A Celebration of Statistics

The ISI Centenary Volume

A Celebration of Statistics

The ISI Centenary Volume

A Volume to Celebrate the Founding of the International Statistical Institute in 1885

Edited by
Anthony C. Atkinson and Stephen E. Fienberg

Springer-Verlag New York Berlin Heidelberg Tokyo

Anthony C. Atkinson
Imperial College of Science and Technology
London SW7 2BZ, England

Stephen E. Fienberg
Carnegie-Mellon University
Pittsburgh, Pennsylvania 15213
U.S.A.

Library of Congress Cataloging in Publication Data
Main entry under title:
A Celebration of statistics.
Includes index.
1. Mathematical statistics—Addresses, essays, lectures.
2. Statistics—Addresses, essays, lectures. I. Atkinson,
A. C. (Anthony Curtis) II. Fienberg, Stephen E.
III. International Statistical Institute.
QA276.16.C45 1985 519.5 84-26825

Typeset by Asco Trade Typesetting Ltd., Hong Kong.
Printed and bound by R.R. Donnelley and Sons, Harrisonburg, Virginia.
Printed in the United States of America.

9 8 7 6 5 4 3 2 1

ISBN 0-387-96111-9 Springer-Verlag New York Berlin Heidelberg Tokyo
ISBN 3-540-96111-9 Springer-Verlag Berlin Heidelberg New York Tokyo

Foreword

The International Statistical Institute was founded in 1885 and is therefore one of the world's oldest international scientific societies. The field of statistics is still expanding rapidly and possesses a rich variety of applications in many areas of human activity such as science, government, business, industry, and everyday affairs. In consequence, the celebration of the Institute's centenary in 1985 is of considerable interest not only to statisticians but also more widely to the international scientific community. As part of its centennial celebration planning the Institute decided to publish a volume of papers representing the immensely wide range of interests encompassed by statistics in its international context, viewed both from a historical and from a contemporary standpoint. We were fortunate in securing the services of Anthony Atkinson and Stephen Fienberg as Editors of this volume: they have worked hard over a period of several years to put together a most fascinating collection of papers. On behalf of the Institute it is my pleasant duty to thank them and the authors for their contributions.

J. DURBIN, President
International Statistical Institute

Preface

The papers in this volume were prepared to help celebrate the centenary of the International Statistical Institute. During the ISI's first 100 years statistics has matured, both as a scientific discipline and as a profession, in ways that the ISI's founders could not possibly have imagined. With this maturation, not only has statistics become the language of science, but statistical terms and concepts have also permeated all aspects of modern society, from the workplace, to our newspapers, and even to the courtroom. Being able to see statistics in action all around us is a source of constant stimulation to us as individual statisticians, and thus to our field. We hope that the papers in this volume reflect the variety, excitement, importance, and international range of our subject.

There should be something here for everyone. When we wrote to prospective authors inviting papers we stated: "The plan of the International Statistical Institute is that the book should be truly international. We hope for a geographical spread of authors and also for a range of subjects that will interest statisticians in all countries, both developed and developing. We are most concerned that the papers should be a way of strengthening communication between groups of statisticians. In some cases an appreciable proportion of survey, exposition, and discussion may be more appropriate than new results. We are especially interested in a discussion of unsolved problems and of future challenges." At the back of our minds was the additional hope that many of the papers would provide a glimpse of what statisticians do—for laymen and for our colleagues in other professions.

A brief outline of our method of working will help explain the contents of the volume. We received the invitation to edit the volume early in 1980. In the process of inviting and refereeing contributions we have had the support of a panel of advisors, listed on the first page, some of whom were chosen explicitly as representatives of the three sections of ISI. They and other members of the ISI were asked to suggest topics and names of authors for papers. As a result of these suggestions we wrote, in the autumn of 1980, to an initial list of 18 authors in 12 countries. Other invitations to contribute followed as the structure and needs of the volume emerged.

Our purpose throughout has been to display the best of statistics, while at the same time achieving a balance across topics and across countries. We attempted to do this while relying upon authors whom we knew could provide interesting and even provocative papers. As the work on the volume progressed, the balance has shifted somewhat, in part by design and in part by happenstance. Some potential authors felt that they could not accept our invitation to contribute. Others accepted but then found that they could not produce the papers they had offered. In these cases, rather than lose the balance across subjects, we sometimes had to turn to someone near at hand to produce a key paper. This process has slightly distorted the national balance for which we strove, but the quality of the papers has been our primary concern.

Looking back over five years of work we are impressed by the quality, the quantity, and the variety of the papers which have been assembled. Some of the most technical papers are among the most readable and some of the less technical present us with the most fascinating challenges for the future. Nonetheless, we are aware of some major gaps which are worth mentioning if only in the hope that someone we did not contact will feel impelled to put pen to paper. One such gap is the difficult subject of the abuse, or perhaps distortion, of statistics by government and of the duties of government statisticians to their employers and to their fellow citizens. It is stated that Mussolini made the trains run on time. He is also said to have reduced unemployment in Italy by decreasing the categories of those counted as unemployed. Governments of many persuasions continue to behave in this manner, or at least occasionally attempt to do so. Some examples from British experience are quoted by Sir Claus Moser in Section 3, "Integrity," of his presidential address to the Royal Statistical Society, published in 1980. Readers in other countries can doubtless provide their own examples. What we sought, but failed to find, was a paper that would give examples of this behavior in an international context. There are other ethical issues going beyond integrity to the statistician's responsibility for the use and interpretation of government statistics, especially those subject to serious technical uncertainties associated with measurement error, in forms that the data do not support. We had hoped to see these issues discussed as well, again from an international perspective.

Other subjects on which we would have liked additional papers relate to statistical activities in developing countries and the transfer of technology. Despite our best efforts, authors from developing countries are underrepresented in this volume. Some reasons for this may be gleaned from the paper by Pearce. We also failed in our efforts to obtain a paper on the transfer of technology from one developed country to another. We had in mind the Japanese effort in the area of quality control and its transfer to the factories of older industrialized nations.

Finally, in our initial round of invitations we attempted to secure papers that reflected the varied areas of application of present-day statistical methodology. As we review the list of papers actually written, we note that we have had some success in that a large fraction do deal with applications. Yet there are missing subject matter areas, three notable ones being: business and management, engineering, and environmental science. There are also methodological topics that have not received much attention. We can only note that the absence of an extended discussion of your favorite statistical method is unintentional.

It may seem strange that in a centenary volume there is so little explicit mention of the history of the International Statistical Institute. One reason for this is that the first 75 years of the Institute are chronicled by J.W. Nixon in *A History of the International Statistical Institute, 1885–1960,* published for the

75th anniversary in 1960. But, in addition, we and our advisors decided that it would be more appropriate, and more fun for our readers, if instead of a history we were to produce a volume on the "stuff" of statistics, and to point to the future of our field instead of to its past.

All of the papers in this volume have been refereed and many have been thoroughly revised, sometimes more than once. We are grateful to the authors, the members of our advisory panel, and the referees for all of their work and cooperation. The result of their efforts is a much improved collection of papers. We believe that there is something here for anyone who is interested in, or curious about statistics. We hope that you agree, and that you take from this volume a sense of the excitement with which we view the future of statistics.

ANTHONY C. ATKINSON
STEPHEN E. FIENBERG

Acknowledgements

We are indebted to a large number of people who assisted in the preparation of this volume. David Cox served as liaison with the ISI, and we turned to him for advice and assistance on numerous occasions. Our Editorial Advisory Panel provided suggestions for authors and topics as we began to get organized, and the members served as primary reviewers of most of the manuscripts. A.L. Finkner gave us special assistance and suggestions in the area of sample surveys. In addition, the following individuals served as referees of one or more papers: R. A. Bailey, O. E. Barndorff-Nielsen, D. J. Bartholomew, J. L. Bell, J. T. Bonnen, C. Cress, B. S. Everitt, V. V. Fedorov, I. Ford, J. Greenhouse, A. M. Herzberg, Wm. Kruskal, N. Laird, J. Lehoczky, R. Mead, M. M. Meyer, H. D. Patterson, J. N. K. Rao, R. L. Smith, C. C. Spicer, J. M. Tanur, E. J. Thompson, and F. Vogel. Finally, we would like to thank Margie Krest for keeping track of everything as work on the volume progressed, and for her assistance in putting the final manuscript together for the publisher.

Contents

List of Contributors

HIROTUGU AKAIKE
Institute of Statistical Mathematics, Tokyo, Japan

ANTHONY C. ATKINSON
Imperial College of Science and Technology, London, England

G. A. BARNARD
Colchester, England

OLE E. BARNDORFF-NIELSEN
University of Aarhus, Aarhus, Denmark

PETTER JAKOB BJERVE
Central Bureau of Statistics, Oslo, Norway

PREBEN BLÆSILD
University of Aarhus, Aarhus, Denmark

N. E. BRESLOW
University of Washington, Seattle, WA, U.S.A.

JOHN CLELAND
International Statistical Research Centre, Voorburg, The Netherlands

TORE DALENIUS
Brown University, Providence, RI, U.S.A.

MORRIS H. DEGROOT
Carnegie-Mellon University, Pittsburgh, PA, U.S.A.

K. DIETZ
University of Tübingen, Tübingen, Federal Republic of Germany

J. DURBIN
London School of Economics and Political Science, London, England

WILLIAM F. EDDY
Carnegie-Mellon University, Pittsburgh, PA, U.S.A.

NEIL R. ERICSSON
Board of Governors of the Federal Reserve System, Washington, D.C., U.S.A.

STEPHEN E. FIENBERG
Carnegie-Mellon University, Pittsburgh, PA, U.S.A.

CHRISTOPHER FIELD
Dalhousie University, Halifax, Nova Scotia, Canada

JAMES E. GENTLE
IMSL Inc., Houston, TX, U.S.A.

T.I. GOLIKOVA
Moscow State University, Moscow, U.S.S.R.

YU. V. GRANOVSKY
Moscow State University, Moscow, U.S.S.R.

MORRIS H. HANSEN
WESTAT Inc., Rockville, MD, U.S.A.

DAVID F. HENDRY
Nuffield College, Oxford, England

C.C. HEYDE
University of Melbourne, Parkville, Australia

JENS LEDET JENSEN
University of Aarhus, Aarhus, Denmark

M.P. JHA
Indian Agricultural Statistics Research Institute, New Delhi, India

JANA JUREČKOVÁ
Charles University, Prague, Czechoslovakia

DAVID G. KENDALL
Statistical Laboratory, Cambridge, England

MILOŠ MACURA
Ekonomski Institut, Belgrade, Yugoslavia

E. MALINVAUD
INSEE, Paris, France

JUAN E. MEZZICH
University of Pittsburgh, Pittsburgh, PA, U.S.A.

V.V. NALIMOV
Moscow State University, Moscow, U.S.S.R.

S.C. PEARCE
University of Kent, Canterbury, England

LUIS RAÚL PERICCHI
Universidad Simón Bolívar, Caracas, Venezuela

R.L. PLACKETT
Newcastle-Upon-Tyne, England

C. Radhakrishna Rao
University of Pittsburgh, Pittsburgh, PA, U.S.A.

Ignacio Rodríguez-Iturbe
Universidad Simón Bolívar, Caracas, Venezuela

D. Schenzle
University of Tübingen, Tübingen, Federal Republic of Germany

Burton Singer
Columbia University, New York, NY, U.S.A.

D. Singh
Council of Scientific and Industrial Research, New Delhi, India

Michael Sørenson
University of Aarhus, Aarhus, Denmark

Judith M. Tanur
State University of New York at Stony Brook, Stony Brook, NY, U.S.A.

Benjamin J. Tepping
Silver Spring, MD, U.S.A.

CHAPTER 1

Prediction and Entropy

Hirotugu Akaike

Abstract

The emergence of the magic number 2 in recent statistical literature is explained by adopting the predictive point of view of statistics with entropy as the basic criterion of the goodness of a fitted model. The historical development of the concept of entropy is reviewed, and its relation to statistics is explained by examples. The importance of the entropy maximization principle as a basis of the unification of conventional and Bayesian statistics is discussed.

1. Introduction and Summary

We start with an observation that the emergence of a particular constant, the magic number 2, in several statistical papers is inherently related to the predictive use of statistics. The generality of the constant can only be appreciated when we adopt the statistical concept of entropy, originally developed by a physicist L. Boltzmann, as the criterion to measure the deviation of a distribution from another.

A historical review of Boltzmann's work on entropy is given to provide a basis for the interpretation of statistical entropy. The negentropy, or the negative of the entropy, is often equated to the amount of information. This review clarifies the limitation of Shannon's definition of the entropy of a probability distribution. The relation between Boltzmann entropy and the asymptotic theory of statistics is discussed briefly.

The concept of entropy provides a proof of the objectivity of the log likelihood as a measure of the goodness of a statistical model. It is shown that this observation, combined with the predictive point of view, provides a simple explanation of the generality of the magic number 2. This is done through the explanation of the AIC statistic introduced by the present author. The use of AIC is illustrated by its application to multidimensional contingency table analysis.

The discussion of AIC naturally leads to the entropy maximization principle which specifies the object of statistics as the maximization of the expected

Key words and phrases: AIC, Bayes procedure, Entropy, entropy maximization principle, information, likelihood, model selection, predictive distribution.

entropy of a true distribution with respect to the fitted predictive distribution. The generality of this principle is demonstrated through its application to Bayesian statistics. The necessity of Bayesian modeling is discussed and its similarity to the construction of the statistical model of thermodynamics by Boltzmann is pointed out. The principle provides a basis for the unification of Bayesian and conventional approaches to statistics. Referring to Boltzmann's fundamental contribution to statistics, the paper concludes by emphasizing the importance of the research on real problems for the development of statistics.

2. Emergence of the Magic Number 2

Around the year 1970, the constant 2 appeared in a curious fashion over and over again in a series of papers. This represents the emergence of what Stone (1977a) symbolically calls the magic number 2.

The number appears in Mallow's C_p statistic for the selection of independent variables in multiple regression, which is by definition

$$C_p = \frac{1}{s^2}\mathrm{RSS}_p - n + 2p,$$

where RSS_p denotes the residual sum of squares after regression on p independent variables, n the sample size, and s^2 an estimate of the common variance σ^2 of the error terms (Mallows, 1973). The final prediction error (FPE) introduced by Akaike (1969, 1970) for the determination of the order of an autoregression is an estimate of the mean squared error of the one-step prediction when the fitted model is used for prediction. It satisfies asymptotically the relation

$$n \log \mathrm{FPE} = n \log S_p + 2p,$$

where n denotes the length of the time series, S_p the maximum-likelihood estimate of the innovation variance obtained by fitting the pth-order autoregression. Both Leonard and Ord (1976) and Stone (1977a) noticed the number 2 as the asymptotic critical level of F-tests when the number of observations is increased.

An explanation of the multiple appearances of this number 2 can easily be given for the case of the multiple regression analysis. The effect of regression is usually evaluated by the value of RSS_p. A smaller RSS_p may be obtained by increasing the number of independent variables p. However, we know that after adding a certain number of independent variables further addition of variables often merely increases the expected variability of the estimate. When the increase of the expected variability is measured in terms of the mean squared prediction error, it will be seen that the increase is exactly equal to the expected amount of decrease of the sample residual variance RSS_p/n. Thus to convert RSS_p into an unbiased estimate of the mean squared error of prediction we must apply *twice* the correction that is required to convert RSS_p into an unbiased estimate of $n\sigma^2$.

The appearance of the critical value 2 for the F-test discussed by Leonard and Ord (1976) is more instructive. The F-test is considered as a preliminary test of significance in the estimation of the one-way ANOVA model where K independent observations y_{jk} ($k = 1, 2, \ldots, K$) are taken from each group j ($j = 1, 2, \ldots, J$). Under the assumption that y_{jk} are distributed as normal with mean θ_j and variance σ_W^2, the F-statistic for testing the hypothesis $\theta_1 = \theta_2 = \cdots = \theta_J$ is given by

$$F = \frac{(J-1)^{-1}S_B^2}{(J(K-1))^{-1}S_W^2},$$

where $S_B^2 = K\sum_j (y_{j\cdot} - y_{\cdot\cdot})^2$ and $S_W^2 = \sum_j \sum_k (y_{jk} - y_{j\cdot})^2$, and where $y_{j\cdot}$ and $y_{\cdot\cdot}$ denote the mean of the jth group and the grand mean, respectively. The final estimate of θ_j is defined by

$$\tilde{\theta}_j = \begin{cases} y_{j\cdot} & \text{if the hypothesis is rejected,} \\ y_{\cdot\cdot} & \text{otherwise.} \end{cases}$$

Consider the loss function $L(\tilde{\theta}, \theta) = \sum(\tilde{\theta}_j - \theta_j)^2$. For the simpler estimates defined by $\tilde{\theta}_j = y_{\cdot\cdot}$ and $\tilde{\theta}_j = y_{j\cdot}$ it can easily be shown that the difference of the risks of these estimates has one and the same sign as that of $E(J(K-1))^{-1}S_W^2(F-2)$, where E denotes expectation. Thus when the sample size K is sufficiently large, the choice of the critical value 2 for the F-test to select $\tilde{\theta}_j$ is appropriate.

The one characteristic that is clearly common to these papers is that the authors considered some predictive use of the models. An early example of the use of the concept of future observation to clarify the structure of an inference procedure is given by Fisher (1935, p. 393). The concept is explicitly adopted as the philosophical motivation in a work by Guttman (1967). In the present paper, the predictive point of view adopts as the purpose of statistics the realization of appropriate predictions.

In the example of the ANOVA model above, if the number of groups, J, is increased indefinitely, the test statistic F converges to 1 under the null hypothesis. Thus the critical value of the F-test for any fixed level of significance must also converge to 1 instead of 2. As is observed by Leonard and Ord, this dramatically demonstrates the difference between the conventional approach to model selection by testing with a fixed level of significance and the predictive approach. Thus the emergence of the magic number 2 must be considered as a sign of the impending change of the paradigm of statistics. However, to fully appreciate the generality of the number, we have first to expand our view of the statistical estimation.

3. From Point to Distribution

The risk functions considered in the preceding section were the mean squared errors of the predictions. Such a choice of the criterion is conventional but quite arbitrary. The weakness of the ad hoc definition becomes apparent when we try to extend the concept to multivariate problems.

A typical example of multivariate analysis is factor analysis. At first sight it is not at all clear how the analysis is related to prediction. In 1971, in an attempt to extend the concept of FPE to solve the problem of determination of the number of factors, the present author recognized that in factor analysis a prediction was realized through the specification of a distribution (Akaike, 1981). This quickly led to the observation that almost all the important statistical procedures hitherto developed were concerned, either explicitly or implicitly, with the realization of predictions through the specification of distributions.

Stigler (1975) remarks that the interest of statisticians shifted from point estimation to distribution estimation towards the end of the 19th century. However, it seems that Fisher's very effective use of the concept of parameter drew the attention of statisticians back to the estimation of a point in a parameter space. We are now in a position to return to distributions, and here the basic problem is the introduction of a natural topology in the space of distributions. The probabilistic interpretation of thermodynamic entropy developed by Boltzmann provides, historically, one of the most successful examples of a solution to this problem.

4. Entropy and Information

The statistical interpretation of the thermodynamic entropy, a measure of the unavailable energy within a thermodynamic system, was developed in a series of papers by L. Boltzmann in the 1870's. His first contribution was the observation of the monotone decreasing behavior in time of a quantity defined by

$$E = \int_0^\infty f(x, t) \log\left[\frac{f(x, t)}{\sqrt{x}}\right] dx,$$

where $f(x, t)$ denotes the frequency distribution of the number of molecules with energy between x and $x + dx$ at time t (Boltzmann, 1872). Boltzmann showed that for a closed system, under proper assumptions on the collision process of the molecules, the quantity E can only decrease. When the distribution f is defined in terms of the velocities and positions of the molecules, the above quantity takes the form

$$E = \iint f \log f \, dx \, d\xi,$$

where x and ξ denote the vectors of the position and velocity, respectively. Boltzmann showed that for some gases this quantity, multiplied by a negative constant, was identical to the thermodynamic entropy.

The negative of the above quantity was adopted by C. E. Shannon as the definition of the entropy of a probability distribution:

$$H = -\int p(x)\log p(x)\,dx,$$

where $p(x)$ denotes the probability density with respect to the measure dx (Shannon and Weaver, 1949).

Almost uncountably many papers and books have been written about the use of the Shannon entropy, where the quantity H is simply referred to as a measure of information, or uncertainty, or randomness. One departure from this definition of entropy, known as the Kullback–Leibler information (Kullback and Leibler, 1951), is defined by

$$I(q;p) = \int q(x)\log\left(\frac{q(x)}{p(x)}\right)dx$$

and relates the distribution $q(x)$ to another distribution $p(x)$.

Much interest has been shown in the use of these quantities as measures of statistical information. However, it seems that their potential as statistical concepts has not been fully evaluated. It seems to the present author that this is due to the neglect of Boltzmann's original work on the probabilistic interpretation of thermodynamic entropy. Karl Pearson (1929, p. 205) cites the words of D. F. Gregory: "... we sacrifice many of the advantages and more of the pleasure of studying any science by omitting all reference to the history of its progress." It seems that this has been precisely the case with the development of the statistical concept of entropy or information.

5. Distribution and Entropy

The work of Boltzmann (1872) produced a demonstration of the second law of thermodynamics, the irreversible increase of entropy in an isolated closed system. In answering the criticism that the proof of irreversibility is based on the assumption of a reversible mechanical process, Boltzmann (1877a) pointed out the necessity of probabilistic interpretation of the result.

At that time Meyer, a physicist, produced a derivation of the Maxwell distribution of the kinetic energy among gas molecules at equilibrium as the "most probable" distribution. Pointing out the error in Meyer's proof, Boltzmann (1877b) established the now well-known identity

$$\text{entropy} = \log(\text{probability of a statistical distribution}).$$

His reasoning was based on the asymptotic equality

$$\log \frac{n!}{n_0!n_1!\cdots n_p!} = -n \sum_{i=0}^{p} \frac{n_i}{n} \log \frac{n_i}{n}, \tag{5.1}$$

where n_i denotes the frequency of the molecules at the ith energy level and $n = n_0 + n_1 + \cdots + n_p$. If we put $p_i = n_i/n$, then the right-hand side is equal to $nH(p)$, where

$$H(p) = -\sum_{i=0}^{p} p_i \log p_i,$$

i.e., the Shannon entropy of the distribution $p = (p_0, p_1, \ldots, p_p)$.

Following the idea that the frequency distribution f of molecules at thermal equilibrium is the distribution which is the most probable under the assumption of a given total energy, Boltzmann maximized

$$H(f) = -\int_0^\infty f \log f \, dx$$

under the constraints

$$\int_0^\infty f \, dx = N \quad \text{and} \quad \int_0^\infty x f(x) \, dx = L,$$

where x denotes the energy level, N the total number of molecules, and L the total energy. The maximization produces as the energy distribution $f(x) = Ce^{-hx}$ with a proper positive constant h. Boltzmann discussed in great detail that this result could be physically meaningful only for a proper definition of the energy level x, a point commonly ignored by later users of the Shannon entropy. Incidentally, we notice here an early derivation of the exponential family of distributions by the constrained maximization of $H(f)$, a technique of probability-distribution generation later named the maximum-entropy method (Jaynes, 1957).

The change in Boltzmann's view of the energy distribution between 1872 and 1877 is quite significant. In the 1872 paper the distribution $f(x, t)$ represented a unique entity. In the 1877b paper the distribution was considered as a random sample and its probability of occurrence was the main subject.

Boltzmann (1878) further extended the discussion of this point. Noting that the probability of geting a sample frequency distribution $(w_0, w_1, \ldots, w_p)$ from a probability distribution $(f_0, f_1, \ldots, f_p)$ is given by

$$\Omega = f_0^{w_0} f_1^{w_1} \cdots f_p^{w_p} \cdot \frac{n!}{w_0! w_1! \cdots w_p!},$$

Boltzmann derived an asymptotic equality

$$l\Omega = w_0 l f_0 + w_1 l f_1 + \cdots + w_p l f_p - w_0 l w_0 - w_1 l w_1 \cdots - w_p l w_p + \text{const}, \tag{5.2}$$

where $n = w_1 + w_2 + \cdots + w_p$ and l denoted the natural logarithm. He pointed out that the former formula (5.1) is a special case of (5.2) where it is assumed that $f_0 = f_1 = \cdots = f_p$. If the additive constant is ignored, the present formula (5.2) can be rearranged in the form

$$l\Omega = -n \sum_{i=0}^{p} g_i l\left(\frac{g_i}{f_i}\right),$$

where $g_i = w_i/n$. Thus to retain the interpretation that the entropy is the log probability of a distribution we have to adopt, instead of $H(p)$, the quantity

$$B(g; f) = -\sum_i g_i \log\left(\frac{g_i}{f_i}\right)$$

as the definition of the entropy of the secondary distribution g with respect to the primary distribution f. When the distributions g and f are defined in terms of densities $f(x)$ and $g(x)$, the entropy is defined by

$$B(g; f) = -\int g(x) \log\left(\frac{g(x)}{f(x)}\right) dx.$$

When it is necessary to distinguish this quantity from the thermodynamic entropy or the Shannon entropy, we will call it the Boltzmann entropy. It is now obvious that $B(g; f)$ provides a natural measure of the deviation of g from f.

The equality of the above quantity to the thermodynamic entropy holds only when the former is maximized under the assumption of a given mean energy for an appropriately chosen "primary distribution" f and then multiplied by a proper constant. Thus it can be seen that the Shannon entropy $H(g) = -\sum g_i \log g_i$ yields the physical meaning of the entropy contemplated by Boltzmann only under very limited circumstances. Obviously $l\Omega$ or $B(g; f)$ is the more fundamental concept. This point is reflected in the fact that in Shannon and Weaver (1949) essential use is made not of $H(f)$ but of its derived quantities taking the form of $B(g; f)$.

The Kullback–Leibler (KL) information number is defined by $I(g; f) = -B(g; f)$. Kullback (1959) describes this quantity as the mean information per observation from $g(x)$ for discrimination in favor of $g(x)$ against $f(x)$ and simultaneously considers $I(f; g)$ to define the Jeffreys divergence:

$$J(g; f) = I(g; f) + I(f; g).$$

Contrary to the formal definition of $I(g; f)$ by Kullback, the present derivation of $B(g; f)$ based on Boltzmann's $l\Omega$ clearly explains the difference of the roles played by g and f. The primary distribution f is hypothetical, while the secondary g is factual. It is the fictitious sampling from $f(x)$ that provides the probabilistic meaning of $B(g; f)$. This may be seen more clearly by the representation

$$B(g; f) = -\int \frac{g(x)}{f(x)} \log\left(\frac{g(x)}{f(x)}\right) f(x)\, dx.$$

Boltzmann (1878) also arrived at a generalization of the exponential family of distributions by maximizing the entropy under certain constraints. These results demonstrate the fundamental contribution of Boltzmann to the science of statistics. A good summary of mathematical properties of the Boltzmann entropy or the Kullback–Leibler information is given by Csiszar (1975).

6. Entropy and the Asymptotic Theory of Statistics

The Boltzmann entropy appears, sometimes implicitly, in many basic contributions to statistics, particularly in the area of asymptotic theory. For a pair of distributions $p(\cdot\,|\,\theta_1)$ and $p(\cdot\,|\,\theta_2)$ from a parametric family $\{p(\cdot\,|\,\theta);\theta\in\Theta\}$ the deviation of the former from the latter can be measured by $B(\theta_1;\theta_2) = B(p(\cdot\,|\,\theta_1);\, p(\cdot\,|\,\theta_2))$. This induces a natural topology in the space of parameters.

When θ_1 and θ_2 are k-dimensional parameters given by $\theta_1 = (\theta_{11}, \theta_{12}, \ldots, \theta_{1k})$ and $\theta_2 = (\theta_{21}, \theta_{22}, \ldots, \theta_{2k})$, under appropriate regularity conditions we have

$$B(\theta_1;\theta_2) = -\tfrac{1}{2}(\theta_2-\theta_1)'E\left[\frac{\partial^2}{\partial\theta'\,\partial\theta}\log p(x\,|\,\theta_1)\right](\theta_2-\theta_1) + o(\|\theta_2-\theta_1\|^2),$$

where $(\partial^2/\partial\theta'\,\partial\theta)\log p(x\,|\,\theta_1)$ denotes the Hessian evaluated at $\theta=\theta_1$, E denotes the expectation with respect to $p(\cdot\,|\,\theta_1)$, and $o(\|\theta_2-\theta_1\|^2)$ is a term of order lower than $\|\theta_2-\theta_1\|^2 = \sum(\theta_{1i}-\theta_{2i})^2$. The quantity $-E[(\partial^2/\partial\theta'\,\partial\theta)\log p(x\,|\,\theta_1)]$ is the Fisher information matrix. The fact that the Fisher information matrix is just minus twice the Hessian of the entropy clearly shows that it is related to the local property of the topology induced by the entropy.

The likelihood-ratio test statistic for testing a specific model, or hypothesis, defined by $\theta=\theta_0$ is given by

$$\lambda_n = \frac{\prod p(x_i\,|\,\theta_0)}{\sup\{\prod p(x_i\,|\,\theta);\theta\in\Theta\}},$$

where $(x_1, x_2, \ldots, x_n)$ denotes the sample. If the true distribution is defined by $p(\cdot\,|\,\theta)$, we expect that

$$T_n = -\frac{1}{n}\log\lambda_n$$

will converge stochastically to $-B(\theta;\theta_0)$ as n is increased to infinity. The result of Bahadur (1976) shows that under certain regularity conditions

$$\lim_{n\to\infty}\frac{1}{n}\log P(T_n > t_n\,|\,\theta_0) = B(\theta;\theta_0),$$

where t_n denotes the sample value of the test statistic T_n for a particular realization $(x_1, x_2, \ldots, x_n)$. This means that if one calculates the probability of the statistic T_n being larger than t_n, assuming that the data have come from the hypothetical distribution $p(\cdot \mid \theta_0)$, it will asymptotically be equal to $\exp(nB(\theta; \theta_0))$, where θ denotes the true distribution.

In a practical application the hypothesis will never be exact, and the above result says that by calculating the P-value of the log likelihood-ratio test we are actually measuring the entropy $nB(\theta; \theta_0)$. It is often argued that the test is logically meaningless, since the falsity of θ_0 is almost always certain. The present observation partially clarifies the confusion in this argument.

The concept of second-order efficiency was introduced by Rao (1961). In that paper he discussed the performance of an estimator obtained by minimizing the Kullback–Leibler information number $\sum \pi_r \log(\pi_r/p_r)$, where π_r denotes the probability of the rth cell in a multinomial distribution, defined as a function of a parameter θ, and p_r the observed relative frequency. This estimator can also be characterized as the one that maximizes $B(\pi; p)$, while the maximum-likelihood estimate maximizes $B(p; \pi)$.

If we carefully follow the derivation of $B(g;f)$, we can see that the primary distribution f is always hypothetical, while the secondary distribution g is factual. It is interesting to note that Rao has shown that the minimum KL number estimator, defined by the entropy with a factual primary distribution and an hypothetical secondary, is less efficient than the maximum-likelihood estimator defined by the more natural definition of the entropy. A similar relation has been observed between the estimators defined by minimizing the chi-square and the modified chi-square that are approximations to $-2B(p; \pi)$ and $-2B(\pi; p)$, respectively. These results suggest that the present interpretation of entropy can produce useful insights not available from the use of Fisher information which does not discriminate between the primary and secondary distributions.

The relation between the entropy and the asymptotic distribution of the corresponding sample distribution function is discussed by Sanov (1957) and Stone (1974). Other standard references on the relation between the entropy and large-sample theory are Chernoff (1956) and Rao (1962).

7. Likelihood, Entropy and the Predictive Point of View

Obviously, one of the most significant contributions to statistics by R. A. Fisher is the development of the method of maximum likelihood. However, there is a definite limitation to the applicability of the idea of maximizing the likelihood.

The limitation can most clearly be seen by the following model selection problem. Consider a set of nested parametric families $\{p(\cdot \mid \theta_k)\}$ ($k = 1, 2, \ldots, K$), defined by $\theta_k = (\theta_{k1}, \theta_{k2}, \ldots, \theta_{kk}, \theta_{0k+1}, \ldots, \theta_{0K})$. In the kth family, only the first k components of the parameter vector θ_k are allowed to vary; the rest are fixed at some preasssigned values $\theta_{0\,k+1}, \ldots, \theta_{0K}$. When data x are given, if we simply maximize the likelihood over the whole family, we always end up with the choice of $p(\cdot\,|\,\theta_K^*)$, where θ_K^* denotes the maximum-likelihood estimate that maximizes $p(x\,|\,\theta_K)$. This means that the method of maximum likelihood always leads to the selection of the unconstrained model. This is obviously against our expectation. If a statistician suggests the choice of the highest possible order whenever fitting a polynomial regression, he will certainly lose the trust of his clients.

Fisher was clearly aware of the limitation of his theory of estimation. After pointing out the necessity of the knowledge of the functional form of the distribution as the prerequisite of his theory of estimation, Fisher (1936, p. 250) admits the possibility of a wider type of inductive argument that would discuss methods of determining the functional form by data. However he also states, "At present it is only important to make clear that no such theory has been established." This clearly suggests the necessity of extending the theory of statistical estimation to the situation where several possible parametric models are involved. Such an extension is possible with a proper combination of the predictive point of view and the concept of entropy.

The predictive point of view generalizes the concept of estimation from that of a parameter to that of the distribution of a future observation. We refer to such an estimate as a predictive distribution. The basic criterion in this generalized theory of estimation is then the measure of the "goodness" of the predictive distribution. One natural choice of such a measure is the expected deviation of the true distribution from the predictive distribution as measured by the expected entropy EB(true; predictive). Here, the expectation E is taken with respect to the true distribution of the data used to define the predictive distribution.

Except for data obtained by an artificial sampling scheme, we do not know exactly what is meant by the true distribution. Indeed, the concept of the true distribution obtains a practical meaning only through the specification of an estimation procedure or a model. The true distribution may thus be viewed as a conceptual construct that provides a basis for the design of an estimation procedure for a particular type of data. Since the concept of the true distribution is quite personal, the validity of an estimation procedure based on such a concept must be judged by the collective experience of its use by human society. In such a circumstance it becomes crucial to find the objectivity of a statistical inference procedure to make it a vehicle for the communication of our experiences.

When a parametric model $\{p(\cdot\,|\,\theta);\ \theta \in \Theta\}$ of the distribution of a future observation y is given, the goodness of a particular model $p(\cdot\,|\,\theta)$ as the predictive distribution of y is evaluated by the entropy

$$B(f; p(\cdot\,|\,\theta)) = -E_y \log\left(\frac{f(y)}{p(y\,|\,\theta)}\right),$$

where E_y denotes the expectation with respect to the true distribution denoted by $f(y)$. Here the true distribution is unspecified; only its existence is assumed. Since it holds that

$$B(f; p(\cdot \mid \theta)) = E_y \log p(y \mid \theta) - E_y \log f(y),$$

we may restrict our attention to $E_y \log p(y \mid \theta)$ for the comparison of possible choices of θ.

We further specify the predictive point of view by assuming that *the future observation y is another independent sample taken from the same distribution as that of the present data x*. The accuracy of our inference is evaluated only in its relation to the prediction of an observation similar to the present one.

One of the important consequences of the present specification of the predictive point of view is that it leads to the observation that the log likelihood, $\log p(x \mid \theta)$, is a natural estimate of $E_y \log p(y \mid \theta)$. Obviously, by the present predictive point of view, *the log likelihood* $\log p(x \mid \theta)$ *provides an unbiased estimate of* $E_y \log p(y \mid \theta)$, *irrespective of the form of the true distribution* $f(y)$. The log likelihood provides an unbiased estimate of the basic criterion $E_y \log p(y \mid \theta)$ to everyone who accepts the concept of the true distribution, irrespective of the form of $f(y)$. In this sense, objectivity is imparted to statistical inference through the use of log likelihoods. We can see that the range of the validity of the concept of likelihood is not restricted to one particular parametric family of distributions. This observation constitutes the basis for the solution of the model selection problem considered at the beginning of this section.

8. Model Selection and an Information Criterion (AIC)

We will first show that our basic criterion, the expected entropy, provides a natural extension of the mean-squared-error criterion. The quality of a predictive distribution $f(y \mid x)$ is evaluated by the expected negentropy defined by

$$-E_x B(f; f(\cdot \mid x)) = E_y \log f(y) - E_x E_y \log f(y \mid x),$$

where $f(y)$ denotes the true distribution of y, which is assumed to be independent of x, and E_x and E_y denote the expectations with respect to the true distributions of x and y, respectively. By Jensen's inequality we have $E_x \log f(y \mid x) \leqslant \log E_x f(y \mid x)$, and we get the additive decomposition

$$-E_x B(f; f(\cdot \mid x)) = \{E_y \log f(y) - E_y \log E_x f(y \mid x)\} + \{E_y \log E_x f(y \mid x) - E_y E_x \log f(y \mid x)\}.$$

The term inside the first braces on the right-hand side represents the amount of increase of the expected negentropy due to the deviation of $f(y)$ from $E_x f(y \mid x)$. This term corresponds to the squared bias in the case of ordinary estimation of a parameter. The term inside the second braces represents the increase of the expected negentropy due to the sampling fluctuation of $f(y \mid x)$

around $E_x f(y|x)$. This quantity corresponds to the variance. The present result shows why the two different concepts, squared bias and variance, can be added together in a meaningful way.

Having observed that the expected negentropy provides a natural extension of the mean-squared-error criterion, we recognize that the main problem is the estimation of the entropy or the expected log likelihood $E_y \log f(y|x)$ of the predictive distribution. In the case of the ANOVA model discussed by Leonard and Ord, the F-test was used for the selection of the model underlying the definition of the final estimate. For the present general model we consider the use of the log likelihood-ratio test. The test statistic for the testing of $\{p(\cdot|\theta_k)\}$ against $\{p(\cdot|\theta_K)\}$ of the preceding section is defined y

$$(-2)\{\log p(x|\theta_k^*) - \log p(x|\theta_K^*)\},$$

where θ_k^* denotes the maximum-likelihood estimate determined by the date x, and the statistic is taken to follow a chi-square distribution with K-k degrees of freedom.

We consider that the test is developed to make a reasonable choice between $p(y|\theta_k^*)$ and $p(y|\theta_K^*)$. From our present point of view this means that the test must be in good correspondence to the choice by $(-2)E_y\{\log p(y|\theta_k^*) - \log p(y|\theta_K^*)\}$. The result of Wald (1943) on the asymptotic behavior of the log-likelihood-ratio test shows that, when x is a vector of observations of independently identically distributed random variables with the likelihood functions satisfying certain regularity conditions, we have asymptotically

$$E_x^*[-2\{\log p(x|\theta_k^*) - \log p(x|\theta_K^*)\}] = \|\theta_k^0 - \theta_K^0\|_I^2 + (K-k),$$

where E_x^* denotes the mean of the limiting distribution, $\| \ \|_I$ is the Euclidean norm defined by the Fisher information matrix, and θ_k^0 denotes the value of θ_k that maximizes $E_x \log p(x|\theta_k)$, where E_x denotes the expectation with respect to the true distribution under the assumption that it is given by $p(x|\theta_K^0)$.

Similarly, from the analysis of the asymptotic behavior of the maximum-likelihood estimates we have asymptotically

$$E_x^*[-2E_y\{\log p(y|\theta_k^*) - \log p(y|\theta_K^*)\}] = \|\theta_k^0 - \theta_K^0\|_I^2 - (K-k),$$

where the restricted predictive point of view is adopted, and x and y are assumed to be independently identically distributed.

From these two results it can be seen that as a measurement of $(-2)E_y\{\log p(y|\theta_k^*) - \log p(y|\theta_K^*)\}$ the log-likelihood-ratio test statistic $(-2)\{\log p(x|\theta_k^*) - \log p(x|\theta_K^*)\}$ shows an upward bias by the amount $2(K-k)$. If we correct for this bias, then we get $\{-2\log p(x|\theta_k^*) + 2k\} - \{-2\log p(x|\theta_K^*) + 2K\}$ as a measurement of the difference of the entropies of the models specified by $p(\cdot|\theta_k^*)$ and $p(\cdot|\theta_K^*)$. This observation leads to the conclusion that the statistic $-2\log p(x|\theta_k^*) + 2k$ should be used as a measure of the badness of the model specified by $p(\cdot|\theta_k^*)$ (Akaike, 1973). The acronym AIC adopted by Akaike (1974) for this statistic is an abbreviation of "an information criterion" and is symbolically defined by

$$\text{AIC} = -2\log(\text{maximum likelihood}) + 2(\text{number of parameters}),$$

where log denotes natural logarithm.

If the log-likelihood-ratio test is considered as a measurement of the entropy difference, then the above observation suggests that from our present point of view *we should choose the model with smaller value of* AIC. If we follow this idea, we get an estimation procedure which simultaneously realizes the model selection and parameter estimation. An estimate thus obtained is called a minimum AIC estimate (MAICE). Now it is a simple matter to see that the critical level 2 of the F-test by Leonard and Ord corresponds to the factor 2 of the second term in the definition of AIC.

One important observation about AIC is that it is defined without specific reference to the true model $p(\cdot \mid \theta_K^0)$. Thus, for any finite number of parametric models, we may always consider an extended model that will play the role of $p(\cdot \mid \theta_K^0)$. This suggests that AIC can be useful, at least in principle, for the comparison of models which are nonnested, i.e., the situation where the conventional log likelihood-ratio test is not applicable.

We will demonstrate the practical utility of AIC by its application to the multidimensional contingency-table analysis discussed by Goodman (1971). Observing the frequency f_{ijkl} in the cell (i, j, k, l) of a 4-way contingency table ($i = 1, 2, \ldots, I; j = 1, 2, \ldots, J; k = 1, 2, \ldots, K; l = 1, 2, \ldots, L$) with $\sum_{ijkl} f_{ijkl} = n$, the basic model is specified by the parametrization

$$\log F_{ijkl} = \theta + \lambda_i^A + \cdots + \lambda_l^D + \lambda_{ij}^{AB} + \cdots + \lambda_{kl}^{CD} + \lambda_{ijk}^{ABC} + \cdots + \lambda_{jkl}^{BCD} + \lambda_{ijkl}^{ABCD},$$

where F_{ijkl} denotes the expected frequency and the λ's satisfy the condition that any sum with respect to one of the suffixes is equal to zero. The characters A, B, C, D symbolically denote the group of parameters that are related to the factors denoted by these characters. Hypotheses are defined by putting some of the parameters equal to zero.

Goodman discussed the application to the analysis of detergent-user data which included information on the following four factors: the softness of the water used (S), the previous use of a brand (U), the temperature of the water used (T), and the preference for one brand over the other (P). In Table 1 the initial portion of Goodman's Table 3 is shown with the corresponding AIC's. In Goodman's modeling, when a higher-order effect is considered, all the corresponding lower-order effects are included in the model.

Goodman asserts that H_1 and H_2 do not fit the data but H_3 and H_4 do, where H_i denotes hypothesis number i. By the present definition of AIC the negative signs of AIC for H_3 and H_4 mean that the corresponding models are preferred to the saturated nonrestricted model. This corresponds to Goodman's assertion. The AIC already suggests that H_4 is an overfit, and Goodman actually proceeds to the detailed analysis of H_3 and arrives at H_5.

The significances of S and T are then respectively checked by comparing H_6 and H_7 with H_5. The hypothesis H_8 is then judged to be an improvement over

Table 1. Goodman's Analysis of Consumer Data

Hypothesis	Estimated group of parameters	Degrees of freedom	$-2x$ log likelihood ratio)	AIC[a]
1	None	23	118.63	72.63
2, (a)[b]	S, P, T, U	18	42.93	6.93
3	All the pairs	9	9.85	-8.15
4	All the triplets	2	0.74	-3.26
5, (b)[b]	PU, S, T	17	22.35	-11.65
6	PU, S	18	95.56	59.56
7	PU, T	19	22.85	-15.15
8	PU, PT	18	18.49	-17.51
9	PT, U	19	39.07	1.07
10, (d)[b]	PU, PT, ST	14	11.89	-16.11
11, (c)[b]	PU, PT, S	16	17.99	-14.01
—[c], (e)[b]	PTU, ST	12	8.4	-15.6
28, (f)[b]	PTU, STU	8	5.66	-10.34

[a] AIC = -2(log likelihood ratio) -2(number of degrees of freedom) = AIC(i) − AIC(∞), where AIC(i) denotes the original AIC of H_i, and AIC(∞) denotes that of the saturated model with all the parameters unrestricted.

[b] Models considered in Table 5-4 of Fienberg (1980, p. 77).

[c] Missing in Goodman (1971). Numbers obtained from Fienberg (1980).

H_7. The effect of PU is then confirmed by comparing H_8 with H_9. Further elaboration of H_8 leads to H_{10}. However, its improvement over H_8 is not considered to be significant, although the effect ST is judged to be significant by the comparison of H_{10} with H_{11}. The path of Goodman's stepwise search is schematically represented by Table 2.

Table 2 shows that we come to the same conclusions as those obtained by Goodman with the choice of 5% as the critical level, simply by choosing models with lower values of AIC. The fact that AIC does not require the table lookup of the chi-squares with different degrees of freedom adds to the significance of this result. Since AIC is defined with a unique scaling unit, it allows easy extraction of useful information from a collection of fitted models. For example, by comparing the difference of AIC's of H_7 and H_5 with that of H_8 and H_{11}, we can clearly see the deteriorating effect of including S in the model. Also the direct comparison of H_6 and H_7, not possible by the log likelihood-ratio test, is now possible by AIC, and the inferiority of H_6, which contains S, is clearly recognizable. The ability of AIC to allow the researcher to extract global information from the result of fitting a large number of models is a characteristic that is not shared by the conventional model selection procedure realized by some ad hoc application of significance tests.

To avoid possible misconceptions some precautions are in order. The fact that AIC allows simple comparison of models does not justify the mechanistic enumeration of all possible models. The selection of the basic set of models must represent the particular way of looking at the data by the researcher. This point is discussed extensively, in relation to the categorical data analysis,

Table 2. The Path of Goodman's Stepwise Search and the Corresponding AIC's[a]

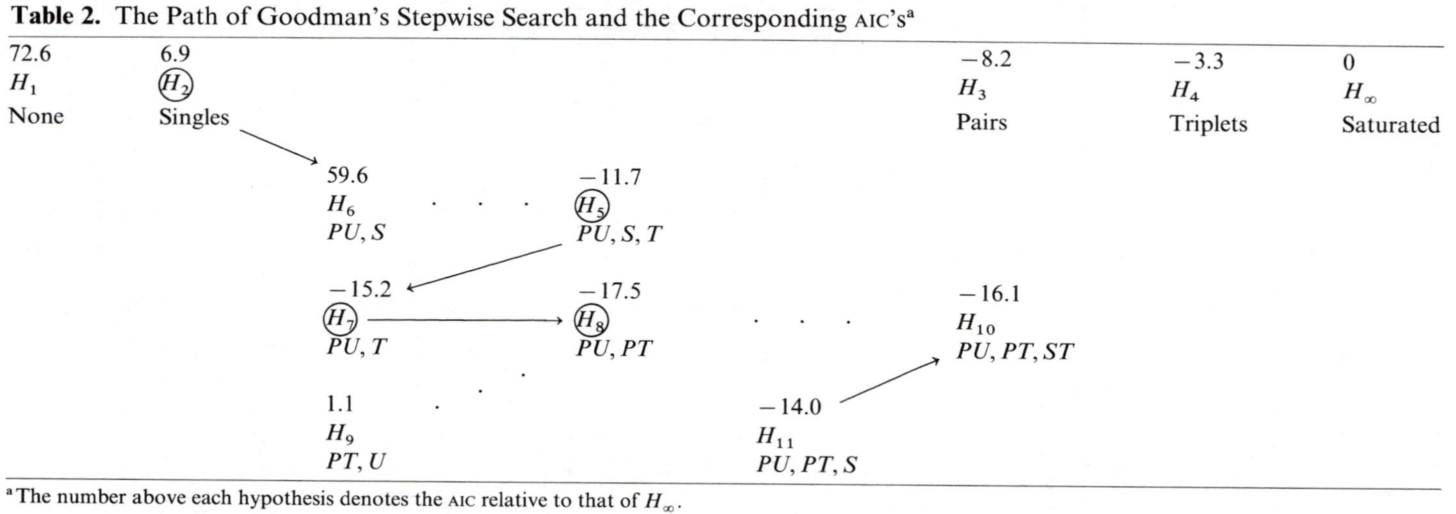

72.6	6.9				−8.2	−3.3	0
H_1	H_2				H_3	H_4	H_∞
None	Singles				Pairs	Triplets	Saturated
		59.6	−11.7				
		H_6	H_5				
		PU, S	PU, S, T				
		−15.2	−17.5		−16.1		
		H_7	H_8		H_{10}		
		PU, T	PU, PT		PU, PT, ST		
		1.1		−14.0			
		H_9		H_{11}			
		PT, U		PU, PT, S			

[a] The number above each hypothesis denotes the AIC relative to that of H_∞.

by Fienberg (1980), who also treats the detergent-user data. In particular, without proper restriction of the basic set of models, the growing number of possible models easily makes the model selection by minimum AIC quite unreliable. The situation is worse with the selection by repeated applications of tests, as the procedure cannot provide the global view of the models equivalent to the one given by the distribution of AIC over the models.

Another possible confusion is to equate AIC to a test statistic. The choice of one and the same model by the testing procedure defined with the critical level 5% and by the minimum AIC procedure in the above example is merely a coincidence. There are situations where judgements resulting from the use of AIC differ drastically from those based on the sequential application of testing procedures. One typical example of this difference is discussed in Akaike (1983b), where good agreement between the judgement by AIC and that by an expert is observed, while the test-based procedure fails to provide a reasonable explanation of the situation.

It seems that AIC has attracted the attention of people in various fields of application of statistics. This can be seen by the fact that the Institute for Scientific Information denoted the 1974 paper (Akaike, 1974) as one of the most frequently cited papers in the area of engineering, technology, and applied sciences (Akaike, 1981). However, there are rather limited number of theoretical works related to AIC. These include the discussion of the asymptotic equivalence of the minimum-AIC procedure to cross-validation, by M. Stone (1977b); modifications of the criterion by Schwarz (1978) and by Hannan and Quinn (1979); discussions of the relation to the Bayes procedure by Zellner (1978), Atkinson (1980), and Smith and Spiegelhalter (1980); and discussions of the optimality of the MAICE procedure by Akaike (1978a), Shibata (1980), and C. J. Stone (1982). Evidence of the inherent relation between the magic number 2 and the predictive point of view can be found in works by Geisser and Eddy (1979) and Leonard (1977).

When the number of possible alternatives is increased, the MAICE procedure may tend to be sensitive to sampling fluctuations. One solution to this problem is to use some averaging procedure, as is discussed in Akaike (1979). However, this brings us closer to the Bayesian modeling approach, which is discussed in the next section.

9. Entropy-Maximization Principle and the Bayes Procedure

The discussion of the concept of true model and its relation to entropy shows that there is no end to the process of statistical model building. All we can do is attempt to produce better models. When we admit this, then it is easy to accept the following very modest, yet very productive view of statistics: *all statistical activities are directed to maximizing the expected entropy of the predictive distribution in each particular application.* We call this the entropy-maximization principle (Akaike, 1977). The minimum-AIC procedure may be considered as a realization of this principle. The generality of this principle can be seen by the following discussion of the Bayesian approach to modeling.

Consider the set of models given by $\{g_k(\cdot);\, k = 1, 2, \ldots, K\}$, where $g_k(y)$ denotes a predictive distribution specified by the parameter k. Assume that we consider the use of a random mechanism for the selection of the predictive distribution. Our preferences with respect to the models is represented by the distribution of probabilities $w_k(x)$ of selecting the kth model, where $w_k(x)$ is specified by combining our knowledge of the problem and the data x. However, irrespective of the form of the true distribution of y, the following relation holds:

$$E_y \log\left\{\sum_{k=1}^{K} g_k(y) w_k(x)\right\} \geqslant \sum_{k=1}^{K} w_k(x) E_y \log g_k(y),$$

where E_y denotes the expectation with respect to the true distribution of y. This means that the entropy of the true distribution with respect to the averaged distribution $\sum g_k(y)w_k(x)$ is always greater than or equal to that with respect to the distribution chosen by the random mechanism. The entropy-maximization principle suggests that we should consider the use of the averaged distribution $\sum g_k(y)w_k(x)$ as our predictive distribution, rather than a distribution chosen by a random mechanism. Taking into account the fact that a conventional model selection procedure corresponds to a particular choice of $w_k(x)$ which takes either the value 0 or 1, the present result suggests the possibility of improved modeling for the purpose of prediction by extending the basic set of models from $\{g_k(\cdot); k = 1, 2, \ldots, K\}$ to $\{\sum g_k(\cdot)w_k;\, w_k \geqslant 0, \sum w_k = 1\}$.

The problem now is how to define $w_k(x)$. Since the distribution $w_k(x)$, which we will call the inferential distribution, is introduced to define a predictive distribution, we consider the more general problem of the selection of a predictive distribution. Assume that the variable x takes a finite number of discrete values $x = 1, 2, \ldots, I$. Before observing the value of x, we consider the selection of the predictive distribution of x, where the possible predictive distributions of x are given by $f_k(x)$. Since x is not available yet, we consider the use of a probability distribution w_k over k, defined independently of x. Thus we are specifying a probability distribution $w_k f_k(x)$ over (k, x).

When the observation produces $x = x_0$, a Bayesian will say that we should follow the Bayes procedure and replace the distribution $w_k f_k(x)$ by the distribution $w(k, x)$ which is defined by

$$w(k, x) = \begin{cases} \dfrac{w_k f_k(x_0)}{\sum\limits_{k} w_k f_k(x_0)} & \text{for } x = x_0, \\ 0 & \text{otherwise.} \end{cases}$$

The common counsel of the subjectivist that a probabilistic structure must be based on the whole set of available information is definitely correct, but it does not imply the use of the Bayes procedure. De Finetti (1972, p. 150) mentions that "according to a criterion of temporal coherency" the posterior probability must be the new probability after the person has observed the data. However, no explanation is given about the criterion. In fact Bayes' theorem does not contain any element of time, and thus its temporal interpretation is arbitrary.

There is an essential analogy between Boltzmann's derivation of the exponential family of distributions for energy and the use of the Bayes procedure. To see this we consider more generally an arbitrary distribution $\pi(k, x)$ over (k, x) and try to find a distribution $w(k, x)$ concentrated on $\{(k, x_0)\}$ and such that the Boltzman entropy with respect to the original $\pi(k, x)$ is maximum. This leads to the maximization of

$$\sum_x \sum_k w(k, x) \{\log \pi(k, x) - \log w(k, x)\} + \lambda \left\{ \sum_k w(k, x_0) - 1 \right\},$$

where λ is the Lagrange multiplier. The solution is given by

$$w(k, x) = \begin{cases} \dfrac{\pi(k, x_0)}{\sum_k \pi(k, x_0)} & \text{for } x = x_0, \\ 0 & \text{otherwise.} \end{cases}$$

This result characterizes the transition from the original distribution to the conditional distribution as the most conservative action that conforms to the observation of the data x_0 yet otherwise maximally retains the structure of the originally assumed distribution. We refer to this particular application of the maximum-entropy method of probability distribution generation as the conditioning principle. That the Bayesian rule of conditionalization is a special case of the principle of minimum information, or of maximum entropy, was also noticed by Williams (1980).

Coming back to Bayesian modeling, we can now see that the assumption of the original distribution $\pi(k, x)$ and the conditioning principle leads to the use of the "posterior distribution" $w(k, x)$ as the inferential distribution $w_k(x)$. That such a definition of the inferential distribution is a reasonable one can be shown as follows. First we assume that when k is given, y and x are independent and the distribution is given by $g_k(y)f_k(x)$. The expected performance of a predictive distribution $h(y \mid x)$ is then evaluated by $E_k E_{x|k} E_{y|k} \log h(y \mid x)$, where E_k denotes the expectation with respect to the distribution w_k, and $E_{x|k}$ and $E_{y|k}$ denote the expectations with respect to $f_k(x)$ and $g_k(y)$, respectively. We have

$$E_k E_{x|k} E_{y|k} \log h(y \mid x) = \sum_x f(x) \sum_y \sum_k g_k(y) w(k \mid x) \log h(y \mid x),$$

where $f(x) = \sum f_k(x) w_k$ and $w(k \mid x) = f_k(x) w_k / f(x)$. This quantity is maximized by putting

$$h(y \mid x) = \sum_k g_k(y) w(k \mid x),$$

which means that, *as long as we assume the validity of the original probabilistic setup*, the use of the posterior distribution $w(k \mid x)$ as the inferential distribution is the best choice. This result was recognized earlier by Kerridge (1961) and Aitchison (1975).

10. Statistical Inference and Bayesian Modeling

What the result of the preceding section has shown is that the conditioning principle leads to the best choice of the inferential distribution *under the assumption of the validity of the Bayesian model defined by* $f_k(y) f_k(x) w_k$. What will happen when we are uncertain about the choice of the "prior distribution" w_k?

Here we recall our basic observation that statistical model building is an unending process. This means that the validity of a model can only be established by a careful analysis of other possibilities. This leads to the situation where we have several alternative prior distributions $w_k^{(i)}$ ($i = 1, 2, \ldots, I$). Here we have to assume a (hyper) prior distribution $\pi(i)$ over these alternatives. When the data x are observed the posterior probability $p(i \mid x)$ of the ith model is given by the relation

$$p(i \mid x) \propto f^{(i)}(x) \pi(i),$$

where $f^{(i)}(x)$ is the likelihood of the ith Bayesian model, defined by

$$f^{(i)}(x) = \sum_k f_k(x) w_k^{(i)}.$$

Thus, even when we do not know how to specify $\pi(i)$, we can see how much relative support is given to each model by the observation x.

Based on the concept of entropy we have demonstrated the objectivity of the log likelihood as a criterion of fit of a probabilistic structure to a set of data. Since each Bayesian model specifies a probabilistic structure for the data, the objectivity holds for the log likelihood $\log f^{(i)}(x)$ as a measure of the goodness of the model. Accordingly, in spite of the firm belief of some strict Bayesians that a probabilistic structure must be constructed whenever there are several possibilities, we may safely insist that the goodness of one Bayesian model relative to another can be evaluated by the difference of the log likelihoods.

Good (1965) calls the procedure of hyperparameter estimation by maximizing the likelihood of a Bayesian model "type II maximum likelihood." The use of the likelihood for the assessment of a Bayesian model is demonstrated in an illuminating paper by Box (1980). The application to the very practical problem of seasonal adjustment is discussed by the present author (Akaike, 1980a).

The discussion of Bayesian modeling will never be complete unless we provide a procedure for the modeling of the situation where no further prior information is available for the modeling. The concept of entropy again finds an interesting application in this type of situation. It has been shown that the well-known Jeffreys ignorance prior distribution (Jeffreys, 1946) can be given an interpretation as the locally or globally impartial prior distribution (Akaike, 1978b). However, this concept is essentially dependent on the continuity of the parameter involved. Recently the present author applied the predictive point of view and the concept of entropy to define a prior distribution that will retain its impartiality for a discrete set of alternatives. For the Bayesian model discussed in the preceding section a minimax-type prior distribution is defined by minimizing

$$\max_k \sum_x f_k(x) \sum_y g_k(y) \log \left(\frac{g_k(y)}{p(y \mid x)} \right),$$

where $p(y \mid x) = \sum g_k(y) f_k(x) w_k / f(x)$ and $f(x) = \sum f_k(x) w_k$. The strict predictive point of view requires us to put $g_k(y) = f_k(y)$. It has been observed by numerical investigation that this definition leads to interesting nontrivial specifications of the prior distribution that exhibits a local uniformity (Akaike, 1983b). Related works in this area are those by Zellner (1977) and Bernardo (1979), based on the earlier work of Lindley (1956), who discussed the use of the Shannon entropy in statistics.

Do these formal procedures for generating prior distributions produce useful results? The answer can be obtained only through the detailed analysis of the final output of each Bayesian model thus obtained. An example of such an analysis is given by Akaike (1980b), where admissibility is proved for a James–Stein type of estimator of a multivariate normal distribution obtained by applying the ignorance prior to the hyperparameter of a prior distribution.

Here again we are reminded of the attitude of Boltzmann, who considered that the justification of the primary distribution used in the derivation of the distribution of the energy could only be obtained through the observation of the validity of the final result. It is the author's view that the use of a Bayesian procedure can only be justified when the procedure produces good results for those data which are "similar" to the present one and for which unequivocal judgment of the results is possible.

11. Conclusion

The predictive point of view, particularly in its strict form, and the concept of entropy can produce a unifying view of statistics. This view is not only conceptually simple and unifying, but also practical and very productive. It leads, for example, to a practical solution to the notoriously difficult problems associated with significance tests involving multiple hypotheses.

The entropy-maximization principle which is obtained by combining the predictive point of view with the concept of entropy clearly states that the search for better models is the purpose of statistical data analysis. From this perspective Bayesian modeling will often be an improvement over non-Bayesian approaches. Nevertheless, the objectivity of the log likelihood, established with the aid of entropy, as an evaluation of a stochastic model provides us with a firm basis for the selection of a Bayesian model when further Bayesian modeling of the situation is impractical.

Statistical models are formulations of our past experiences, and only new interesting problems can stimulate the development of useful models. The fundamental contribution by Boltzmann came from the deep study of one particular real problem. Thus we can see that for the development of statistics the main emphasis should be placed on the search for important practical problems.

Acknowledgements

The author is grateful to A. P. Dawid and T. Leonard for helpful comments. The reference to the work by Williams on Bayesian conditionalization was made possible by the comment of Dawid. The presentation of the paper has been significantly improved by the comments of the editors and reviewers. This work was partly supported by the United States Army under Contract No. DAAG29-80-C-0041 at the Mathematics Research Center, University of Wisconsin—Madison, and by the Ministry of Education, Science and Culture, Grant-in-Aid No.58450058 at the Institute of Statistical Mathematics.

Bibliography

Aitchison, J. (1975). "Goodness of prediction fit." *Biometrika*, **62**, 547–554.

Akaike, H. (1969). "Fitting autoregressive models for prediction." *Ann. Inst. Statist. Math.*, **21**, 243–247.

Akaike, H. (1970). "Statistical predictor identification." *Ann. Inst. Statist. Math.*, **22**, 203–217.

Akaike, H. (1973). "Information theory and an extension of the maximum likelihood principle." In B. N. Petrov and F. Csaki (eds.), *Second International Symposium on Information Theory*. Budapest: Akademiai Kiado, 267–281.

Akaike, H. (1974). "A new look at the statistical model identification." *IEEE Trans. Automat. Control*, **AC-19**, 716–723.

Akaike, H. (1977). On entropy maximization principle. In P. R. Krishnaiah, (ed.), *Applications of Statistics*. Amsterdam: North-Holland, 27–41.

Akaike, H. (1978a). "A Bayesian analysis of the minimum AIC procedure". *Ann. Inst. Statist. Math.*, **30A**, 9–14.

Akaike, H. (1978b). "A new look at the Bayes procedure". *Biometrika*, **65**, 53–59.

Akaike, H. (1979). "A Bayesian extension of the minimum AIC procedure of autoregressive model fitting." *Biometrika*, **66**, 237–242.

Akaike, H. (1980a). "Seasonal adjustment by a Bayesian modeling." *J. Time Series Anal.*, **1**, 1–13.

Akaike, H. (1980b). "Ignorance prior distribution of a hyperparameter and Stein's estimator." *Ann. Inst. Statist. Math.*, **33A**, 171–179.

Akaike, H. (1981). "Abstract and commentary on 'A new look at the statistical model identification'." *Current Contents*, Engineering, Technology and Applied Sciences, **12**, No. 51, 22.

Akaike, H. (1983a). "On minimum information prior distributions." *Ann. Inst. Statist. Math.*, **34A**, 139–149.

Akaike, H. (1983b). "Information measures and model selection." In *Proceedings of the 44th Session of ISI*, **1**, 277–291.

Atkinson, A. C. (1980). "A note on the generalized information criterion for choice of a model." *Biometrika*, **67**, 413–418.

Bahadur, R. R. (1967). An optimal property of the likelihood ratio statistic. In L. M. LeCam and J. Neyman (eds.), *Proc. 5th Berkeley Symp. Math. Statist. and Probab.*, **1**. Berkeley: Univ. of California Press, 13–26.

Bernardo, J. M. (1979). "Reference posterior distributions for Bayesian inference (with discussion)." *J. Roy. Statist. Soc. Ser. B*, **41**, 113–147.

Boltzman, L. (1872). "Weitere Studien über das Wärmegleichgewicht unter Gasmolekülen." *Wiener Berichte*, **66**, 275–370.

Boltzman, L. (1877a). "Bemerkungen über einige Probleme der mechanischen Wärmetheorie." *Wiener Berichte*, **75**, 62–100.

Boltzman, L. (1877b). "Über die Beziehung zwischen dem zweiten Hauptsatze der mechanischen Wärmetheorie und der Wahrscheinlichkeitsrechnung respective den Sätzen über das Wärmegleichgewicht." *Wiener Berichte*, **76**, 373–435.

Boltzmann, L. (1878). "Weitere Bemerkungen über einige Plobleme der mechanischen Wärmetheorie." *Wiener Berichte*, **78**, 7–46.

Box, G. E. P. (1980). "Sampling and Bayes' inference in scientific modelling and robustness." *J. Roy. Statist. Soc. Ser. A*, **143**, 383–430.

Chernoff, H. (1956). "Large sample theory—parametric case." *Ann. Math. Statist.*, **27**, 1–22.

Csiszar, I. (1975). "*I*-divergence geometry of probability distributions and minimization problems." *Ann. Probab.*, **3**, 146–158.

Fienberg, S. E. (1980). *Analysis of Cross-classified Categorical Data* (2nd ed.). Cambridge, MA: M.I.T. Press.

de Finetti, B. (1972). *Probability, Induction and Statistics*. London: Wiley.

Fisher, R. A. (1935). "The fiducial argument in statistical inference." *Ann. Eugenics*, **6**, 391–398. Paper 25 in *Contributions to Mathematical Statistics* (1950). New York: Wiley.

Fisher, R. A. (1936). "Uncertain inference." *Proc. Amer. Acad. Arts and Sciences*, **71**, 245–258.

Geisser, S. and Eddy, W. F. (1979). "A predictive approach to model selection." *J. Amer. Statist. Assoc.*, **74**, 153–160.

Good, I. J. (1965). *The Estimation of Probabilities*. Cambridge, MA: M.I.T. Press.

Goodman, L. A. (1971). "The analysis of multidimensional contingency tables: Stepwise procedures and direct estimation methods for building models for multiple classifications." *Technometrics*, **13**, 33–61.

Guttman, I. (1967). "The use of the concept of a future observation in goodness-of-fit problems." *J. Roy. Statist. Soc. Ser. B*, **29**, 83–100.

Hannan, E. J. and Quinn, B. G. (1979). "The determination of the order of an autoregression." *J. Roy. Statist. Soc. Ser. B*, **41**, 190–195.

Jaynes, E. T. (1957). "Information theory and statistical mechanics." *Phys. Rev.*, **106**, 620–630; **108**, 171–182.

Jeffreys, H. (1946). "An invariant form for the prior probability in estimation problems." *Proc. Roy. Soc. London Ser. A*, **186**, 453–461.

Kerridge, D. F. (1961). "Inaccuracy and inference." *J. Roy. Statist. Soc. Ser. B*, **23**, 184–194.

Kullback, S. (1959). *Information Theory and Statistics*. New York: Wiley.

Kullback, S. and Leibler, R. A. (1951). "On information and sufficiency." *Ann. Math. Statist.*, **22**, 79–86.

Leonard, T. (1977). "A Bayesian approach to some multinomial estimation and pretesting problems." *J. Amer. Statist. Assoc.*, **72**, 869–876.

Leonard, T. and Ord, K. (1976). "An investigation of the *F*-test procedure as an estimation short-cut." *J. Roy. Statist. Soc. Ser. B*, **38**, 95–98.

Lindley, D. V. (1956). "On a measure of the information provided by an experiment." *Ann. Math. Statist.*, **27**, 986–1005.

Mallows, C. L. (1973). "Some comments on C_p." *Technometrics*, **15**, 661–675.

Pearson, K. (1929). "Laplace, being extracts from lectures delivered by Karl Pearson." *Biometrika*, **21**, 202–216.

Rao, C. R. (1961). "Asymptotic efficiency and limiting information." In J. Neyman, (ed.), *Proc. 4th Berkeley Symp. Math. Statist. and Probab.*, **1**. Berkeley: Univ. of California Press, 531–548.

Rao, C. R. (1962). "Efficient estimates and optimum inference procedures in large samples." *J. Roy. Statist. Soc. Ser. B*, **24**, 46–72.

Sanov, I. N. (1957). "On the probability of large deviations of random variables." (in Russian). *Mat. Sbornik N.S.*, **42**, No. 84, 11–44. English transl., *Selected Transl. Math. Statist. Probab.*, **1** (1961), 213–244.

Schwarz, G. (1978). "Estimating the dimension of a model." *Ann. Statist.*, **6**, 461–464.

Shannon, C. E. and Weaver, W. (1949). *The Mathematical Theory of Communication*. Urbana: Univ. of Illinois Press.

Shibata, R. (1980). "Asymptotically efficient selection of the order of the model for estimating parameter of a linear process." *Ann. Statist.*, **8**, 147–164.

Smith, A. F. M. and Spiegelhalter, D. J. (1980). "Bayes factors and choice criteria for linear models." *J. Roy. Statist. Soc. Ser. B*, **42**, 213–220.

Stigler, S. M. (1975). "The transition from point to distribution estimation." In *Proceedings of the 40th ISI Meeting*, **2**, 332–340.

Stone, C. J. (1982). "Local asymptotic admissibility of a generalization of Akaike's model selection rule." *Ann. Inst. Statist. Math.*, **34A**, 123–133.

Stone, M. (1974). "Large deviations of empirical probability measures." *Ann. Statist.*, **2**, 362–366.

Stone, M. (1977a). "Asymptotics for and against cross-validation." *Biometrika*, **64**, 29–35.

Stone, M. (1977b). "Asymptotics equivalence of choice of models by cross-validation and Akaike's criterion." *J. Roy. Statist. Soc. Ser. B*, **39**, 44–47.

Wald, A. (1943). "Tests of statistical hypotheses concerning several parameters when the number of observations is large." *Trans. Amer. Math. Soc.*, **54**, 426–482.

Williams, P. M. (1980). "Bayesian conditionalization and the principle of minimum information." *Brit. J. Philos. Sci.*, **31**, 131–144.

Zellner, A. (1977). "Maximal data information prior distributions." In A. Aykac and C. Brumat (eds.), *New Developments in the Applications of Bayesian Methods*. Amsterdam: North-Holland, 211–232.

Zellner, A. (1978). "Jeffreys–Bayes posterior odds ratio and the Akaike information criterion for discriminating between models." *Economic Letters*, **1**, 337–342.

CHAPTER 2

Statistical Developments in World War II: An International Perspective

Stephen E. Fienberg

Abstract

The impact of World War II on the practice and theory of statistics was substantial. Major new statistical developments occurred as part of the wartime activities, especially in the United Kingdom and the United States. The paper provides some background to these developments and serves as an introduction to a longer and more detailed description by Barnard and Plackett on *Statistics in the United Kingdom, 1939–45*.

In the United States, the 40th anniversary of the June, 1944 Normandy invasion brought forth numerous accounts of activities surrounding World War II, but most of these relate to the battlefield. There are some descriptions that recount the activities of the physicists at Los Alamos, but virtually none chronicle the dramatic development of ideas in other areas of the sciences as part of the war effort. In many ways, statistics as a field was one of the great beneficiaries of war-related scientific activities, especially in the United Kingdom and the United States. The following article by Barnard and Plackett describes the ways in which statistics flourished in the United Kingdom as part of the war effort. This brief note is intended as a backdrop to their paper, pointing the interested reader to accounts of parallel activities in the United States, and summarizing some of what is known about the wartime activities of statisticians in other countries.

The contrast between World Wars I and II, as far as statistical development goes, is enormous. Kendall (1978), in his description of the history of statistics, associates the beginning of the modern theory of statistics with the work of Galton, Edgeworth, Pearson, Weldon, and Yule that began about 1890. Yet, at the beginning of World War I, statistics as a field had not really been developed; e.g. "the then current ideas on estimation from a sample were intuitive and far from clear" (Kendall, 1978). There appears to have been little or no attempt to use statistical ideas as part of the wartime effort of either side during World War I, and statisticians in the United Kingdom especially had considerable difficulty pursuing their scholarly activities. R. A. Fisher, for

Key words and phrases: History of statistics, quality control, weapons research example, after being rejected by the army because of his eyesight, became a school teacher "in the wilderness," and wartime difficulties delayed substantially the publication in *Biometrika* of Karl Pearson's "Cooperative Study" on the distribution of the correlation coefficient (Box, 1978, Chapter 2, 3).

The activities of the International Statistical Institute itself reflected the state of official statistics at the time of World War I:

> The war of 1914–1918 dealt a severe blow to the Institute as to many other international organizations but thanks to the creation of the Permanent Office in 1913, the Institute continued to function as such at its seat in the Hague. No biennial sessions or elections of members or officers were possible and the Secretary-General, Director of the Permanent Office was able to keep in touch with only a few of the members, but the work of publishing volumes of the Annuaire International de Statistique continued. The number of members fell from 204 in 1913 to about 150 in 1919 and of these nearly 90% were Europeans. The division of Europe into two hostile camps resulted in a similar division among the members and officers of the Institute.... National sentiments were strong and there was even some talk of disbanding the Institute. (Nixon, 1960, pp. 27–28)

Perhaps it was the lack of demand for technical support in World War I, or perhaps statistics was simply not well enough developed at the time, for others to realize how it might be used as part of wartime activities. Whatever the explanation, it is clear that statistical work was slowed by World War I, but flourished between the wars.

The situation at the outbreak of World War II was dramatically different. By 1939, as Barnard and Plackett note, a considerable amount of basic statistical theory had been firmly established, and with this work as a foundation, the practical application of statistics to military and naval problems could realistically be pursued. And as with statistical work for all real problems, these new applications served to stimulate substantial new theoretical developments.

In the United States, World War II saw the development of several major groups of statisticians, especially at Columbia and Princeton Universities. W. A. Wallis (1980) describes many of the activities (and publications) of the Columbia group, which under his direction included such distinguished statisticians and economists as A. Bowker, C. Eisenhart, M. Friedman, M. Girshick, H. Hotelling, E. Paulson, L. J. Savage, G. Stigler, H. Solomon, A. Wald and J. Wolfowitz. At Princeton, under the direction of S. S. Wilks, there were R. L. Anderson, T. W. Anderson, W. G. Cochran, P. McCarthy, A. Mood, and D. Votaw (see R. L. Anderson's 1982 recollections), and the Princeton group also had a branch working in New York City under the direction of John Williams that included F. Mosteller. In addition, there was another Princeton research team doing fire-control work that included G. W. Brown, W. J. Dixon, P. S. Dwyer, I. E. Siegal, A. Tucker, J. W. Tukey, and C. Winsor (Harshbarger, 1976).

The output of these groups was substantial, as was their subsequent influence. But they did not encompass all of the U.S. wartime statistical researchers. At the University of California, Berkeley, J. Neyman organized a group, initially under subcontract from Princeton, for work on bombing research that included E. Fix and E. Scott, and subsequently E. Lehmann and others (Reid, 1982, pp. 180–199). At MIT, N. Wiener's work on the design of fire-control apparatus for antiaircraft guns led to his preparation for restricted circulation in 1942 of his memorandum, *The Extrapolation, Interpolation, and Smoothing of Time Series*. And Wm. Kruskal (1980) describes the statistical work at the U.S. Naval Proving Ground in Dahlgren, Virginia, and how it was influenced by Hotelling. There were also some significant developments at the Allegheny Ballistics Laboratory, in particular in connection with the "up-and-down method" of sensitivity testing for probit analysis. In a quite different setting, S. Kullback applied statistical ideas to cryptanalysis.

In addition, W. E. Deming, M. H. Hansen, F. F. Stephan, W. G. Madow, and others worked at the development and implementation of sampling theory in the U.S. Bureau of the Census and the government statistical agencies, and Deming collaborated with Wilks and W. Shewart to initiate a nationwide program in statistical-quality-control instruction. Although ideas on quality control had been around for at least ten years, the introduction of these ideas to those involved in industrial production was clearly a wartime phenomenon—virtually every industry felt it had to have its own quality-control expert. Out of this program grew the American Society for Quality Control, and Deming's postwar efforts in Japan.

Besides these activities in the United States, there were others that involved statistics, but under different labels, e.g. the work by S. Stouffer, F. Stephan, and others on social statistics, which led after the war to the publication of *The American Soldier* (Stouffer, (1949). Also, there was the work of various operations-research and operations-analysis groups, much of which was statistical in nature. The most important of these groups was the one with the 8th Air Force, which included J. Youden; and the one with the 20th Air Force in Guam, which included J. Hodges, E. Lehmann, and G. Nicholson.

Statistical research continued to be published during the war in the U.S. in journals such as the *Annals of Mathematical Statistics*, but many of the statistical publications growing out of the war-related activities mentioned above did not appear until several years after the war. For an extreme example, there is Wald's wartime work on vulnerability, which was not published until 1984 (see Mangel and Samaniego, 1984).

In Australia, by contrast, World War II put a virtual stop to most statistical work. E. J. G. Pitman (1982) notes that in addition to his teaching duties at the University of Tasmania, he had heavy administrative duties and he worked for the Royal Australian Air Force as Education Officer. "I had no time or energy for research." H. O. Lancaster spent World War II as a Medical Officer in the Australian Imperial Force, and it was as part of an epidemiological investigation of intestinal protozoal infections, in 1942, that he was first exposed to statistical problems in the analysis of $2 \times 2 \times 2$ contingency tables—these provided the focus of much of his research following the war. In China, P-L. Hsu struggled to continue his statistical work, first at Peking University, and later at the Southwestern Associated University at Kunming, Yunnan. As Reid (1982, p. 194) notes, "Conditions in China were chaotic, the country divided into three parts—one under the Japanese, a second under the Communists, and a third under the Nationalists. At one point Hsu was doing his scientific research in a cave."

In my own home country of Canada, wartime statistical developments were limited, and the major activities parallelled some of those in the U.S. For example, a physical chemist, A. E. R. Westman, led a quality-control effort (which ended shortly after the war), and N. Keyfitz introduced sampling theory into government data collection. These sampling activities also had their counterpart in India, where groups under the direction of P. C. Mahalanobis and P. V. Sukhatme developed methods for forecasting agricultural production as well as for large-scale social surveys. Other important work on statistical theory by R. C. Bose, C. R. Rao, and S. N. Roy and others during this period did not stem specifically from war-related efforts.

In Europe outside the United Kingdom there were some major statistical developments during World War II. For example, although the deputy head of the Central Statistical Office in Norway was a leading member of the resistance, under instructions from the German occupying power the Office set up a system of national financial accounts which was quite similar to that set up in England. The purpose of this system was to present an account for reparations to the British Government after its anticipated defeat. In the end, of course, an account based on the same system was presented to the German government. But elsewhere life was more difficult. In France, although M. Fréchet developed the general form of the information inequality now known as the Cramér–Rao inequality, a senior member of the statistical office, M. Carmille, died in a concentration camp after refusing to give information about the number of Jews living in various districts of the country. Kerrich in Denmark and H. Hamaker in Holland passed the long hours of internment by carrying out randomization experiments which subsequently led to interesting research papers. In the U.S.S.R., A. N. Kolmogorov and others worked on statistical and mathematical problems associated with anti-aircraft gunnery.

The statistical situation in Germany itself is a little more difficult to put together, but as Hoeffding (1982) notes, prior to the war "Probability and statistics were very poorly represented in Berlin at the time. Hitler had become Chancellor of the Reich in 1933 and Richard von Mises, who was Jewish, had already left the university." Hoeffding goes on to describe some of his experiences at the Berlin actuarial institute during the War, but they reveal little about possible statistical activity in Germany at the time.

The political chaos in Europe was also reflected in official statistics activities, especially at the ISI:

> The outbreak of the second World War in 1939, like that of the first World War, created a crisis in the affairs of the Institute, but whereas the first crisis was overcome in due course and the Institute then continued to function, more or less on the same lines as before and under the same statutes, the second led to fundamental changes in the organization, constitution and aims of the Institute.... The Prague session of the Institute in September 1938 closed ... prematurely on the second day owing to the political events of that momentous month.
>
> The occupation of the Netherlands in May 1940 by the German forces put an end to all negotiations on [an invitation to hold a session in Washington]. The offices of the Institute were in due course evacuated and the Secretary General, Director of the Permanent Office (Methorst) transferred the office to the Peace Palace where under great difficulties, he managed to carry on its work.... Both Spain and the U.S.A. had offered hospitality to the Permanent Office but the Secretary General considered it advisable to carry on in the Hague. In 1941 and 1942, German statisticians approached the Secretary General and other statisticians in the occupied territories concerning the formation of a European Statistical Institute, but their *démarches* remained fruitless. Contact with most of the members of the Institute was no longer possible though the Secretary General was able to maintain some contact with the President (Julin) in Belgium and a vice-president (Huber) in France. (Nixon, 1960, pp. 39–40)

Finally, as H. Wold (1982) notes: "Sweden remained a quiet corner during the war, and in 1942 as Professor of Statistics at Uppsala I could begin to explore demand analysis and related problems both extensively and in depth."

Among all of the countries of Western Europe, Britain was unique in being within the war zone, and yet having a continuity of government and social organization to allow for a truly remarkable set of statistical developments. And Britain, too, was the home of Edgeworth, Fisher, Galton, Gosset, Jeffreys, the Pearsons, and Yule—the pioneers of the foundations of statistical methodology and theory. It was for these reasons that A.C. Atkinson and I invited George Barnard and Robin Plackett to prepare their paper "*Statistics in the United Kingdom, 1939–1945*" for inclusion in this centenary volume.

The picture I have painted of war-related statistical activities, even within the United States, is far from complete. Unlike Barnard and Plackett, I have had to rely in large part on secondary sources. I am sure that there must have been a number of people, all over the world, who did important statistical work during World War II, but I was unable to embark on anything close to a proper survey outside the United States. Such an effort would be important for the profession, and it becomes more difficult with the passage of time, as we have fewer and fewer first-hand accounts of actual activities and their impact. For that reason I invite readers to send me information on omitted people and the important statistical things they did during the war. Even more interesting than a comprehensive listing of war-related statistical work, however, would be a study of why statistical ideas and techniques were used effectively in some settings, and not used or misused in others. Someone needs to answer this deeper question.

Acknowledgements

The preparation of this paper was partially supported by the Office of Naval Research Contract N00014-80-C-0637 at Carnegie-Mellon University, and by the Guggenheim Foundation while the author was a Fellow at the Center for Advanced Study in the Behavioral Sciences. Reproduction in whole or part is permitted for any purpose of the United States Government. I am indebted to G. A. Barnard for much of the account of the European activities described here, and to many others who provided comments on an early draft: A. C. Atkinson, D. DeLury, M. Hansen, Wm. Kruskal, E. Lehmann, F. Mosteller, R. Plackett, C. R. Rao, S. Stigler, J. Tanur, and W. A. Wallis.

Bibliography

Anderson, R. L. (1982). "My experience as a statistician: From the farm to the university." In J. Gani (ed.) *The Making of Statisticians*, New York: Springer, 130–148.

Barnard, G. A. and Plackett, R. L. (1985). "Statistics in the United Kingdom, 1939–45." In this volume, Chapter 3.

Box, J. F. (1978). *R. A. Fisher, The Life of a Scientist*. New York: Wiley.

Harshbarger, B. (1976). "History of early developments of modern statistics in America (1920–1944)." In D. B. Owen (ed.) *On the History of Statistics and Probability*. New York: Marcel Dekker, 133–145.

Hoeffding, W. (1982). "A statistician's progress from Berlin to Chapel Hill." In J. Gani (ed.), *The Making of Statisticians*. New York: Springer, 100–109.

Kendall, M. G. (1978). "Statistics: The history of statistical method." In Wm. H. Kruskal and J. M. Tanur (eds.) *International Encyclopedia of Statistics*, **2**. New York: Free Press, 1093–1101.

Kruskal, Wm. H. (1980). "Comment: First interactions with Harold Hotelling; testing the Norden bombsight." J. *Amer. Statist. Assoc.*, **75**, 331–333.

Lancaster, H. O. (1982). "From medicine through medical to mathematical statistics: Autobiographical notes." In J. Gani (ed.) *The Making of Statisticians*. New York: Springer, 236–252.

Mangel, M. and Samaniego, F.J. (1984). "Abraham Wald's work on aircraft survivability (with discussion)." J. *Amer. Statist*. Assoc., **79**, 259–271.

Nixon, J. W. (1960). *A History of the International Statistical Institute, 1885–1960*. The Hague: International Statistical Institute.

Pitman, E. J. G. (1982). "Reminiscences of a mathematician who strayed into statistics." In J. Gani (ed.) *The Making of Statisticians*. New York: Springer, 112–125.

Reid, C. (1982). *Neyman—from Life*. New York: Springer.

Stouffer, S. A. et al. (1949). *The American Soldier. Studies in Social Psychology in World War II*. Vol. 1 and 2. Princeton U. P.

Wallis, W. A. (1980). "The Statistical Research Group, 1942–1945 (with discussion)." J. *Amer. Statist. Assoc.*, **75**, 320–335.

Wold, H. (1982). "Models for knowledge." In J. Gani (ed.), *The Making of Statisticians*. New York: Springer, 190–212.

CHAPTER 3

Statistics in the United Kingdom,* 1939–45

G. A. Barnard and R. L. Plackett

Abstract

During the period between the two world wars, the scope of statistical methods was extended, and the foundations of modern statistical theory were laid. Statistics in Britain continued a gradual process of change. When war came, the pace of change increased. The official statistics were reformed by the creation of a central department. Statistical problems arose in different aspects of the war effort, and statistical advice was widely sought. Many young people were trained to join the few with experience. When peace returned, the statistical scene in Britain had been completely transformed.

1. Introduction

This paper presents an account of statistical activities in the United Kingdom during the Second World War, which began in 1939 and ended in 1945. Some of these activities were completely new, but others extended those already existing, and Section 2 therefore describes the statistical resources of the country between the First World War and the Second.

During this period, the scope of statistical methods was extended, and the foundations of modern statistical theory were laid. The long career of Karl Pearson drew to a close, the powerful techniques of R. A. Fisher were introduced, quality-control charts were developed by W. A. Shewhart, and the mathematical basis of probability theory was established by A. N. Kolmogorov. Structural changes of the magnitude achieved by the pioneers from different nations could not be assimilated in five or ten years, and in 1939 the great expansion in the theory and applications of statistics and probability still lay ahead.

Before the war, little progress was made in reforming the official statistics of

*The United Kingdom consists of Great Britain and Northern Ireland. Great Britain comprises the island consisting of England, Scotland, and Wales, together with some smaller islands. The United Kingdom is sometimes erroneously referred to as England, but this terminology does grave injustice to nations equally old which make up the whole. When convenient, Britain is used as a shorthand for the United Kingdom.

Key words and phrases: computing, cryptanalysis, experimental design, operational research, quality control, Royal Statistical Society, sample surveys, sampling inspection, sequential analysis.

the United Kingdom. Notwithstanding considerable efforts by individuals like A. L. Bowley who saw the need for change, and by the Royal Statistical Society which supported these efforts, successive Governments never accepted the proposed improvements. When war came, those in responsible positions saw at once that reform of the official statistics was one of their first important tasks. The account of the wartime period begins by explaining in Section 3 what was accomplished, and how the postwar structure came to be set up.

The Department of Statistics at Rothamsted Experimental Station had for long been consulted by outside bodies on the statistical methods appropriate to the analysis of widely varying problems. It was therefore a natural consequence that a great deal of time of the Department should be devoted to the study of urgent war problems, where accurate information was demanded as a basis for immediate administrative action. The resulting activities are outlined in Section 4.

Production of war materials began at a low level, but in the course of time grew to a massive scale, and the need for statistical methods of quality control and sampling inspection became apparent. A large group was formed for this purpose, mostly young and at first inexperienced, but with a leaven of trained statisticians. This group was part of the scientific research division of the Ministry of Supply, and eventually became known as SR17. The main account continues in Section 5 by describing how SR17 came to be established, what functions were performed, what problems addressed, and what members involved, as well as personal anecdotes to give a flavor of those far-off days.

Statistical problems arose in many different aspects of the war effort, and statistical advice was widely sought. Trained statisticians were soon absorbed by the new governmental organizations that came into being, and their experience was used in a diversity of novel fields, but their numbers were wholly inadequate to meet the demand. Some of the best intellects in biology, mathematics, and physics turned their attention to statistical matters when required in the course of widely ranging duties. They achieved outstanding successes in operational research and cryptanalysis, especially during 1940–1942 when the country was isolated and under continuous threat. Many small groups were formed, and regrouping continued throughout the whole period. The main account concludes in Section 6 with a survey of their activities, and names many of the investigators concerned.

Consequences of the war for statistics in Britain are outlined very briefly in Section 7.

2. Between the Wars

The First World War ended in November 1918, and the Second began in September 1939. During the period which intervened, a huge progress in statistical methodology was achieved by a small number of people in several nations. But statistics in Britain continued the process of gradual change which had been taking place since the time of Graunt. The background at the start of this period together with the knowledge and experience acquired before the close are outlined in what follows.

Up to 1939, the principal interests of the Royal Statistical Society remained those defined in its Charter, and the number of Fellows was almost constant at about one thousand. According to the Charter, the Society was established "to collect, arrange, digest and publish facts illustrating the conditions and prospect of society in its material, social and moral relations." Nearly all the papers given at the main meetings of the Society were concerned with these traditional matters. Developments in statistical theory and methodology came to be regarded as legitimate fields of interest only slowly, although discussed at meetings of the Study Group, formed in 1928. The crest of the Society was originally a wheatsheaf with the motto *Aliis exterendum* (to be threshed out by others). Although the motto was removed from the crest in 1857, both the wheatsheaf and the feeling behind the motto survived.

However, the Society had always been active within the area defined by the Charter, and especially with the official statistics of Government. For fifty years, members of the Society had from time to time called attention to the imperfection of official statistics, and when the First World War ended, another attempt was made to reform them. The chief instigator this time was A. L. Bowley. He saw a need to extend Government responsibility for statistics and to coordinate the work of different departments. Accordingly, he proposed a Central Thinking Office of Statistics and the creation of a trained statistical officer class in the Civil Service. Many of his ideas were embodied in a formal petition made in 1919 by the Royal Statistical Society, but Government considered the proposals to be impracticable.

The traditional view of statistics was changed largely by the efforts of a few outstanding individuals directed towards the solution of problems with a practical basis. Karl Pearson had for many years been using statistical methods to study evolution and heredity, but his considerable energies were devoted to war work from 1914 to 1919. The Department of Applied Statistics at University College London was formally opened in 1920, when he was sixty-three years old, and he remained as active Head until his retirement in 1933 from the Galton Chair of Eugenics. Between 1894 and 1930, University College London was the only place in the United Kingdom where statistical theory and its applications were taught to any depth, and graduate students, whether from this country or abroad, who wished to learn more of the techniques developed by the Biometric School, came there year after year. Among those who came from abroad was J. Neyman, whose work with Egon S. Pearson on the theory of testing statistical hypotheses takes pride of place in the story of the Department up to 1939. Throughout the whole of the period under review, *Biometrika* published the research of the Department of Applied Statistics and of many other individuals.

In 1919, R. A. Fisher was appointed to the staff of Rothamsted Experimental Station in order to reexamine by "modern statistical methods" the data accumulated from long-term field trials. His response to the statistical problems of biological and agricultural research workers was to introduce and develop many original ideas, both in the theory of statistical inference, and in the design and analysis of experiments. *Statistical Methods for Research Workers*, first published in 1925, made the new techniques available to biologists. Rothamsted became a centre of statistical research, and Fisher made statisticians think about fundamental ideas. Some who had received their first training at University College London went to Rothamsted for further study, notably J. O. Irwin and J. Wishart. Many "voluntary workers" came from overseas, particularly North America and India, to work with him for varying periods. In 1933, Fisher left Rothamsted to go to University College London, where he had a similar stream of students and more senior statisticians. He continued his interest in agricultural problems, and *The Design of Experiments* was published in 1935. His successor as head of the Department of Statistics was F. Yates, who, assisted by W. G. Cochran until 1939, continued and extended the activities which Fisher had begun.

From 1919 to 1936, the only Chair of Statistics was occupied by Bowley at the London School of Economics, where he had been teaching since 1895. He was a pioneer in two areas: what is now econometrics, and sampling techniques in social surveys. Under his influence, the teaching of statistics advanced greatly in the 1930's and by 1938–1939 included a theoretical treatment on modern lines. His book *Elements of Statistics* went through six editions from 1901 to 1937. The London and Cambridge Economic Service, founded in 1923, was edited by Bowley from the beginning until 1945.

At Edinburgh, the mathematical laboratory founded by E. T. Whittaker in 1914 continued under his successor A. C. Aitken to have a strong association with the actuaries in that center of insurance. *The Calculus of Observations* (Whittaker and Robinson, 1924) was subtitled *A Treatise on Numerical Mathematics*, but *Statistical Mathematics* (Aitken, 1939) reflected a change in outlook.

Statistical methods in British industry were introduced by W. S. Gosset at the Dublin brewery of Arthur Guinness Son & Co. in the early years of this century. Another pioneer was B. P. Dudding, who applied statistical methods in British manufacturing industry after 1919 at the Research Laboratories of the General Electric Company. He was concerned to study the manufacturing processes of lamps and compare the quality of lamps made by new methods with those made by the usual factory routine. In 1925, L. H. C. Tippett began his long career with the British Cotton Industry Research Association (The Shirley Institute), where he helped the scientists to apply the statistical approach to their experimental work on spinning and weaving. His book *The Methods of Statistics* appeared in 1931.

A great advance in the presentation of data from industrial mass production was made when W. A. Shewhart developed quality-control charts at Bell Telephone Laboratories in the United States. *Economic Control of Quality of*

Manufactured Product, published in 1931, was a landmark in the industrial applications of statistics. In the following year, Shewhart gave three lectures at University College London on the role of statistical method in industrial standardization. There were two important consequences: the formation in 1933 of the Industrial and Agricultural Research Section of the Royal Statistical Society, and the publication in 1935 by the British Standards Institution of BS 600—*The Application of Statistical Methods to Industrial Standardization and Quality Control.* Egon Pearson played a leading role in these activities, and BS 600 appeared under his name.

The papers read before the Industrial and Agricultural Research Section in the years 1934–1939 all placed the field of application first and statistical methodology second. During this period, statistical appointments were made in the research departments of trade and industry, for example in Imperial Chemical Industries (O. L. Davies, M. S. Bartlett), Boots Pure Drug Company (E. C. Fieller), the Forest Products Research Laboratory (E. D. van Rest), and the Wool Industries Research Association (H. E. Daniels).

In the 1930s a great deal was done to develop statistical techniques for use in economic and social enquiries. There were teaching posts at Cambridge, Manchester, and Liverpool, for example, as well as at the London School of Economics. Discussions took place in the Study Groups of the Royal and Manchester Statistical Societies. The Oxford Institute of Statistics was founded, and there were the beginnings of the National Institute of Economic and Social Research. As regards medical statistics, M. Greenwood became the first professor of Epidemiology and Vital Statistics when the London School of Hygiene was founded in 1928. *Principles of Medical Statistics* by his colleague, A. Bradford Hill, appeared in 1937, and an 11th edition is currently in preparation.

But university teaching of advanced theoretical statistics was very limited. When Karl Pearson retired in 1933, Fisher succeeded him in the Chair of Eugenics, the Department of Applied Statistics was split into two, and Egon Pearson became Reader in a separate Department of Statistics. When Bowley retired in 1936, his Chair remained unfilled until 1944. At Cambridge, G. U. Yule was Lecturer and latterly Reader in Statistics from 1912 to 1931. His *Introduction to the Theory of Statistics* went through ten editions between 1911 and 1932, and continued as the version rewritten by M. G. Kendall. However, Yule's appointment was in the Faculty of Agriculture, he taught economists and biological students, and his successor Wishart was single-handed until M. S. Bartlett was appointed to the Faculty of Mathematics in 1938. The last year of peace was graced by the appearance of *Theory of Probability* (H. Jeffreys).

Statistical work in other countries has always been important for statistics in Britain. Thus, A. Quetelet helped to found the Statistical Society of London, and K. Pearson took note of work by A. Bravais, E. Czuber, E. L. DeForest, and T. N. Thiele. After 1930, these influences began to grow strongly; for example, Neyman's visit drew attention to research in Poland, notably through his 1934 paper on the role of statistical theory in sample surveys; and A. N. Kolmogorov continued the long tradition of Russian advances in probability theory with *Grundbegriffe der Wahrscheinlichkeitsrechnung* (1933). But the most diverse—and ultimately the most influential—activities took place in the United States and Canada. They included: the foundation by H. C. Carver of *The Annals of Mathematical Statistics* in 1930 and the foundation of a Statistical Laboratory at Iowa State College in 1933; the work on sample surveys by C. L. Dedrick and M. H. Hansen, on medical statistics by R. Pearl and L. T. Reed, on statistical theory by H. Hotelling and S. S. Wilks, and on probability theory by N. Wiener and J. L. Doob; and the books by H. L. Rietz (*Mathematical Statistics*, 1927), T. C. Fry (*Probability and Its Engineering Uses*, 1928), G. W. Snedecor (*Statistical Methods Applied to Experiments in Agriculture and Biology*, 1937), W. E. Deming (*Statistical Adjustment of Data*, 1938), and C. H. Goulden (*Methods of Statistical Analysis*, 1939).

3. Official Statistics

Before the war, there was no central office to connect the various statistical departments of Government. The statisticians were men of erudition and integrity, but they came from the Administrative or Executive classes of the Civil Service and had no professional statistical training. They were few in number, and had little scope to develop the statistics that the nation needed. The preparation of estimates of national income and expenditure, changes in total consumption, and changes in stocks were undertaken mainly by private individuals. Those who found themselves responsible for the conduct of the country in rapidly changing conditions had very little factual information to help them. Woolton (1946) gives an example.

> When I was called upon to clothe the Army in May 1939, the War Office had no statistical evidence to assist me. Here there could be no doubt that to guess was to endanger the chance of victory and the security of the State. I had the greatest difficulty in arriving at any figures that would show how many suits of uniform and how many boots were involved. I found that the estimates of the potential loss due to various forms of military reverse—such as attacks on supply dumps by gas, or on troops in action—were little more than guesses, and in those urgent months before the war I called on the services of Professor Greenwood ... to help me to create a statistical basis on which the operations of this aspect of the war effort could be built.

In 1938, Lord Stamp was summoned back to England to deal with railway problems of mobilization and evacuation. During the summer of 1939 he joined Sir Henry Clay and Sir Hubert Henderson in a three-man team to survey the financial and economic plans for the war. The Central Economic Information Service was formed around December 1939 to assist the Stamp Survey, and the staff included H. Campion, J. Jewkes, and Austin Robinson.

They found that the statistical services of the Government had to be increased, new statistical series devised, and the speed of collection changed.

When war broke out on 3 September 1939, Winston S. Churchill became First Lord of the Admiralty, a post he had occupied during the First World War. He immediately asked a trusted friend and confidant of twenty years to advise him on scientific matters. This was Professor F. A. Lindemann, always known to his friends as "the Prof." In October 1939, Churchill asked Lindemann to form a Statistical Branch, and he in turn asked R. F. Harrod to suggest an economist. It was thus that G. D. A. MacDougall came to be appointed, and he was soon followed by another batch of Oxford economists consisting of G. L. S. Shackle, Helen Makower, and H. W. Robinson. Later recruits included Harrod himself and D. G. Champernowne. The duties of S Branch, as it was called, are recorded by Churchill (1948, p. 580). When he became Prime Minister in May 1940, S Branch was transformed into the Prime Minister's Statistical Section. This was a department of size twenty which was concerned with all aspects of the war effort and which remained in existence until July 1945. The continuing task of the Section was the construction of graphs in pen and ink or color, to keep Churchill informed of the progress of events. Skill was necessary in devising the right mode of presentation for matters concerned with economics, defence, equipment, manpower, machine tools, imports, and so on. The Section was always a personal advisory service to Churchill, and never the central statistical organization of the Government.

After the fall of France in June 1940, there were big changes in Ministerial structure. Sir John Anderson became Lord President of the Council, and more of the best economists in the country were gathered to form the personnel of the War Cabinet Office and other Departments. At the end of 1940, the group in the War Cabinet Office was divided into two, thus forming the Central Statistical Office and the Economic Section. The first head of the Central Statistical Office was H. Campion, and the first Director of the Economic Section was J. Jewkes. A minute circulated to the War Cabinet on 27 January 1941 included the following:

> A Central Statistical Office is being established whose duty will comprise the collection from Government Departments of a regular series of figures on a coherent and well-ordered basis covering the development of our war effort. The Prime Minister has directed that the figures so collected should form an agreed *corpus* which will be accepted and used without question, not only in interdepartmental discussions, but in the preparation of documents submitted to Ministers for circulation to the War Cabinet and to War Cabinet Committees. This section, which will take over the work of issuing Statistical Digests hitherto performed by the Economic Section of the War Cabinet Secretariat, will form part of the staff of the War Cabinet Offices.

The Central Statistical Office served all Ministers and all Departments. There were very few people with statistical training, and actuaries were recruited from the insurance companies. The real dilemma was that people with mathematical training were ideal, except that they had no experience of the subject. They were in any case in short supply—at the beginning of the war physicists were a reserved occupation, kept out of army service because their special training was needed elsewhere. As an afterthought, mathematicians were told to register as physicists, but this message was not received by all of them. Statistician too became a reserved occupation. To spread the available talent as economically as possible an infrastructure was built up where staff occupied key positions in the different Ministries, so as to provide focal centers for information. Matters were complicated by the dispersal of these organizations for safety reasons: thus Customs and Excise was in Buxton, and the Ministry of Food in Colwyn Bay. A system was created where the figures were agreed on step by step and published in the form of Digests, with the Ministry responsible named at the foot of each Table. On this basis the consumption of raw materials and the pattern of imports could be controlled. According to Harrod (1969), the Central Statistical Office supplied figures without comment other than purely statistical, whereas the Economic Section made comments and tendered advice. Nevertheless, statisticians became involved with questions of administration and the formation of policy. During the latter part of the war, the realization that a devolved system like the Central Statistical Office could be adapted to postwar conditions led to the creation of the Government Statistical Service, with staff who had professional statistical training and could function in any part of the organization. In 1945, the Central Statistical Office was made permanent, and subsequently prepared a *Statistical Digest of the War*.

Collaboration with the United States preceded the entry of that country into the war on 7 December 1941. At the end of 1940, R. G. D. Allen was posted to Washington to serve in the British Supply Council during 1941–1942, a vital period of the financing of the British war effort. From 1942 to 1945, he was British Director of Research and Statistics in the Combined Production and Resources Board, directly concerned with most aspects of lend–lease and reciprocal aid, and with military procurement in particular. A vast range of statistics arising from a variety of sources on the combined war effort was assembled, analysed, and commented on.

Many individuals were eventually involved with statistical work for the Government, and they moved from post to post as the situation changed. A key reference is the series of essays collected by Chester (1951). The following distinguished men, now dead, illustrate only a small fraction of the manifold activities. They were primarily economists by training, with a few survivors from Karl Pearson's Biometric School. Lord Woolton was appointed Minister of Food in 1940, and his control of the food supply by rationing is remembered by every citizen who lived through those times. Sir William Elderton was a statistical adviser in the Ministry of Shipping, which was later absorbed into the Ministry of War Transport. He had performed similar functions during the period 1917–1920. Lord Beveridge was chairman of committees which made surveys of manpower resources and requirements, and of the use of skilled men by the armed forces. His influential report on social insurance appeared in 1942, and his book on *Full Employment in a Free*

Society in 1944. At the Ministry of Labour, E. C. Ramsbottom was Director of Statistics and R. B. Ainsworth Deputy Director. They were concerned with the statistics of wages, hours of work, and working-class budgets, and their activities ultimately led to the replacement of Bowley's "Cost of Living Index" by the interim index of retail prices. At the Board of Trade, H. Leak was Director of Statistics from 1932 to 1948. At the Ministry of Agriculture and Fisheries, C. Oswald George became Director of Statistics in 1942. He was responsible for the Agricultural Censuses and the data required to administer food and agricultural programmes. Sir Arnold Plant was concerned with the allocation of materials. He was also the first Director of the Wartime Social Survey in the Ministry of Information. This unit carried out social surveys of public opinion and behavior on a variety of topics of interest of government departments. R. F. George was Statistician to the Air Ministry, responsible for essential figures of personnel and numbers and movement of aircraft. E. C. Snow, the Director of the United Tanners' Federation, was the obvious choice for Leather Controller. Sir Paul Chambers was appointed in 1940 to the post of Director of Statistics and Intelligence to the Board of Inland Revenue, where he was responsible both for the Budget estimates and for budgetary policy in the Inland Revenue field. In 1942 he became a Commissioner of Inland Revenue, and was primarily responsible for devising the PAYE system which played a crucial part in raising the revenue needed for the war effort. D. H. Robertson was a senior advisor at the Treasury working in overseas finance and lend-lease matters including economic statistics. At the end of the war, Lord Keynes negotiated a loan of $3.75 billion from the United States. This was his culminating achievement, firmly based on statistical evidence both of the war effort of the United Kingdom and of the economic position to which the country had been reduced.

4. Rothamsted

The methods evolved during this period by the Department of Statistics at Rothamsted Experimental Station were mainly developments of well-known principles, but they also included the introduction of fractional replication by D. J. Finney and the steady extension of ideas on sampling and sample surveys by F. Yates.

Early in 1940 Yates was invited by S. Zuckerman to contribute an article on food to an anonymous Penguin Special *Science in War*. The article emphasised the importance of efficient production, in particular by increasing the yield per acre through the use of fertilizers. Proposals were made for a reexamination of existing experimental results and for rapid sample surveys to see what was actually happening on farms. Yates recalls subsequent events:

> Stimulated by the disaster at Dunkirk and the fall of France, a memorandum summarising all fertilizer trials done in the U.K. since 1900 was produced, with much help from members of the Rothamsted departments, under my and Crowther's name, in the very short time of ten weeks (Crowther was then head

> of the Chemistry Department, and a leading authority on fertilizers). It should have been welcomed by the Ministry of Agriculture, but they had decided that this was to be a non-scientific war and that they had all the information that they required! However the Fertilizer Control (which was responsible for the supply side and with which Crowther had personal contacts) greatly welcomed it.... The outcome was that Crowther was co-opted to the Ministry's Fertilizer Committee by the Fertilizer Control, and the whole fertilizer rationing policy was changed. Throughout the war Crowther and myself were increasingly involved in this work, including providing data for such problems as balancing food and fertilizer imports in the light of available shipping.

The survey of fertilizer practice which followed in 1942 yielded useful and timely information, for example that much acid and phosphate-deficient ploughed-out grassland was not receiving lime or phosphate, with resultant crop failure. A qualitative survey of all farms of over five acres in England and Wales was carried out by the Ministry of Agriculture during 1941–42, and the Rothamsted statisticians used variable sampling fractions for the different size groups in order to speed analysis. Other surveys of national importance were concerned with the assessment of wireworm infestation when large amounts of old grass were turned over to arable cropping, and with the assessment of timber supplies, where a "guesstimate" based on a survey only half complete was combined with information dating from 1923 to give a volume estimate within 5% of that obtained from a survey on a proper random basis made two years later.

From the winter of 1940 onwards, Yates devoted about half his time to operational research at home and overseas, concerned to work with Zuckerman on the effects of air raids, and on the effective use of air forces. As regards the first, the concept of "vulnerable area" or "standardized casualty rate," based on injuries caused at increasing distances from the point of impact of a bomb, was evolved in 1938. This became the standard method of assessing the effects of bombs of various types and the protective quality of various types of shelter. The effect on industrial production of bombing raids on German cities was assessed from aerial photographs. At first, the percentage of buildings destroyed in the various towns was obtained (at great labor), but this statistic was largely meaningless, because by and large it was the centers of towns that were destroyed and factories were on the outskirts. Yates was instrumental in compelling the use of line sampling to determine quickly the area of damage to factories, and this statistic could be converted into loss of production. The estimated loss turned out to be quite small, and investigations in Germany after the war showed that production had in fact increased.

This is an appropriate place to review Fisher's wartime activities. When war was declared, University College London notified all departmental staff that research work would cease at once and the workers disband. Fisher protested vigorously, and the difficulties were eventually settled by arranging for the removal of the Department of Eugenics to Rothamsted in October 1939. The staff gradually drifted away until only two remained. Fisher got some financial support from the Agricultural Research Council and elsewhere, and worked on genetics and blood-group data until appointed in 1943 to the Arthur Balfour Chair of Genetics at Cambridge. The reasons why he was not invited to do war work remain a matter for conjecture.

5. SR 17

Standardization and reliability were essential requirements for war materials produced on a massive scale, and such qualities could be assured by the use of statistical methods. In 1941, Sir Charles Darwin went on a scientific liaison mission to the United States, and became convinced that these methods were important. Meetings were arranged at which publicity was given to statistical methods in industrial standardization, and there were consequent demands from various departments of the Ministry of Supply and the Ministry of Production for advice on problems of quality control and sampling inspection. Proposals made by F. Smithies and J. R. Womersley led to the formation of the Ministry of Supply Advisory Service on Statistical Method and Quality Control in the summer of 1942. The new unit, part of the scientific research division of the Ministry, was known first as SR 1e and after 1943 as SR 17. A small proportion of the staff had some previous knowledge of the industrial applications of statistics. The remainder consisted largely of actuaries and batches (in 1942 and 1943) of young persons who had up to then been studying mathematics at Cambridge. By the end of the war, the staff numbered about forty.

Womersley was appointed as the first head of the Advisory Service. His interest in the application of statistics to industrial problems began during his time at the Shirley Institute in Manchester, where he had come under Tippett's influence. He was later concerned with ballistics research, where he initiated a statistical investigation of cordite proof records, and from that field he moved to SR 1e. Womersley provided the new unit with a sense of direction, firm support, and the warmth of his personality. During the periods when he was absent, E. D. van Rest gave kind and friendly leadership. SR 17 was well served by both men, who contributed much to the strong *esprit de corps*.

Many of the staff were recruited at a mysterious address near the Strand. A typical starting salary was £230 per annum. The unit occupied a floor in Berkeley Court, a block of flats opposite Baker Street Station. By a happy coincidence, the clerical side was in the hands of Mr. Baker, who was also responsible for the fire-fighting arrangements. Although raids by manned aircraft had virtually ceased by 1942, the flying bomb (V1) and rocket (V2) came in 1944. There were dives for cover when the sound of V1 appeared, whereas V2 arrived silently and allowed no evasive action.

At first, most members of SR 17 knew little about quality control. There were no copies of BS600 to help them because the standing type had been destroyed by enemy action. However, a partial revision BS600R written by

Dudding and W. J. Jennett was issued in 1942. Knowledge and experience grew with the applications, made in visits to munition factories and manufacturing industry. Such information was embodied in *A First Guide to Quality Control for Engineers*, written by E. H. Sealy and completed in July 1943. This was a practical counterpart for the more theoretical BS600R. In the Royal Ordnance Factories (ROF), W. T. Hale, G. H. Jowett, and G. V. Owen set up a workable system to control the size and density of explosives pressed into pellets. Another project to install acceptance sampling schemes for ROF stores was perhaps less successful. Elsewhere, the same methods found a variety of applications, and small groups went to centers of production to show how quality control worked. For varying periods, H. C. H. Carpenter, J. G. Day, L. Edwards, W. A. Pridmore, and P. Stanley were attached to the Ministry of Production Quality Control Panel in Birmingham, where demonstration showplaces included major firms in the West Midlands. Another member of the Panel was D. J. Desmond of Joseph Lucas Ltd, an enthusiast for "compressed" limits which would reject ten to fifteen per cent of production when under control. SR 17 representatives who visited the Lucas plant noted that out of five thousand fraction defective charts, only a handful used compressed limits. Thus they learned that quality could be improved merely by placing charts in a public place, and the full statistical treatment reserved for a few special cases. At least one of the schemes thus installed stood the test of time: S. N. Collings recalls that a rectifying sequence scheme he set up in 1944 was still working in 1972.

Before long, statistical problems were met for which the handbooks had no solution to offer, and the activities of SR 17 were widened in order to deal with the analysis of data collected during production and the design of experiments. Questions which seemed to lie beyond the statistical knowledge of the time began to arise. An early example was the problem of finding the distribution of the mean deviation from the sample mean in normal samples. This was tackled by Sealy and C. D. Bates, who obtained an empirical solution using the table of random samples published in *Sankhyā* in the autumn of 1939. The theoretical solution was subsequently determined by H. J. Godwin. A new impetus was given to these developments by the creation of a specialist research group under G. A. Barnard late in 1942. The discussion of statistical problems was raised to a new level. Members of SR 17 began to study textbooks which might assist in the wider tasks which confronted them. *Statistical Methods for Research Workers* was found difficult to read because the theory was withheld and the examples were all biological. The year 1943 saw the appearance of *The Advanced Theory of Statistics* (M. G. Kendall) and *Mathematical Statistics* (S. S. Wilks). They supplied a logical development absent from the training of the Cambridge mathematicians, who had encountered statistics only in short courses likely to deal a speedy death to any nascent interest. The new books were carefully studied, and a long list of errors drawn up.

During the spring and summer of 1943, the work of the research group was directed towards economy in sampling. Development work on new fuses uncovered faults, and modifications were made in the hope of eliminating them. Trials of modified versus unmodified fuses, giving results in the form

	Faulty	Not faulty	Total
Unmodified	a	b	$m = a + b$
Modified	c	d	$n = c + d$

were exceedingly expensive in skilled manpower. The question was put: supposing the fault has been eliminated (so $c = 0$), what is the best allocation of $m + n$ trials in order to give evidence at the 5% level that a real improvement has occurred? The answer turned out to be that m should be chosen so that a should equal 3 approximately (and hence exactly, since it had to be a whole number). The obvious procedure was to continue trying unmodified fuzes until 3 had failed, and then switch to the modified fuzes. It was then recognized that this meant regarding the sample size as variable and the number of failures as fixed; and it soon became apparent that by generalizing this idea considerable savings in testing and in inspection costs could be achieved. Economy in sampling meant the imposition of stopping rules on standard schemes, and properties of the modified schemes were worked out. The analogies with games between two players, studied by classical authors in the seventeenth century, were recognized. At this stage, draft copies of the report by A. Wald on sequential analysis arrived from the United States and aroused much interest. His work showed that the procedure could be reversed by specifying requirements and deriving the sequential probability-ratio test (SPRT) which satisified them, to very good approximation in most cases. But earlier preoccupations served the group well because the exact properties of some SPRT were worked out and compared with the Wald approximations. The main results, obtained by J. P. Burman and published in 1946, were later rediscovered by Dvoretzky, Kiefer, and Wolfowitz (1953). The analogy with a game between two players was used as a basis for practical application in the booklet *Target-Handicap Charts for Sampling Inspection*. Wald's sequential t-test was investigated by P. Armitage with respect to the possibility of getting different answers using different weight functions for the parameter σ. A letter (Barnard, 1945b) was later sent to *Nature* to put on record the stage reached in the development of these ideas up to the arrival of Wald's report.

Long after the war ended it became clear that likelihood ratios, the basis of the SPRT, had been introduced by A. M. Turing into his work at Bletchley Park on decoding enemy communications, and his use of them amounted effectively to a sequential test procedure. Precise details of this work remain subject to military secrecy, but a good account of what is publicly known can be found in the biography of Turing by Andrew Hodges (1983). I. J. Good, who was working with Turing at the time, had known Barnard from student days at Cambridge, and recalls discussing related theoretical issues, though not their practical application, with Barnard. This was during the war, but recollections differ as to the date.

The practical side of these theoretical advances in sequential analysis is illustrated by the following experience of D. Newman.

> An inspection scheme was needed for certain textile components used in the filling of shells and because the tests used were destructive and the staff available to carry them out limited, the most economical sample possible had to be used.
>
> The scheme was worked out in London and I took it to Lancashire to install. I remember that we were all very apprehensive about the ability of the operatives to understand the scheme and run it properly without the supervision of a statistician. I left the instruction in the inspection shop overnight with the object of going in in the morning to train the inspection staff in the method of using it. Statisticians, of course, got in much later than the inspectors, who worked normal shift hours. When I arrived I found that the staff had found the instructions, had started to work the scheme, and what is more were working it perfectly. I took the next train back to London.
>
> I think that this illustrates an aspect of the application of statistical methods to industry that I learned during that period. The methods of statistical quality control and inspection may often have appeared complicated to us but we never produced any scheme which the ordinary worker involved could not cope with. I always found that the real menace was the technician or scientist with no understanding of statistics.

Another field of activity for the research group concerned the operation of a weapon which contained twenty-two components. A realistic number of treatment combinations was obtained by taking each factor at two levels and assuming that all the interactions were zero. The problem posed by Barnard was to choose the treatment combinations in such a way as to minimize the variances of factorial effects. This was solved by Burman and R. L. Plackett. When the results were presented at a meeting of the London Mathematical Society, J. Bronowski pointed out a connection with Hadamard matrices. An experiment was carried out on the lines proposed, and the interpretation of the data was found to be somewhat difficult.

Computing facilities gradually became more sophisticated, but consisted always of desk-top machines. The first calculators were Brunsvigas and Facits, for which the motive power was supplied by the operator. They were followed by Marchants and Fridens, where the teeth and cogs remained but were now driven by an electric motor. Much could be achieved with such equipment, as Armitage recalls:

> In the computation of exact properties of sampling inspection schemes it was necessary to have binomial coefficients with high indices, and over a period of a year or more we filled in odd moments by constructing a large Pascal's triangle which when completed filled most of a wall. I think it got up to about $n = 125$. Each new entry contained enormous numbers of digits, and was quite a tricky feat with the usual sort of electric calculators. Dennis Lindley, who was regarded as having the neatest handwriting, was given the task of copying the new entries onto the sheet.

The preface to the Royal Society *Table of Binomial Coefficients* records thanks to D. V. Lindley for lending a copy of Armitage, Godwin, and Lindley (MS) for comparison with the proofs.

Life in SR 17 was never entirely concerned with statistical matters. The need to relax, the alternation of idle periods and busy ones, the generally low age and often high spirits of the participants all had their inevitable consequences. Those who enjoyed music attended concerts: in London, the Amadeus Quartet in RAF uniform played at lunchtime in a sandbagged National Gallery; and the City of Birmingham Orchestra performed under George Weldon. A series of talks was organized in which members spoke on the topics that interested them: F. J. Anscombe on Ezra Pound, A. Brown on the New River (London's water supply), A. H. J. Baines on the poetry of William Barnes, and A. M. Walker on the piano works of Bartók with illustrations. On the lighter side, records were kept of individual times of arrival; the variance was computed using Sheppard's correction, so that those with regular habits achieved an imaginary standard deviation. A powerful stimulus was applied to SR 17 by J. Taylor, who always kept the three buttons of his jacket done up. The Worshipful Order for the Unbuttoning of Taylor was established, with appropriate Officers, e.g. Keeper of the Middle Button, and a journal: *The Independent Variable*. Bill Pridmore puts these matters into perspective:

> The months during which the Worshipful Order flourished may have seemed like a waste of time. I have to report however that both Denis Ward and myself in later years met as advisers to the Department of Prices and Consumer Protection, and as UK representatives in Brussels to the European equivalent of the CBI in dealing with weights and measures legislation. We are quite sure that our experience with the Worshipful Order in our SR 17 days was an excellent preparation for this activity.

SR 17 was disbanded late in 1945, when the members returned to their former occupations or began new lives elsewhere.

6. Specialist Groups

At the Admiralty, H. L. Seal was first appointed in 1939 as a Statistical Officer to the Director of Air Materiel, where he was joined by R. E. Beard. Their work was concerned with the provision of naval aircraft to meet the demand for them. There was little time to analyse problems in depth, and practical approximations were sought. In 1942, Seal moved to the Civil Establishments Branch, where he was asked to plan the postwar civil establishment based on orderly promotion through the civilian grades. He was assisted by S. Vajda, who also analysed casualties to ships. H. R. Pitt worked part time for the Admiralty while still at Aberdeen University. He clarified Wiener's results on prediction from past behavior of random functions and sequences. In his later work at the Air Ministry, he proved that the replacement of stock is better done by fixed amounts at irregular intervals than by irregular amounts at constant intervals.

At Bletchley Park, the Government Code and Cypher School was concerned with breaking ciphers based on an electromechanical machine, the Enigma. The daily keys were eventually recovered with the help of high-speed testing on a cryptanalytic machine, the Bombe. Statistical ideas useful in this connection included sequential analysis, weight of evidence, and a form of empirical Bayes. They were introduced by A. M. Turing, who was assisted by I. J. Good. Turing was later an adviser to M. H. A. Newman in the construction of a series of special-purpose electronic computers, the Colossus machines, which were used to break messages too important for Enigma. The magnitude of his contribution to the survival of the United Kingdom has only slowly been realized.

The Chemical Defence Experimental Station at Porton Down was concerned with what to do if poison gas were used against the military or the civilian population. A young laboratory assistant, G. E. P. Box, suggested that statistical methods were needed, and since nobody knew any, he was assigned to look into the matter. Books by Fisher and Goulden provided ideas, on the basis of which he designed experiments on animals and humans using randomized blocks, latin squares, and factorial designs. The results indicated, for example, that the best treatment for mustard-gas blister was to cover it with gauze and let nature go ahead. At a later stage, problems involving quantal response arose, and it was suggested that Fisher be consulted. However, a visit by Sergeant Box to Professor Fisher was not in accordance with protocol, and so the fiction was devised of a visit to Cambridge in order to collect a horse. Fisher was very helpful, and when Box remarked that he had no books on the theory of statistics, he was presented with Fisher's copy of M. G. Kendall's *Advanced Theory of Statistics*, Vol. 1.

In the Ministry of Aircraft Production, the Air Warfare Analysis Section under L. B. C. Cunningham included H. E. Daniels, W. R. B. Hynd, H. R. Pitt, and W. Rudoe. They worked on problems related to radar navigation and position finding, and the theory of combat for aircraft. The group was also able to determine the location of German radar stations by fitting best lines to bomb plots and producing back to Northern France.

Many aspects of enemy bombing were investigated by the Experiments Branch of the Ministry of Home Security, housed in the Forest Products Research Laboratories at Princes Risborough. During the earlier years of the war, the incidence and nature of civilian casualties was studied by A. Bradford Hill, assisted by Florence N. David and E. D. van Rest. Sir Austin Bradford Hill contributes these recollections:

> One main problem was to measure the relative safety of shelters and parts of the ordinary domestic home and whether, as I would say, it was safer to be "upstairs, downstairs, or in my lady's chamber". What were the frequency and nature of injuries and causes of death? Inevitably it was virtually impossible to get appropriate denominators to match the casualties, i.e. what were the numbers exposed to risk in defined locations when the bomb fell? One could not get these figures in advance of the events since people did not necessarily stay put in their position in relation to the severity of the raid. And one had to be careful in a

> post-raid inquiry that one did not infuriate shocked and perhaps homeless persons with what might well appear to be damn-fool questions. But we did, I think, succeed in building up some kind of a statistical picture of the events.

Other aspects of enemy bombing were considered by F. Garwood and, from 1942, J. Bronowski. The amount of destruction was estimated theoretically by dropping circles at random, and an air raid on Coventry was simulated by dropping rice on a large scale map, the damage from each grain being assessed by S. Zuckerman. Members of the group counted the bomb distribution over the country—the Bomb Census. Late in 1944, Neyman worked in England for three months on bomb damage and collaborated with Bronowski.

In the latter years of the war, Bradford Hill moved to the RAF to work on research problems in the department of the Director-General of Medical Services (housed in Kingsway, London) and to serve as consultant in medical statistics to the Institute of Aviation Medicine:

> Here many problems arose in relation to aircrew—was the accident rate increased by very long sorties; what were the early physical and mental signs of the strains to which they were subject; how could one measure night vision and could it be improved by training; by psychological testing could the embryo bomber pilot be distinguished from the fighter. The late Professor D. D. Reid was associated with me in much of this work.
>
> I had a room high up in the office block in Kingsway. When I had a visitor to consult me I would first tell him that if a flying bomb made itself heard we would stop talking and listen to it. If its engine cut off I should go under my desk, my Wing Commander assistant would go under his and the rest of the room was at his (my visitor's) disposal. We had never to put it to the test though when out at lunch one day a bomb entirely destroyed a bus by Bush House below us.

In the Directorate of Filling Factories of the Ministry of Supply, A. W. Swan, assisted by G. Herdan, gave advice on applied statistical techniques of inspection. These routine activities were transformed into something quite different by the lively and stimulating character of Albert Swan.

At the Directorate of Ordnance Factories (Explosives), K. A. Brownlee was appointed to introduce statistical quality control, and he spent some time seconded to SR 17 in order to become acquainted with the newer techniques being developed. He soon realized that the more advanced methods there in use—analysis of variance, multiple regression, and experimental design—were more appropriate techniques for bulk chemicals. The success of these applications led to his book on industrial experimentation.

When war broke out, the staff of the Department of Statistics at University College London became attached to the Ordnance Board, Ministry of Supply. They worked first in Oxford, then in Sidcup, and finally in South Kensington. The transfer arose partly from personal contact between E. S. Pearson and his University College colleague, A. V. Hill, who was already involved with the preparation of a Central Register of Scientific and Technical Personnel. Other members of the group were D. J. Bishop, N. L. Johnson, and B. L. Welch, and also Florence N. David until she left for the Ministry of Home Security. Later recruits were D. F. Mills and L. S. Vallance. That part of the work concerned with the lethal effectiveness of antiaircraft fire is described by Pearson (1963). He outlines the problem, gives a simple mathematical model, and explains how knowledge of positioning errors, fragmentation characteristics of the shell, and target vulnerability were necessary in reaching a solution. The back of the problem was broken by 1944, by which time recommendations could be made with confidence on many of the relevant matters. E. A. Milne also worked for the Ordnance Board on probabilistic problems. He produced a solution to an "optimum distribution of fire" problem, one noteworthy feature being the recognition that an optimum solution could imply a whole distribution.

The Projectile Development Establishment of the Ministry of Supply was originally located in Kent but was evacuated to Aberporth in Wales after the fall of France in 1940. This group included M. S. Bartlett, F. J. Anscombe, D. G. Kendall, and P. A. P. Moran. A considerable part of the time was spent in assessing the effectiveness of proposed rocket weapons, and this involved the statistical design and analysis of firing trials. Other problems concerned the optimum distribution of shell fragment sizes, and the mean area of projection of shell fragments. When Anscombe and Bartlett left for activities elsewhere, Kendall was told by his chief "to learn statistics":

> He was kind enough to give me a week in which to do this, and what I did was to go to London and stay in Anscombe's digs and talk to him until the small hours every night, as during the day he was busy with other things. That is the only statistical instruction I have ever received but I have found it more than sufficient for most of my needs.

At the Wool Industries Research Association, H. E. Daniels worked until 1942 on a project for impregnating uniform cloth with carbon particles so as to absorb odors from wounds.

The ideas of statistics and probability are important in operational research, and brief comments on the use of OR methods should be made before concluding this account. During the period immediately before the war, OR was applied to radio direction finding, later known as radar, and the group concerned was subsequently transferred to Fighter Command. Throughout 1940–1942, P. M. S. Blackett was invited to establish OR groups successively in Anti-Aircraft Command, Coastal Command, and the Admiralty. Some of the best biologists in the country worked in OR. Their success often seemed due to the way they had been trained to handle experimental data rather than to their rapid mastery of statistical technique.

7. Consequences

Peace finally returned, and the statistical scene in the United Kingdom had been completely transformed by the changes which had come in response to the immense demand of war. No other method would have produced these changes in only six years, such is the stability—some would say conservatism—of all British institutions.

For the first time, the official statistics of Government were unified by the creation of a Central Statistical Office, and the importance of the subject was recognized by the formation of a Government Statistical Service. There was a large increase in the number of people who knew that statistics was an interesting subject, and they had been given an excellent training free of charge. Many of them decided to continue working in statistics. The provision of statistical advice for industry and agriculture had slowly grown between the wars and now expanded more rapidly. Those in charge of higher education realized that an important field had in the past been largely neglected, and during the next twenty years the situation was remedied. The Royal Statistical Society, now much larger in size, developed new Sections and Local Groups. An expressed need for professional status led to the formation of the Association of Incorporated Statisticians, later to become the Institute of Statisticians. The difficult questions then posed about the nature of the statistical profession have still to be resolved. Operational research came into being as a new scientific discipline applied to management problems.

But the details of what happened in the postwar statistical history of the United Kingdom, and the subsequent careers of leading figures in the new wave, must be deferred to another occasion.

Acknowledgements

The following persons kindly supplied information, but are not responsible for any defects that may be found in the foregoing account: Sir Roy Allen, Vanessa Allinson, F. J. Anscombe, P. Armitage, M. S. Bartlett, C. D. Bates, R. E. Beard, Joan F. Box, G. E. P. Box, K. A. Brownlee, Sir Harry Campion, D. G. Champernowne, S. N. Collings, Sir William Cook, D. R. Cox, R. Croasdale, H. E. Daniels, D. J. Finney, F. Garwood, H. J. Godwin, I. J. Good, A. H. R. Grimsey, W. T. Hale, H. C. Hamaker, R. Henstock, Sir Austin Bradford Hill, N. L. Johnson, D. G. Kendall, Sir Maurice Kendall, D. V. Lindley, P. A. P. Moran, S. J. Morrison, D. Newman, J. B. Parker, Sir Harry Pitt, W. A. Pridmore, Constance Reid, Florence A. Rigg, C. A. Rogers, L. Rosenhead, W. Rudoe, P. J. K. Salmon, H. L. Seal, F. Smithies, Sir Richard Stone, L. H. C. Tippett, C. J. Tranter, S. Vajda, D. J. van Rest, Sir Arthur Vick, and F. Yates. Their letters, and manuscript notes of their recollections, will be deposited in the archives of the Royal Statistical Society.

Bibliography

Ainsworth, R. B. (1960). "Edmund Cecil Ramsbottom, 1881–1959." *J. Roy. Statist. Soc. Ser. A*, **123**, 86–87.

Aitken, A. C. (1939). *Statistical Mathematics*. Edinburgh: Oliver and Boyd.

Allen, R. G. D. and George, R. F. (1957). "Professor Sir Arthur Lyon Bowley, November 6th 1869–January 21st 1957." *J. Roy. Statist. Soc. Ser. A*, **120**, 236–241'

Anon. (1954). "Sir Henry Clay." *J. Roy. Statist. Soc. Ser. A*, **117**, 500–501.

Bartlett, M. S. (1956). "John Wishart, D.Sc., F.R.S.E." *J. Roy. Statist. Soc. Ser. A*, **119**, 492–493.

Bartlett, M. S. (1981). "Egon Sharpe Pearson, 1895–1980." *Biographical Memoirs of Fellows of the Royal Society*, **27**, 425–443.

Bartlett, M. S. (1982). "Chance and change." In J. Gani (ed.), *The Making of Statisticians*. New York: Springer, 41–60.

Bartlett, M. S. and Tippett, L. H. C. (1981). "Egon Sharpe Pearson, 1895–1980. Appreciation." *Biometrika*, **68**, 1–12.

Benjamin, B. and Douglas, Iris (1970). "R. F. George." *J. Roy. Statist. Soc. Ser. A*, **133**, 128–129.

Birkenhead, The Earl of (1961). *The Prof in Two Worlds: The Official Life of Professor F. A. Lindemann, Viscount Cherwell*. London: Collins.

Blackett, P. M. S. (1962). *Studies of War: Nuclear and Conventional*. Edinburgh: Oliver & Boyd.

Bowley, A. L. (1941). "Lord Stamp, G.C.B., G.B.E., F.B.A., D.Sc., LL. D." *J. Roy. Statist. Soc.*, **104**, 193–196.

Box, Joan F. (1978). *R. A. Fisher: The Life of a Scientist*. New York: Wiley.

Cairncross, A. (1968). "Professor Ely Devons." *J. Roy. Statist. Soc. Ser. A*, **131**, 459–460.

Chester, D. N. (ed.) (1951). *Lessons of the British War Economy*, National Institute for Economic and Social Research, Economic and Social Studies, X. Cambridge U.P.

Chester, D. N. (1963). "Lord Beveridge, 1879–1963." *J. Roy. Statist. Soc. Ser. A*, **126**, 618–620.

Churchill, W. S. (1948). *The Second World War, Vol. 1: The Gathering Storm*. London: Cassell.

Clark, R. W. (1965). *Tizard*. London: Methuen.

Cockfield, Lord (1982). "Sir Paul Chambers." *The Times*, January 5, 1982.

Daniels, H. E. (1982). "A tribute to L. H. C. Tippett." *J. Roy. Statist. Soc. Ser. A*, **145**, 261–263.

Dudding, B. P. (1943). "The Industrial Applications Group of the Royal Statistical Society: First session, 1942–43." *J. Roy. Statist. Soc.*, **106**, 64–67.

Dudding, B. P. (1944). "The Industrial Applications Group of the Royal Statistical Society: Second session, 1943–44." *J. Roy. Statist. Soc.*, **107**, 60–63.

Dudding, B. P. (1952). "The introduction of statistical methods to industry." *Appl. Statist.*, **1**, 3–20.

Dvoretzky, Kiefer, J., and Wolfowitz, J. (1953). "Sequential Decision Problems for processes with continuous time Parameter Testing Hypotheses." *Ann. Math. Statist.* **24**, 254–264.

Eisenhart, C. (1974). "Karl Pearson." In *Dictionary of Scientific Biography*, **10**, 447–473.

Finney, D. J. (1982). "A tribute to Frank Yates." *J. Roy. Statist. Soc. Ser. A*, **145**, 259–260.

Finney, D. J. and Yates, F. (1981). "Statistics and computing in agricultural research." In G. W. Cooke (ed.), *Agricultural Research 1931–1981* Agricultural Research Council, 219–236.

Fowler, R. F. (1972). "R. B. Ainsworth, C.B.E., M.C." *J. Roy. Statist. Soc. Ser. A*, **135**, 295.

Good, I. J. (1979a). "Studies in the history of probability and statistics XXXVII. A. M. Turing's statistical work in World War II." *Biometrika*, **66**, 393–396.

Good, I. J. (1979b). "Early work on computers at Bletchley." *Ann. Hist. Comput.*, **1**, 38–48.

Good, I. J. (1981). "Some comments on Rejewski's paper on the Polish decipherment of the Enigma." *Ann. Hist. Comput.*, **3**, 232–234.

Greenwood, M. (1941–1943). "Medical statistics from Graunt to Farr." *Biometrika*, **32**, 101–127, 203–225; **33**, 1–24. Reprinted in Pearson and Kendall (1970), 47–120.

Greenwood, M. (1949). "Karl Pearson." In *Dictionary of National Biography, 1931–40*, 681–684.

Harris, José (1977). *William Beveridge: A Biography*. Oxford: Clarendon.

Harrod, R. F. (1951). *The Life of John Maynard Keynes*. London: Macmillan.

Harrod, R. F. (1959). *The Prof: A Personal Memoir of Lord Cherwell*. London: Macmillan.

Hill, A. B. and Butler, W. (1949). "Major Greenwood." *J. Roy. Statist. Soc. Ser. A*, **112**, 487–489.

Hill, A. B. (1965). "The Earl of Woolton." *J. Roy. Statist. Soc. Ser. A*, **128**, 462–463.

Hilts, V. (1978). "Aliis Exterendum, or the origins of the Statistical Society of London." *Isis*, **69**, 21–43.

Hinsley, F. H., Thomas, E. E., Ransom, C. F. G., and Knight, R. C. (1979, 1981, 1984). *British Intelligence in the Second World War*, Vols. 1, 2, 3 pt. I London: HMSO.

Hodges, A. (1983). *Alan Turing: The Enigma*. New York: Simon & Schuster.

Irving, D. (1964). *The Mare's Nest*. London: Kimber.

Irwin, J. O. and van Rest, E. D. (1961). "Edgar Charles Fieller, 1907–1960." *J. Roy. Statist. Soc. Ser. A*, **124**, 275–277.

Jones, R. V., org. (1975). "A discussion on the effects of the two world wars on the organization and development of science in the UK." *Proc. Roy. Soc. London Ser. A*, **342**, 439–586.

Kendall, M. G. (1952). "George Udny Yule, 1871–1951." *J. Roy. Statist. Soc. Ser. A*, **115**, 156–161. Reprinted in Pearson and Kendall (1970), 419–425.

Kendall, M. G. (1963). "Ronald Aylmer Fisher, 1890–1962." *Biometrika*, **50**, 1–15. Reprinted in Pearson and Kendall (1970), 439–453.

Kendall, Sir Maurice and Plackett, R. L. (ed.) (1977). *Studies in the History of Statistics and Probability*, Vol II. London: Griffin.

Kruskal, W. (1980). "The significance of Fisher: A review of *R. A. Fisher: The Life of a Scientist*." *J. Amer. Statist. Assoc.*, **75**, 1019–1030.

Lander, M. (1948). "War and the actuary." In *Proc. Centenary Assembly Inst. Actuaries*, 291–297.

Lovell, Sir Bernard (1976). *PMS Blackett. A biographical memoir*. London: Royal Society.

MacDougall, G. D. A. (1951). "The Prime Minister's Statistical Section." Chapter IV in Chester (1951).

MacKenzie, D. A. (1981). *Statistics in Britain, 1865–1930. The Social Construction of Scientific Knowledge*. Edinburgh U.P.

McKinlay, P. L. (1951). "Major Greenwood, 1880–1949." *Biometrika*, **38**, 1–3.

Maunder, W. F. (1972). "Sir Arthur Lyon Bowley (1869–1957)." Inaugural lecture, Univ. of Exeter. Reprinted in Kendall and Plackett (1977), 459–480.

Menzler, F. A. A. (1962). "Sir William Palin Elderton, 1877–1962." *J. Roy. Statist. Soc. Ser. A*, **125**, 669–672.

Morrison, S. J. (1981). *International Comparative Quality Assurance: A Pilot Study*. Univ. of Hull, Dep. of Operational Research.

Owen, D. B. (ed.) (1976). *On the History of Statistics and Probability*. New York: Marcel Dekker.

Paine, T. (1982). "Sir Paul Chambers, 1904–1981." *J. Roy. Statist. Soc. Ser. A*, **145**, 374–375.

Pearson, E. S. (1938). *Karl Pearson: An Appreciation of Some Aspects of His Life and Work*. Cambridge U.P.

Pearson, E. S. (1939). "William Sealy Gosset: 'Student' as a statistician." *Biometrika*, **30**, 210–250. Reprinted in Pearson and Kendall (1970), 360–403.

Pearson, E. S. (1957). "John Wishart, 1898–1956." *Biometrika*, **44**, 1–8.

Pearson, E. S. (1962). "William Palin Elderton, 1877–1962." *Biometrika*, **49**, 297–303.

Pearson, E. S. (1963). "A statistician's place in assessing the likely operational performance of Army weapons and equipment." In *Proc. 8th Conference on the Design of Experiments in Army Research Development and Testing*, ARO-D Report 63-2, 1–15. Reprinted in Pearson (1966a), 314–323.

Pearson, E. S. (1966a). *The Selected Papers of E. S. Pearson*. Cambridge U.P.

Pearson, E. S. (1966b). "The Neyman–Pearson story: 1926–1934. Historical sidelights on an episode in Anglo–Polish collaboration." In F. N. David (ed.), *Research Papers in Statistics. Festschrift for J. Neyman*. Chichester: Wiley, 1–23. Reprinted in Pearson and Kendall (1970), 455–477.

Pearson, E. S. (1967). "Some notes on W. A. Shewhart's influence on the application of statistical methods in Great Britain." *Industrial Quality Control*, **24**, 81–83.

Pearson, E. S. (1970). "David Heron, 1881–1969." *J. Roy. Statist. Soc. Ser. A*, **133**, 287–291.

Pearson, E. S. (1974a). "Some historical reflections on the introduction of statistical methods in industry." *The Statistician*, **22**, 165–179.

Pearson, E. S. (1974b). "Memories of the impact of Fisher's work in the 1920's." *Internat. Statist. Rev.*, **42**, 5–8.

Pearson, E. S. and Kendall, M. G. (eds.) (1970). *Studies in the History of Statistics and Probability*. London: Griffin.

Reid, Constance (1982). *Neyman from Life*. New York: Springer.

Pridmore, W.A. (1976). "Albert Swan 1892–1975." *J. Roy. Statist. Soc. Ser. A*, **139**, 416.

Randell, B. (ed.) (1982). *The Origins of Digital Computers: Selected Papers*, 3rd edn. Berlin: Springer.

Robinson, H. W. (1974). "C. Oswald George." *J. Roy. Statist. Soc. Ser. A*, **137**, 457–458.

Rothamsted Experimental Station. *Report for the Years 1939–1945*.

Royal Statistical Society (1934). *Annals of the Royal Statistical Society 1834–1934*. London: Royal Statistical Society.

Schlapp, R., Copson, E. T., Kendall, D. G., Miller, J. C. P., and Ledermann, W. (1968). "A. C. Aitken, D.Sc., F.R.S." *Proc. Edinburgh Math. Soc. (2)*, **16**, 151–176.

Science in War (1940). Penguin Special S 74.

Silverstone, H. (1968). "Alexander Craig Aitken, M.A., D.Sc., LLD., F.R.S.E., F.R.S. (1895–1967)." *J. Roy. Statist. Soc. Ser. A*, **131**, 259–261.

Smithies, F. (1959). "John Ronald Womersley." *J London Math. Soc.*, **34**, 370–372.

Smithies, F (1974). *Inspiration or Perspiration?* St John's College Cambridge Lecture. Univ. of Hull.

Stone, Richard (1980). "Keynes, political arithmetic and econometrics." In *Proc. Brit. Acad.*, **64** (1978). Oxford U.P.

Temple, G. (1956). "Edmund Taylor Whittaker, 1873–1956." *Biographical Memoirs of Fellows of the Royal Society*, **2**, 299–325.

Thomson, Sir George (1958). "Frederick Alexander Lindemann, Viscount Cherwell." *Biographical Memoirs of Fellows of the Royal Society*, **4**, 45–71.

Tippett, L. H. C. (1955). "Philip Lyle." *J. Roy. Statist. Soc. Ser. A*, **118**, 497–498.

Tippett, L. H. C. (1972). "Annals of the Royal Statistical Society, 1934–71." *J. Roy. Statist. Soc. Ser. A*, **135**, 545–568.

van Rest, E. D. (1946). "Applied statistics in England." *Statistica Neerlandica*, **1**, 5–11.

Waddington, C. H. (1973). *OR in World War 2. Operational Research against the U-boat*. London: Elek Science.

Wallis, W. A. (1980). "The Statistical Research Group, 1942–1945 (with discussion)." *J. Amer. Statist. Assoc.*, **75**, 320–335.

Welchman, G. (1982). *The Hut Six Story: Breaking the Enigma Codes*. London: Allen Lane.

White, G. R. (1960). "Ernest Charles Snow, C.B.E., D.Sc., 1886–1959." *J. Roy. Statist. Soc. Ser. A*, **123**, 355–356.

Whittaker, E. T. and Robinson, G. (1924). *The Calculus* of *observations*. London: Blackie & Son.

Whittaker, J. M. and Bartlett, M. S. (1968). "Alexander Craig Aitken." *Biographical Memoirs of Fellows of the Royal Society*, **14**, 1–14.

Wilson, H. (1977). "Hector Leak, 1887–1976." *J. Roy. Statist. Soc. Ser. A*, **140**, 111.

Wishart, J. (1939). "Some aspects of the teaching of statistics (with discussion)." *J. Roy. Statist. Soc.*, **102**, 532–564.

Woolton, Lord (1946). "Some aspects of the use of statistics in Government with special reference to the human budget (with discussion)." *J. Roy. Statist. Soc.*, **109**, 1–10.

Yamey, B. S. (1980). "Professor Sir Arnold Plant, 1898–1978." *J. Roy. Statist. Soc. Ser. A*, **143**, 92.

Yates, F. (1951). "The influence of "Statistical Methods for Research Workers" on the development of the science of statistics." *J. Amer. Statist. Assoc.*, **46**, 19–34.

Yates, F. (1952), "George Udny Yule, 1871–1951." *Obituary Notices of Fellows of the Royal Society*, **8**, 309–323.

Yates, F. (1968), "Theory and practice in statistics." *J. Roy. Statist. Soc. Ser. A*, **131**, 463–475.

Yates, F. (1981), *Sampling Methods for Censuses and Surveys*, 4th edn. London: Griffin.

Yates, F. and Mather, K. (1963). "Ronald Aylmer Fisher, 1890–1962." *Biographical Memoirs of Fellows of the Royal Society*, **9**, 91–129.

Zuckerman, S. (1978). *From Apes to Warlords*. London: Hamish Hamilton.

Publications Arising from War Work

Allen, R. G. D. (1946). "Mutual aid between the US and the British Empire, 1941–45 (with discussion)." *J. Roy. Statist. Soc.*, **109**, 243–277.

Anscombe, F. J. (1946). "Linear sequential rectifying inspection for controlling fraction defective." *Suppl. J. Roy. Statist. Soc.*, **8**, 216–222.

Anscombe, F. J. (1949). "Tables of sequential inspection schemes to control fraction defective." *J. Roy. Statist. Soc. Ser. A*, **112**, 180–206.

Anscombe, F. J. (1960). "Notes on sequential sampling plans." *J. Roy. Statist. Soc. Ser. A*, **123**, 297–306.

Anscombe, F. J., Godwin, H. J., and Plackett, R. L. (1947). "Methods of deferred sentencing in testing the fraction defective of a continuous output." *Suppl. J. Roy. Statist. Soc.*, **9**, 198–217.

Armitage, P. (1947). "Some sequential tests of Student's hypothesis." *Suppl. J. Roy. Statist. Soc.*, **9**, 250–263.

Balchin, N. (1943). *The Small Back Room.* London: Collins.

Barnard, G. A. (1945a). "A new test for 2 × 2 tables." *Nature*, **156**, 177, 783–784.

Barnard, G. A. (1945b). "Economy in sampling." *Nature*, **156**, 208.

Barnard, G. A. (1946). "Sequential tests in industrial statistics (with discussion)." *Suppl. J. Roy. Statist. Soc.*, **8**, 1–26.

Barnard, G. A. (1947). "Significance tests for 2 × 2 tables." *Biometrika*, **34**, 123–138.

Bartlett, M. S. (1946a). "The large sample theory of sequential tests." *Proc. Cambridge Philos. Soc.*, **42**, 239–244.

Bartlett, M. S. (1946b). "A modified probit technique for small probabilities." *Suppl. J. Roy. Statist. Soc.*, **8**, 113–117.

Bartlett, M. S. and Kendall, D. G. (1946). "The statistical analysis of variance-heterogeneity and the logarithmic transformation." *Suppl. J. Roy. Statist. Soc.*, **8**, 128–138.

Bayley, G. V. and Hammersley, J. M. (1946). "The 'effective' number of independent observations in an auto-correlated time-series." *Suppl. J. Roy. Statist. Soc.*, **8**, 184–197.

Beard, R. E. (1947). "Statistical problems of naval aircraft provisioning." *J. Inst. Actuaries Students' Soc.*, **6**, 144–148.

Beveridge, W. (1943). "Social security: Some trans-Atlantic comparisons (with discussion)." *J. Roy. Statist. Soc.*, **106**, 305–332.

Box, Kathleen, and Thomas, G. (1944). "The wartime social survey." *J. Roy. Statist. Soc.*, **107**, 151–177.

Bronowski, J. and Neyman, J. (1945). "The variance of the measure of a two-dimensional random set." *Ann. Math. Statist.*, **16**, 330–341.

Brownlee, K. A. (1945). *Industrial Experimentation.* Ministry of Supply, Directorate of Royal Ordnance Factories (Explosives).

Burman, J. P. (1946). "Sequential sampling formulae for a binomial population." *Suppl. J. Roy. Statist. Soc.*, **8**, 98–103.

Crowther, E. M. and Yates, F. (1941). "Fertilizer policy in war-time: The fertilizer requirements of arable crops." *Emp. J. Exp. Agric.*, **9**, 77–97.

Cunningham, L. B. C. and Hynd, W. R. B. (1946). "Random processes in problems of air warfare." *Suppl. J. Roy. Statist. Soc.*, **8**, 62–85.

Daniels, H. E. (1941). "A method of improving certain routine measurements." *Suppl. J. Roy. Statist. Soc.*, **7**, 146–150.

Daniels, H. E. (1951). "The theory of position finding (with discussion)." *J. Roy. Statist. Soc. Ser. B*, **13**, 186–207; corr., **14**, 246.

Dudding, B. P. and Jennett, W. J. (1942). *BS 600R. Quality Control Charts.* British Standards Inst.

Dudding, B. P. and Jennett, W. J. (1944). *Quality Control Chart Technique when Manufacturing to a Specification.* London: General Electric Co. Ltd. of England.

Garwood, F. (1947). "The variance of the overlap of geometrical figures with reference to a bombing problem." *Biometrika*, **34**, 1–17.

Godwin, H. J. (1945). "On the distribution of the estimate of mean deviation obtained from samples from a normal population. *Biometrika*, **33**, 254–256.

Grimsey, A. H. R. (1946). "Ultimate risks in sampling inspection." *Suppl. J. Roy. Statist. Soc.*, **8**, 244–250.

Harding, E. W. (1946). "Statistical control applied to high duty iron production." *Suppl. J. Roy. Statist. Soc.*, **8**, 233–243.

Kempthorne, O. (1946). "The use of a punched-card system for the analysis of survey data, with special reference to the analysis of the national farm survey." *J. Roy. Statist. Soc.*, **109**, 284–295.

Kempthorne, O. and Boyd, D. A. (1946). "The stock-carrying capacity of farms." *J. Roy. Statist. Soc.*, **109**, 469–475.

Lindley, D. V. (1946). "On the solution of some equations in least squares." *Biometrika*, **33**, 326–327.

Moran, P. A. P. (1944). "Measuring the surface area of a convex body." *Ann. Math.*, **45**, 793–799.

Pitt, H. R. (1946). "A theorem on random functions with applications to a theory of provisioning." *J. London Math. Soc.*, **21**, 16–22.

Plackett, R. L. (1947). "Limits of the ratio of mean range to standard deviation." *Biometrika*, **34**, 120–122.

Plackett, R. L. and Burman, J. P. (1946). "The design of optimum multifactorial experiments." *Biometrika*, **33**, 305–325.

Seal, H. L. (1945). "The mathematics of a population composed of k stationary strata each recruited from the stratum below and supported at the lowest level by a uniform annual number of entrants." *Biometrika*, **33**, 226–230.

Silvey, R. J. E. (1944). "Methods of listener research employed by the British Broadcasting Corporation (with discussion)." *J. Roy. Statist. Soc.*, **107**, 190–230.

Stockman, C. M. and Armitage, P. (1946). "Some properties of closed sequential schemes." *Suppl. J. Roy. Statist. Soc.*, **8**, 104–112.

Stocks, P. (1941). "Diptheria and scarlet fever incidence during the dispersal of 1939–40 (with discussion)." *J. Roy. Statist. Soc.*, **104**, 311–345.

Stocks, P. (1942). "Measles and whooping-cough incidence before and during the dispersal of 1939–41 (with discussion)." *J. Roy. Statist. Soc.*, **105**, 259–291.

Tippett, L. H. C. (1944). "The control of industrial processes subject to trends in quality." *Biometrika*, 33, 163–172.

Vajda, S. (1947). "Statistical investigation of the casualties suffered by certain types of vessels (with discussion)." *Suppl. J. Roy. Statist. Soc.*, **9**, 141–175.

Vernon, P. E. (1946). "Statistical methods in the selection of navy and army personnel (with discussion)." *Suppl. J. Roy. Statist. Soc.*, **8**, 139–153.

Winsten, C. B. (1946). "Inequalities in terms of mean range." *Biometrika*, **33**, 283–295.

Wishart, J. (1947). "Statistical aspects of demobilisation in the Royal Navy (with discussion)." *J. Roy. Statist. Soc.*, **110**, 27–50.

Yates, F., Boyd, D. A., and Mathison, I. (1944). "The manuring of farm crops: Some results of a survey of fertilizer practice in England. *Emp. J. Exp. Agric.*, **12**, 163–176.

CHAPTER 4

The Fascination of Sand

Ole E. Barndorff-Nielsen, Preben Blæsild, Jens Ledet Jensen, and Michael Sørensen

Abstract

The physics of wind-blown sand poses a variety of intriguing problems whose proper solution seems to require statistical ideas and methods, partly new. The main traits of the processes of transport and sorting of sand particles by wind or by water are described, and a review is given of results obtained and of subjects for further study. Various aspects of general statistical interest are also discussed; in particular, properties and applications of hyperbolic and certain related distributions are outlined. Other, nontechnical ramifications are considered, and this part of the discussion includes a brief biography of R. A. Bagnold, the founder of the physics of blown sands as a scientific field.

1. Introduction

The properties of sand appeal to children and adults alike, to the layman and the scientist. The distinct and remarkably regular shapes of the various types of dunes and ripples, as observed in the desert, on the shore, in the offshore sea bed, and at the bottom of alluvial streams, raise immediate and intriguing questions as to their generation and stability, and a closer study of how wind or water transports and sorts the sand grains brings a wealth of further interesting problems to the fore. Many of the observed phenomena could, it seems, best be understood from a statistical or stochastic viewpoint, but work in this direction has been very limited. The field, however, holds promise of giving rise to interesting new types of statistical questions and of stochastic model building in particular. Focusing on wind-blown sand, we aim in this paper to indicate what has been done and to delineate some of the most intriguing problems, as we see them, in the hope of attracting the attention of statisticians and probabilists to the field. Our basis for doing so derives from a joint research project between the Institute of Geology and the Department of Theoretical Statistics at Aarhus University. This project comprises field and laboratory investigations and wind-tunnel experiments, in a 15-m-long

Key words and phrases: earth magnetization, effective sample size, hyperbolic distributions, Long Range Desert Group, modified profile likelihood, particle sizes, physics of blown sand, sand sorting, shape triangle, stochastic models for sand transport.

wind tunnel, as well as fluid dynamical and mathematical-statistical studies. Some of the geological questions have given the cue to the development and investigation of various new parametric families of distributions, and we shall also briefly review the definitions and properties of these families and exemplify their application to problems other than those of sand transport.

In Section 3 we describe some basic physical aspects of the process of particle transport by wind or by water, emphasizing similarities as well as differences between these two types of transport. This then forms the background for the discussion in Section 4 of findings and investigations on the sorting and distributions of sand particles according to size and on the relation of these phenomena to the velocity of the fluid in which the particles are embedded.

The founder of the physics of blown sands as a field of study is the English scientist and soldier R. A. Bagnold. He has also contributed significantly to the study of alluvial sand transport. His life is an intriguing subject in itself, and we present a very brief biography of Bagnold in Section 2. In Section 5 we turn to the various statistical ramifications indicated above. Section 6 consists of some concluding remarks by which we delineate two main lines of possible further studies. In the rest of the present section we mention a number of points of general interest concerning sand.

About one quarter of the Earth's land surface is covered by desert, and this area is constantly increasing. The study of deserts and desertification has been attracting much new interest in recent years, and our understanding of the deserts has been helped by remote sensing measurements and by satellite observations. A proper understanding of the morphology of the deserts evidently requires a detailed knowledge of the local wind regimes, but extensive data of the kind required have only become available in recent years, by use of remote sensing devices. An account of much of the recently acquired information on the regional structure of the main deserts and on the prevailing wind regimes in the areas concerned is given in *Global Sand Seas* (McKee, 1979). See also Mainguet and Callot (1978) and Bagnold (1951).

The main dune forms are classified into longitudinal dunes and transverse dunes, relative to the prevailing wind direction, and within both groups there is a more or less well-defined subclassification.

The most important type of longitudinal dune is the seif (which means sword in Arabic). The typical seif dune meanders about a straight line, and the crest curve rises and falls, i.e., there are peaks and saddles along the dune. The lengths of the longitudinal dunes vary from a few hundred meters to a few hundred kilometers, their heights from five to fifty meters, and their widths, correspondingly, from tens of meters to more than a hundred meters. In the great Saharan desert, Arabian desert, and Australian desert such dunes run close and parallel, at nearly equal distances. A substantial investigation of the dynamics of longitudinal dunes has been made by Tsoar (1978).

The most intriguing of the transverse dunes are the star dunes and, particularly, the barchans (crescent-shaped dunes). The barchan dunes may have

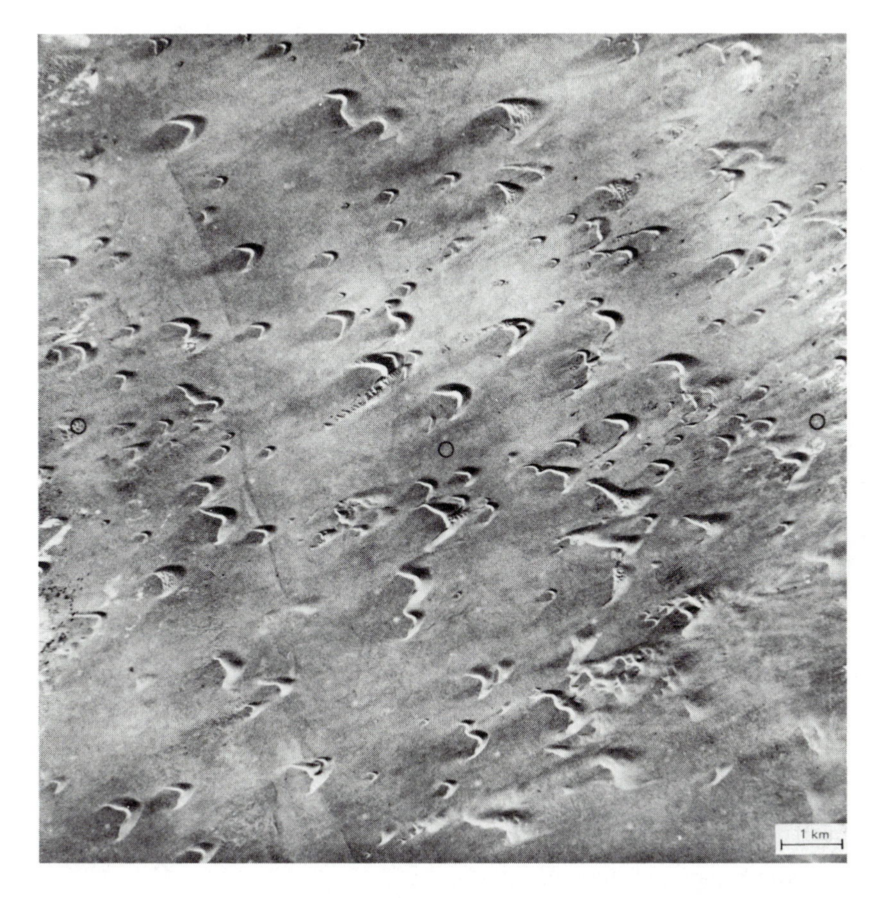

Figure 1. A field of barchan dunes in Tchad. Aerial photo. Note the scale indicated. [Reproduced with permission (aletorisation No. 70-0463) from *Photo Interpretation 10.2* (1971), Editions Technip, Paris. © IGN-Paris 1971.]

width of several hundred meters and may rise to a height of ten or more meters. Great fields of barchan dunes are found in several of the main desert areas of the earth, and Figure 1 shows an aerial photo of such a field, in Tchad. However, the most striking examples of barchan dunes occur in the arid coastal areas of Peru and Chile, as illustrated in Figure 2.

Lettau and Lettau (1978) report on a comprehensive study of part of these areas, and that paper contains much statistical information on the barchan dunes. Thus, for instance, a diagram is given which shows how the individual members of a field of barchans have moved over the period 1955–1964; the diagram is reproduced here as Figure 3.

The investigations of aeolean material transport carried out on this planet have turned out to be of crucial importance for the study and understanding of the data and pictures sent back to Earth from Mars and Venus by recent spacecraft; see e.g. Greeley and Iversen (1984), Greeley, Leach, White,

Figure 2. Three barchan dunes in the Pampa de La Joya. Note expedition's tent and cars in lee of leftmost dune. [Photograph courtesy of H. H. Lettau and K. Lettau; from Lettau and Lettau (1978).]

Iversen, and Pollack (1980), Iversen, Greeley, and Pollack (1976), Iversen, Greeley, White, and Pollack (1975), Iversen, Pollack, Greeley, and White (1976), and Sagan and Bagnold (1975).

A curious phenomenon, still only incompletely understood, is that of "singing" or "booming" sand. This has constantly intrigued travelers of the deserts; see for instance Marco Polo's travel account from about 1300, *Il Milione* (Polo, 1939, Chapter 36) and *The Gobi Desert* by Cable and French (1942). (We are indebted to David G. Kendall for these two references.) The sounds referred to vary according to circumstances from slight squeaks to thunderlike roars and can occur in wet as well as dry sands. They are generated when sand grains move against each other in a shearing fashion, as when part of a dune avalanches down the lee side of the dune. For scientific discussions of the phenomenon, see Bagnold (1941, Chapter 17) and Bagnold (1966b).

Artistic uses of sand and other granular materials have been discussed by Miller (1979).

2. Bagnold

The foundations of the physics of blown sands have been laid primarily by one man, Ralph Alger Bagnold, whose path-breaking monograph (Bagnold, 1941) is still the main reference in the field. Bagnold, a Fellow of the Royal

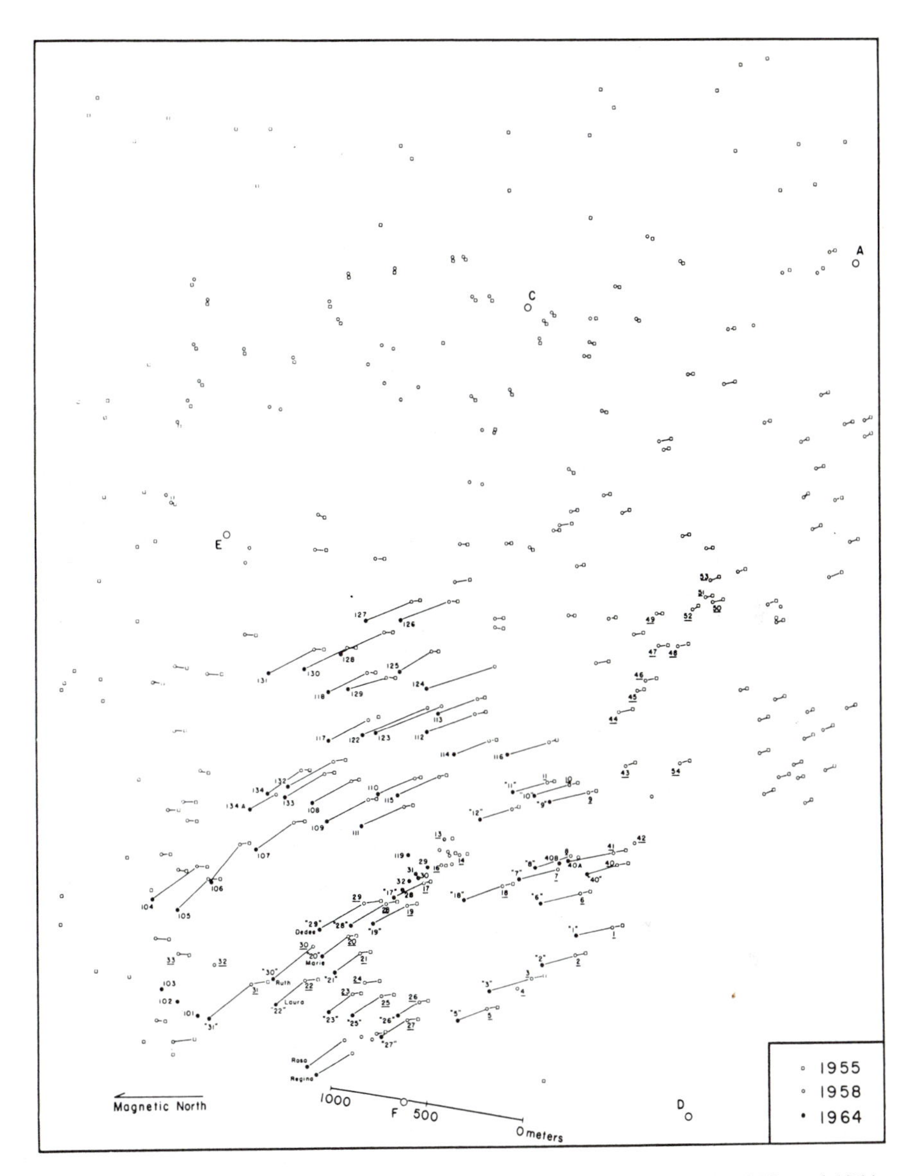

Figure 3. Map showing individual positions of barchan dunes in 1955, 1958, and 1964. The dune field is located in the Pampa de La Joya region of Chile. [Reproduced from Chapter 5 of Lettau and Lettau (1978).]

Society since 1944, has had extraordinary military and scientific careers, but has never been affiliated in any ordinary, permanent way with an institution of research and higher education. Born in 1896, out of a family with military engineering traditions—both his father and his grandfather were engineers in the British Army in India, and his father also participated in the rescue expedition 1884–1885 to save General Gordon in Khartoum—Bagnold volunteered for service in the First World War in 1914 and earned a number of military distinctions in that war.

In 1921 Bagnold took a degree in engineering from Cambridge University, and thereafter he served again with the British Army. As army officer he was in Egypt 1926–1929 and on the North-West Frontier of India 1929–1931. While in Egypt, the desert and its challenges caught Bagnold's imagination and he began, with some friends, to explore the deserts both east and west of the Nile, during weekends and whenever opportunity arose. Some of his experiences are described in the book *Libyan Sands* (Bagnold, 1935). This charming work is a fine example of Bagnold's writing skills. The literary talent exhibited there was paralleled in the life of his sister, the well-known English novelist and playwright Enid Bagnold (1889–1981) whose colourful life is vividly portrayed in her autobiography (Bagnold, 1969).

From the purely adventurous and exploratory, Bagnold's interest gradually turned to the questions of the laws of Nature that govern the formation of the various distinct types of sand dunes and, more generally, the transportation, sorting, and sedimentation of sand particles by the wind. He made many observations in the desert, and having left the army in 1935 with the rank of Major, he built on his own a wind tunnel and performed extensive experiments with that. These studies resulted in a number of path-breaking and fundamental papers, and the work culminated in 1941 with the publication of *The Physics of Blown Sands and Desert Dunes*.

During the same period Bagnold participated in the expedition sent out jointly by the Royal Astronomical Society and the Royal Society to Japan in order to take observations during the total solar eclipse in 1936. The expedition was led by Professor F. J. M. Stratton, who as a Fellow of Gonville and Caius College had known Bagnold, and also R. A. Fisher, during their years as students at that college. Stratton, in fact, was instrumental to some degree in bringing Fisher's very first paper (Fisher, 1912), in which he introduced the concept of likelihood, to publication (see Fisher's acknowledgement at the end of the paper).

In 1939 Bagnold was called up for service and posted to East Africa. On the way there the troopship he was on collided with another ship in the convoy, and thus it happened that he came to spend some days in Cairo. A journalist recognized him in the street and mentioned it in the Egyptian Gazette, referring to Bagnold's expertise in desert navigation and self-contained travel, and commending the authorities for having put a square peg in a square hole. By chance General Wavell heard about this and had Bagnold transferred to his command, under which Bagnold conceived, raised, and commanded the Long Range Desert Group, still renowned and often referred to in factual and fictional literature; see e.g. Ambler (1967), Deighton (1976) and Lyall (1980). Drawing on Bagnold's intimate knowledge of the desert and using a suncompass for navigation in the desert, invented by Bagnold in connection with his explorations of the Libyan desert in the Twenties, the Long Range Desert Group was capable of making raids, with trucks, of up to 2000 km. The

LRDG made a great number of daring and important raids, often far behind enemy lines, during the whole of the Desert War 1940–42. The history of the LRDG is described in Shaw (1945) and Owen (1980); see also Bagnold (1945). Bagnold was appointed Brigadier in 1943 and released from the army in 1944.

After the war Bagnold turned his attention to problems of sediment transport in water, and through the fifties and sixties he published a number of papers on this subject, which are of an importance comparable to that of his work on aeolean sediment transport. From 1956 onwards he worked as consultant on movements of sediments by wind and water, and in this capacity he has assisted most of the Arabic countries and also the U.S. Geological Survey. He has been honoured by a considerable number of military distinctions and dispatches as well as scientific prizes and degrees.

A number of his most important papers are listed in the bibliography (Bagnold 1956, 1966a, 1973, 1980).

3. Modes of Particle Motion

In this section we shall describe, in broad terms, the transport process and the modes of particle motion in wind-driven sand, in particular the saltation mode. In addition, we shall point out some essential differences between sand transport in water and in air. These dissimilarities are important to note when seeking inspiration in the literature on sand transport in water, which is much larger than that on aeolean transport. Finally, we shall explain a method for studying saltation. More comprehensive expositions of the physics of wind-blown sand can be found in Bagnold (1941) and P. R. Owen (1980).

The most important mode of sand transport in air is *saltation*, where the grains are jumping along the sand bed. A similar transport mode exists for sand in water. Saltation can be defined as the motion of a solid which, while generally moving above a boundary, gets most of its upward lift by occasional contact with the boundary. It will thus make a succession of jumps. Bagnold (1973) mentions as a commonplace example that the wheel of a wheelbarrow saltates when the barrow is pushed fast across the furrows of a ploughed field. The turbulence of the fluid flow may or may not affect the motion of a saltating sand grain. The movement of a saltating grain is initiated by some vertical force, which acts transiently. In water, the upward motion of the grain will cease shortly after the initiation, at a height of only a few grain diameters above the bed, whereupon the grain will approach the bed, relatively gently. Often, the grains start upwards again when they approach or strike the bed, and so perform a series of jumps. Since the vertical velocity is small, this cannot be due to an elastic jump, but it can be explained by a negative pressure gradient in the vicinity of the bed; see Engelund (1970). In air, on the other hand, the grain will attain considerable height in the flow due to its large relative inertia. Typically, it will reach a height of the order of a hundred grain diameters—some grains will even jump as high as a few thousand diameters. During the jump the grain will obtain a horizontal velocity, which will be a considerable fraction of the wind speed at this height. When the grain hits the sand surface again, it will do so very violently, and as a consequence the grain will often either rebound or hurl one or more other grains up into the air flow. Therefore, once the saltation is initiated, a chain reaction is released, implying that the sand transport can be maintained by a weaker wind than the one which initiated it. Although saltation in air and water are both covered by the general definition of saltation given above, the nature of the forces involved is different in the two cases. This has led some authors to avoid the name saltation in the case of sand transport in water. On the other hand, Bagnold's general definition has enabled him to analyze a broad variety of transport phenomena from the viewpoint of general physics; see Bagnold (1956, 1966a, 1973).

We have not yet given account of how all of the horizontal momentum attained by the grains during the saltation is spent when the grains hit the bed surface. Some of it is used in the initiation of saltating grains, but most of it is spent to push the grains in the sand bed in the direction of the wind. This transport mode, which is typical of aeolean sand transport, is called *surface creep*. It constitutes approximately 25 per cent of the total transport. There is no generally accepted definition of surface creep, and the demarcation between this transport mode and saltation may vary from one experimental study to another. It seems sensible to base a precise definition on the dynamics of the two transport modes by defining surface creep as the motion of a grain which does not lose contact with the bed during the motion. Thus surface creep is a transport mode where the frictional resistance between the grain and the sand bed plays an important role. The material moving in this way consists mainly of grains that are being pushed forward by the impact of a saltating grain or by the direct action of the wind, but includes also, for instance, grains that are rolling forward because the grains supporting them have been removed. Typically, the sand grains transported as surface creep are coarser than the saltating grains. In particular, grains that are considerably larger than the largest capable of being set in saltation can be pushed forward by the impact of a saltating grain.

When sand is transported in either wind or water, some of the grains will be *suspended* in the fluid in question, i.e., upward currents of the turbulence will keep them aloft. Thus they will be transported over large distances without the frequent contact with the bed which is characteristic of saltation.

In water, suspension will usually play an essential role, while it is of minor importance in air for all but fine dust. Bagnold (1941) states on the basis of theoretical calculations that the suspended material does not constitute more than 5 per cent of the total transport for a sand with a mean diameter of 250 μm, even for large wind speeds. In the case of very fine sand, however, the situation in air approaches that in water, in that for sand of a typical diameter equal to 80 μm more than 50 per cent of the transported material is suspended.

The velocity of the fluid flow in the layer of saltation is affected by the saltating grains that are continuously being thrown up into the flow with a negligible horizontal momentum. This effect implies that the shear stress (i.e. force per unit of area) which the fluid flow exerts on the bed is controlled by the concentration of sand grains in the saltation layer. However, an important consequence of the difference between the densities of air and water is that a sand grain moving at the same speed as the wind will have a momentum which is 2000 times that of the same volume of air, whereas the ratio is only 2.65 in the case of water. Therefore, in the case of saltation in air an equilibrium is established under most conditions where the number of grains dislodged per second equals the number hitting the bed, whereas a grain saltating in water will extract much less momentum from the flow, and an equilibrium between the shear stress at the bottom and the transport rate by saltation can exist for low mean velocities only. If the mean velocity is large enough compared to the depth of the flow, the bed is unstable, and irregular dunes occur, which will cause an additional shape friction. The force on the far smaller grains in the bed is therefore reduced, and a new equilibrium is established. If the mean velocity is increased further, suspension will become more important. Part of the material in the dunes will be suspended, and eventually the dunes will give way to a stable equilibrium with a plane bed. This sequence of events is usually not found in the case of sand transport in air. However, Bagnold (1941, p. 165) reports that small regular ripples are formed on a bed of very fine sand at a friction velocity less than 30 cm/s, whereas these ripples are replaced by much larger and more irregular ripples at higher fluid velocities. This resembles the pattern found in water, and Bagnold explains it by the fact that at friction velocities over 30 cm/s such a large fraction of the transported material is suspended that the remaining saltation is not sufficient to maintain an equilibrium between shear stress and saltation transport rate. Therefore, an equilibrium can only be established by the formation of large ripples. A review of mathematical analyses of the stability of a sand bed in a water flow can be found in Engelund and Fredsøe (1982). A more detailed mathematical study of the shape of dunes in rivers, as this depends on the flow, is presented in Fredsøe (1982).

The transport rate, defined as the mass of sand transported per second across one unit of length perpendicular to the wind direction, is obviously an important quantity. Bagnold (1980) has found empirical evidence that the saltation transport rate in water is proportional to $(\omega - \omega_0)^{3/2}$, where ω is the stream power defined as the product of the shear stress at the bed and the mean velocity of the water flow, and ω_0 is the threshold value of ω at which sediment begins to be moved. This relationship was found to hold over a very wide range of flow conditions including flume experiments as well as large natural rivers. In air the transport rate tends to be proportional to $\tau^{3/2}$, where τ denotes the shear stress at the bed. The quantity $\tau^{3/2}$ can be interpreted as a stream power. However, quite large departures from the latter relationship have been found for grains which are far from being spherical; see Willetts,

Rice, and Swaine (1982). These departures can be explained by the fact that the height to which a grain jumps on impact with the bed depends on its shape; see Jensen and Sørensen (1982b).

We now outline a method proposed by Jensen and Sørensen (1982b, 1983) for analyzing detailed data on aeolean saltation, where the transport rate at various heights above the sand bed has been measured for each particle size class. Data of this kind were obtained by Williams (1964). The transport rates are usually measured by means of traps.

In the analysis a ballistic–stochastic model of the trajectories of a saltating sand grain is used. Given the initial velocity v of a dislodged sand grain, we calculate the trajectory by assuming that the only forces exerted on the grain are those of drag and gravity. Thus the turbulent eddies are disregarded, and the only stochastic element of the model is the liftoff speed. The unknown probability distribution of this quantity is approximated by supposing it to be discrete and concentrated at the points corresponding to the centers of the traps. Now, the point probabilities and the dislodgement rates for each size class are estimated by the values for which the amount of sand actually caught in each trap equals the amount calculated by means of the model.

From these estimates several interesting quantities, such as the transport rate for each size class, the force exerted on the bed by the descending grains (the so-called grain-borne shear stress), and the angle of descent, can be calculated by means of the model.

For details, see Jensen and Sørensen (1982b). A number of other recent theoretical and experimental works on sand transport by water and wind can be found in Kikkawa and Iwasa (1980) and Sumer and Müller (1983).

4. Sorting and Size Distributions

The results of both wind-tunnel experiments and observations in the field carried out by Bagnold (see Bagnold, 1941) show that aeolian sand deposits laid down under steady conditions have a distinctive size distribution. The graph of the log probability density function of the logarithm of the grain diameter has the shape of a hyperbola. This is illustrated in Figure 4. A sand deposit of this kind is termed a regular sand, and the size distribution of a regular sand is called log hyperbolic. The mathematics of the hyperbolic distributions is discussed in Section 5.

Bagnold and Barndorff-Nielsen (1980) give a number of further examples of sand size distributions of the hyperbolic shape, from a variety of alluvial and aeolean regimes.

Bagnold conducted a number of experiments on the changes which take place in the size grading when sand is removed, transported, and redeposited

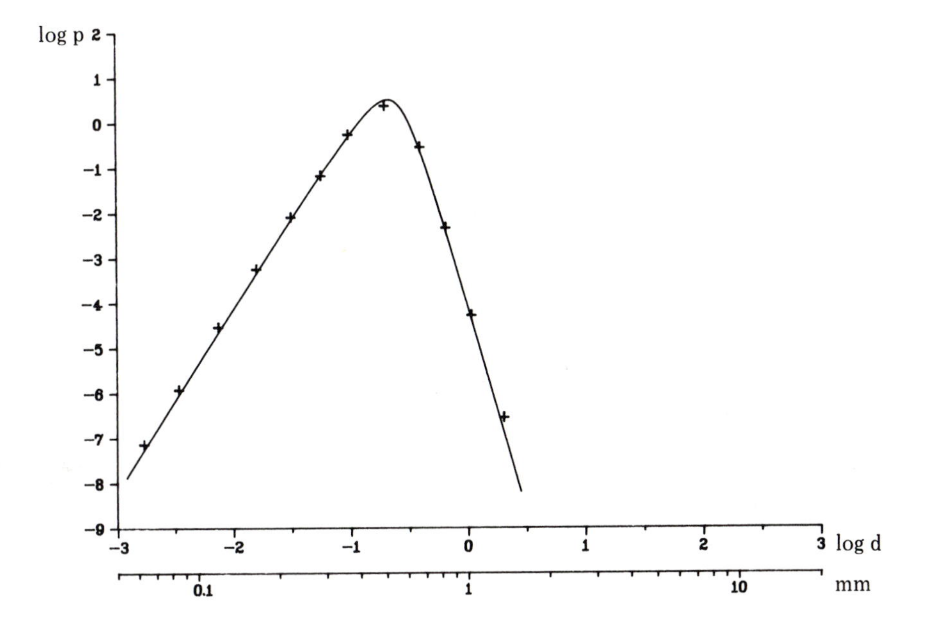

Figure 4. The histogram of a mass–size distribution of a sand sample [data from Bagnold (1941, p. 114)] together with the estimated hyperbolic distribution in a double-logarithmic plot. (Only the top midpoints of the histogram slabs are shown.) Here, d and p denote particle diameter and probability density, respectively. [For details and values of the estimated parameters, see Barndorff-Nielsen (1977).]

by the wind. In one series of experiments he used a wind tunnel, the floor and the roof of which were parallel in the upwind half of the tunnel whereas the roof was inclined in the downwind half, so that the tunnel expanded in the wind direction. The upwind half was covered with a regular sand, and a very thin layer of the same sand was laid over the floor of the downwind half in order to make the whole surface texture uniform. Between the two parts of the tunnel a device for trapping a sample of surface creep was positioned. When the wind was blowing, saltating sand was deposited in the downwind half of the tunnel because the expansion implied a slowing up of the wind and a consequent decrease in the wind's ability to transport sand.

Bagnold found that the sand deposited in subsections of the expanding part of the tunnel was regular in each section but with different hyperbolic distributions. The slope of the coarse-side asymptote of the hyperbola was close to 9 in all subsections, irrespective of the wind speed and of the size distribution of the initial regular sand. Similarly, the slope of the fine-side asymptote had the same value in all sections. This asymptote was steeper than that of the parent

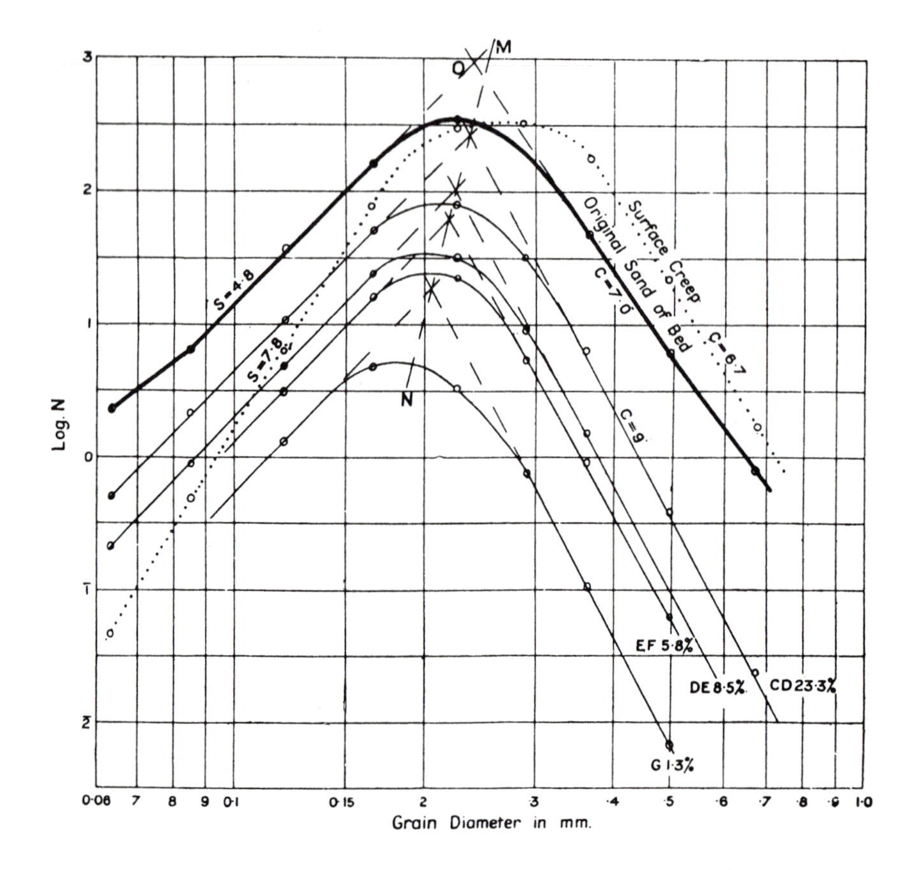

Figure 5. Double-logarithmic plot [reproduced, with permission, from Bagnold (1941, p. 137)] of mass–size distributions of sand samples from a wind-tunnel experiment. The original sand was spread out in the upwind section *AB* of the tunnel, and after the wind had blown, samples were taken at *B* (surface creep) and from the subsequent sections *CD*, *DE*, *EF*, and *G* of the tunnel. The weights of the various samples are shown on the figure in per cent of the total weight of the samples.

sand, but approached the latter as the wind speed was increased. The mode point of the size distributions in the subsections decreased downwind, and (as would be expected) the mass of sand deposited in each subsection decreased downwind too. Intriguingly, however, Bagnold found a linear dependence between the logarithm of the mass deposited per unit of area and the logarithm of the peak diameter; cf. Figure 5 and Bagnold (1941, Section 10.3). The surface creep was also a regular sand, but all the removed sand taken together was not regular.

In another type of experiment in the same wind tunnel Bagnold artificially made a regular sand nonregular by adding more sand to the extreme fine size class and to the extreme coarse class. This nonregular sand was used as parent sand. It turned out that the distortion in the fine end of the size distribution almost disappeared in the surface creep but hardly at all in the saltation deposits. On the other hand, the distortion in the coarse tail was slightly reduced in the case of surface creep, whereas it disappeared entirely in the saltation deposits at the end of the tunnel. This experiment illustrates that the processes which produce the hyperbolic size distributions are different for the fine and coarse tails of the size distribution.

In trying to explain the changes in the size grading when sand is transported by the wind and, in particular, the generation of hyperbolic size distributions, it is essential to determine the relation between the transport velocity and the size of a sand grain under various conditions. The motion of the individual sand grains has often been modeled by a compound Poisson process in the case of sand transport in a water flow; see e.g. Einstein (1937), Hubbell and Sayre (1964), and Todorović and Nordin (1975). In the case of aeolean sand transport a compound Poisson process appears likewise to be a reasonable model, and this type of model was used by Barndorff-Nielsen, Jensen, and Sørensen (1983) to gain information about the relation between transport velocity and size of a sand grain, from an experiment using coloured sand grains.

In this type of experiment a ridge of coloured heterogeneous sand is eroded and transported down through a wind tunnel by the wind, over a uniform and uncoloured bed of the same kind of sand. When the wind is turned off, samples of the uppermost layer of sand are taken at a number of locations along the tunnel. A very thin sample can be taken by means of an adhesive spray. Each sample is divided into size classes by sieving. For each size class a subsample is taken, and the proportions of coloured and uncoloured grains in the subsamples are determined by counting, using a microscope. Let δ_{ij} denote the ratio between the number of coloured grains and the number of uncoloured grains in the ith size fraction at location j. The data reported by Barndorff-Nielsen, Jensen, and Sørensen (1983) suggested that

$$\log \delta_{ij} = \beta_i - \alpha x_j, \tag{4.1}$$

where $\alpha > 0$ and where x_j denotes the distance between the sample location and the initial position of the ridge. It should at this point be emphasized that the experiment had the character of a pilot experiment. For instance, the calculation of δ_{ij} involved an adjustment in order to correct for an insufficiently accurate sampling method. On combining (4.1) with the assumption that the size distribution is the same everywhere in the sampling region, one obtains that the probability density function of the total distance traveled by a coloured sand grain of size s that took part in the transport process is

$$\pi(x; s) = \alpha\{\log(1 + \varphi(s)^{-1})(1 + \varphi(s)e^{\alpha x})\}^{-1}, \tag{4.2}$$

where s denotes log grain diameter and $\varphi(s) = e^{-\beta(s)}$. The moments of the distribution (4.2) can be found from its Laplace transform by expressing the latter in terms of a hypergeometric function.

Now, suppose that the motion of the individual sand grains can be modeled by an additive stochastic process. Actually, it would not be an essential restriction to assume that the process is a compound Poisson process, since this is equivalent to assuming that the process has an atom at the origin. Let $\varphi(\theta; s, t)$ denote the Laplace transform of the distance traveled by a grain of size s in the time interval $[0, t]$. Provided the process is additive, $\varphi(\theta; s, t)$ equals $\varphi(\theta; s, 1)^t$, and the corresponding mean and variance are of the form $t\mu(s)$ and $t\sigma^2(s)$. Furthermore, we shall suppose that the ridge of coloured sand is eroded in such a way that the sand in the interior of the ridge becomes part of the layer in which the sand motion takes place, at a constant rate during the entire experiment. Under these assumptions the relation between the Laplace transform $\varphi_\pi(\theta; s)$ of the distribution (4.2) and $\varphi(\theta; s, 1)$ is easily found to be

$$\varphi_\pi(\theta; s) = \frac{\varphi(\theta; s, 1)^T - 1}{T \log \varphi(\theta; s, 1)},$$

where T denotes the duration of the experiment. This equation cannot be solved explicitly with respect to $\varphi(\theta; s, 1)$. However, being less ambitious, we can under the assumptions given above calculate $\mu(s)$ and $\sigma^2(s)$ from the mean $m(s)$ and the variance $v(s)$ of (4.2), since

$$\mu(s) = \frac{2m(s)}{T}$$

and

$$\sigma^2(s) = 2\frac{v(s) - m(s)^2/3}{T}.$$

When the estimates of α and $\varphi(s)$ obtained in Barndorff-Nielsen, Jensen, and Sørensen (1983) are used in these formulae it is found, somewhat unexpectedly, that $\mu(s)$ increases as s is increased. [The standard deviation $\sigma(s)$ increases too, but at a much slower rate.] This result can obviously be valid only for a limited range of grain sizes. However, similar observations have been made in the case of alluvial sediment transport; for references, see Barndorff-Nielsen, Jensen, and Sørensen (1983).

An entirely different approach has been used by Deigaard and Fredsøe (1978) to explain the longitudinal grain sorting in rivers. From a formula for the total transport rate in an alluvial channel, developed by Engelund and Fredsøe (1976), they derive a formula for the transport rate of the individual size fractions. The idea behind these formulae for transport rates is also applicable to aeolean sand transport, as indicated by calculations carried out in Jensen and Sørensen (1982a, 1983). By keeping account of the mass of sand in the individual size fractions that enter and leave suitably chosen sections of the river, Deigaard and Fredsøe were able to follow the development of the size distribution of the sediment in a numerical study. Their analysis yielded a longitudinal variation in the mean grain diameter in reasonable accordance

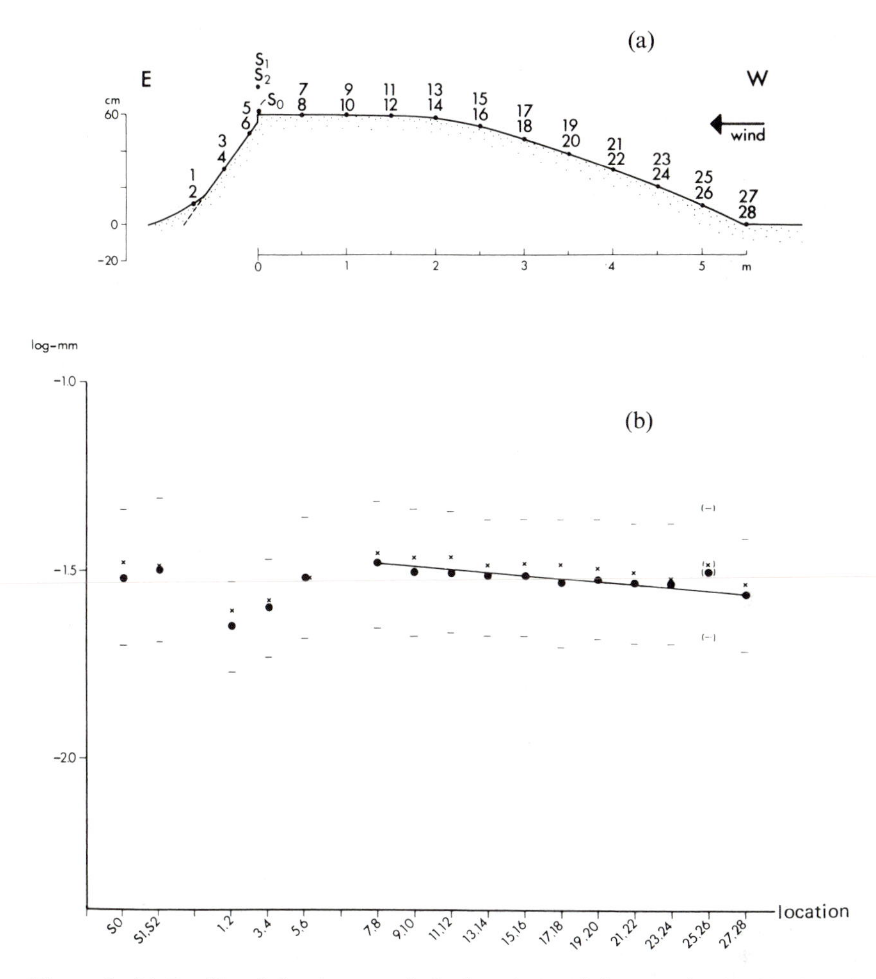

Figure 6. (a) Profile of the dune, and the locations of the samples (1–28, S_0–S_2). (b) Summarizing the variation in particle size distribution over the dune. The graph shows the typical log grain size $\overline{v}$ at the various locations (●), and the associated values of $\overline{\mu}$ (×) and $\overline{v} \pm \overline{\tau^{-1}}$. (The bars indicate average value for duplicate samples.) [Reproduced, with permission, from Barndorff-Nielsen *et al.* (1982).]

with the variation found in flume experiments as well as in nature. Interestingly, the assumed lognormal distribution of the incoming sand at the start of the river turned out to change towards a hyperbolic shape in the downstream direction.

As a final example of the intriguing regularities found in the present field of study, we reproduce in Figure 6(a) and (b) some results from Barndorff-Nielsen, Dalsgaard, Halgreen, Kuhlman, Møller, and Schou (1982). Duplicate sand samples were collected from a small barchanoid dune, as indicated in Figure 6(a). Each sample was fitted with a hyperbolic distribution by maximum likelihood estimation, and some of the estimated parameters, averaged over duplicate samples, are shown in Figure 6(b). The parameters in question are the mode υ of the hyperbolic distribution, the abscissa μ of the point of intersection of the asymptotes of the hyperbola, and the square root τ of the curvature of the hyperbola at the mode point υ. Here υ and μ indicate the typical grain size, while τ is a measure of the spread of the distribution in the neighbourhood of the mode. (For a normal distribution the analogous definition of τ yields $\tau = \sigma^{-1}$, where σ denotes the standard deviation.) One notes, in particular, how the typical grain size υ increases linearly with distance up the windward side of the dune.

5. Hyperbolic Distributions; Theory, Applications, and Related Distributions

The r-dimensional hyperbolic distribution is characterized by the fact that the graph of the log probability function of the distribution is a hyperboloid (hyperbola when $r = 1$).

The one-dimensional hyperbolic distribution has four parameters, and its probability density function may be written as

$$p(x; \mu, \delta, \alpha, \beta) = \frac{\kappa}{2\alpha\delta K_1(\delta\kappa)} e^{-\alpha\sqrt{\delta^2+(x-\mu)^2}+\beta(x-\mu)}, \qquad x \in R. \tag{5.1}$$

Here K_1 is a Bessel function and $\kappa = \sqrt{\alpha^2 - \beta^2}$, where α and β are parameters such that $\alpha > |\beta|$. The location parameter μ is the abscissa of the point of intersection between the linear asymptotes of the hyperbola, while δ is a scale parameter. The mode of the distribution is $\upsilon = \mu + \delta\beta/\kappa$, and the distribution is symmetric for $\beta = 0$. The slopes of the asymptotes are $\varphi = \alpha + \beta$ and $-\gamma = \beta - \alpha$, respectively, and replacing (α, β) in (5.1) by (φ, γ) one obtains

$$p(x, \mu, \delta, \varphi, \gamma) = a(\delta, \varphi, \gamma) e^{-\frac{1}{2}\varphi(\sqrt{\delta^2+(x-\mu)^2}-(x-\mu))-\frac{1}{2}\gamma(\sqrt{\delta^2+(x-\mu)^2}+(x-\mu))},$$

where

$$a(\delta, \varphi, \gamma) = \frac{\sqrt{\varphi\gamma}}{\delta(\varphi+\gamma)K_1(\delta\sqrt{\varphi\gamma})}.$$

This representation of the hyperbolic distribution is very suitable in connection with the study of mass–size distributions of sand samples, φ and γ being the theoretical counterparts of the small-grade coefficient and the coarse-grade coefficient, respectively.

Estimation of all four parameters of (5.1) from a single sample of ungrouped observations generally requires a rather large sample size, say a hundred observations or more. An effective algorithm for determining the maximum likelihood estimates is described in Jensen (1983). This emerged only after a considerable number of approaches had been tried out, with varying success. The main point of the algorithm is to switch between iteration in δ and in the complementary three parameters, μ, α, and β.

In studies of particle size distributions of sand samples, the observational procedure based on sieving yields grouped observations. The underlying sample size, measured by the number of single grains in a typical sample of about 30 g, is of the order of 10^6, but is never known precisely. Moreover, the random variation exhibited by the measured size distribution of a sand sample, such as that shown in Figure 4, depends not only on the number of grains, but also on the sampling and sieving procedures. An interesting question is then: is it possible, somehow, to describe the random variation by specifying an "actual" or "effective," but fictitious, sample size—N say—for empirical distributions of this kind? This would enable us to describe the uncertainty of the estimates of the hyperbolic parameters, obtained (Barndorff-Nielsen, 1977a, b) by maximizing the function

$$\sum r_i \ln p_i(\mu, \delta, \alpha, \beta) \tag{5.2}$$

or, equivalently, by minimizing the function

$$I = I(\mu, \delta, \alpha, \beta) = \sum r_i \ln \left\{ \frac{r_i}{p_i(\mu, \delta, \alpha, \beta)} \right\}, \tag{5.3}$$

where r_i is the relative weight of the grains in the ith sieving interval and where p_i is the probability mass of the hyperbolic distribution (5.1) in the same interval, size being measured on the logarithmic scale. We term the estimates obtained in this way maximum likelihood estimates even though strictly speaking a likelihood function for the parameters is not available, due to the indefiniteness of the sample size. However, the function I in (5.3) may be conceived of as being proportional to a log likelihood function l for $(\mu, \delta, \alpha, \beta)$, the constant of proportionality being minus the reciprocal of the quasi sample size N. If l and N were real, with N large, we would have that $2N\hat{I} = 2NI(\hat{\mu}, \hat{\delta}, \hat{\alpha}, \hat{\beta})$ followed, approximately, a χ^2 distribution on $k - 1 - 4$ degrees of freedom, where k denotes the number of sieving intervals. This suggests that, assuming hyperbolic variation, the quasi sample size N may be sensibly defined as

$$N = (k - 3)(2\hat{I})^{-1}, \tag{5.4}$$

$(k - 3)^{-1}$ being the mode point of the distribution of the reciprocal of a χ^2 variate on $k - 5$ degrees of freedom. (Admittedly, there is an element of arbitrariness in this choice of definition. Thus, for instance, one might instead have taken N so that $2N\hat{I}$ was the mode point of the χ^2 distribution on $k - 5$ degrees of freedom.) We can then adopt $-NI$ as a quasi log likelihood function for $(\mu, \delta, \alpha, \beta)$, and in particular we may use the formation matrix derived from this [i.e. the inverse of the matrix of second-order derivatives of NI, evaluated at the estimate $(\hat{\mu}, \hat{\delta}, \hat{\alpha}, \hat{\beta})$] to specify the actual uncertainty and correlations of the estimates.

Table 1. Maximum Likelihood Estimates of Hyperbolic Parameters, Together with Their Estimated Standard Deviations and Correlations, for the Mass–Size Distribution of the Sand Sample Shown in Figure 4.

			Correlations			
	Estimate	Standard deviation	υ	δ	φ	γ
υ	−0.676	.00294		−.484	−.657	.149
δ	0.197	.00923			.824	.689
φ	4.11	.0569				.436
γ	9.53	.2301				

Applying this procedure to the data for Figure 4, we obtain $N = 37{,}519$, the corresponding estimated values, standard deviations, and correlations for the parameters υ, δ, φ, and γ being set out in Table 1.

The above idea of a quasi sample size would seem to be more broadly applicable, to situations where random effects of secondary importance have a nonnegligible influence on the fit of and inference under a model, as is quite often the case with large sample sizes. While ordinary tests of the model specification may indicate gross significance, one may for various good reasons—such as previous experience, theoretical background, incisiveness of analysis, and ease of interpretation—wish to uphold the model. Ignoring the lack of fit will, however, imply an exaggeration of the precision of the parameter estimates. This may be avoided by a formal adjustment of the sample size to a quasi sample size, as above.

The distribution (5.1) has been found to fit a great variety of empirical distributions from geology, biology, astronomy, fluid mechanics, economics, etc.; see Barndorff-Nielsen and Blæsild (1981, 1983) and the references given there. As yet another instance we exhibit in Figure 7 the distribution of the log intensity of primary magnetization in 2163 samples from lava flows on Iceland. The original data, which have kindly been put at our disposal by Dr. L. Kristjansson, are discussed in Kristjansson and McDougall (1982). Of the total number of 2163 observations, 1067 are from the normal polarity state of the earth magnetization and 1096 from the reverse polarity state. From a comparison of the grouped distributions of, respectively, normal- and reverse-state observations Kristjansson and McDougall (1982) concluded, cogently, that the strength of the earth magnetization does not depend on the polarity. A sharper test of this is obtained by comparing the normal and reverse samples separately with the hyperbolic distribution fitted to the combined data, that is, the fitted hyperbolic distribution shown in Figure 7. This comparison is carried out in Figure 8, which gives the quantile diagrams for each of the two data sets. In both cases there is remarkable agreement with the hyperbolic distribution estimated from the total data. If, for the combined data, we apply formula (5.4) to obtain an estimated "effective" or quasi

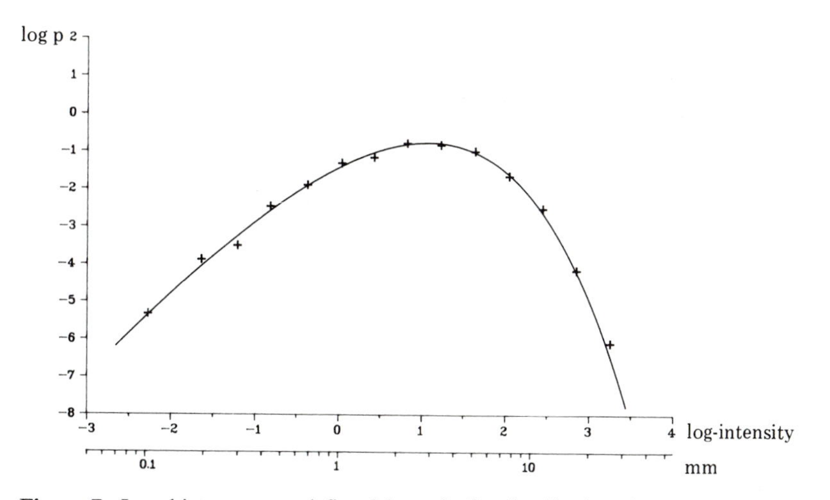

Figure 7. Log histogram and fitted hyperbolic distribution for the log intensity of primary magnetization of 2163 samples from lava flows on Iceland. [Data courtesy of Dr. L. Kristjansson; cf. Kristjansson and McDougall (1982).]

sample size N, we obtain $N = 2142$, in agreement with the actual sample size of 2163.

Further examples of empirical distributions, of discrete as well as continuous type, with exponentially decreasing tails are presented in Bagnold (1983), and in that paper the author makes an interesting comparison of the distributions by means of a standardization which eliminates the dependence on location and scale.

A general theoretical argument, like the central limit law for the normal distribution, which could explain the ubiquity of the hyperbolic distributional form has not so far been found. However, the one-dimensional hyperbolic distribution has been shown (Halgreen, 1979) to be self-decomposable, i.e., it occurs as the limit law for sequences of random variables y_n of the form $y_n = b_n^{-1}(x_1 + x_2 + \cdots + x_n) - a_n$, where $x_1, x_2, \ldots, x_n, \ldots$ are independent random variables and a_n, b_n are constants. Furthermore, the hyperbolic distribution of arbitrary dimension r, as defined at the outset of the present section, is a variance–mean mixture of normal distributions (Barndorff-Nielsen, 1977a, b). More specifically,

$$H_r(\alpha, \beta, \mu, \delta, \Delta) = N_r(\mu + w\beta\Delta, w\Delta) \underset{w}{\wedge} G\left(\frac{r+1}{2}, \delta^2, \alpha^2 - \beta\Delta\beta^*\right) \quad (5.5)$$

where $H_r(\alpha, \beta, \mu, \delta, \Delta)$ denotes the r-dimensional hyperbolic distribution with a certain parametrization, $N_r(\xi, \Sigma)$ is the r-dimensional normal distribution

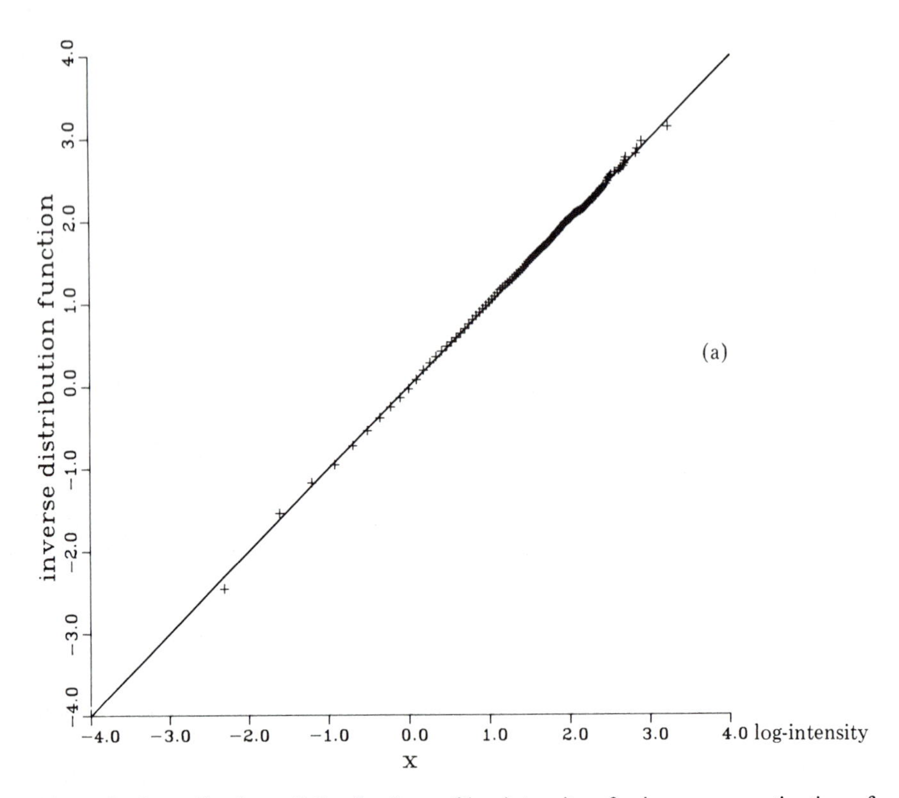

Figure 8. Quantile plots of distributions of log intensity of primary magnetization of lava flow samples: (a) normal earth magnetization; sample size 1067; (b) reverse earth magnetization; sample size 1096. Both distributions are compared with the hyperbolic distribution determined by maximum likelihood estimation from the combined sample.

with mean vector ξ and variance matrix Σ, and $G(\lambda, \chi, \psi)$ is the generalized inverse Gaussian distribution with index λ whose probability density function is given by

$$\frac{(\psi/\chi)^{\lambda/2}}{2K_\lambda(\sqrt{\chi\psi})} w^{\lambda-1} e^{-\frac{1}{2}(\chi w^{-1}+\psi w)}, \qquad w > 0. \tag{5.6}$$

For a comprehensive account of the generalized inverse Gaussian distributions, see Jørgensen (1982). Also, we refer to Embrechts (1983) for an application of these distributions in risk theory, and to Letac and Seshadri (1983), who have shown that for λ negative (5.6) is the limit distribution of continued fractions in independent gamma variates.

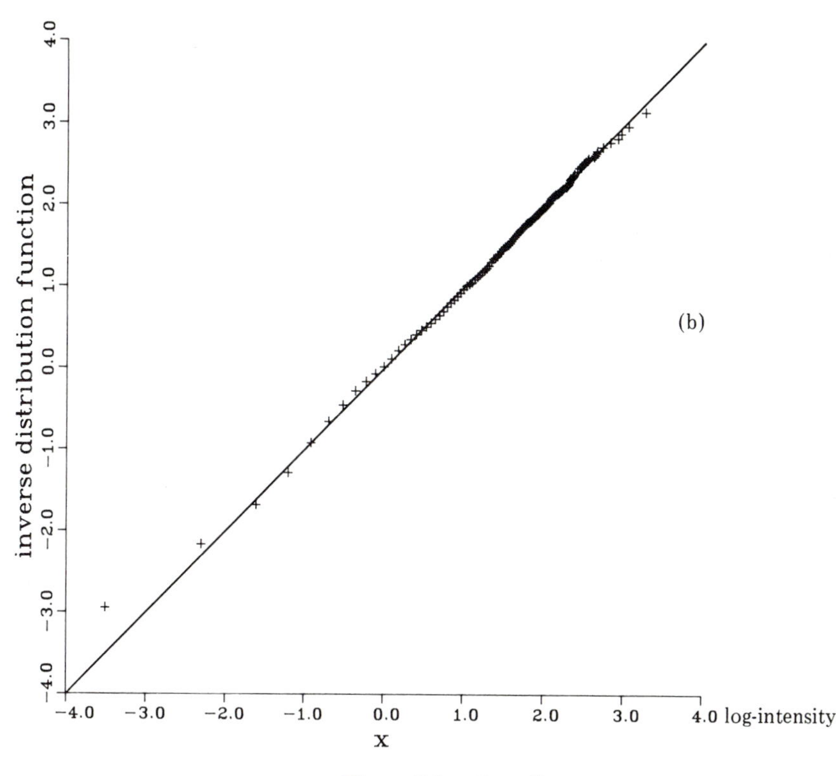

Figure 8 (continued).

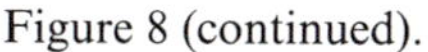

In the special context of relativistic statistical mechanics the three-dimensional hyperbolic distribution occurs in a basic way as the distributional law of the momentum vector of a single particle in an ideal gas. This result derives from the Boltzmann–Gibbs law $ae^{-\lambda E}$. In classical, i.e. Newtonian, physics the analogous derivation leads to the normal distribution. For a brief exposition, with references to the relevant physics literature, see Barndorff-Nielsen (1982).

A one-parameter extension of the class of hyperbolic distributions is obtained by replacing $(r + 1)/2$ with an arbitrary real number λ on the right-hand side of (5.5). The resulting class of generalized hyperbolic distributions, which has been discussed by Barndorff-Nielsen (1977a, b; 1978), Blæsild (1981), and Blæsild and Jensen (1981), is closed under conditioning, marginalization, and affine transformations. It includes, corresponding to $\lambda = (r - 1)/2$, the so-called hyperboloid distributions (Barndorff-Nielsen, 1978; Jensen, 1981), which have properties analogous to those of the von Mises–Fisher distributions, but relate to hyperboloids instead of spheres.

For more detailed surveys of the properties of distributions of hyperbolic type the reader is referred to Barndorff-Nielsen and Blæsild (1981, 1983), and

Blæsild and Jensen (1981). Atkinson (1982) gives algorithms for simulation of the associated random variables.

We conclude this section with a discussion of the usefulness of the hyperbolic distribution as an error law.

As is evident from the discussion in this and the previous section, the hyperbolic distribution is capable of describing a considerable range of long-tailed and asymmetrical, as well as symmetrical, empirical distributions. A numerical expression of this is obtained by specifying the domain of joint variation of the skewness $\gamma_1 = \kappa_3/\kappa_2^{3/2}$ and the kurtosis $\gamma_2 = \kappa_4/\kappa_2^2$, where κ_i denotes the ith cumulant of (5.1). This was done in Barndorff-Nielsen and Blæsild (1981), and it was found that the domain is approximately given by

$$0 \leq \gamma_1^2 < \tfrac{2}{3}\gamma_2 < 2 + \tfrac{1}{2}\gamma_1^2 < 4.$$

Unfortunately, the exact formulae for γ_1 and γ_2 are rather complicated expressions in terms of the quotient of Bessel functions K_2/K_1; cf. Barndorff-Nielsen and Blæsild (1981). However, using well-known asymptotic expressions for the Bessel functions, we find

$$\gamma_1 \sim \begin{cases} \dfrac{6}{\sqrt{2}}\rho\{1 + O(\rho^2)\} & \text{as } \rho \to 0,\ \zeta \to 0, \\ 3\zeta^{-\frac{1}{2}}\rho\{1 + O(\rho^2)\} & \text{as } \rho \to 0,\ \zeta \to \infty \end{cases} \tag{5.7}$$

and

$$\gamma_2 \sim \begin{cases} 3\{1 + O(\rho^2)\} & \text{as } \rho \to 0,\ \zeta \to 0, \\ 3\zeta^{-1}\{1 + O(\rho^2)\} & \text{as } \rho \to 0,\ \zeta \to \infty. \end{cases} \tag{5.8}$$

Here

$$\zeta = \delta\sqrt{\alpha^2 - \beta^2} \quad \text{and} \quad \rho = \beta/\alpha.$$

Note that ρ and ζ are invariant under location–scale transformations of the hyperbolic distribution. The results (5.7) and (5.8), and further properties to be discussed below, suggest that the shape of the hyperbolic distribution, up to location–scale transformations, may be well represented by the invariant parameters

$$\xi = (1 + \zeta)^{-\frac{1}{2}} \quad \text{and} \quad \chi = \rho\xi.$$

In particular, (5.7) and (5.8) indicate that for not too large values of ρ we have roughly $\gamma_1 \simeq 3\chi$ and $\gamma_2 \simeq 3\xi^2$. The joint domain of variation of the parameters ξ and χ is given by

$$0 \leq |\chi| < \xi < 1,$$

and this domain, which we refer to as the shape triangle, is depicted in Figure 9. The estimated values of (ξ, χ) for the data of Figures 4 and 7 and for a further data set to be mentioned below are indicated in the figure. Moreover,

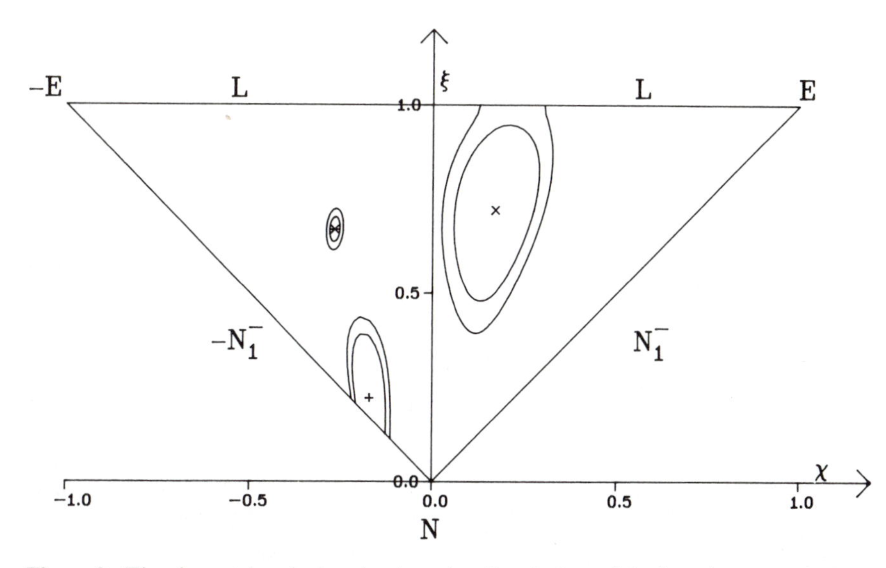

Figure 9. The shape triangle, i.e. the domain of variation of the invariant parameters ξ and χ of the hyperbolic distribution. The letters N, N_1^-, L, and E at the boundaries indicate how the normal distribution, generalized inverse Gaussian distribution with $\lambda = 1$, Laplace distribution (symmetrical of skew), and exponential distribution are limits of the hyperbolic. Also indicated are the estimated values of (χ, ξ) for an aeolian sand sample (*; cf. Figure 4), the lava magnetization data (+, cf. Figure 7), and Pierce's telegraph-key data (×), together with the contours of the (modified profile likelihood approximation to the) marginal likelihood function, corresponding to 3 and 5 log likelihood units.

the symbols N, N_1^-, L, and E at the boundaries of the domain indicate that the normal distribution, generalized inverse Gaussian distribution with index $\lambda = 1$, Laplace distribution (symmetric or skew), and exponential distribution are all limit distributions of the hyperbolic distribution.

More specifically, to obtain the above limit distributions we choose δ such that the variance of the hyperbolic distribution is of order 1. The variance is asymptotically given by

$$Vx \sim \begin{cases} 2\dfrac{\delta^2(1+\rho^2)}{\zeta^2(1-\rho^2)} & \text{for } \zeta \to 0, \\[2ex] \dfrac{\delta^2}{\zeta(1-\rho^2)} & \text{for } \zeta \to \infty, \end{cases}$$

and we therefore take

$$\delta = (\sqrt{1+\zeta} - 1)\sqrt{1-\rho^2} = (\xi^{-1} - 1)\sqrt{1 - (\chi/\xi)^2}.$$

If, moreover, we arrange for the mode point of the distribution to be at 0, the density (5.1) takes the form

$$\psi(x; \xi, \chi) = a(\xi, \chi)e^{-b(\xi, \chi)\{\sqrt{1-(\chi/\xi)^2+z^2}-(\chi/\xi)z\}}, \tag{5.9}$$

where

$$z = \frac{x}{\xi^{-1}-1} + \frac{\chi}{\xi}$$

and

$$a(\xi, \chi) = \{2(\xi^{-1}-1)K_1(\xi^{-2}-1)\}^{-1}, \qquad b(\xi, \chi) = \frac{\xi^{-2}-1}{1-\chi/\xi}.$$

If we now let ξ tend to 0 we obtain the normal density

$$\frac{1}{\sqrt{2\pi}}e^{-\frac{1}{2}x^2},$$

with a relative error of the order ξ (or, equivalently, $\zeta^{-\frac{1}{2}}$), uniformly in ρ and uniformly in x for x belonging to a compact set. In fact, for $\rho = 0\ (=\chi)$ the relative error is of order ξ^2. If instead we let ξ tend to 1 we obtain the Laplace distribution

$$\exp\left(-\frac{2}{1-\chi^2}(|x| - \chi x)\right);$$

in particular, for $\chi = 0$ we have the Laplace distribution

$$e^{-2|x|}.$$

For x not belonging to an open interval around 0 the relative error is uniformly of the order $1 - \xi$, and for $x = 0$ the relative error is of the order $\sqrt{1-\xi}$. Finally, we consider the limit operation χ tending to ξ. If we relocate the distribution by introducing $y = x + \xi^{-1} - 1$, we obtain the limiting density

$$\begin{cases} \dfrac{1}{2(\xi^{-1}-1)K_1(\xi^{-2}-1)}e^{-\frac{1}{2}(\xi^{-2}-1)\{(\xi^{-1}-1)y^{-1}+(\xi^{-1}-1)^{-1}y\}} & \text{for } y > 0, \\ 0 & \text{for } y \le 0, \end{cases}$$

which is a generalized inverse Gaussian distribution with index $\lambda = 1$; cf. (5.6). The relative error is exponentially small for $y < 0$, and uniformly of the order $\xi - \chi$ for y in a compact set of positive real numbers.

Suppose we have a set of independent observations $y_1, \ldots, y_n$ and we specify a linear model with hyperbolic errors for these, i.e., we assume that $y = (y_1, \ldots, y_n)$ is of the form

$$y = \mu D + \delta x, \tag{5.10}$$

where μ is a $1 \times d$ parameter vector, D is a $d \times n$ design matrix of rank d, δ is a positive scale parameter, and the coordinates $x_1, \ldots, x_n$ of x are independent and identically distributed with the hyperbolic probability density function

$\psi(x; \xi, \chi)$ given by (5.9). Since $\log \psi$ is a concave function in x for fixed (ξ, χ), we have (cf. Burridge, 1980) that the maximum likelihood estimate of (μ, δ) exists, is unique, and is also the unique solution of the likelihood equations for (μ, δ), provided only that y does not belong to the linear subspace $R^d D$ of R^n.

Under this model, inference on (ξ, χ) should in principle be drawn from the marginal model for the statistic u that is maximal invariant relative to the group of affine transformations generating the structure (5.10). In the simplest case, of a pure location–scale model with $d = 1$ and $D = (1, \ldots, 1)$, u is Fisher's configuration statistic

$$u = \left(\frac{y_1 - \bar{y}}{s}, \ldots, \frac{y_n - \bar{y}}{s}\right), \tag{5.11}$$

where $\bar{y} = n^{-1}(y_1 + \cdots + y_n)$ and $s^2 = (n-1)^{-1} \sum (y_i - \bar{y})^2$. An explicit expression for the distribution of (5.11) is not available, but the marginal likelihood function for (ξ, χ) based on u may be expressed as

$$L(\xi, \chi) = \int_{-\infty}^{\infty} \int_{0}^{\infty} \delta^{-n} \prod_{i=1}^{n} \psi\left(\frac{x_i - \mu}{\delta}; \xi, \chi\right) \delta^{-1} \, d\delta \, d\mu; \tag{5.12}$$

see, for instance, Barndorff-Nielsen (1983). Except for the cases $\xi = 1$ and $\xi = 0$, corresponding, respectively, to the double exponential distribution and the normal distribution, this double integral cannot be calculated explicitly and numerical integration may be very time consuming, especially for large values of n. [For $\xi = 1$ and $\chi = 0$ an explicit form of (5.12) is available from Uthoff (1973).]

This difficulty may be avoided by considering an approximation to the marginal likelihood, the so-called modified profile likelihood, introduced by Barndorff-Nielsen (1983). Let $\tilde{L}$ denote the profile likelihood for (ξ, χ) corresponding to the sample $y_1, \ldots, y_n$, i.e.

$$\tilde{L}(\xi, \chi) = \sup_{(\mu, \delta)} L(\mu, \delta, \xi, \chi) = L(\hat{\mu}(\xi, \chi), \hat{\delta}(\xi, \chi), \xi, \chi),$$

where $\hat{\mu}(\xi, \chi)$ and $\hat{\delta}(\xi, \chi)$ are the maximum likelihood estimates of μ and δ for fixed (ξ, χ), and let

$$g_{(\xi, \chi)}(x) = -\log \psi(x; \xi, \chi)$$

and

$$a_i(\xi, \chi) = \frac{x_i - \hat{\mu}(\xi, \chi)}{\hat{\delta}(\xi, \chi)}.$$

The modified profile likelihood $\tilde{\tilde{L}}$ for (ξ, χ) is then given by [Barndorff-Nielsen, 1983, formula (3.3)]

$$\tilde{\tilde{L}}(\xi, \chi) = \hat{\delta}(\xi, \chi) D_{(\xi, \chi)}^{-\frac{1}{2}} \tilde{L}(\xi, \chi), \tag{5.13}$$

where

$$D_{(\xi,\chi)} = \left\{\sum_i g''_{(\xi,\chi)}(a_i(\xi,\chi))\right\}\left\{n + \sum_i a_i(\xi,\chi)^2 g''_{(\xi,\chi)}(a_i(\xi,\chi))\right\}$$
$$-\left\{\sum_i a_i(\xi,\chi) g''_{(\xi,\chi)}(a_i(\xi,\chi))\right\}^2.$$

The approximation (5.13) to (5.12) has been used in Figure 9 to plot contours of the marginal likelihood surface for (ξ,χ), both for the lava magnetization data discussed above and for a sample observed by Pierce (1870).

In this paper "On the theory of errors of observations," Pierce discussed the suitability of the normal distribution for describing the random variation of experimental data, and in particular he reports and considers in detail an extensive set of observations which he introduces as follows:

> But there was one series of experiments which deserves particular description. I employed a young man about eighteen years of age, who had had no previous experience whatever in observations, to answer a signal consisting of a sharp sound like a rap, the answer being made upon a telegraph-operator's key nicely adjusted. Five hundred observations were made on every weekday during a month, twenty-four days' observations in all. ... It was found that after the first two or three days the curve differed very little from that derived from the theory of least squares. It will be noticed that on the first day, when the observer was entirely inexperienced, the observations scattered to such an extent that I have been obliged to draw the curve upon a different scale from that adopted for the other days.

From this set of observations we have selected for consideration those of the second day of the experiment. The number of these observations is 490 (not 500), and their empirical distribution may be described as hyperbolic, the fit being as good as those seen in Figure 8 for the lava data.

It is evident from the likelihood contour plots in Figure 9 that the data from Pierce (1870), as well as the lava magnetization data, are far from being normally distributed.

6. Concluding Remarks

In the search for a deeper and more comprehensive understanding of how the wind transports and sorts sand and how the sand particles in transit interact with the sand bed there are two main lines of study that hold promise, both of which involve stochastic considerations and statistical methods in an essential way.

One approach is based on studying the movements and impacts of single particles, by means of tracers and of photographic and electronic equipment. Each of the three main phases of the movement, i.e. initiation of the movement, the trajectory flight of a saltating grain, and the impingement of the grain on the sand bed, requires separate specialized observational techniques.

It seems possible that laser–Doppler anemometry can be helpful in the investigation of initiation of the movement. In the study of the grain trajectories and of what happens when a descending grain hits the bed surface photographic techniques may be used. Some first experiments of this kind with artificially projected grains have recently been performed by Dr. B. B. Willetts of the University of Aberdeen, and an investigation of the trajectories of saltating glass spheres by means of high-speed photography was reported by White and Schulz (1977). Also the turbulence of the saltation layer, and in particular the mean velocity profile, must be investigated more closely. Here laser–Doppler anemometry is an obvious technique, but specialized hot-wire probes may be useful too. The difficulties of obtaining sufficiently detailed and reliable measurements of the kind indicated are, however, very considerable, due, in particular, to the high velocity of the sand grains and the problem of following a single particle in a whole population of particles. For example, when registering trajectories in a cloud of saltating grains, it becomes increasingly difficult to distinguish a single trajectory as the bed is approached, due to the increasing density of grains. Also, in order by this route to obtain sufficient information on the behavior of the sand particles, for each fixed set of experimental conditions a great number of repetitions of the experiment are required. If such experimental material becomes available, one may try to build a mathematical model for the motion of a single grain and, eventually, for the entire transport process. Perhaps reasoning more or less similar to that used in statistical mechanics can be applied for this. The question is then what observational quantities should be taken as analogous to the temperature, pressure, energy, etc., of classical statistical mechanics. Some candidates are wind speed and transport rate or stream power. One would hope to be able to derive the hyperbolic form of the particle size distributions within such a theory.

The other approach consists in attempting to obtain relevant statistical information by studying a substantial subpopulation of the sand particles. The experiment described briefly in Section 4, where colouring of a sand sample was used to obtain information on the dependence of grain velocity on grain size, is an example of this. Trapping sand grains at various heights in the saltation layer is an element in another such technique; see Section 3. Furthermore, it is not unlikely that the technique of laser–Doppler anemometry will soon have reached a state of development that will allow comprehensive, simultaneous measurements of particle size and velocity in the various parts of the saltation layer. On the whole this second approach seems to pose much lesser experimental problems than the first and to lead more directly to an understanding of important parts of the sand transport process. In the case of sand transport in water, Tsujimoto and Nakagawa (1983) have, for instance, discussed how the spectral density of the stochastic process describing the surface of the sand bed develops when sand starts to move over an initially flat bed. Their derivation is based on experimentally determined relations between certain physical parameters and the parameters of a stochastic process, of the kind mentioned in Section 4, modeling the one-dimensional behavior of a single sand grain. The result of the analysis is that under many circumstances a single frequency will after a while dominate the spectrum, and the authors interpret this as ripple formation.

Each of the two experimental approaches will clearly give insights that the other is not capable of providing. Moreover, they support each other. The interpretation of experiments of one kind will often need results from experiments of the other kind. For instance, results on the motion of a single grain were necessary in the analysis of sand grains trapped at various heights explained in Section 3. Another example is that since surface creep presumably plays an important role in the formation of aeolian ripples, a model of the ripple formation must in this case involve information about what happens when a grain hits the bed surface.

Acknowledgements

We are grateful to R. A. Bagnold and B. B. Willetts for critical readings of the manuscript for this paper. Part of the work of M. Sørensen was done with financial support, in the form of a stipend, from the Danish Natural Science Research Council.

References

Ambler, E. (1967). *Dirty Story*. London: Bodley Head.

Atkinson, A. C. (1982). "The simulation of generalized inverse Gaussian and hyperbolic random variables." *SIAM J. Sci. Statist. Comput.*, **3**, 502–515.

Bagnold, E. (1969). *Enid Bagnold's Autobiography*. London: Heinemann.

Bagnold, R. A. (1935). *Libyan Sands*. London: Travel Book Club.

Bagnold, R. A. (1941). *The Physics of Blown Sands and Desert Dunes*. London: Methuen. Second edition (1976), London: Chapman and Hall.

Bagnold, R. A. (1945). "Early days of the Long Range Desert Group." *Geogr. J.*, **105**, 30–46.

Bagnold, R. A. (1951). "Sand formations in Southern Arabia." *Geogr. J.*, **117**, 78–85.

Bagnold, R. A. (1956). "The flow of cohesionless grains in fluids." *Philos. Trans. Roy. Soc. London Ser. A*, **249**, 235–297.

Bagnold, R. A. (1966a). *An Approach to the Sediment Transport Problem from General Physics*. U.S. Geol. Surv. Prof. Pap. 422-I.

Bagnold, R. A. (1966b). "The shearing and dilatation of dry sand and the singing mechanism." *Proc. Roy. Soc. London Ser. A*, **295**, 219–232.

Bagnold, R. A. (1973). "The nature of saltation and bed-load transport in water." *Proc. Roy. Soc. London Ser. A*, **332**, 473–504.

Bagnold, R. A. (1980). "An empirical correlation of bedload transport rates in flumes and natural rivers." *Proc. Roy. Soc. London Ser. A*, **372**, 453–473.

Bagnold, R. A. (1983). "The nature and correlation of random distributions." *Proc. Roy. Soc. London Ser. A*, **388**, 273–291.

Bagnold, R. A. and Barndorff-Nielsen, O. (1980). "The pattern of natural size distributions." *Sedimentology*, **27**, 199–207.

Barndorff-Nielsen, O. (1977a). "Exponentially decreasing distributions for the logarithm of particle size." *Proc. Roy. Soc. London Ser. A*, **353**, 401–419.

Barndorff-Nielsen, O. (1977b). "Contribution to the discussion of Cox, D. R.: The role of significance tests." *Scand. J. Statist.*, **4**, 49–70.

Barndorff-Nielsen, O. (1978). "Hyperbolic distributions and distributions on hyperbolae." *Scand. J. Statist.*, **5**, 151–157.

Barndorff-Nielsen, O. (1982). "The hyperbolic distribution in statistical physics." *Scand. J. Statist.*, **9**, 43–46.

Barndorff-Nielsen, O. (1983). "On a formula for the distribution of the maximum likelihood estimator." *Biometrika*, **70**, 343–365.

Barndorff-Nielsen, O. and Blæsild, P. (1981). "Hyperbolic distributions and ramifications: Contributions to theory and applications." In C. Taillie, G. P. Patil and B. A. Baldessari (eds.), *Statistical Distributions in Scientific Work*, **4**. Dordrecht: Reidel, 19–44.

Barndorff-Nielsen, O. and Blæsild, P. (1983). "Hyperbolic distributions." In *Encyclopedia of Statistical Sciences*, **3**. New York: Wiley.

Barndorff-Nielsen, O., Dalsgaard, K., Halgreen, C., Kuhlman, H., Møller, J. T., and Schou, G. (1982). "Variation in particle size distribution over a small dune." *Sedimentology*, **29**, 53–65.

Barndorff-Nielsen, O., Jensen, J. L., and Sørensen, M. (1983). "The relation between size and distance travelled for wind-driven sand grains—results and discussion of a pilot experiment using coloured sand." In B. Mutlu Sumer and A. Müller (eds.), *Mechanics of Sediment Transport*. Rotterdam: Balkema, 55–64.

Blæsild, P. (1981). "On the two-dimensional hyperbolic distribution and some related distributions; with an application to Johannsen's bean data." *Biometrika*, **68**, 251–263.

Blæsild, P. and Jensen, J. L. (1981). "Multivariate distributions of hyperbolic type." In C. Taillie, G. P. Patil, and B. A. Baldessari (eds.), *Statistical Distributions in Scientific Work*, **4**, Dordrecht: Reidel, 45–66.

Burridge, J. (1981). "A note on maximum likelihood estimation for regression models using grouped data." *J. Roy. Statist. Soc. B*, **43**, 41–45.

Cable, M. and French, F. (1942). *The Gobi Desert*. London: Hodder and Stoughton.

Deigaard, R. and Fredsøe, J. (1978). "Longitudinal grain sorting by current in alluvial streams." *Nordic Hydrology*, **9**, 7–16.

Deighton, L. (1976). *Twinkle, Twinkle Little Spy*. London: Jonathan Cape.

Einstein, H. A. (1937). *Der Geschiebetreib als Wahrscheinlichkeits-Problem*. Zürich: Rascher. Reprinted in English in H. W. Shen (ed.), *Sedimentation Symposium to Honor Professor H. A. Einstein*. Fort Collins, Colorado (1972), Appendix C.

Embrechts, P. (1983). "A Property of the Generalized Inverse Gaussian Distribution with Some Applications." *J. Appl. Prob.*, **20**, 537–544.

Engelund, F. A. (1970). "A note on the mechanics of sediment suspension." *Basic Res. Prog. Rep., Hydraulic Lab., Tech. Univ. Denmark*, **21**, 7.

Engelund, F. A. and Fredsøe, J. (1976). "A sediment transport model for straight alluvial channels." *Nordic Hydrology*, **7**, 293–306.

Engelund, F. A. and Fredsøe, J. (1982). "Sediment ripples and dunes." *Ann. Rev. Fluid Mech.*, **14**, 13–37.

Fisher, R. A. (1912). "On an absolute criterion for fitting frequency curves." *Messenger of Mathematics*, **41**, 155–160. Reprinted in J. H. Bennett (ed.), *Collected Papers of R. A. Fisher*, **I**, Univ. of Adelaide, 53–58.

Fredsøe, J. (1982). "Shape and dimensions of stationary dunes in rivers." *J. Hyd. Div. ASCE*, **108**, No. HY8, 932–947.

Greeley, R. and Iversen, J. D. (1984). *Wind as a Geological Process. Earth, Mars, Venus and Titan.* Cambridge: Cambridge U. P.

Greeley, R., Leach, R., White, B., Iversen, J., and Pollack, J. (1980). "Threshold windspeeds for sand on Mars: Wind tunnel simulations." *Geophys. Res. Letters*, **7**, 121–124.

Halgreen, C. (1979). "Self-decomposability of the generalized inverse Gaussian and hyperbolic distributions." *Z. Wahrsch. Verw. Gebiete*, **47**, 13–18.

Hubbell, D. W. and Sayre, W. W. (1964). "Sand transport studies with radioactive tracers." *J. Hydraul. Div. ASCE*, **90**, No. HY3, 39–68.

Iversen, J. D., Greeley, R., and Pollack, J. B. (1976). "Windblown dust on Earth, Mars and Venus." *J. Atmosph. Sci.*, **33**, 2425–2429.

Iversen, J. D., Greeley, R., White, B. R., and Pollack, J. B. (1975). "Eolian erosion of the Martin surface, Part 1: Erosion rate similitude." *Icarus*, **26**, 321–331.

Iversen, J. D., Pollack, J. B., Greeley, R., and White, B. R. (1976). "Saltation threshold on Mars: The effect of interparticle force, surface roughness, and low atmospheric density." *Icarus*, **29**, 381–393.

Jensen, J. L. (1981). "On the hyperboloid distribution." *Scand. J. Statist.*, **8**, 193–206.

Jensen, J. L. (1983). *Maximum Likelihood Estimation of the Hyperbolic Parameters from Grouped Observations.* Research Report 91. Dept. Theoret. Statist., Aarhus Univ.

Jensen, J. L. and Sørensen, M. (1982a). *A Model for Saltation Trajectories—Evaluated by Means of Trap Data.* Research Report 85. Dept. Theoret. Statist., Aarhus Univ.

Jensen, J. L. and Sørensen, M. (1982b). "A reanalysis of Williams' aeolean saltation data." *Sedimentology* (to appear).

Jensen, J. L. and Sørensen, M. (1983). "On the mathematical modelling of aeolian saltation." In B. Mutlu Sumer and A. Müller (eds.), *Mechanics of Sediment Transport.* Rotterdam: Balkema, 65–72.

Jørgensen, B. (1982). *Statistical Properties of the Generalized Inverse Gaussian Distribution.* Lecture Notes in Statistics, **9**. Springer-Verlag, New York.

Kikkawa, H. and Iwasa, Y. (eds.) (1980). *Proceedings of the Third International Symposium on Stochastic Hydraulics.* Tokyo: Japan Society of Civil Engineers.

Kristjansson, L. and McDougall, I. (1982). "Some aspects of the late Tertiary geomagnetic field in Iceland." *Geophys. J. Roy. Astron. Soc.*, **68**, 273–294.

Letac, G. and Seshadri, V. (1983). "A characterization of the generalized inverse Gaussian distribution by continued fractions." *Z. Wahrsch. Verw. Gebiete*, **62**, 485–490.

Lettau, H. H. and Lettau, K. (eds.) (1978). *Exploring the World's Driest Climate.* IES Report 101. Center for Climatic Research, Institute for Environment Studies, Univ. of Wisconsin—Madison.

Lyall, G. (1980). *The Secret Servant.* New York: Ballantine.

Mainguet, M. and Callot, Y. (1978). *L'Erg de Fachi-Bilma. (Tchad-Niger.)* Paris: Editions du Centre Nat. Rec. Sci.

McKee, E. D. (ed.) (1979). *A Study of Global Sand Seas.* Geological Survey Professional Paper 1052. Washington: U.S. Government Printing Office.

Miller, R. M. (1979). "Kinetic art: Sculpture displaying strata and changing forms in granular materials." *Leonardo*, **12**, 271–274.

Owen, D. L. Lloyd (1980). *Providence Their Guide*. London: Harrap.

Owen, P. R. (1980). *The Physics of Sand Movement*. Lecture Notes. Workshop on Physics of Desertification, Trieste, November 10–28, 1980.

Pierce, C. S. (1870). "*On the Theory of Errors of Observations*." In *The United States Coast Survey Report*, Appendix No. 21.

Polo, M. (1939). *Travels of Marco Polo*. No. 306. Everyman's Library. (Reprinted from 1908 English edition of Marco Polo's *Travels*, known as "Il Milione," from about 1300.)

Sagan, C. and Bagnold, R. A. (1975). "Fluid transport on Earth and aeolian transport on Mars." *Icarus*, **26**, 209–218.

Shaw, W. B. Kennedy (1945). *Long Range Desert Group*. London: Collins. French edition 1948: *Patrouilles du Désert*. Paris, Flammarion.

Sumer, B. M. and Müller, A. (eds.) (1983). *Mechanics of Sediment Transport*. Proceedings of Euromech 156. Rotterdam: Balkema.

Todorović, P. and Nordin, C. F. (1975). "Evaluation of stochastic models describing movement of sediment particles on riverbeds." *J. Res. U.S. Geol. Survey*, **3**, 513–517.

Tsoar, H. (1978). *The Dynamics of Longitudinal Dunes*. Technical Report. Dept. of Geography, Ben-Gurion Univ. of the Negev.

Tsujimoto, T. and Nakagawa, H. (1983). "Sand wave formation due to irregular bed load motion." In B. Mutlu Sumer and A. Müller (eds.), *Mechanics of Sediment Transport*, Rotterdam: Balkema, 109–117.

Uthoff, V. A. (1973). "The most powerful scale and location invariant test of the normal versus the double exponential." *Ann. Statist.*, **1**, 170–174.

White, B. R. and Schulz, J. C. (1977). "Magnus effect in saltation." *J. Fluid Mech.*, **81**, 497–512.

Willetts, B. B., Rice, A. A., and Swaine, S. E. (1982). "Shape effects in aeolian grain transport." *Sedimentology*, **29**, 409–417.

Williams, G. (1964). "Some aspects of the eolian saltation load." *Sedimentology*, **3**, 257–287.

CHAPTER 5

International Trends in Official Statistics

Petter Jakob Bjerve

Abstract

The paper describes trends in official statistics in areas of concern for the top management of statistical services. Major trends in functions, organization, and policy of these services are dealt with.

1. Scope and Focus

During the period since the ISI was established, profound changes have taken place in official statistics. On the demand side, new groups of users and new kinds of use have emerged. New government planning units and research groups have an almost insatiable need for data, and new groups of analysts at universities and private research institutes, in organizations of various kinds, and in large enterprises represent important users of official statistics. Advances made in subject matter theories, statistical theory, methodology, and technology have improved tremendously the possibilities for users to extract knowledge from statistical data and have thereby stimulated new users. Similarly, on the supply side, methodological and technological advances have augmented to a much higher level the possibility of producing the official statistics demanded, and the resources allocated for production of these statistics have multiplied. In particular, since World War II an almost explosive expansion of both demand for and supply of official statistics has taken place in most countries. One hundred years ago, similar statistics, if available at all, were produced by single persons, or by small and perhaps unknown units within the government. Today statistical agencies, particularly in developed countries, constitute relatively large, competent, and highly independent government units with a considerable degree of prestige.

The scope of this chapter is confined to selected aspects of this development, viz. to major trends that have taken place, primarily during the last decades. The presentation focuses on the description of trends that are likely to continue in the future and to be of major concern for the top leadership of statistical agencies. It also points at problems connected with the various

Key words and phrases: archival dissemination, microdata collection, national accounts and similar frameworks, role of statistical services, statistical coordination, statistical organization.

trends. Most of the trends occur in all countries, but they are particularly pronounced where the degree of development is highest. Trends that may be limited to centrally planned economies are not dealt with. Moreover, the presentation does not at all cover all important aspects of official statistics. For instance, trends in the relationships of statistical agencies to data suppliers, in the data legislation, and in the education and training of statistical personnel are not dealt with, although interesting developments have taken place also in these fields.

It would have been preferable if this review were based on information collected by a survey covering the countries concerned. But since such information was not available, personal observations had to be relied on, and examples for illustration had to be limited to the countries which are best known to the author.[1] Moreover, the countries named in various parts of this paper often demonstrate special features, so that one is oversimplifying complex arrangements by citing them as examples.

The presentation is organized by areas where the most important international trends appear to have occurred. In Section 2 trends in data collection, in Sections 3–5 trends as regards the statistical output, and in Section 6 trends in the dissemination of official statistics are discussed. In Sections 7 and 8 trends in the organization and role of statistical services are dealt with. The great technological improvement that has taken place is not separately described, but its impact is pointed out where relevant.

2. The Collection of Microdata

Shortly after the establishment of the ISI a possible application of stratified and partly random sampling was discussed at the biannual sessions (Seng, 1951). In fact, this method of data collection was also, to some extent, applied during the subsequent decades. However, in most cases when a full count was not feasible, microdata were collected from the most important units only or from an intentionally selected, so-called "representative," sample of units. For a comprehensive description of the development of sample survey theory and methods, see the paper by M. H. Hansen, T. Dalenius and B. J. Tepping, (1985, in this volume), and also papers by Kruskal and Mosteller (1979a, b, c).

After World War II statistical agencies began utilizing stratified random sampling to an increasing degree. Progress in the theory of sampling had convinced many statistical agencies that random sampling could be preferable for a part of their data collection. The successful application of this method by some agencies, especially in the U.S. and India, gradually stimulated others to follow suit. There was an increase in the number of sample surveys of both

[1] I acknowledge assistance from the United Nations Statistical Office and from Norwegian friends in my efforts to specify such examples.

establishments and persons. In addition, particularly for the collection of sociodemographic data, statistical agencies began applying trained interviewers, and some of them established a permanent field staff of interviewers located in such a manner that they could collect data for sample surveys as economically as possible. This contributed to increasing the number of subject matter fields in which sample surveys could be applied and to strengthening the international trend towards greater application of sampling.

One implication of this development is that today most statistical agencies cover a considerably wider spectrum of subject matter areas and collect data on many more characteristics than they did a few decades ago. Moreover, statistical agencies in developing countries are applying sample surveys on an increasing scale, *inter alia*, because the administrative data available are not, as a rule, suitable for the production of official statistics. In accordance with this trend, the UN Statistical Office has initiated a National Household Capability Program (NHCP) to promote the application of sample surveys in developing countries (United Nations, 1980b).

The usefulness for official statistics of microdata collected for administrative purposes was an another subject for discussion at early sessions of the ISI. The interest in this subject was revived when administrative bodies in many countries during the 1960s began computerizing some of their activities. One aspect discussed was the possible consequences for the organization of statistical work within the government, which is dealt with in Section 7, but the main reason why the statistical services took a renewed interest in administrative data was the improved prospects that computerization provided for their statistical utilization. Computerization meant that the administrative bodies defined more precisely the data which they collected, improved the reliability of these data by means of machine editing, and used them for more purposes than before, all of which contributed to improving the quality of administrative data. By taking part in the establishment of computerized administrative data systems, official statisticians could also succeed in persuading the administrators to apply concepts and classifications that were analytically more relevant than they otherwise would have done. Finally, when the administrative agencies had entered their data on machine-readable media, the statistical agencies could get copies and save resources which otherwise would have been absorbed for own data entry.

As a consequence, the statistical services began utilizing computerized administrative microdata more and more for the processing of official statistics, partly in addition to and partly instead of data previously received from administrative bodies on manual records; in some cases they could even leave the data processing to the administrative bodies and get statistical tables from them ready for publishing. To some extent this allowed statistical agencies to discontinue their own direct collection of microdata from persons or establishments and thereby to reduce the burden on data suppliers. For instance, an increasing part of agricultural statistics has been processed on the basis of administrative data, and recently a large part of the population census statistics in some countries (in Denmark the whole of them) were derived from administrative systems, such as population registers, income tax registers, etc., by linking data from these systems to census data or to one another.

This trend has been carried quite far in administratively advanced countries, while most developing countries still are lagging behind. The most serious obstacle to the broader utilization of administrative data is still conceptual. Some administrative data systems are based on concepts that have to differ from those which are analytically most relevant. In such cases bridges between the concepts need to be built. For other administrative data systems the concepts may be modified so as to satisfy both administrative and analytical purposes. In such cases official statistics based on administrative data systems may be improved by better cooperation between statistical and administrative agencies. In addition, the coverage and reliability of the administrative data are not always satisfactory, but these weaknesses are likely to diminish when the data are used more frequently for administrative decision making. In countries where the administrative bodies of the government have not as yet succeeded in collecting data with sufficient coverage and reliability, such as in the less developed countries, computerization would only mean "garbage in, garbage out." This means that for a considerable period of time ahead the trend towards increased utilization of computerized administrative data may be confined primarily to the more developed countries.

The application of sample surveys and the statistical utilization of administrative data are intended to serve different groups of users and different uses. The primary purpose of sample surveys in official statistics is to obtain aggregated data for the country as a whole or subaggregates for major parts of the country. By statistical utilization of administrative data it may be possible to obtain geographical and subject matter information which is more detailed than that which sample surveys can provide. For this purpose the only alternative is a direct collection of such data by the statistical agency itself, i.e. by census taking. Since computerized administrative data systems are not likely to satisfy all needs for official statistics, the trend towards increased utilization of such systems cannot be expected to make direct data collection by sample surveys or censuses superfluous.

3. The System of National Accounts and Similar Frameworks

During the twentieth century considerable changes have taken place in the composition of official statistics. From the end of World War I until the 1950s statistical services in most countries placed primary emphasis upon extending and improving economic statistics. This emphasis conformed with the increased use of such statistics by governments for economic policy and by researchers for economic analyses. Then a shift of government interest towards social policy was accompanied by a more rapid development of social statistics relative to economic statistics. From the end of the 1960s onward this shift was supplemented by efforts at developing statistics needed for environmental policy, following an international trend in government policies that put higher weight than before on action in this field and on the primary statistics required to inform such action. In addition to these changes by broad subject matter areas, the official statistics both within each area and as a whole benefited so much from the system of national accounts and similar frameworks for sociodemographic and environmental statistics that their impact requires a separate discussion.

From World War II on, the system of national accounts had a profound impact on the development of economic statistics. Gradually, national accounts data were estimated and published by most countries, though with varying comprehensiveness and reliability. International comparability of these data was to a large degree promoted by the publication of a United Nations System of National Accounts (SNA) in 1952 and by a detailed revision of this system completed in 1968.

The transformation of economic statistics into such a system made the existing basic data much more useful, primarily for macroeconomic analysis and economic planning. In addition, in the longer term the national accounts in various ways had an important effect on the basic data. In the first place, they provided a framework for further development of the economic statistics and in particular stimulated extension of data collection to industrial and institutional sectors for which previously few or no data were available. Secondly, the data work on national accounts changed the attitude of statistical services from confining themselves to aggregation of incomplete microdata collected, to supplementing this aggregation by making the estimates required to arrive at totals for the entire country. In both ways, the comprehensiveness of economic statistics was promoted. Thirdly, the cross-checking implied in double bookkeeping improved the reliability of these statistics and made possible the computation of so-called residuals for entries which could not be directly estimated, e.g. for investment in inventories. Finally, the system of national accounts had an important feedback on the consistency of economic statistics, a point to which we shall return in Section 4.

During the postwar period the governments of more and more countries began using the national accounts as a basis for the elaboration and implementation of quantitative plans for the economic policy. This required, *inter alia*, provisional data for the preplan year at the very beginning of the plan preparation, and similarly, provisional data for the plan year as a basis for monitoring the plan implementation. Such provisional national accounts data must either be provided by the national statistical service, or be calculated by the planning agency itself. During the last decades an increasing number of governments preferred the former solution. Consequently, more and more national statistical services adopted the practice of preparing first one set of provisional data with a high degree of timeliness at the expense of accuracy, and later another set of final data with less timeliness, but satisfying the higher degree of accuracy desired. Gradually, this practice was applied to the basic data also.

The beneficial impact of national accounting on the production and use of economic data created the hope that similar accounting systems could be useful also in other areas of official statistics. Attempts were made in the 1960s at developing a system of sociodemographic accounts and in the 1970s at developing an accounting system for natural resources (United Nations, 1975b, 1979, 1982). However, the usefulness of the first kind of system turned out to be disappointing, and the second kind is too new as yet to be properly evaluated. At the same time, work on less ambitious systems for these areas of statistics was initiated by the United Nations. So-called frameworks of sociodemographic statistics and of environment statistics were devolped, consisting of a set of standard concepts, classifications, and definitions especially designed for users who want to carry out macroanalyses based on a comprehensive and consistent set of data.

Parallel with the trend in social statistics a so-called "social indicator movement" emerged. Its main idea was to measure the total social welfare through the construction of indicators of its components. This movement was initiated by sociologists, mainly working at universities and research institutes, but gradually the idea was also discussed among official statisticians. To date no international agreement has been reached on the kind of indicators to be developed; however, the social indicator movement has helped to clarify the needs for social statistics. In addition, as explained in Section 6, the movement has had an important influence on the presentation of official statistics in this field.

4. Coordination and Integration

Modern users of official statistics to a large extent want to apply in combination data from a variety of sources for a single analytical purpose. This requires a high degree of consistency between these data sources. The data, as far as possible, need to be based on the same concepts, classifications, and definitions—over time, across subject matter areas, and, as regards international analyses, for different countries (see the paper by Malinvaud, 1985, in this volume). Gradually, statistical agencies have recognized this need and have developed several tools for coordinating and integrating their data within coherent logical systems. In fact, this was one of the reason why the International Statistical Institute was established, and why at a later stage the League of Nations and subsequently the United Nations gradually took over the function of promoting the international comparability of statistics.

This international work, in conformity with the national efforts at achieving statistical integration, began with the elaboration of standards for concepts, classifications, and definitions. Today, such standards exist for most subject matter fields, and practically all national statistical services are applying these international standards in more or less modified form. (United Nations, 1975a) In particular, the statistical services of small countries have developed their own standards mainly by adjusting existing international standards so as to conform with their own needs.

One of the conditions for a successful implementation of statistical standards is that all forms used for data collection be designed in conformity with these standards. Both the questionnaires issued by statistical agencies and the forms used for collection of administrative data that can be utilized statistically must be so designed. The organizational arrangements made to ensure this differ among countries (see Section 7).

After World War II more and more statistical services gradually began using their systems of national accounts as a means of improving the consistency of economic statistics. These systems contain a large number of related concepts and classifications that are defined so as to be mutually consistent. Consequently, a logical integration of economic statistics can be achieved, to a large extent automatically, if the data are produced so as to fit into the national accounts system. Similar frameworks for sociodemographic and environment statistics were developed at a later stage, *inter alia*, for the same purpose (see Section 3). While the systems of national accounts are being extensively used as a means of coordination and integration, it is not as well known to what extent the other frameworks are being successfully used for this purpose.

The tools described above can be used for coordination and integration at the macro level only. They cannot ensure that each individual statistical unit is given the same codes in all areas of statistics where these frameworks apply. For instance, though all subject matter units in a statistical service may apply the same standard classification of economic activities, they may in some cases interpret the standard definition of an establishment differently and even give the same establishment different industrial codes, size codes, etc. Experience has shown that doubt frequently arises about which activity some establishments (and persons occupied by them) belong to. Inconsistencies in these cases can be avoided if a central list or register is available with names, addresses, and identification numbers of the establishments and with the codes to be applied, provided that all producers of statistics are required to apply this register for relevant mailing and coding purposes (United Nations, 1969). Such an arrangement can secure that each individual statistical unit is defined and classified in the same manner in all relevant subject matter statistics. Thereby, consistency is promoted at the level of the individual statistical unit and consequently also at the macro level. The same result can be achieved by means of similar multipurpose central registers for other statistical units, e.g. by a register for persons and a register of buildings.

To date only a few national statistical services, notably in the Nordic countries, have begun applying such registers as a means of coordination and integration. However, the successful experience of these services suggests that more countries are likely to adopt the same practice in the future. Computerization of the registers in the form of data bases to which all users have direct access will facilitate such a development.

The establishment and maintenance of central multipurpose registers involve difficult problems, which explain why progress in this field has been slow in many countries. Although census data and available administrative data can be used as a basis for the establishment of such registers, their maintenance requires currently available information. If this information is lacking, the desired data must be collected for all individual units included. In addition to the burdens of data supply which such a collection implies, both the establishment and the maintenance of registers require considerable resources. Furthermore, within the national statistical services several units producing subject matter statistics have, by tradition, maintained their own register of establishments; for instance, the units for manufacturing statistics, internal trade statistics, and transport statistics have often used separate and more or less overlapping registers as a basis for their data collection, and frequently a unit producing wage statistics for the same industries has maintained an additional separate register. Thus, the transfer of this function to a central register unit means that the subject matter units must refrain from maintaining a register of their own. Resistance to this seems to have delayed the establishment of a central multipurpose register in some countries. Finally, since such registers may be useful for both statistical and other purposes, their location within the government may become an issue. This problem is dealt with in Section 7.

Though multipurpose central registers of statistical units appear necessary to achieve full consistency of statistical aggregates, they are not entirely sufficient. If different methods of aggregation are used in different areas of statistics, the resulting macrodata may be inconsistent even if the microdata are fully consistent; for instance, if the nonresponse problem is differently solved, inconsistencies may remain. Consequently, consistency requires the application of standard methods in the conduct of sample surveys, in adjustments for missing data, in seasonal adjustments, etc. This need does not seem to be widely recognized as yet. However, the problem gradually has been acknowledged by official statisticians, and the United Nations and other international organizations have during the last decades contributed substantially to its solution by developing recommendations on methodology.

5. The Archival Function

The availability of modern data processing equipment, with very great potential for storing data and for retrieving and reprocessing them at relatively low cost, from the 1950s gradually stimulated national statistical agencies to invest considerable resources in systematized archives of microdata and macrodata together with the metadata required for prospective utilization of the information stored. Thereby the archival function of these agencies could become much more important than before.

Statistical agencies always used to store a part of the microdata collected after having used them, in the beginning by archiving questionnaires and later on by archiving punched cards used for censuses and surveys. Some of these data sets, particularly the population census returns, were taken advantage of by historians, but they were never (or seldom) reprocessed, i.e. used for the preparation of additional statistics. Reprocessing turned out to be difficult even for data stored on punched cards. With the entrance of the computer the possibility of reprocessing microdata and of storing aggregated data for prospective retrieval was radically improved, and new perspectives were opened for utilization of such data.

Also with less advanced technology, and particularly after having put punched card equipment to use, the statistical services produced statistical tables with more details than could be published and made them available for interested users by copying, frequently carried out by the users themselves. However, the computer made the supply of such unpublished data possible on an ever increasing scale by processing either simultaneously with the published statistics or at a later date. In the latter case, the supply required systematized data archives, where computer programs had to be well documented, stored data easily accessible and well described, information on definitions and classifications fully available, etc. The development of such metadata requires both time and resources, and this may explain why the establishment of systematized data archives did not until recently become an international trend in official statistics.

Statistical agencies with systematized archives of data in machine-readable form were gradually able to utilize the data collected more intensively to the benefit of users. In the first place, many more macrodata could be stored and, on subsequent requests, retrieved and supplied to users at low cost. Secondly, stored microdata could be reprocessed to supply "tailor-made statistics" to satisfy the demand of individual users. In both ways the supply of unpublished statistics appears to have increased considerably in recent years. In addition, the data archives enabled statistical agencies to utilize data stored for the processing of more statistics for publication, if desired. To date this seems to have been done primarily by linking of data from different sources or from the same source at different periods or points of time.

Linking of microdata appears to have been practiced primarily in the Nordic countries (See, e.g., Ohlsson, 1967). Already, a considerable part of the official statistics in these countries is based on such linking, and this part is likely to increase in the future. However, in many other countries restrictions are imposed on this kind of data linkage, even if performed by the national statistical service, and it is still uncertain if linkage of microdata will become a widespread international trend. Linking of aggregated data is being practiced to a larger extent and is likely to be more widespread in the future.

Systematized data archives also represent preparedness for disseminating data to users who want them for their own data processing. This and other aspects of dissemination related to such archives are dealt with in Section 6.

6. Dissemination of Official Statistics

The rapidly expanded volume of official statistics since World War II has been disseminated to users by means of an increasing number and variety of media designed to satisfy the needs of particular user groups. During the first couple of decades publications of various kinds, including press releases, represented the dominant means of data dissemination, but from the 1960s other media, such as computer printouts, magnetic tapes, and microfilm, gained in importance.

In the 1970s the dissemination of official statistics was brought to a focus in international discussions among official statisticians. A widespread consensus emerged that statistical agencies ought to strengthen dissemination relative to other functions. Despite the rapid expansion that had taken place, in both the number of uses and the number of users of official statistics, not enough had been done to ensure that the data produced were used widely and properly. Several aspects of dissemination were discussed, such as the identification of users and their needs, the design of statistical outputs to satisfy these needs, the promotion of the use of these outputs, *inter alia*, by more direct contact with users, and the education of users to improve their ability to apply the statistics available. In conformity with this consensus, statistical agencies in many countries began allocating more resources for dissemination and strengthened this function also in other ways, but the main emphasis was placed on providing more user-oriented media of dissemination and on promoting their use.

Among the many new publications presented by statistical agencies in recent decades, three kinds need to be briefly described, viz. the regional publications, the so-called social reports, and the compendia on environment statistics.

Increased interest in regional planning and policy after World War II augmented the need for regional and small area data and for new ways of disseminating such statistics. To ensure their timeliness more and more countries began publishing data for individual administrative units before national data could be produced. This meant that, in particular, census data could be used for local and regional purposes at a much earlier date than otherwise would be possible. In addition, more and more statistical agencies began presenting separate publications for particular regions and major administrative areas containing a selection of previously published statistics, which thereby could be made more easily accessible.

The first modern social report was the United Kingdom's *Social Trends,* published annually since 1970. In subsequent years other countries published a number of more or less similar volumes, and by now about thirty nations publish them on a regular basis, annually or periodically. The idea of publishing social statistics in the form of such reports originated from the social indicator movement (discussed in Section 3), which envisaged presenting the social indicators to be developed in a separate, new kind of publication. However, pending the development of such indicators and the completion of a framework of sociodemographic statistics (also discussed in Section 3), an increasing number of statistical agencies began publishing in one volume social statistics that were already available and previously presented in different subject matter publications. Some countries merely assembled the most important existing statistics on individuals and households with the primary aim of illuminating conditions of living, the social structure, and social concerns. Other countries applied a more problem-oriented approach, restricting the selection of data to more direct indicators of social welfare.

The first compendium on environmental statistics was published by Finland in 1972, and similar volumes were later presented periodically by several other European countries as well as by Australia, Japan, and the USA. To date, such compendia have been issued in fifteen countries; their major purpose is considered to be integration of widely dispersed and not easily accessible environmental data. In countries where major efforts have been devoted to developing a framework of environmental statistics, attempts have also been made at applying their approaches to the presentation of such data. The compendia on environmental statistics have been well received by users, and more countries are likely to issue such publications in the future.

The dissemination of statistics by means of media other than publications began at the time when computers made large scale production of data at low cost possible. Users currently needing more data than it was possible to publish, primarily detailed data on foreign trade and manufacturing production, could now be offered such data on computer printouts by subscription, and a growing number of business enterprises took advantage of this. Furthermore, the demand for macrodata or for particular tabulations of microdata stored in the data archives could be accommodated, for a charge, to an increasing extent. Moreover, users with their own terminals could get access to stored macrodata, either directly on line or by way of copies through machine-readable media, and to some extent this even allowed the users to process the data received. The extent to which statistics are disseminated in these ways varies greatly from country to country, and in many countries this trend has not even started yet. However, the dissemination of statistics by means of media other than publications is likely to grow in importance and to become more and more international in the future.

In debates among statisticians, both at the national and the international level, it has been predicted that statistical publications are likely to be replaced by other media of dissemination, and particularly by direct access to archives of macrodata from user terminals. However, to date the number and variety of publications have continued to grow, and no signs are indicating a prospective decline.

In recent years the statistical services of some countries, for instance, the U.S. and the Nordic countries, on certain conditions have begun making available to particular users unidentifiable sets of data on persons and households for processing by means of computing equipment. Most of these sets have been disseminated by magnetic tapes, but there also exist arrangements for direct access to microdata. This trend is still in its initial stage. Cheaper computers and more user-friendly programs may rapidly increase its importance in the future, but confidentiality problems may restrict this trend and limit its international character.

Increased supply of detailed statistics, in particular to business enterprises and to users of small area data, has augmented the risk of disclosure and consequently imposed upon statistical agencies the need for improving their methods of safeguarding confidentiality. Direct access of users to statistics stored on machine-readable media has added to this need. The storage of microdata on machine-readable media has also compelled the statistical agencies to take new physical and administrative measures in order to prevent unauthorized use of such data. These problems are likely to become even more serious in the future.

In recent years statistical agencies have attempted in various ways to promote the utilization of their data. Thus, in many countries they have

instructed their field staff to acquaint local users with the official statistics, set up seminars and conferences for users,

arranged press conferences and other meetings for representatives of the mass media,

distributed data directories and dictionaries facilitating effective utilization of data stored on tables or tapes,

published lists of publications recently released or expected to be released in the near future,

assisted in establishing small libraries of statistical publications for particular user groups such as ministries and other national or subnational authorities, universities, and research institutes,

presented programs on radio and television, articles in newspapers and professional periodicals, brochures, and other materials aimed at promoting the utilization of statistics (in some countries also educational material, describing how statistics could be used, has been issued),

approached individual groups of users and provided them with information on what kind of statistics they in particular can utilize and how to go about using them, offering them subscriptions to individual packages of publications and other dissemination media fitting their needs.

Combined with such "marketing" activities, many statistical agencies have strengthened their reproduction and distribution services in order to ensure that all users receive as quickly as possible the statistics they request.

The use of official statistics has also been promoted by international organizations. The number and variety of international publications presenting such statistics have multiplied, and their international dissemination by means of other media is now rapidly growing. Even private companies have engaged in the dissemination of statistics by means of computerized data bases. This trend involves the risk that more or less conflicting statistics will be disseminated by different suppliers, and imposes increasing burdens on the statistical agencies supplying the original data.

7. Statistical Organization

Both the organization of statistical work within the government and the organizational structure within individual statistical agencies of the government have changed considerably in recent years (United Nations, 1980a). A number of governments have centralized the management and operation of their statistical programs so that today there are many more countries than some decades ago that have a strong central statistical agency. Moreover, a centralization of functions within each agency has also taken place.

Centralization of statistics within the government in particular occurred in small and medium sized countries such as in the Nordic countries, the Netherlands, Australia, Canada, Kenya, Zimbabwe, the Yemen Arab Republic, and Panama. In the developing world also, some large countries such as Pakistan and Bangladesh centralized their official statistics to a large degree. This international trend gained strength at the time when the advantages of large and very expensive electronic computers became apparent. Even where the statistical service is still to a high degree decentralized, as in the United States and the United Kingdom, substitutes for a far-reaching centralization have been introduced, mainly by concentrating statistical work in a few large agencies and charging one of them (or, alternatively a separate centrally located unit) with the responsibility for integration and coordination. In this sense, these countries are also characterized by a trend towards centralization.

At the time when government agencies in many countries (as described in Section 2) began computerizing their administrative routines, and thereby became better equipped for producing statistics based on the microdata which they collected (mainly for control purposes), the question arose whether decentralization of statistical work within the government would follow, i.e., whether official statistics based on administrative data would be processed and presented by the data-collecting bodies instead of by the central statistical agency. It is not clear to what extent such a development actually has taken place, but it appears that when the processing of official statistics was taken over by an administrative body (at least in some countries, e.g. Norway), the central statistical agency maintained its dissemination function.

The centralization of functions within the statistical agencies has been almost universal. Specialization of activities, such as personnel management, office administration, library services, etc., and centralization of these activities in new administrative units has been justified by statistical agencies for the same reason as by other large establishments, viz. the expectation of gaining increased efficiency. Moreover, the increased emphasis on integration and coordination of statistics and the technological development is a major factor contributing to centralization of functions such as the development and implementation of statistical standards, the central control of forms, the administration of multipurpose registers of statistical units, the dissemination of statistics, the analysis of statistics, and methodological work.

Statistical standards are frequently developed by those subject matter units within the statistical service where the need for them is felt most urgent, and in a central statistical agency the Director General or one of his deputies often looks after their implementation. However, in some countries where statistics were to a high degree decentralized or where the central statistical agency was large, new administrative units were established to take charge of the development and implementation of statistical standards. Such units also, as a rule, were made responsible for seeing that relevant questionnaires and forms conform with existing statistical standards, and in addition perhaps for promoting a technically proper design of forms and for preventing duplication of the data collection. In Japan and the United States, for instance, separate units of this kind exist while in Australia, Canada, Indonesia, Peru, the USSR, and several other countries the central statistical agency is responsible for statistical standards and the control of questionnaires. Irrespective of who is developing the statistical standards, more and more countries have in recent decades established advisory committees for this task, with representation of both users and producers of statistics.

Central registers of establishments, persons,and other statistical units, such as those dealt with in Section 4, can be used for administrative as well as statistical purposes. If the achievement of both is desired, the question arises where in the government the registers should be located. Undoubtedly, the statistical needs can be best satisfied if the central statistical agency is in charge; however, the agency must in this case perform an administrative function which from the point of view of the data suppliers may appear to be in conflict with its confidentiality obligations. Even though the statistical agency may be authorized to perform this function by a separate law, the data suppliers may not understand that the collection of data is based on this law and not on the statistical law. Consequently, they may believe that the confidentiality requirement of the latter law is violated. For this and other reasons, in most countries central registers are located outside the statistical service, but there are examples of countries where the statistical service is in charge in spite of the confidentiality problems implied. In Denmark, for instance, both the central register of establishments and the central register of persons are located outside the statistical service, but are fully utilized for statistical purposes; in Norway the Central Bureau of Statistics is in charge of both, being authorized by a special law on the population register to provide the name, identification number, and address of persons to government bodies needing such information (as well as the date of birth and address of persons to others if they can supply the name and prove the need for this information), but is prohibited from providing any information included in the register of establishments for use other than statistical; and in France a register of enterprises is located in the central statistical agency, which can provide all information included for nonstatistical use in accordance with a special law. As yet, it is uncertain which location will be preferred by countries establishing similar multipurpose registers in the future.

The increased emphasis on the dissemination of official statistics pointed out in Section 6 has motivated the establishment of a central information unit in the statistical services of several countries. Furthermore, in some countries the extended role of the statistical service described in Section 7 has involved the establishment of central units within the service for, respectively, statistical methodology and analysis of statistics. For instance, countries such as Brazil, Canada, India, the USA, the United Kingdom, and Sweden have a relatively strong methodological unit, and in the statistical agency of Norway one of its five major departments (with approximately 50 professionals out of 200 for all departments) is engaged entirely in analytical work. However, while thus the methodological work has been centralized in many countries, there appears to be fewer countries who have centralized work on the analysis of statistics.

In general, the international trend towards centralization of functions is likely to become still wider in the future. However, recently a trend in the opposite direction has emerged as regards some functions related to electronic data processing. When punch card equipment was brought into use, it seemed rather obvious that a central unit should take charge of data entry (punching), and when computers were introduced, editing and tabulation were also, to a large degree, transferred from the subject matter divisions to functional service units, frequently centrally located in a data processing department. But the widespread availability of video-display computer terminals and of mini and micro computers at low cost, together with an increased supply of user-friendly computer programs, now motivates a retransfer of these functions to the subject matter divisions concerned, and in fact such a decentralization is already taking place. Within the subject matter units as well, a similar decentralization and integration of functions occurs. During the next decade computer technology may well change radically the organizational structure of statistical agencies.

8. The Role of Statistical Services

Simultaneously with the expansion of the national statistical services and the international trends described with respect to their functions and organization, the role and statistical policy of these services has changed to a considerable degree. As examples of such changes it may be mentioned that almost universally statistical services have extended their activities

from producing data mainly for government use to providing statistics to other users as well,
from confining themselves to aggregation of incomplete microdata to supplementing this activity by making estimates for the entire country, and
from maintaining archives of questionnaires and punched cards, which practically could not be used for statistical reprocessing, to establishing systematized archives of machine-readable media which make quick data retrieval and the processing of additional statistics possible.

There are also examples of changes

from presenting statistics in tabular form only, at most supplemented by some "commodity description," to engaging also in analyses of statistics,
from producing data for past periods only, perhaps supplemented by preliminary estimates shortly after the end of these periods, to making projections for the future, and
from relying on methods developed outside the statistical service to making one's own efforts at improving the statistical methodology.

Many statistical services do not as yet engage in such activities, and international debates in recent years reflect different views on their future course. Nevertheless, these discussions suggest that more and more statistical services will be carrying out analyses, projections, and methodological work, or will extend present activities in these areas. Therefore, in conclusion, it may be appropriate to deal with these activities. In addition, a few observations will be made on the trends as regards organization of the cooperation between national and international statistical services.

Analytical activities enable statistical agencies to learn more about weaknesses of the statistical output, to get a better sense of its relevance, and to acquire an improved ability to set priorities. In addition, producers of official statistics can perform some kinds of analyses more efficiently than other government agencies.

Several statistical agencies appear to place priority on the description of economic and sociodemographic trends. Such analysis, which *inter alia* is performed in connection with the publication of economic and social surveys, provides a test of the quality of time series and promotes production of the proper kind of such series. It also stimulates efforts at making them consistent. A few statistical services, e.g. those of Canada, France, and Norway, have engaged themselves quite heavily in developing numerical mathematical models, and in applying such models to explanatory macroeconomic analyses. However, the majority of services have so far preferred to leave this activity entirely to others. The question of whether or to what degree a statistical agency should engage in the construction of numerical models is frequently discussed among official statisticians. Those in favor argue that statistical agencies, because of their easy access to data and a high competence in data processing, are particularly effective in constructing such models, at least models containing a large number of data. Those against mainly argue that analysis of economic or social problems, with or without models, may endanger the objectivity of statistical services.

Since World War II an increasing number of statistical services have begun preparing and publishing projections of population by means of more and more refined numerical models. The United Nations Population Division has supplemented these efforts and is now publishing projections of the world population by sex, age, region, and country at regular intervals. In Norway, similar projections are being made for the prospective flows of students through the educational system and of the prospective supply of labor by various categories; in addition, experiments are being made with projections of other human phenomena, such as the prospective number of patients at hospitals.

To date, most national statistical services have been reluctant to make economic projections. However, in countries where statistical agencies have developed economic models, projections of economic trends have also been made, mainly for the use of policy makers. Nevertheless, it does not seem likely that the elaboration of such projections will become an international trend among statistical agencies in the foreseeable future. Macroeconomic models have also been successfully used for the estimation of up to date provisional national accounts data. More statistical agencies may in the future consider this kind of use to be a part of their role. At present, the methods of making provisional estimates in general do not appear to be advanced beyond simple regression models, and in many cases intuitive procedures, not to say guesswork, are applied. Therefore, methodological improvements are urgently needed.

It is becoming more and more widely recognized that to produce statistics properly one has to take advantage of the available statistical theory and methods. However, a wide gap still exists between statisticians working in statistical agencies and theorists employed by universities and colleges. The two groups of statisticians as a rule are not even able to communicate with one another professionally. One way by which the statistical agency can bridge this gap is to employ more statisticians who have sufficient theoretical knowledge to communicate with the academic statisticians and who know the work of official statistics well enough that they also can communicate with the other staff members of the agency (Bjerve, 1975). By allowing such statisticians to engage actively in methodological work, the statistical service can improve its ability to utilize methodological advances made by others and can focus on those methodological problems which others cannot be expected to solve, such as the building of conceptual bridges between administrative and statistical data, the editing of microdata, the linkage of data from different sources or for different periods or points of time, the elimination of the risk of disclosures, the evaluation of the quality of its data, and improvement of the timeliness of statistics.

The methodological work of statistical services during the last decades has to a considerable degree been focused on sampling methods, at the same time as the application of sample surveys has gained in importance, as explained in Section 2. However, the statistical services of quite a few countries have recently extended their activities into other areas also. In conformity with this trend, the Conference of European Statisticians in 1979 adopted statistical methodology as a new project on its work program, and has to date arranged three meetings on this subject.[2]

The role of national statistical services have become more and more closely connected with that of the international statistical agencies, primarily the statistical offices of the United Nations and its specialized agencies.[3] These and the statistical offices of the various regional organizations such as the European Community (EC), the Organization for Economic Cooperation and Development (OECD), the Organization of American States (OAS), and the Council for Mutual Economic Assistance (CMEA), have in recent decades to an increasing degree supplemented the national statistical services in disseminating their data. Moreover, to make national data as useful as possible for international analysis, they modify them so as to fit into an international frame. To some extent they also aggregate national data for the world as a whole and for particular regions of the world. In this sense they carry out production of international statistics in addition to performing collection and dissemination functions. Some international organizations even carry out statistical surveys covering selected countries. The World Fertility Survey of the International Statistical Institute represents an outstanding example of such an activity (see the paper by Macura and Cleland, 1985, in this volume). Finally, some organizations devote considerable resources to harmonization of national data and to improving the methods and techniques applied in the production of official statistics of member countries. These activities are likely to become still more important in the future.

During the postwar years the national statistical agencies gained considerable influence over the statistical work within the United Nations family of organizations. To an increasing extent they were invited by these organizations to make comments on their plans for collection of new data and on the harmonization and development work to be carried out by them. The national agencies also took part in an increasing number of international meetings on the activities mentioned above. The professional benefit of these meetings for

[2] The first two meetings dealt with methodological problems relating to household surveys, and the third with linkage of data from different sources and interdependences of survey data and other statistics, methodology of panel surveys, nonresponse problems, and guidelines for quality presentations.

[3] For a comprehensive review of the statistical work and responsibilities of some 16 organizations within and outside the United Nations System carrying out substantial activities in international statistics, see the *Directory of International Statistics*.

the staff members participating should not be underrated. In this manner, international cooperation on statistics considerably extended the professional milieu of official statisticians (Bjerve, 1982). The statisticians of small countries in particular benefited from this international trend in official statistics.

9. Summary

During the past century increased application of sampling was the major international trend in the collection of microdata for official statistics, primarily after World War II, when the success of improved sampling methods had been clearly demonstrated. The statistical utilization of data collected for administrative purposes, which originally was the main basis for current official statistics, was augmented when the computerization of governmental routines began in the 1960s, but to date this trend has been confined to the more developed countries. From World War II onwards, the system of national accounts had in several ways a profound impact on the economic statistics, making them much more useful as a basis for macroeconomic analysis and planning, and providing a framework for integrating such data. At a later stage similar frameworks were developed for sociodemographic and environmental statistics. These three kinds of frameworks, together with more specific and detailed standards for concepts, classifications, and definitions, the development of which began more than a century ago, increasingly benefited those users of official statistics who wanted to combine data from a variety of sources for a single analytical purpose. During the last decades several national statistical services established multipurpose registers of persons, establishments, and other statistical units, which they applied as tools for integrating statistics at the micro level. Another and wider international trend was the establishment of systematized archives of micro- and macrodata, together with the metadata required for prospective utilization of the information stored. In the 1970s the dissemination of official statistics was brought to a focus in international discussions among official statisticians. In conformity with the wide agreement attained that a strengthening of this function was desirable, statistical agencies in many countries began allocating more resources for dissemination and applying more user-oriented means of dissemination. The trends described had an important impact on the organization of statistical work. In many countries a centralization of statistical work within the government and, almost universally, a centralization of functions within each agency took place. However, an opposite trend has recently emerged as regards functions related to electronic data processing. Finally, the rapid expansion of the statistical output and new activities engaged in by the statistical services has meant that their role has extended and become even more important.

Bibliography

Bjerve, P. J. (1975). "Presidential address presented to the 40th session of the International Statistical Institute." *Proc. Internat. Statist. Inst.*, **XLVI**, Book 1, 41–48

Bjerve, P. J. (1982). "Three decades of the Conference of European Statisticians: past achievements and perspectives for the future." *Statist. J. U.N. ECE,* 3–27.

Hansen, M. H., Dalenius, T., and Tepping, B. J. (1985). "The development of sample surveys of finite populations." In this volume, Chapter 13.

Kruskal, W. and Mosteller, F. (1979a). "Representative sampling. I: Non-scientific literature." *Internat. Statist. Rev.,* **47**, 13–24.

Kruskal, W. and Mosteller, F. (1979b). "Representative sampling, II: Scientific literature, excluding statistics." *Internat. Statist. Rev.*, **47**, 111–127.

Kruskal, W. and Mosteller, F. (1979c). "Representative sampling, III: The current statistical literature." *Internat. Statist. Rev.*, **47**, 245–265.

Macura, M. and Cleland, J. (1985). "Reflections on the world fertility survey." In this volume, Chapter 18.

Malinvaud, E. (1985). "Economic and social statistics for comparative assessments." In this volume, Chapter 9.

Ohlsson, I. (1967). "Merging of data for statistical use." *Proc. Internat. Statist. Inst.*, **XLII**, Book 2, 750–766.

Seng, Y. P. (1951). "Historical survey of the development of sampling theories and practice." *J. Roy. Statist. Soc.*, **114**, 214–231.

United Nations (1969). *Methodology and Evaluation of Population Registers and Similar Systems.* Series F, No. 15. New York.

United Nations (1975a). *Directory of International Statistics.* New York, 199–266.

United Nations (1975b). *Towards a System of Social and Demographic Statistics.* Series F, No. 18. New York.

United Nations (1979). *Studies in the Integration of Social Statistics: Technical Report.* Series F, No. 24. New York.

United Nations (1980a). *Handbook of Statistical Organization, Vol. 1, Studies in Methods.* Series F, No. 28. New York.

United Nations (1980b). *The National Household Survey Capability Programme—Prospectus.* DP/UN/INT-79-020/1. New York.

United Nations (1982). *Survey of Environment Statistics: Frameworks, Approaches and Statistical Publications.* Statistical papers series M, No. 73. New York.

CHAPTER 6

Cohort Analysis in Epidemiology

N. E. Breslow

Abstract

Epidemiologic cohort studies typically involve the follow-up of large population groups over many years to ascertain the effects of environmental exposures on the outbreak of illness and the age and cause of death. An efficient method of analysis is to fit Poisson regression models to grouped data consisting of a multidimensional classification of disease cases and person-years of observation by discrete categories of age, calendar period, and various aspects of exposure. Extension of these models for use with disease rates and exposure variables that vary continuously with age or time leads to the well-known proportional hazards model. Incorporation of external standard rates is more likely to improve the estimates of exposure effects in additive or excess risk models than in multiplicative or relative risk situations. Examples are provided of the maximum likelihood fitting of such models to data from cohort studies of British doctors and Montana smelter workers. The discussion considers the choice between models and certain problems that may arise when attempting to fit nonmultiplicative relationships.

1. Introduction

Epidemiology has been defined as the study of the distribution of disease in human populations and of the search for determinants of disease encountered in different population groups (MacMahon and Pugh, 1970). Popular conceptions of the "shoe leather" epidemiologist, so aptly portrayed by Roueché (1967), conjure up a public health sleuth sent out to investigate and control outbreaks of infectious disease or episodes of acute poisoning. As communicable diseases have been brought gradually under control through improvements in living standards, hygiene, immunization programs, and antibiotic therapy, however, the attention of the epidemiologist has turned increasingly towards cancer and the cardiovascular diseases that today account for the majority of deaths in industrialized nations. Study of the determinants of these chronic diseases is more difficult, since they may have their origins in an interaction between genetic predispositions and environmental exposures that occur decades before the diagnosis. Moreover, the relevant exposures are

Key words and phrases: efficiency, excess risk, healthy worker effect, Poisson regression, proportional hazards, relative risk, standardized mortality ratio.

often widely distributed in the home, the workplace, or the community. Statisticians have played an important role in the development of the conceptual and methodological tools, in the design and execution of the observational studies, and in the analysis and interpretation of the data needed to explore such relationships.

Two distinct study designs are currently used in the search for risk factors of chronic disease. The case-control design bases sampling on the disease outcome. Cases of a specific disease are ascertained as they arise from population based disease registers or lists of hospital admissions. Controls are sampled either as disease-free individuals from the population at risk, or as hospitalized patients having a diagnosis other than the one under study. Since the relevant exposure histories of cases and controls are obtained by interview or other retrospective means, these studies may be carried out relatively quickly and at moderate cost. They facilitate the investigation of multiple interacting risk factors. Major limitations relate to the accuracy of the exposure histories and uncertainty about the appropriateness of the control sample. Extensive discussions of statistical issues that arise in the design and analysis of case-control studies are provided by Mantel and Haenszel (1959), Breslow and Day (1980), Kleinbaum, Kupper, and Morgenstern (1982), Schlesselman (1982), and Breslow (1982).

The cohort study design focuses attention on a particular exposure rather than a particular disease. Groups of individuals are assembled and classified with regard to the exposures of interest and are followed forward in time to determine which diseases develop.[1] Advantages of such a longitudinal approach include the opportunity for more accurate measurement of the exposure history and a careful examination of the time relationships between exposure and disease. However, in order to collect sufficiently many cases of rare diseases, cohort studies may require that tens of thousands of individuals be followed for many years. Although it is sometimes possible to use existing records to define "retrospective cohorts" of groups assembled in the past, and then to trace the cohort members via national health and welfare systems or death registers, even then the expense of data collection is considerable.

Cohort studies have had a major impact on public awareness of health risks. The twenty year follow-up of British doctors (Doll and Hill, 1964; Doll and Peto, 1976; see also Tables 1 and 2 below) and the American Cancer Society study of 1,000,000 men and women (Hammond, 1966) are widely cited with regard to the dose–time–response relationships between smoking and lung

[1] Use of the word "cohort" to describe such a design was made as early as 1935 by Frost (1939), who studied mortality rates from tuberculosis in successive birth cohorts (generations) using the rudiments of what has come to be called age–period–cohort analysis (e.g., Fienberg and Mason, 1985). The term was also introduced during a planning session for statistical surveys held that same year in England. R. A. Fisher noted that the minimum subgroup size of 100 men and 100 women in each of three social classes led to a total sample of 600 persons, and Beveridge remarked that this was the same size as the unit of the Roman army known as a cohort (Wall and Williams, 1970).

cancer or heart disease. The life span study of over 100,000 atom bomb survivors in Hiroshima and Nagasaki (Beebe, 1981) is still the primary source of information on the epidemiologic effects of ionizing radiation. Selikoff et al.'s (1980) investigation of mesothelioma, lung cancer, and asbestosis occurring among 18,000 members of an international asbestos and insulation workers' union has been used as a basis for legal claims by tens of thousands of affected workers.

This paper reviews some methods of analysis of cohort data that are based on statistical models for disease incidence or mortality rates. A primary objective is to show that the classical epidemiologic methods both follow from and are extended by the modeling approach. Illustrative analyses of both grouped and continuous data are provided.

2. Selection of Variables for Analysis

The most relevant epidemiologic measure of disease occurrence is the incidence rate, defined as the number of new cases of disease diagnosed during a given period divided by the number of person-years of observation time accumulated by the population at risk. Accurate measurement of incidence requires an active surveillance mechanism such as an established disease register. Good registers are rare, however, and it is more common for large scale cohort studies to use as their primary endpoint death from a specific disease instead of its diagnosis. Death rates are reasonably satisfactory measures for rapidly fatal diseases because there is little opportunity for the competing risks of death from other causes to operate after diagnosis. They are less satisfactory for lingering illnesses, both because of competing risks and because a sick person is likely to modify his exposures. The interpretation of exposure–death rate associations may be seriously complicated by the fact that heavily exposed persons are selected to be in good health during the time that their exposures are received.

Mortality rates for different diseases are influenced by a large number of demographic and personal factors. Knowledge of these relationships is essential for the proper design and analysis of cohort studies so that measurements are taken on relevant variables and adjustments are made for their confounding effects. Careful quantitative study of how the dose and timing of environmental exposures modifies the natural or spontaneous disease rates is needed to further our understanding of disease mechanisms and in order to formulate policy for the regulation of causative agents.

Nuisance factors that explain the variation in background rates include age, calendar year, sex, race, and place of residence. These are not so much causative factors *per se* as they are surrogates for nonspecific exposures. Thus, while the data analyst needs to be aware of their effects, for example the sharp rise in most cancer rates with age, he or she is usually not particularly interested in making precise statistical inferences about them. Attention is more properly focused on the intensity and duration of exposures, e.g. to cigarette smoke or radiation, on the age at which such exposures started or stopped, and on the "latent interval" between exposure and disease.

3. Elementary Analysis of Grouped Data

Disease rates calculated from grouped data defined by broad categories of exposure, age, and time are easily displayed and examined graphically. When only a few exposure factors need to be considered, such analyses provide access to most of the useful information in the data. Presentation of results for grouped data is a desirable prelude to the development of more elaborate models for continuous data. In this section we introduce some simple epidemiologic methods for calculating disease rates and measuring the possible effects of exposure.

Suppose there are J strata defined by age and other demographic variables, and K exposure groups. Each individual may contribute observation time to several of the JK cells during the course of the study. Transitions from one cell to another could occur, for example, on designated birthdays, at the start of every fifth calendar year, at specified anniversaries of the date of hire, or at defined levels of cumulative exposure. Clayton (1982) has developed a computer algorithm to perform the calculations. It produces summary data consisting of the numbers d_{jk} of disease cases or deaths and the totals n_{jk} of person-years of observation time accumulated in the (j, k)th cell. The ratios $\hat{\lambda}_{jk} = d_{jk}/n_{jk}$ are viewed as estimates of true but unknown rates λ_{jk} that, technically speaking, are assumed to remain constant within each cell. Models that allow the disease rates to vary continuously with time or age are considered in Section 5.

3.1. The British Doctors Study

Data from the British doctors study provide a simple example with which to illustrate our notation. In 1951, Doll and Hill sent a questionnaire to all men on the British Medical Register, inquiring as to their smoking habits. Nearly 35,000 replies were received, which represented about 70% of such men then alive. Follow-up for mortality was carried out chiefly through the UK Registrars-General, who provided certificates for all deaths of medical practitioners. Causes of death were assigned on the basis of the death certificate diagnosis, supplemented in some cases by consultation with the doctor who signed the certificate. Table 1 shows person-years of observation and deaths from coronary artery disease accumulated during the first ten years of the study (Doll and Hill, 1966). There are $J = 5$ ten-year age groups and $K = 2$ smoking categories for a total of $JK = 10$ cells.

Table 1. Death Rates from Coronary Disease Among British Male Doctors[a]

Age	Person-years		Coronary deaths		Death rates[b]		Rate difference[b]	Rate ratio
	Nonsmokers	Smokers	Nonsmokers	Smokers	Nonsmokers	Smokers		
j	n_{j1}	n_{j2}	d_{j1}	d_{j2}	$\hat{\lambda}_{j1}$	$\hat{\lambda}_{j2}$	$\hat{\lambda}_{j2} - \hat{\lambda}_{j1}$	$\hat{\lambda}_{j2}/\hat{\lambda}_{j1}$
35–44	18,790	52,407	2	32	0.11	0.50	0.60	5.73
45–54	10,673	43,248	12	104	1.12	2.40	1.28	2.14
55–64	5,710	28,612	28	206	4.90	7.20	2.30	1.47
65–74	2,585	12,663	28	186	10.83	14.69	3.86	1.36
75–84	1,462	5,317	31	102	21.20	19.18	−2.02	0.90
Totals	39,220	142,247	101	630	2.57	4.43	1.86	1.72

[a] Source: Doll and Hill (1966) as quoted by Rothman and Boice (1979).

[b] Per 1000 person-years.

Examination of the last two columns of the table shows that the age-specific coronary death rates for smokers exceed those for nonsmokers in all but the highest age category. The differences in rates for smokers vs. nonsmokers generally increase with advancing age while the rate ratios decline, and it is not entirely clear whether the difference or the ratio provides the best summary measure of the effect of smoking. One goal of the analysis will be to determine which of these scales is the most appropriate for expressing the exposure effect.

Data on lung cancer diagnoses from the same study, but with follow-up continued through 1971, are shown in Table 2 (Doll and Peto, 1978). Here there are $J = 8$ five-year age groups and $K = 9$ categories of cigarette consumption. The authors restricted their analyses to regular smokers or non-smokers by eliminating from further study those who reported a change in their smoking habits. Observations for the heaviest smokers (over 40 cigarettes per day) and oldest ages (80 years or more) were excluded on the ground that they were particularly unreliable.

3.2. Models for Grouped Data

In order to make statistical inferences about the effects of exposure on disease rates, some structure must be imposed on the underlying rates λ_{jk}. In what follows they are assumed to be functions of a set of background rates λ_{j0}, which nominally represent the disease incidence for stratum j in the absence of exposure, and a set of linear predictors $\boldsymbol{\beta}\mathbf{z}_{jk}$. The $\mathbf{z}_{jk}$ are vectors of covariables that represent the quantitative effects of exposure and their interactions with age and other demographic factors, and $\boldsymbol{\beta}$ is a vector of regression coefficients. In our example, the first component of $\mathbf{z}$ is chosen to be a binary indicator of smoking status (1 for smokers, 0 for nonsmokers), and a second component represents the quantitative interaction of smoking and age.

Three functional relationships that have been used to describe the influence of the linear predictor on the underlying disease rates are multiplicative, multiplicative with additive relative risk, and additive with additive excess risk. The model equations are given by

$$\lambda_{jk} = \lambda_{j0} \exp(\boldsymbol{\beta}\mathbf{z}_{jk}) = \exp(\alpha_j + \boldsymbol{\beta}\mathbf{z}_{jk}), \tag{3.1}$$

$$\lambda_{jk} = \lambda_{j0}(1 + \boldsymbol{\beta}\mathbf{z}_{jk}), \tag{3.2}$$

$$\lambda_{jk} = \lambda_{j0} + \boldsymbol{\beta}\mathbf{z}_{jk}, \tag{3.3}$$

respectively, where $\alpha_j = \log \lambda_{j0}$ is the log background rate in stratum j. The log-linear model (3.1) defines a regular exponential family of distributions (Lehman, 1959) and has the most desirable statistical properties due to the regularity in the shape of its likelihood surface. However, (3.2) is often invoked, since it allows for a linear relationship between a quantitative exposure variable and lifetime (cumulated) disease risk (Berry, 1980; Thomas,

Table 2. Numbers of Lung Cancers (OBS) and Person-Years of Observation (PY) by Age and Smoking Level Among British Male Doctors[a]

Cigarettes per day	Av. no. smoked		Age (years) 40–44	45–49	50–54	55–59	60–64	65–69	70–74	75–79
0	0	OBS	0	0	1	2	0	0	1	2
		PY	17846.5	15832.5	12226	8905.5	6248	4351	2723.5	1772
1–4	2.7	OBS	0	0	0	1	1	0	1	0
		PY	1216.0	1000.5	853.5	625.0	509.5	392.5	242.0	208.5
5–9	6.6	OBS	0	0	0	0	1	1	2	0
		PY	2041.5	1745	1562.5	1355	1068	843.5	696.5	517.5
10–14	11.3	OBS	1	1	2	1	1	2	4	4
		PY	3795.5	3205	2727	2288	1714	1214	862	547
15–19	16.0	OBS	0	1	4	0	2	2	4	5
		PY	4824	3995	3278.5	2466.5	1829.5	1237	683.5	370.5
20–24	20.4	OBS	1	1	6	8	13	12	10	7
		PY	7046	6460.5	5583	4357.5	2863.5	1930	1055	512
25–29	25.4	OBS	0	2	3	5	4	5	7	4
		PY	2523	2565.5	2620	2108.5	1508.5	974.5	527	209.5
30–34	30.2	OBS	1	2	3	6	11	9	2	2
		PY	1715.5	2123	2226.5	1923	1362	763.5	317.5	130
35–40	38.0	OBS	0	0	3	4	7	9	5	2
		PY	892.5	1150	1281	1063	826	515	233	88.5

[a] Source: Doll and Peto (1978).

1981). Also, some epidemiologists consider that the additive scale is the best one for assessing the causal effects of two or more distinct exposures (Rothman, 1974). The model (3.3) allows for the additive combination of effects not only among the exposures, but also between these and the background rates. Considerable flexibility is available with each of these models through the choice of exposure variables and their interactions with age and time. Provided that a reasonably good fit can be achieved, the multiplicative or log-linear model is the first choice for routine data analysis, since it is the easiest to work with. Other structures may be needed in specialized situations. For example, dose–response functions of the form $(\beta_0 + \beta_1 z + \beta_2 z^2) \times \exp(-\gamma_1 z - \gamma_2 z^2)$ are used in radiation studies to account for the sterilizing effects of higher doses.

Information regarding the background age-specific rates is often available from vital statistics bureaus, specialized disease registers, or theoretical models of the disease process. Incorporation of such standard rates into the model could conceivably result in more accurate estimation of the regression parameters for the exposure variables. On the other hand, if the standard rates do not apply to the population under study, their use may lead to serious bias (Yule, 1934). Sometimes it helps to introduce an unknown scale factor $\theta = e^{\alpha}$ to adjust the standard rates so that they more nearly represent the true background rates. Thus we have three corresponding models

$$\lambda_{jk} = \lambda_j^* \exp(\alpha + \boldsymbol{\beta}\mathbf{z}_{jk}), \tag{3.1$'$}$$

$$\lambda_{jk} = \lambda_j^* \theta(1 + \boldsymbol{\beta}\mathbf{z}_{jk}), \tag{3.2$'$}$$

$$\lambda_{jk} = \lambda_j^* \theta + \boldsymbol{\beta}\mathbf{z}_{jk}, \tag{3.3$'$}$$

and where the λ_j^* denote known standard rates.

For purposes of making statistical inferences, it is customary to assume that the deaths d_{jk} have independent Poisson distributions with means $E(d_{jk}) = \lambda_{jk} n_{jk}$ (Armitage, 1971). Even though the person-year denominators n_{jk} are more properly regarded as random variables rather than fixed constants, this assumption works because the kernel of the Poisson likelihood function is the same as that based on a more realistic model. [See Section 4.3, and also Holford (1980), Whitehead (1981), and Laird and Olivier (1981).]

3.3. The Standardized Mortality Ratio and the Healthy Worker Effect

Table 3 presents results from Fox and Collier's (1976) study of some 7000 British workers exposed to vinyl chloride monomer who were followed several years to determine whether their mortality patterns differed from those of the general population. The far right column shows totals of observed and expected deaths from specific causes, the expected deaths having been derived by applying national death rates to age- and calendar-year-specific person-years

Table 3. Observed and Expected Deaths Among Vinyl Chloride Workers by Cause and Years Since Entering the Industry[a]

Cause		Years since entering the industry				Total
		1–4	5–9	10–14	15+	
All causes	OBS	34	55	74	230	393
	EXP	91.00	87.45	98.47	244.30	521.22
	SMR (%)	37.4	62.9	75.1	94.2	75.4
All cancers	OBS	9	15	23	68	115
	EXP	20.33	21.25	24.48	60.81	126.77
	SMR (%)	44.5	70.6	94.0	111.8	90.7
Circulatory disease	OBS	7	25	38	110	180
	EXP	32.49	35.56	44.87	121.26	234.18
	SMR (%)	21.5	70.3	84.7	90.7	76.9
Respiratory disease	OBS	2	4	4	32	42
	EXP	9.59	10.31	12.76	34.43	67.09
	SMR (%)	20.9	38.8	31.3	93.0	62.6

[a] Source: Fox and Collier (1976).

of observation. This data display is typical of those used with industrial cohort studies.

As a first approach to the analysis of such data, consider the multiplicative model (3.1′) with external standard rates. In the simplest situation there is no subdivision by exposure category ($K = 1$) and the model implies that the age- and time-specific cohort rates are a constant multiple of the standard, i.e., $\lambda_{j1} = \theta\lambda_j^*$, where $\theta = e^\alpha$. Application of the Poisson assumption with OBS $= \sum_j d_{j1}$ and EXP $= \sum_j \lambda_j^* n_{j1}$ denoting the total numbers of cases observed and expected yields the maximum likelihood estimate $\hat{\theta} = \text{OBS}/\text{EXP}$. Known to epidemiologists as the standardized mortality or morbidity ratio (SMR), $\hat{\theta}$ is often calculated for a large number of causes of death or disease diagnoses in an initial screening of the data. The efficient score test (Rao, 1965) of the null hypothesis $H_0: \theta = 1$ takes the familiar form $\chi^2 = (\text{OBS-EXP})^2/\text{EXP}$. Formal application of the multiplicative model thus yields the classical estimate and test statistic of epidemiologic analysis (Armitage, 1971).

For all causes of death we find $\hat{\theta} = 0.754$, $\chi_1^2 = 31.5$, and conclude that the vinyl chloride workers have an overall death rate that is substantially less than that of the general population. This phenomenon, known as the "healthy worker" effect, is most likely a consequence of a selection factor whereby workers are necessarily in good health at their time of entry into the workforce. Note the attenuation of this effect with the passage of time, so that the cancer rates even show a slight excess after 15 years. Vinyl chloride exposures are known to induce a rare form of liver cancer, and they may also increase rates of brain cancer (Beaumont and Breslow, 1981).

Once hypotheses have been established that relate the exposures to specific disease entities, further analysis is undertaken to identify subgroups at particularly high risk, to explore dose–time–response relationships, or to validate model assumptions. For example, one may wish to test the homogeneity of SMRs calculated separately for each of K exposure categories. Denoting by $\text{OBS}_k = \sum_j d_{jk}$ and $\text{EXP}_k = \sum_j \lambda_j^* n_{jk}$ the observed and expected (from standard rates) numbers of cases in exposure group k, the score test for the hypothesis $H_0: \beta_1 = \cdots = \beta_K$ in the model $\lambda_{jk} = \lambda_j^* \exp(\alpha + \beta_k)$ is

$$\chi^2_{K-1} = \sum_k \frac{(\text{OBS}_k - \text{EXP}_k^*)^2}{\text{EXP}_k^*}, \tag{3.4}$$

where $\text{EXP}_k^* = \text{EXP}_k(\text{OBS}/\text{EXP})$ is an expected number of cases adjusted so that $\sum_k \text{EXP}_k^* = \text{OBS}$. If the K exposure categories correspond to levels of a quantitative exposure variable with values z_k, the score statistic for testing $\beta = 0$ in the model $\lambda_{jk} = \lambda_j^* \exp(\alpha + \beta z_k)$ is

$$\chi^2_1 = \frac{[\sum_k z_k(\text{OBS}_k - \text{EXP}_k^*)]^2}{\sum_k z_k^2 \text{EXP}_k^* - (\sum_k z_k \text{EXP}_k^*)^2/\text{OBS}} \tag{3.5}$$

(Armitage, 1955; Tarone, 1982).

We apply these formulas to the $K = 4$ categories in Table 3, defined by years since first exposure, in order to illustrate some further features of the healthy worker effect. The linear trend test for a regression variable coded $z_k = k$ accounts for most of the heterogeneity in the SMRs for all causes of death; the statistics (3.4) and (3.5) yield values $\chi^2_3 = 8.73$ and $\chi^2_1 = 8.68$, respectively. The usual interpretation of this finding is that the effects of the initial selectivity gradually wear off, so that eventually the workers, some of whom may have retired, are not much healthier than other persons of the same age. Nevertheless, there is a continuing selection bias for those who remain employed. The SMR of 230/244.30 = 94.2% for the period from 15 years since hire decomposes into a ratio of 75/101.36 = 74.0% for those who were still employed at the 15 years anniversary, vs. 155/142.94 = 108.4% for those who were not. Testing for the significance of the difference between these two SMRs yields $\chi^2_1 = 7.47$. Thus the healthy worker effect is continuously operative throughout the study, in the sense that persons who continue employment, and therefore continue to accumulate exposures, have lower relative mortality rates due to their presumably better health. Gilbert (1983) suggests lagging exposure variables by two years or so, and eliminating the first few years of follow-up from analysis, as a means of coping with this problem.

3.4. A Simple Test and Estimate for the Common Rate Ratio

The data in Table 1 exemplify the situation where there is a single category of exposure. The key parameters under the model (3.1) are then the rate ratios $\psi_j = \lambda_{j2}/\lambda_{j1}$ for exposed vs. nonexposed. Suppose that z_k is a binary indicator that distinguishes exposed ($z_2 = 1$) from nonexposed ($z_1 = 0$) and that x_j is a quantitative value associated with the jth stratum, for example the midpoint of the jth age interval. Several hypotheses of interest are $H_0 : \lambda_{jk} = \exp(\alpha_j)$, i.e. $\psi_j = 1$, the global null null hypothesis; $H_1 : \lambda_{jk} = \exp\{\alpha_j + \beta z_k\}$, i.e. $\psi_j = \psi = e^\beta$, the hypothesis of a common rate ratio; $H_2 : \lambda_{jk} = \exp\{\alpha_j + \beta z_k + \gamma x_j z_k\}$, i.e. $\psi_j = \exp(\beta + \gamma x_j)$, the alternative of (log-linear) trend; and $\mathrm{H}_3 : \lambda_{jk}$ or ψ_j unrestricted, the general alternative. The usual goal of the statistical analysis is to test the null hypothesis, estimate the rate ratio assuming it is common to all strata, and test this latter hypothesis against alternatives of trend or heterogeneity. The stratum parameters α_j play no essential role and may be eliminated from consideration by conditioning on the total number of deaths $D_j = d_{j1} + d_{j2}$ in each stratum. Conditionally, the d_{j2} have independent binomial distributions with denominators D_j and probabilities $p_j = \psi_j n_{j2}/(n_{j1} + \psi_j n_{j2})$.

The numerator of the score test of H_0 vs. H_1 contrasts the number of deaths among the exposed to that expected if the rates for both exposure groups were equal within each stratum. The equivalent normal deviate is

$$\chi = \frac{\sum_{j=1}^{J} \left(d_{j2} - \frac{D_j n_{j2}}{N_j}\right)}{\left\{\sum_{j=1}^{J} \frac{n_{j1} n_{j2} D_j}{N_j^2}\right\}^{1/2}}, \tag{3.6}$$

where $N_j = n_{j1} + n_{j2}$ denotes the total number of person-years in stratum j. A simple noniterative estimate of the common rate ratio $\psi = e^\beta$ under H_1 is

$$\hat{\psi}_{\mathrm{MH}} = \frac{\sum_{j=1}^{J} R_j}{\sum_{j=1}^{J} S_j} = \frac{\sum_{j=1}^{J} \frac{d_{j2} n_{j1}}{N_j}}{\sum_{j=1}^{J} \frac{d_{j1} n_{j2}}{N_j}}, \tag{3.7}$$

where R_j and S_j are defined by the numerator and denominator expressions, respectively. The statistics (3.6) and (3.7) are adaptations of the famous Cochran–Mantel–Haenszel test statistic (Cochran, 1954; Mantel and Haenszel, 1959) and Mantel–Haenszel estimator, originally proposed for combining data on relative risks (odds ratios) from a series of 2×2 tables. Both are widely used by epidemiologists (Rothman and Boice, 1979).

A robust asymptotic variance for the Mantel–Haenszel estimator is derived by writing $\hat{\psi}_{\mathrm{MH}} - \psi = \sum_j (R_j - \psi S_j)/\sum_j S_j$ and noting that $E(R_j) = \psi E(S_j)$ under H_1. It follows that $\mathrm{Var}_{\mathrm{A}}(\hat{\psi}_{\mathrm{MH}}) = \sum_j \mathrm{Var}(R_j - \psi S_j)/\{\sum_j E(S_j)\}^2$. The standard error of the log Mantel–Haenszel estimator $\hat{\beta}_{\mathrm{MH}} = \log \hat{\psi}_{\mathrm{MH}}$ may thus be written

$$\mathrm{SE}(\hat{\beta}_{\mathrm{MH}}) = \hat{\psi}_{\mathrm{MH}}^{-1} \mathrm{SE}(\hat{\psi}_{\mathrm{MH}}) = \frac{\left\{\sum_{j=1}^{J} \frac{n_{j1} n_{j2} D_j}{N_j^2}\right\}^{1/2}}{\hat{\psi}_{\mathrm{MH}}^{1/2} \sum_{j=1}^{J} \frac{n_{j1} n_{j2} D_j}{N_j (n_{j1} + \hat{\psi}_{\mathrm{MH}} n_{j2})}}, \tag{3.8}$$

it being desirable to work on the log scale because $\hat{\beta}$ has a more nearly normal distribution than does $\hat{\psi}$. The Mantel–Haenszel estimator is known to have near-asymptotic optimality relative to the iterative maximum likelihood estimator considered next (Hauck, 1979; Breslow, 1981, 1984).

For the data in Table 1 we find $\chi = 3.32$ ($p = 0.001$, two sided), $\hat{\psi}_{MH} = 1.425$ or $\hat{\beta}_{MH} = 0.3540$, and $\text{SE}(\hat{\beta}_{MH}) = 0.1074$. Thus there is little doubt that the smoking effect is real, with coronary death rates estimated to increase by 42.5% at each age among smokers. However, as already noted, there are some doubts about the constancy of the rate ratios and whether the multiplicative model is appropriate for these data. This question is examined further below.

4. Fitting Models to Grouped Data

Although these elementary techniques have served epidemiologists well for many years, and undoubtedly will continue to do so in the future, they are of limited help in discriminating between different statistical models or in examining the simultaneous effects of several exposure variables in a multivariate setting. Such goals are more readily achieved by the explicit maximum likelihood fitting of models (3.1)–(3.3) or (3.1′)–(3.3′). Berry (1983) and Frome (1983) provide excellent introductory accounts of this approach.

4.1. Testing for a Trend in Rate Ratios

The right hand columns of Table 4 (a) show the results of fitting the multiplicative model (3.1) to the data in Table 1. There are five parameters α_j that represent the log death rates (per 1000 person-years) among nonsmokers in different age groups, and a single parameter $\beta = \log \psi$ that represents the log rate ratio for smokers vs. nonsmokers. Note the numerical closeness of the maximum likelihood estimate and standard error, $\hat{\beta}_{ML} = 0.3545 \pm 0.1073$, to the Mantel–Haenszel estimate and standard error found in Section 3.4.

Comparing the fitted values $\hat{d}_{jk} = n_{jk} \exp(\hat{\alpha}_j + \hat{\beta} z_k)$ for the multiplicative model [Table 4 (b)] with the observed values (Table 1), it is apparent that the model does not fit well. The usual chi-square goodness-of-fit statistic $\chi^2 = \sum_j \sum_k (d_{jk} - \hat{d}_{jk})^2 / \hat{d}_{jk}$ yields 11.15 on 4 degrees of freedom ($p = 0.026$), while the likelihood ratio goodness-of-fit criterion (deviance) yields $\chi_4^2 = 12.13$. We generally expect the chi-square measure to be smaller than the likelihood ratio for moderate sized samples (Fienberg, 1980, Appendix IV).

Even stronger evidence for the lack of fit is obtained by exploiting the decreasing trend in rate ratios with increasing age (Table 1) or, equivalently, the trend in the deviations $d_{j2} - \hat{d}_{j2}$. Fienberg (1980) discusses various tests that may be used. We prefer the score test of H_1 vs. H_2 (see Section 3.4), which is a modification of the usual test for a trend in proportions (Armitage, 1955).

Table 4. Parameter Estimates, Fitted Values, and Regression Diagnostics for Three Statistical Models for the Data in Table 1

		Statistical model[a]					
		Additive ($\rho = 1$)		Power ($\rho = 0.55$)		Multiplicative ($\rho = 0$)	
Age group	Parameter	Nonsmokers	Smokers	Nonsmokers	Smokers	Nonsmokers	Smokers
		(a) Parameter estimates ± SE[b]					
35–44	α_1	0.084 ± 0.066		0.276 ± 0.092		-1.012 ± 0.192	
45–54	α_2	1.556 ± 0.214		0.839 ± 0.100		0.472 ± 0.130	
55–64	α_3	6.219 ± 0.454		2.180 ± 0.121		1.616 ± 0.115	
65–74	α_4	13.440 ± 0.963		3.583 ± 0.173		2.338 ± 0.116	
75–84	α_5	19.085 ± 1.704		4.487 ± 0.253		2.688 ± 0.125	
(Smoking)	β	0.591 ± 0.125		0.493 ± 0.098		0.355 ± 0.107	
		(b) Fitted values[c]					
35–44		1.59	35.37	1.81	32.53	6.83	27.17
45–54		17.51	96.50	13.00	102.56	17.12	98.88
55–64		35.99	197.25	29.26	204.49	28.74	205.26
65–74		34.96	178.73	30.12	183.61	26.81	187.19
75–84		28.03	105.07	24.97	108.65	21.51	111.49
	χ^2_4[d]	7.43		2.14		12.13	

Table 4. (continued)

		Statistical model[a]					
		Additive ($\rho = 1$)		Power ($\rho = 0.55$)		Multiplicative ($\rho = 0$)	
Age group	Parameter	Nonsmokers	Smokers	Nonsmokers	Smokers	Nonsmokers	Smokers
(c) Regression diagnostics (h_{jk})							
35–44		0.98	0.93	0.67	0.86	0.25	0.81
45–54		0.31	0.77	0.42	0.85	0.29	0.88
55–64		0.19	0.82	0.28	0.85	0.38	0.91
65–74		0.18	0.83	0.22	0.84	0.36	0.91
75–84		0.22	0.78	0.24	0.79	0.34	0.87
(d) One step change in β coefficient for smoking							
35–44		0.855	0.855	0.026	0.026	−0.060	−0.060
45–54		−0.058	−0.058	−0.021	−0.021	−0.072	−0.072
55–64		−0.020	−0.020	−0.010	−0.010	−0.012	−0.012
65–74		−0.008	−0.008	−0.010	−0.010	0.018	0.018
75–84		0.002	0.002	0.023	0.023	0.138	0.138

[a] Exponent (ρ) of power function relating death rates and linear predictor.

[b] Person-years denominators expressed in units of 1000.

[c] See Table 1 for observed values.

[d] Goodness of fit (deviance).

The formula is

$$\chi^2 = \frac{\left\{\sum_{j=1}^{J} x_j(d_{j2} - \hat{d}_{j2})\right\}^2}{\sum_{j=1}^{J} \frac{x_j^2 \hat{d}_{j1} \hat{d}_{j2}}{D_j} - \left(\sum_{j=1}^{J} \frac{x_j \hat{d}_{j1} \hat{d}_{j2}}{D_j}\right)^2 \Big/ \left(\sum_{j=1}^{J} \frac{\hat{d}_{j1} \hat{d}_{j2}}{D_j}\right)}, \tag{4.1}$$

and yields $\chi_1^2 = 10.30$ ($p = 0.001$) when the x_j are equally spaced. We conclude that most of the heterogeneity in the (log) rate ratios is explained by their linear trend in the ages x_j. The deviance for the interaction model H_2 is $\chi_3^2 = 1.44$ ($p = 0.69$).

4.2. Choosing between Additive and Multiplicative Models

In view of the lack of fit of the multiplicative model, the next step is to determine if an additive structure might not give a better summary of the data. The left hand columns of Table 4 show results of fitting the model (3.3) to the same set of rates. Smoking is estimated to increase the age-specific coronary death rates by 0.591 deaths per 1000 person-years of observation. While this model apparently fits better, with a deviance of $\chi_4^2 = 7.43$ instead of 12.13, several of the individual deviations are still rather large and there is now a generally increasing trend in the additive smoking effect with increasing age (Table 1). This suggests that some intermediate scale for estimating the smoking effect may give better results.

One method of formally discriminating between additive and multiplicative models is to imbed them in a parametric family that contains both as special cases. Thus, following Aranda-Ordaz (1983), we consider the family of models

$$\lambda_{jk}^{\rho} = \alpha_j + \boldsymbol{\beta}\mathbf{z}_{jk} \tag{4.2}$$

that relate the disease rates to the linear predictor by means of the power transform with exponent ρ. The additive model corresponds to the case $\rho = 1$ and, in view of the relation $\lim_{\rho\to 0} (\lambda^{\rho} - 1)/\rho = \log \lambda$, the multiplicative model occurs in the limit as ρ tends towards zero.

Figure 1 graphs the deviances obtained by fitting a range of power models with $-0.2 \leq \rho \leq 1.2$ to the data in Table 1. Substantial improvement is obtained with $\rho \approx \frac{1}{2}$ over both additive and multiplicative structures. Using a trial and error procedure, we found the minimum deviance was obtained with $\rho = 0.55$; Table 4 presents parameter estimates and fitted values based on this transform. The fit is significantly better than that obtained with either additive or multiplicative structures. Its main drawback is that expressing the smoking effect on a square root scale ($\rho = \frac{1}{2}$) may appear slightly unnatural. Note that the standard errors do not account for the asymptotic correlation between $\hat{\rho}$ and $(\hat{\alpha}, \hat{\beta})$. In typical practice ρ would not be estimated explicitly; rather the goodness-of-fit would be examined for a range of ρ and the final model selected partially on the basis of a *priori* considerations.

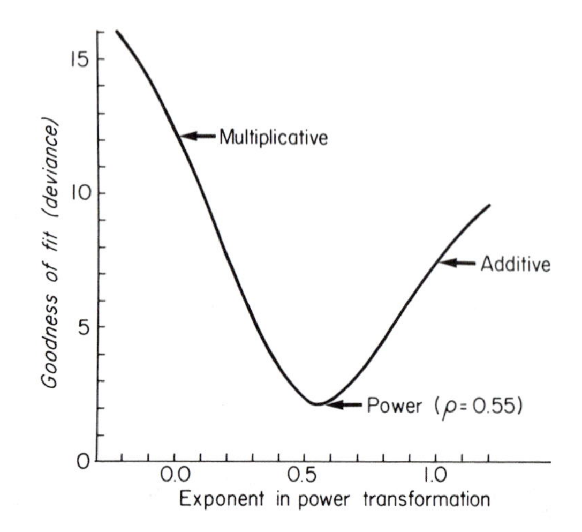

Figure 1. Goodness-of-fit statistics (deviances) for the power family of transformations used to relate smoking and age effects on coronary artery mortality rates in the British doctors study.

4.3. Examining the Influence of Individual Data Points

Frome (1983) notes that the maximum likelihood fitting of generalized linear regression models for Poisson rates may be recast as a problem in iterated reweighted least squares (IRLS) and thus programmed on a model-specific basis in any system that facilitates IRLS calculations. Following Pregibon (1981), he demonstrates how the diagonal elements of the "hat" matrix obtained as a by-product of such fitting may be used to identify observations d_{jk} that have an especially great influence on the parameter estimates. The calculations follow the method developed by Nelder and Wedderburn (1972) that is incorporated in the program GLIM (Baker and Nelder, 1978). Let X denote the design matrix incorporating the stratum effects, $\boldsymbol{\gamma}' = (\boldsymbol{\alpha}', \boldsymbol{\beta}')$ the unknown coefficients, W the diagonal matrix of iterated weights, and $\boldsymbol{\eta} = X\hat{\boldsymbol{\gamma}} + W^{-1}(\mathbf{d} - \hat{\mathbf{d}})$ the "working vector." The parameter estimates are found by recursive solution of $\hat{\boldsymbol{\gamma}} = (X'WX)^{-1}X'W\boldsymbol{\eta}$. At convergence, the diagonal elements h_{jk} of the matrix

$$H = W^{1/2}X(X'WX)^{-1}X'W^{1/2} \tag{4.3}$$

identify observations that have a particularly large influence on the overall fit.

The third part of Table 4 presents the regression diagnostics h_{jk} for each of the three models fitted to the data in Table 1. These are generally larger for smokers, which is not surprising, since smokers are more numerous and account for most of the deaths. It is instructive that the younger age groups have a substantially greater influence on the fit of the additive model than they do for the multiplicative model. This reflects the fact that large numbers of events are required for accurate determination of ratios of rates, whereas large numbers of person-years are of greater help in estimating absolute rates and rate differences.

These diagnostics measure the relative contribution of each observation to the overall fit rather than to any particular regression coefficient. Since the key regression parameter in our example is that for smoking, a more sensitive indication of the influence of any particular data point is to examine the change in the last component of $\hat{\gamma}$ occasioned by its deletion from the sample. An excellent "one-step" approximation to the change in $\hat{\gamma}$ from deleting the (j, k) observation is given by

$$\Delta_{jk}\hat{\gamma} \approx (X'WX)^{-1}\frac{\mathbf{x}_{jk}(d_{jk} - \hat{d}_{jk})}{1 - h_{jk}}, \tag{4.4}$$

where $\mathbf{x}_{jk}$ denotes the jk row of the design matrix (Pregibon, 1981).

Part (d) of Table 4 presents the approximate changes in the value of the smoking coefficient β that follow the deletion of each datum in Table 1. Since a separate stratum parameter is estimated for each age group, deleting the observation for smokers (or for nonsmokers) at any specified age has the same effect as deleting all the data for that age. The coefficient in the additive model is highly sensitive to the presence of the youngest age group, whereas deletion of successively older age groups in the multiplicative model leads to progressively greater estimates for the smoking rate ratio. Only the coefficient for the power model is relatively unchanged by such deletions, which confirms the goodness of fit of this model from yet another perspective.

4.4. The Additive Relative Risk Model

In contrast to the situation with coronary disease, there is no doubt at all that smoking effects on lung cancer rates are multiplicative rather than additive. The last two lines of Table 5 (a) examine the fit of the additive model (3.3), in which the effect of smoking is expressed as an excess risk (ER) added to the background lung cancer rate, to the data in Table 2. Whether this excess risk is allowed to vary arbitrarily with cigarette dose group ($\chi^2_{56} = 184.7$), or is assumed to be linear in dose (ER $= \beta z$, $\chi^2_{63} = 205.8$), the deviances are so enormous as to render the additive model untenable. On the other hand, as shown in the first line of the table, the multiplicative model (3.1) with separate relative risks (RR) for each dose group fits quite well ($\chi^2_{56} = 45.74$). Further analyses therefore considered the relative risk as a quantitative function of z = number of cigarettes smoked per day.

Use of z itself in the model (3.1), which implies an exponential rise in the age-specific risks with amount smoked, is unsatisfactory. A much better fit is obtained using $\log(1 + z)$ in place of z, which specifies a power relationship RR $= (1 + z)^{\beta}$ for the relative risk. The additive relative risk model (3.2) with RR $= 1 + \beta z$ also fits reasonably well. The fact that the quadratic model

Table 5. Results of Fitting Several Models to the Data in Table 2

Equation for excess risk (ER) or relative risk (RR) as a function of daily cigarettes (z)	DF	Deviance	Parameter estimates	Standard errors
(a) Age effects estimated from the data				
Separate RR each dose group	56	45.74	(See Figure 3)	
RR $= \exp(\beta z)$	63	68.91	0.0853	0.0063
RR $= \exp(\beta z + \gamma z^2)$	62	51.87	0.1802 −0.00226	0.0263 (β) 0.00059 (γ)
RR $= (1 + z)^\beta$	63	55.87	1.187	0.123
RR $= 1 + \beta z$	63	58.36	1.130	0.510
RR $= 1 + \beta z + \gamma z^2$	62	51.03	0.4105 0.0237	0.2880 (β) 0.0116 (γ)
Separate ER each dose group	56	184.7	—	—
ER $= \beta z$	63	205.8	0.0000437	0.0000055
(b) Age effects assumed proportional to (AGE-22.5)$^{4.5}$				
Separate RR each dose group	63	47.13	—	—
RR $= \exp(\beta z)$	70	70.29	0.0854	0.0063
RR $= \exp(\beta z + \gamma z^2)$	69	53.26	0.1802 −0.00226	0.0261 (β) 0.00059 (γ)
RR $= (1 + z)^\beta$	70	57.35	1.192	0.122
RR $= 1 + \beta z$	70	60.00	1.141	0.516
RR $= 1 + \beta z + \gamma z^2$	69	52.43	0.409 0.0239	0.286 (β) 0.0116 (γ)
Separate ER each dose group	63	200.0	—	—
ER $= \beta z$	70	225.2	0.0000421	0.0000059

RR $= 1 + \beta z + \gamma z^2$ improved the fit even further was interpreted by Doll and Peto (1978) as consistent with the notion that two or more stages in the carcinogenetic process are influenced by cigarette smoking. Using an earlier summary of the lung cancer data from the British doctors study (Doll and Hill, 1966), however, Frome (1983) finds no evidence for a quadratic effect, a conclusion also reached by Whittemore and Altshuler (1976). Part of the discrepancy between Frome's results and those shown here and in Doll and Peto (1978), in which there is distinct upward curvature in the relative risk as a function of dose, is due to the restriction on subjects contributing data to

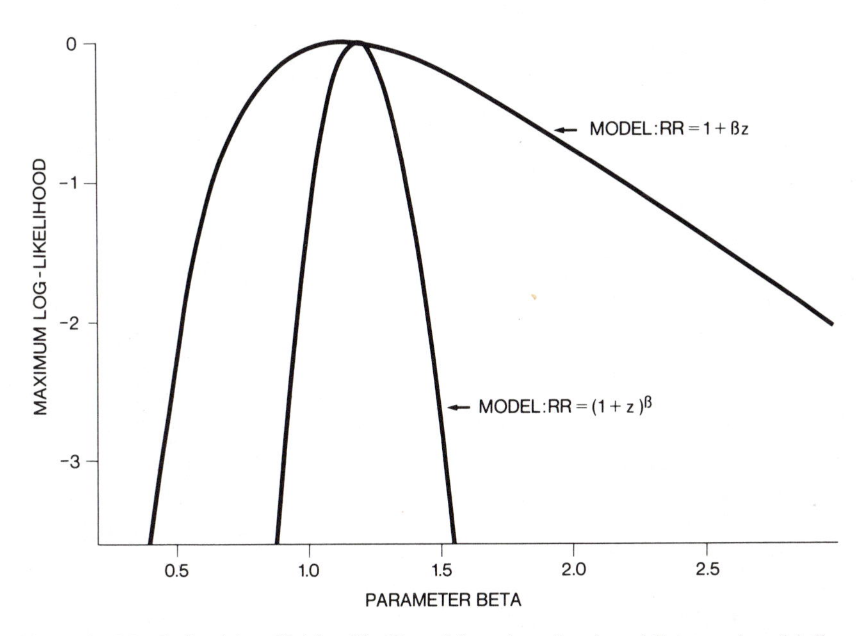

Figure 2. Maximized (profile) log likelihood functions for the additive and multiplicative relative risk models fitted to lung cancer rates from the British doctors study.

Table 2, particularly the elimination of a few doctors who reported smoking over 40 cigarettes per day for whom the relative risk was paradoxically low.

The statistical properties of the additive relative risk models (3.2) are not nearly so convenient as for the log linear models (3.1). In the first place, the fact that relative risks must be positive restricts the range of parameter values. Consequently, the profile log likelihood, i.e. the log likelihood considered as a function of β alone after maximization over the nuisance parameters α_j, may be highly asymmetric, whereas that for the multiplicative model behaves more like a normal (quadratic) log likelihood. As an example, Figure 2 contrasts the profile log likelihoods for the models shown in lines 4 and 5 of Table 5. The difference in the degree of curvature of the maximum is due in part to the fact that for a typical value of z, $1 + \beta z$ is less affected by small changes in β than is $(1 + z)^{\beta}$. The figure suggests that the β estimate from the model (3.2) is rather unstable and that its standard error is not helpful in assessing the degree of uncertainty. Thomas (1981) demonstrates that such behavior may also be expected when fitting additive relative risk functions to data from matched sample case-control studies and suggests that, whenever possible, the additive model be reparametrized as $\text{RR} = 1 + e^{\beta}z$ in order to achieve a better-behaved likelihood function.

One unfortunate property of the additive relative risk model is evident from the estimated standard error for $\hat{\beta}$ shown in Table 5. Comparing deviances for models with and without smoking effects, there is no doubt that the statistical significance of the increases in lung cancer rates with increasing numbers of cigarettes smoked is enormous. This is reflected also in t statistics based on the multiplicative model, for instance $t = 0.0833/0.0063 = 13.5$ from line 2 of the table or $t = 1.187/0.123 = 9.7$ from line 4. However, the t statistic based on the additive relative risk model in line 5 is only $t = 1.130/0.510 = 2.2$. Numerical results from the fitting of this model with external standard rates (see below) suggest that much of the problem is the extreme dependence of $\hat{\beta}$ on the estimates of the nuisance parameters α_j.

Another drawback of the additive relative risk models is that second derivatives of the log likelihood depend on the data. Substantial differences exist in practice between observed and expected information. Our experience with matched sample case-control studies indicates that score statistics for testing the significance of variables in the regression equation are adequate if calculated using expected information, but not with observed information (Storer, Wacholder, and Breslow, 1983). One would expect reasonably good performance from such score tests, since they share the property of the likelihood ratio tests of invariance under transformations of the parameter space (Cox and Hinkley, 1974).

The power function $\text{RR} = (1 + z)^\beta$ shares with $\text{RR} = 1 + \beta z$ the property of linearity with slope β at the origin, and both could be used for purposes of low dose linear extrapolation. Since these two curves have similar shapes in the range of interest, the power law might well be selected for routine work because of its statistical advantages, with the additive relative risk model being reserved for those situations where strong *a priori* considerations dictate its use.

4.5. Possible Efficiency Gains from the Use of External Standard Rates

We remarked earlier that knowledge of the background rates λ_{j0}, at least up to an unknown multiplicative constant, could theoretically improve the estimates of the relative risk parameters. We can measure the increase in efficiency by computing the information for $\boldsymbol{\beta}$ estimation under the models (3.1) and (3.1′), respectively. Suppose for simplicity that z is univariate, and define $I_1(\beta)$ to be the inverse of the appropriate element of the inverse $(J + 1) \times (J + 1)$ dimensional matrix of negative second partial derivatives of the log likelihood for the model (3.1). We find

$$I_1(\beta) = \sum_j \sum_k \mu_{jk} z_{jk}^2 - \sum_j \frac{\sum_k \mu_{jk} z_{jk}^2}{\sum_k \mu_{jk}},$$

where $\mu_{jk} = E(d_{jk}) = n_{jk} \exp(\alpha_j + \beta z_{jk})$. Writing $E(\text{OBS}) = \sum_j \sum_k \mu_{jk}$ for the expected number of cases under the full model, this may be expressed

$$I_1(\beta) = E(\text{OBS}) \times E[\text{Var}(Z \mid S)], \tag{4.5}$$

where $E[\mathrm{Var}(Z \mid S)]$ denotes the expected within stratum variance of a random variable Z which is drawn from the finite distribution putting masses proportional to μ_{jk} at each of the values z_{jk}. A similar expression holds for a vector $\mathbf{z}$ with $\mathrm{Var}(\mathbf{Z} \mid S)$ denoting the expected within stratum covariance matrix.

Slightly easier calculations show that the β information under the model (3.1′) is

$$I_1'(\beta) = \sum_j \sum_k \mu_{jk} z_{jk}^2 - \frac{\left(\sum_j \sum_k \mu_{jk} z_{jk}\right)^2}{\sum_j \sum_k \mu_{jk}} \tag{4.6}$$

$$= E(\mathrm{OBS}) \times \mathrm{Var}(Z),$$

where now $\mu_{jk} = n_{jk} \lambda_j^* \exp(\alpha + \beta z_{jk})$ and $\mathrm{Var}(Z)$ represents the total variability in Z as defined above. In view of the relation $\mathrm{Var}(Z) = E[\mathrm{Var}(Z \mid S)] + \mathrm{Var}[E(Z \mid S)]$, therefore, substantial gains in efficiency are achieved by knowledge of the standard rates only if the exposure and stratification variables are highly correlated.

In our example, standard rates $\lambda_j^* = 10^{-10} \times (\mathrm{AGE}_j - 22.5)^{4.5}$ were obtained not from vital statistics but rather from the multistage theory of carcinogenesis (Armitage and Doll, 1961): $\mathrm{AGE}_j - 22.5$ represents the estimated average duration of smoking in the jth age group and the exponent 4.5 a compromise between 5 and 6 stages in the carcinogenic process. (Although the value 4.5 was originally estimated from the data, this fact is ignored here.) Comparing parts (a) and (b) of Table 5, however, only a slight advantage seems to accrue to the use of the extra assumptions. We would expect to see a gain in efficiency if smoking and age were closely related, and while there is some suggestion from Table 2 that the older doctors smoked less, this relationship is evidently not strong enough to give much advantage to the use of the "known" rates.

Continuing the efficiency calculations under the null hypothesis for the additive model (3.3), we find for a univariate exposure z that the inverse of the corresponding element in the inverse (expected) information matrix is

$$I_3'(\beta) = E(\mathrm{OBS}) \times E[\mathrm{Var}(Z/\Lambda) \mid S]$$

and for the model (3.3′)

$$I_3'(\beta) = E(\mathrm{OBS}) \times \mathrm{Var}(Z/\Lambda), \tag{4.7}$$

where Z/Λ is used to represent a discrete random variable taking values z_{jk}/λ_{j0} with probabilities proportional to $\mu_{jk} = n_{jk}\lambda_{j0}$. Thus the efficiency of $\hat{\beta}$ if the stratum specific rates must be estimated internally, relative to that when they are assumed proportional to standard rates, is given by the ratio $E[\mathrm{Var}(Z/\Lambda \mid S)]/\mathrm{Var}(Z/\Lambda)$. The corresponding ratio for the multiplicative model was $E[\mathrm{Var}(Z \mid S)]/\mathrm{Var}(Z)$. In situations where the background rates Λ depend strongly on stratum, even though the exposures Z do not, one would therefore expect a larger gain in efficiency for the additive model.

One way of writing the additive relative risk model (3.3′) is $E(d_{jk}) = \theta\lambda_j^* n_{jk} + \psi\lambda_j^* n_{jk} z_{jk}$. For the data in Table 2 we estimate $\hat{\theta} = 0.8371\text{E}-1 \pm 0.3560\text{E}-1$, $\hat{\psi} = 0.9549\text{E}-1 \pm 0.7408\text{E}-2$, and $\text{Cov}(\hat{\theta},\hat{\psi}) = -0.6855\text{E}-4$. A t test of the smoking effect is thus $\hat{\psi}/\text{S.E.}(\hat{\psi}) = 12.9$, of the same order of magnitude as with the multiplicative models. However, the regression parameter of interest is $\beta = \psi/\theta$, which shows that a major source of instability in the additive relative risk model is the extreme dependent of β on the fitted background rates. Although the minimum deviance of 60.0 occurs at $\hat{\beta} = \hat{\psi}/\hat{\theta} = 1.141$ (Table 5), virtually identical fits are obtained for a wide combination of (θ, β) values. The test based on a comparison of $\hat{\beta}$ with its standard error of

$$\frac{[\text{Var}(\hat{\theta})\,\hat{\beta}^2 - 2\text{Cov}(\hat{\theta},\hat{\psi})\,\hat{\beta} + \text{Var}(\hat{\psi})]^{1/2}}{\hat{\theta}} = 0.5155$$

yields a normal deviate of only 2.2.

4.6. Interpretation of Relative Risk Functions

Figure 3 graphs four of the best-fitting models for the lung cancer data with background rates estimated internally [Table 5 (a)]. The points • correspond to relative risks estimated separately for each category of smokers relative to nonsmokers (RR = 1.0). These are obtained from β parameters for the model shown in the first line of the table. The other three curves correspond to the quantitative relative risk models shown in lines 4–6 of the table, all of which fit the observed data reasonably well.

A peculiarity of Figure 3 is the fact that the relative risk function specified by the power model $\text{RR} = (1 + z)^\beta$ fails to overlap the others. One must remember that there is an implied intercept that represents the baseline lung cancer rate for nonsmokers. Although this baseline rate is estimated for each model, and indeed may vary substantially among them, it is not represented at all in the relative risk functions, as these are constrained to equal 1.0 at $z = 0$. The apparent discrepancy is aggravated by the choice of nonsmokers ($z = 0$) as a reference category. Most of the lung cancer deaths occur among smokers, and thus most of the information for β estimation comes from comparisons among categories of smokers, e.g. 10 cigarettes vs. 20 cigarettes per day, rather than of smokers with nonsmokers. The relative risk curves would have appeared to be in better agreement if smokers of 10 cigarettes per day had been chosen as the reference category so that baseline rates were estimated for this group instead of nonsmokers and all the relative risk functions were constrained to have RR = 1.0 at $z = 10$. However, use of $z = 0$ as a reference category is standard practice and more easily explained to nonspecialists.

4.7. Notes on Computing with GLIM

The Poisson regression models considered above were all fitted with the Royal Statistical Society's program for generalized linear interactive modeling (GLIM) described by Baker and Nelder (1978). The degree of effort

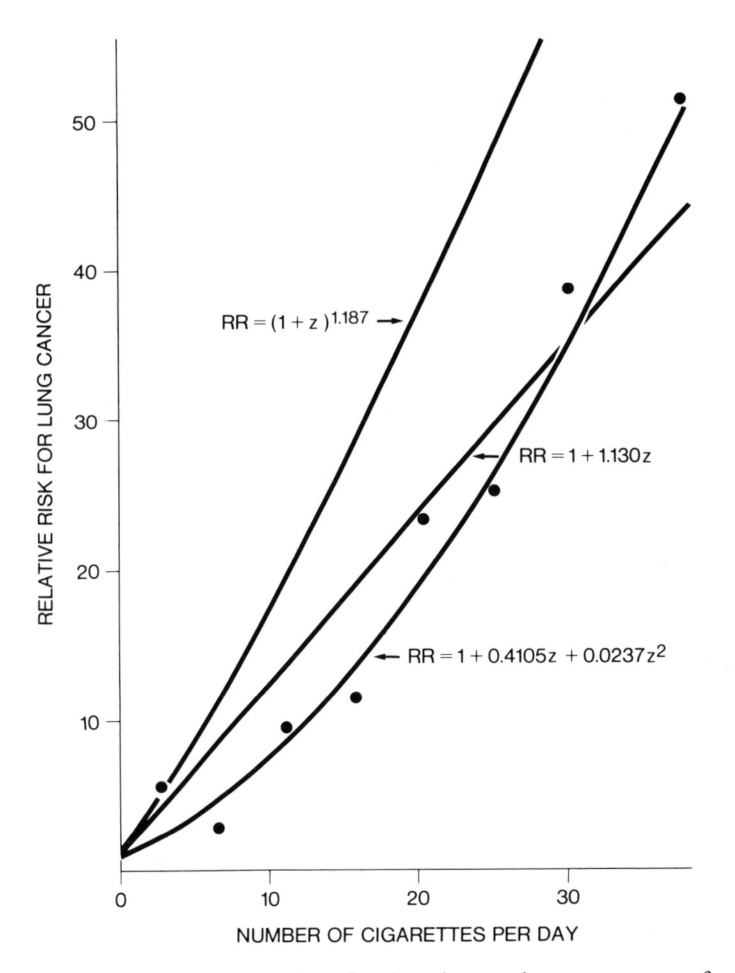

Figure 3. Three relative risk functions fitted to data on lung cancer rates from the British doctors study.

demanded from the user varies considerably with the model. Easiest to fit are the multiplicative models (3.1) or (3.1′), since these may be expressed in the form $\log E(d_{jk}) = \log n_{jk} + \alpha_j + \boldsymbol{\beta}\mathbf{z}_{jk}$ or $\log E(d_{jk}) = \log(n_{jk}\lambda_j^*) + \alpha + \boldsymbol{\beta}\mathbf{z}_{jk}$, respectively. Using GLIM terminology, the log transform "links" the mean values $E(d_{jk})$ to the linear predictors $\alpha_j + \boldsymbol{\beta}\mathbf{z}_{jk}$ or $\log(n_{jk}\lambda_j^*) + \alpha + \boldsymbol{\beta}\mathbf{z}_{jk}$. The log link is the standard (default) option for Poisson variables. Since they enter the model with a known coefficient, the terms $\log n_{jk}$ or $\log(n_{jk}\lambda_j^*)$ are declared as "offset."

The additive models (3.3) or (3.3′) and the intermediate models based on the power transform may also be fitted using the GLIM option for the identity or power functions as links. In this case, the "offset" procedure is not available. Instead the user must construct transformed regression variables $z_{jk}^* = n_{jk}^\rho z_{jk}$, where ρ is the exponent of the power transform, and similarly multiply the "dummy" regression variables associated with the nuisance parameters α_j or α by n_{jk}^{ρ} or $(n_{jk}\lambda_j^*)^{\rho}$.

The additive relative risk models (3.2) define generalized linear models having composite link functions (Berry, 1980; Thompson and Baker, 1981). Here the GLIM facility for user-defined models is invoked. Writing the model equation in the form $\log E(d_{jk}) = \log n_{jk} + \alpha_j + \log(1 + \boldsymbol{\beta}\mathbf{z}_{jk})$, the macros are constructed just as they would be for a Poisson model with log link. However, at each cycle of iteration the exposure variables $\mathbf{z}_{jk}$ are replaced by $\mathbf{z}_{jk}/(1 + \boldsymbol{\beta}\mathbf{z}_{jk})$. The additive relative risk models (3.2′) that incorporate standard rates are simply a reparametrization of the additive models (3.3′) and may be fitted accordingly. Depending upon the particular data being analyzed, substantial difficulties in achieving convergence may be anticipated with any but the log-linear models.

The basic quantities needed to compute regression diagnostics are also provided by GLIM. Specifically, the diagonal elements of the matrix $X'(X'WX)^{-1}X$ needed to find the h_{jk} defined by (4.3) are available as the variances of the linear predictor, while the matrix $(X'WX)^{-1}$ used in the calculation of the approximation (4.4) to $\Delta\hat{\gamma}$ is simply the covariance matrix for the parameter estimates.

5. Analysis of Continuous Cohort Data

Original data records from cohort studies usually consist of a time sequence of observations on individuals, starting with their entry into the study and ending with death, loss to follow-up, or termination of the study. These features are present for data from a cohort of copper smelter workers that we use to illustrate some of the analytic techniques for continuous data.

5.1. The Montana Smelter Workers Study

Lee and Fraumeni (1969) assembled records on 8047 white males who worked at a Montana copper smelter for at least one year between 1937 and 1956. Although some of the older workers had been employed as early as 1885, they were not entered on study until 1938 or upon completion of one year of employment if hired later. Results reported below include follow-up through 1977 for 8014 males of the original cohort, relevant records having been lost for the others. During this period there were 276 respiratory cancer deaths that occurred under 80 years of age, whereas 130.1 would have been expected on the basis of age- and calendar-year-specific rates for U.S. white males (SMR = 2.12).

A major goal of the analysis was to determine the extent to which the excess respiratory cancer mortality could be explained by arsenic exposure. To this end, work areas were classified as having light, moderate, or heavy levels of arsenic. From records of job changes within the plant, time-dependent covariables were constructed for each worker to represent the cumulative number of years exposed to either moderate or heavy arsenic concentrations. Selection bias of the type shown in Table 3 was not felt to play a major role, since few respiratory cancer deaths occurred during the first 15 years of employment and since the rapidly fatal course of the disease meant that any selection was unlikely to be operative for more than a year or so before death. Nevertheless, the exposure variables were lagged two years so that the exposure category at death was based on the worker's status at a time that was presumably closer to the onset of disease.

Grouped data analyses were conducted using several factors considered *a priori* to have possible effects on the age and year specific respiratory cancer death rates. Date of employment was split at 1925, the year in which introduction of a selective flotation process supposedly reduced airborne arsenic concentrations (Lee–Feldstein, 1983). Date of employment, birthplace (U.S. *vs*. foreign), and arsenic exposure were all found to have strong and independent effects in a grouped data analysis conducted using the models (3.1) and (3.1′) with 0–1 regression variables representing different factor levels. However, the effect of years since first employment was largely secondary to that of data of employment. Neither age at employment nor years since termination was found to have much influence on relative risk, though they do affect excess risk (Brown and Chu, 1983).

Grouping of the data into a multidimensional classification of deaths and person-years by categories of age, calendar period, cumulative exposure, and other fixed or time-varying factors is a convenient way to reduce the often voluminous mass of data to a form suitable for statistical analysis. This approach has the advantage of simplicity and of allowing the data analyst to "see" and plot the data. It facilitates model checking via the analysis of residuals and the exploration of interactions between exposure and age or period effects and is less sensitive than continuous variable methods to cases having extremes of exposure. The assumption that underlies this approach, namely that disease rates are constant within each of the cells that make up the classification, can be relaxed by refining certain categories, for example by using five year intervals of age and calendar time rather than ten year intervals. Of course, such refinements may ultimately increase the number of cells and the number of nuisance parameters to the point that the technique breaks down. If this is a serious concern, an alternative method of analysis is available in which the disease rates, and possibly also the exposure variables, are modeled as continuous functions of a continuous time variable.

5.2. Models for Continuous Data

The theory of survival analysis (Kalbfleisch and Prentice, 1980), and in particular the proportional hazards regression model of Cox (1972), provides a suitable framework for the statistical modeling of continuous cohort data.

Under this theory the instantaneous disease rates for the ith individual are written as a function $\lambda_i(t)$ of a continuous time parameter t. Likewise, the exposures accumulated prior to t are expressed in a vector $\mathbf{z}_i(t)$ of covariables. Examples of covariable definitions that express such concepts as latency and repair of accumulated damage are provided by Knox (1973), Berry et al. (1979), Lundin et al. (1979), and Thomas (1982). Time-dependent strata, indexed by an integer function $s = s(t)$, may also be introduced to accommodate the effects of time-varying nuisance factors. The rates $\lambda_i(t)$ are modeled as a function of the linear predictors $\boldsymbol{\beta}\mathbf{z}_i(t)$ and a set of background rates $\lambda_s(t)$, where s denotes the stratum in which a given subject finds himself at time t.

Several alternatives are available for the choice of the fundamental time variable t. The most natural of them from the viewpoint of survival analysis is for t to represent duration of time on study, analogous to the situation in clinical trials, for then everyone starts his observation period at $t = 0$. The effects of age and calendar year would be accounted for by time-dependent stratification, just as in the grouped data models. However, time on study is probably not a good choice for t in cohort studies, for two reasons. First, since the background rates for cancer and other chronic diseases vary rapidly with age, more accuracy is obtained if the age effects are modeled continuously rather than in strata. Second, cumulative exposure is often highly correlated with time on study, so that inclusion of the nuisance functions $\lambda_s(t)$ in the model may mask the very effects that one is attempting to quantify. The illustrative analyses presented here use $t =$ age as the fundamental time variable, with the effects of calendar year controlled by age-dependent stratification into five or ten year calendar periods. For the most part, the effects of time since hire and other measures of cumulative exposure are explicitly modeled in age-dependent covariables $\mathbf{z}(t)$.

With these preliminaries, models analogous to those considered already for grouped data are

$$\lambda_i(t) = \lambda_s(t)\exp\{\boldsymbol{\beta}\mathbf{z}_i(t)\}, \tag{5.1}$$

$$\lambda_i(t) = \lambda_s(t)\{1 + \boldsymbol{\beta}\mathbf{z}_i(t)\}, \tag{5.2}$$

$$\lambda_i(t) = \lambda_s(t) + \boldsymbol{\beta}\mathbf{z}_i(t). \tag{5.3}$$

Under the multiplicative models (5.1) and (5.2), the instantaneous death rates are the product of one factor representing background and another the specific effect of the exposures. Under the additive model (5.3), in contrast, the quantities that factor into background and exposure components are the net probabilities $P_i(t) = \exp\{-\int_0^t \lambda_i(u)\,du\}$ of remaining disease-free for an interval of time. External standard rates are incorporated by assuming that the background rates $\lambda_s(t)$ are proportional to known standard rates $\lambda_s^*(t)$. Denoting by $\theta = e^\alpha$ the unknown constant of proportionality, there are thus three models

$$\lambda_i(t) = \lambda_s^*(t)\exp\{\alpha + \boldsymbol{\beta}\mathbf{z}_i(t)\}, \tag{5.1'}$$

$$\lambda_i(t) = \lambda_s^*(t)\theta\{1 + \boldsymbol{\beta}\mathbf{z}_i(t)\}, \tag{5.2'}$$

$$\lambda_i(t) = \lambda_s^*(t)\theta + \boldsymbol{\beta}\mathbf{z}_i(t) \tag{5.3'}$$

analogous to (5.1)–(5.3).

5.3. Likelihood Inference for Continuous Data

Statistical inference about the parameters of interest in the models for continuous data requires construction of an appropriate likelihood function. Denote by t_i^0 the age at which the ith subject enters the study, and by t_i his age at the end. Heuristic considerations suggest that the likelihood contribution for the ith subject is

$$\lambda_i^{\delta_i}(t_i)\exp\left\{-\int_{t_i^0}^{t_i} \lambda_i(u)\,du\right\}, \tag{5.4}$$

where the exponential term represents the probability of observing someone disease-free between ages t_i^0 and t_i, $\lambda_i(t_i)$ represents the conditional probability of death or diagnosis at t_i, and δ_i is 1 or 0 according as death or diagnosis has or has not occurred. A rigorous derivation of this likelihood requires consideration of the product integral of the instantaneous probabilities of death or disease at each age, conditional on past history (Kalbfleisch and Prentice, 1980; Johansen, 1981).

For models that incorporate standard rates, the only unknowns are the scalar α and the vector $\boldsymbol{\beta}$. In this case the log likelihood function for the entire set of cohort data may be written

$$L(\alpha, \boldsymbol{\beta}) = \sum_i \left[\delta_i \log \lambda_i(t_i; \alpha, \boldsymbol{\beta}) - \int_{t_i^0}^{t_i} \lambda_i(u; \alpha, \boldsymbol{\beta})\,du\right], \tag{5.5}$$

where $\lambda_i(t; \alpha, \boldsymbol{\beta})$ is specified by any of the models (5.1′)–(5.3′). A formal proof that the maximum likelihood estimates have the usual properties of consistency and asymptotic normality, with covariances estimable from the inverse information matrix, may be based on the large sample theory of counting processes (Aalen, 1978; Gill, 1980). These methods have been implemented for the multiplicative model (5.1′) by Breslow et al. (1983), who approximate the integral shown in (5.5) and its first and second derivatives by a summation in which the time-dependent covariates are evaluated once each year.

For the models (5.1)–(5.3) the likelihood (5.4) involves an unknown nuisance function $\lambda_s(t)$ whose presence considerably complicates estimation of $\boldsymbol{\beta}$. Cox (1972, 1975) suggested that the problem could be solved for the multiplicative model (5.1) by construction of an appropriate "partial likelihood." Suppose, for example, that the ith individual is known to have died (or been diagnosed) at age t_i in calendar period s_i. Denote by R_i the set of all subjects who were "at risk" of death at age t_i in period s_i, i.e. those alive and under observation just prior to that time. The conditional probability that the ith subject died, given one death from among those in the risk set R_i, is $\lambda_i(t_i)/\{\sum_{j \in R_i} \lambda_j(t_i)\}$. Summing the logarithms of such contributions over all subjects who die or develop disease yields the log partial likelihood

$$L(\boldsymbol{\beta}) = \sum_i \boldsymbol{\beta}\mathbf{z}_i(t_i) - \log \sum_{j \in R_i} \exp\{\boldsymbol{\beta}\mathbf{z}_j(t_i)\}. \tag{5.6}$$

Andersen and Gill (1982) have shown that the usual large sample likelihood calculations based on differentiation of (5.6) yield asymptotically normal estimates whose variances and covariances may be estimated from the observed information. Prentice and Self (1983) derive analogous results for the model (5.2), in which the relative risk function is given by $r(\boldsymbol{\beta}\mathbf{z}) = 1 + \boldsymbol{\beta}\mathbf{z}$, and for other more general models. Satisfactory methods of estimation in the additive model (5.3) have evidently not yet been developed for continuous data.

5.4. Efficiency Gains from the Use of Standard Rates

The relative risks of exposure, as expressed in the regression coefficients $\boldsymbol{\beta}$ in the multiplicative model, may be estimated either from the parametric likelihood (5.5), which required knowledge of the external standard rates, or from the partial likelihood (5.6). Oakes (1977, 1981) has derived the asymptotic relative efficiency of the two estimators. His results may be obtained from the results already derived for grouped data by making a heuristic limiting argument. We first note that the log partial likelihood (5.6) may be viewed as a profile log likelihood $L(\boldsymbol{\beta}) = \sup_{\boldsymbol{\alpha}} L(\boldsymbol{\alpha}, \boldsymbol{\beta})$ for a log–linear Poisson model that attaches a separate α parameter to each risk set (Breslow, 1972). The information about $\boldsymbol{\beta}$ in the analogous grouped data model, for which there are J α's, is given by (4.5). Letting J and K approach infinity so that the age intervals used for stratification become infinitesimally small and the age-dependent covariables continuous valued, $\mathrm{Var}(\mathbf{Z} \mid S)$ is interpreted in the limit as the variance of the continuous exposure variables of a diseased case, conditional on the exact age at death or diagnosis as well as on any other demographic features used to define strata. The fact that $\mathrm{Var}(\mathbf{Z} \mid S)$ is also the information for the profile likelihood follows from results of Richards (1961), who shows that the Fisher information is the same for full likelihoods and profile likelihoods. Similarly, the information from the parametric likelihood (5.5) is found by taking the limit in (4.6) so that $\mathrm{Var}(\mathbf{Z})$ becomes the total variance of a continuous exposure variable for a randomly selected case.

These theoretical results suggest that substantial gains in efficiency can only be expected when the covariables $\mathbf{z}$ vary markedly from one risk set to another. This was confirmed empirically by an earlier analysis of the Montana smelter data using follow-up data through 1963 (Breslow et al., 1983). A components of variance analysis showed that the ratio of between to between plus within risk set variances was 15.9% for a binary indicator of foreign birth versus 4.5% and 1.7% for heavy and moderate arsenic exposure variables. Correspondingly, the ratio of standard errors for partial vs. parametric likelihoods was 0.92 for foreign-born, but 0.98 and 0.97 for the arsenic variables.

5.5. Comparison of Grouped and Continuous Data Analyses

The Montana smelter workers data with follow-up through 1977 were analyzed using multiplicative models for both grouped and continuous data as specified in equations (3.1) and (5.1), respectively. Data for the grouped analysis consisted of respiratory cancer deaths and person-years classified in six dimensions: (i) age in four ten-year intervals from 40–49 through 70–79; (ii) calendar year in four periods 1938–1949, 1950–1959, 1960–1969, 1970–1977; (iii) date of employment divided into before or after 1925; (iv) birthplace divided into U.S. versus foreign; (v) years moderate arsenic exposure in four categories <1, 1–4, 5–14, and 15+; and (vi) years heavy arsenic exposure in three categories <1, 1–4, and 5+. The last two time-varying exposure variables were lagged two years. Of the $4 \times 4 \times 2 \times 2 \times 4 \times 3 = 768$ possible cells, person-years of observation were only recorded for 458, and this was the number of data records entered into GLIM for analysis.

For the continuous data analysis, 155 risk sets were formed on the basis of distinct integral ages for deaths between 40 and 79 years that occurred during one of the nine calendar periods 1938–1939, 1940–1949, ..., 1973–1977. The sizes of the risk sets ranged from 42 to 882, and the numbers of deaths in each from one to six. A direct comparison with results for grouped data was made possible by defining binary (0–1) covariables that expressed the qualitative effects of factors (iii) through (vi). The regression coefficients and standard errors shown for these covariables in Table 6 are quite comparable, notwithstanding the fact that substantially more computer time was required to fit the same model using the partial likelihood approach than was required with the grouped data analysis.

Table 6 also shows results when the two arsenic exposure variables were analyzed in their original, continuous form rather than by grouping. A somewhat imperfect comparison of the continuous and grouped data analysis of continuous covariables was effected by assigning scores to the grouped exposure categories of 0.02050, 2.219, and 16.669 for heavy arsenic and 0.05775, 2.272, 8.746, and 29.740 for moderate arsenic, these being the average exposure times within each exposure category based on a sample of 20 controls from each risk set. There is also good agreement between the estimated coefficients in this case.

These results confirm the strong and independent effects on respiratory cancer mortality of employment data, birthplace, and arsenic exposure. There is some evidence that the arsenic effect is causal in that the greatest

Table 6. Regression Coefficients and Standard Errors from Multiplicative Models Fitted to Grouped and Continuous Data from the Montana Smelter Workers Study

	Method of analysis			
			Case-control	
Regression variables	Grouped	Continuous	20 controls	5 controls
	(a) All covariables binary (0–1)			
Employed before 1925	0.444 ± 0.151	0.405 ± 0.153	0.410 ± 0.163	0.349 ± 0.189
Foreign born	0.445 ± 0.153	0.484 ± 0.154	0.539 ± 0.168	0.452 ± 0.205
Heavy arsenic[a]				
1–4 years	0.193 ± 0.305	0.170 ± 0.312	0.411 ± 0.337	0.482 ± 0.405
5+ years	1.069 ± 0.230	1.088 ± 0.232	1.228 ± 0.262	1.303 ± 0.346
Moderate arsenic[a]				
1–4 years	0.600 ± 0.166	0.601 ± 0.166	0.639 ± 0.181	0.594 ± 0.216
5–14 years	0.259 ± 0.242	0.261 ± 0.243	0.211 ± 0.258	0.187 ± 0.292
15+ years	0.684 ± 0.206	0.674 ± 0.207	0.611 ± 0.227	0.585 ± 0.269
Deviance or twice log likelihood	282.1	−3167.0	−1432.82	−794.79
	(b) Continuous arsenic variables			
Employed before 1925	0.441 ± 0.151	0.403 ± 0.153	0.378 ± 0.163	0.379 ± 0.188
Foreign born	0.432 ± 0.153	0.473 ± 0.153	0.554 ± 0.167	0.430 ± 0.202
Years heavy arsenic[a]	0.0662 ± 0.0138	0.0664 ± 0.0139	0.0538 ± 0.0189	0.0489 ± 0.0275
Years moderate arsenic[a]	0.0222 ± 0.0067	0.0218 ± 0.0068	0.0159 ± 0.0075	0.0140 ± 0.0089
Deviance or twice log likelihood	292.1	−3177.0	−1448.78	−809.43

[a] Lagged two years.

relative increase in risk is associated with heavy exposure for five or more years. However, no dose–response trend is evident for the duration of moderate exposures, and these are not well represented by a simple linear term in the log–linear model.

5.6. Sampling from the Risk Sets

A major drawback to the continuous variable technique, whether based on parametric or partial likelihoods, is that the computing expense effectively rules out exploratory analyses of the data. Substantial reduction in computational costs can be achieved at the expense of some loss in efficiency by applying the partial likelihood analyses to reduced risk sets consisting of the cases in each one plus a sample of "controls" drawn at random and without replacement from the remainder (Thomas, 1977). With this procedure the time-dependent covariables need be computed only once for each subject, and the number of data records which need to be processed is typically reduced by one or more orders of magnitude. A variety of exploratory analyses may be performed, including the fitting of additive or multiplicative relative risk functions and the detection of risk sets that have a major influence on the estimated coefficients (Storer, Wacholder, and Breslow, 1983; Moolgavkar, Lustbader, and Venzon, 1984).

Breslow et al. (1983) present some examples using the Montana smelter data with follow-up through 1963. They find that twenty or more controls may be required to obtain estimates that are reasonably comparable to those based on the full data, at least when one is trying to estimate large relative risks associated with rare exposures. Fewer controls are needed when the exposures are more common. The last two columns of Table 6 update these analyses for the 1938–1977 follow-up period. For these data, at least, it seems clear that the grouped data analysis provides a better approximation to the full partial likelihood analysis than does the case-control technique.

6. Suggestions for Further Research

A few problems in cohort analysis on which additional work could usefully be done are suggested by the preceding. We have already remarked that the excess risk model for continuous data is in need of further development. While external standard rates would seem to offer little advantage in the multiplicative framework, certainly when one considers the risk of bias if the model fails, our efficiency calculations for grouped data suggest that they may make a more substantial contribution to the determination of excess risk.

The widely held notion that four or five controls per case suffice for case-control analyses needs to be reevaluated in view of both numerical and theoretical results which suggest that more controls are of value for estimating large relative risks associated with rare exposures. Further work is needed to determine the extent to which these results, obtained for single binary risk factors, extend to multiple and continuous variables. It seems likely that varying the case-control ratio depending on whether or not the case is exposed would increase the efficiency of matched analyses in the same way that selecting subjects on the basis of both disease and exposure history improves the design efficiency of unmatched studies (White, 1982). Details on how to perform the control selection and subsequent analyses need to be worked out.

Acknowledgements

This work was supported in part by USPHS grant 1-K07CA-00723 and an award from the Alexander von Humboldt Foundation. The technical assistance of Bryan Langholz and the helpful comments of the editors and refees are gratefully acknowledged. Portions of this paper were presented during the Forum Lectures at the 1984 European Meeting of Statisticians.

Bibliography

Aalen, O. (1978). "Nonparametric inference for a family of counting processes." *Ann. Statist.*, **6**, 701–726.

Andersen, P. K. and Gill, R. D. (1982). "Cox's regression model for counting processes: A large sample study." *Ann. Statist.*, **10**, 1100–1120.

Aranda-Ordaz. F. J. (1983). "An extension of the proportional hazards model for grouped data." *Biometrics*, 39, 109–117.

Armitage, P. (1955). "Tests for linear trend in proportions and frequencies." *Biometrics*, **11**, 375–386.

Armitage, P. (1971). *Statistical Methods in Medical Research*. Oxford: Blackwell.

Armitage, P. and Doll, R. (1961). "Stochastic models for carcinogenesis." In *Proceedings of the Fourth Berkeley Symposium on Mathematical Statistics and Probability*, **4**, 19–38.

Baker, R. J. and Nelder, J. A. (1978). *The* GLIM *System: Release 3*. Oxford: Numerical Algorithms Group.

Beaumont, J. J. and Breslow, N. E. (1981). "Power considerations in epidemiologic studies of vinyl chloride workers." *Amer. J. Epidemiology*, **114**, 725–734.

Beebe, G. W. (1981). "The atomic bomb survivors and the problem of low dose radiation effects." *Amer. J. Epidemiology*, **114**, 761–783.

Berry, G. (1980). "Dose–response in case-control studies." *J. Epidemiology and Community Health*, 34, 217–222.

Berry, G. (1983). "The analysis of mortality by the subject-years method." *Biometrics*, **39**, 173–184.

Berry, G., Gilson, J. C., Holmes, S., Lewinsohn, H. C., and Roach, S. A. (1979).

"Asbestosis: A study of dose–response relationship in an asbestos textile factory." *British J. Industrial Medicine*, **36**, 98–112.

Breslow, N. E. (1972). "Contribution to discussion of paper by D. R. Cox." *J. Roy. Statist. Soc. Ser. B*, **34**, 216–217.

Breslow, N. (1981). "Odds ratio estimators when the data are sparse." *Biometrika*, **68**, 73–84.

Breslow, N. (1982). "Design and analysis of case-control studies." *Annual Rev. Public Health*, **3**, 29–54.

Breslow, N. (1984). "Elementary methods of cohort analysis." *Internat. J. Epidemiology*, **13**, 112–115.

Breslow, N. E. and Day, N. E. (1980). *Statistical Methods in Cancer Research I: The Analysis of Case-Control Studies*. Lyon: International Agency for Research on Cancer.

Breslow, N. E., Lubin, J. H., Marek, P., and Langholz, B. (1983). "Multiplicative models and cohort analysis." *J. Amer. Statist. Assoc.*, **78**, 1–12.

Brown, C. C. and Chu, K. C. (1983). "Implications of the multistage theory of carcinogenesis applied to occupational arsenic exposure." *J. Natl. Cancer Inst.*, **70**, 455–463.

Clayton, D. G. (1982). "The analysis of prospective studies of disease aetiology." *Comm. Statist. A—Theory Methods*, **11**, No. 19, 2129–2155.

Cochran, W. G. (1954). "Some methods for strengthening the common χ^2 tests." *Biometrics*, **10**, 417–451.

Cox, D. R. (1972). "Regression models and life tables (with discussion)." *J. Roy. Statist. Soc. Ser. B*, **34**, 187–220.

Cox, D. R. (1975). "Partial likelihood." *Biometrika*, **62**, 269–276.

Cox, D. R. and Hinkley, D. V. (1974). *Theoretical Statistics*. London: Chapman and Hall.

Doll, R. and Hill, A. B. (1964). "Mortality in relation to smoking: Ten years' observations on British doctors." *British Medical J.*, **1**, 1399–1410, 1460–1467.

Doll, R. and Hill, A. B. (1966). "Mortality of British doctors in relation to smoking: Observations on coronary thrombosis." *Natl. Cancer Inst. Monogr.*, **19**, 205–268.

Doll, R. and Peto, R. (1976). "Mortality in relation to smoking: 20 years' observations on male British doctors." *British Medical J.*, **2**, 1525–1536.

Doll, R. and Peto, R. (1978). "Cigarette smoking and bronchial carcinoma: Dose and time relationships among regular smokers and lifelong non-smokers." *J. Epidemiology and Community Health*, **32**, 303–313.

Fienberg, S. E. (1980). *The Analysis of Cross-Classified Data*, 2nd edn. Cambridge, MA: MIT Press.

Fienberg, S. E. and Mason, W. M. (1985). "Specification and implementation of age, period and cohort models." In W. M. Mason and S. E. Fienberg (eds.), *Cohort Analysis in Social Research: Beyond the Identification Problem*. New York: Springer-Verlag, 45–88.

Fox, A. J. and Collier, P. F. (1976). "Low mortality rates in industrial cohort studies due to selection for work and survival in the industry." *British. J. Preventive and Social Medicine*, **30**, 225–230.

Frome, E. L. (1983). "The analysis of rates using Poisson regression models." *Biometrics*, **39**, 665–674.

Frost, W. H. (1939). "The age selection of mortality from tuberculosis in successive decades." *Amer. J. Hygiene A*, **30**, 91–96.

Gilbert, E. S. (1983). "An evaluation of several methods for assessing the effects of occupational exposure to radiation." *Biometrics*, **39**, 161–171.

Gill, R. (1980). *Censoring and Stochastic Integrals*. Amsterdam: Mathematical Centre Tracts.

Hammond, E. C. (1966). "Smoking in relation to the death rates of one million men and women." *Natl. Cancer Inst. Monogr.*, **19**, 127–204.

Hauck, W. W. (1979). "The large sample variance of the Mantel–Haenszel estimator of a common odds ratio." *Biometrics*, **35**, 817–820.

Holford, T. R. (1980). "The analysis of rates and of survivorship using log–linear models." *Biometrics*, 36, 299–306.

Johansen, S. (1981). "Discussion of paper by D. Oakes." *Internat. Statist. Rev.*, **49**, 258–262.

Kalbfleisch, J. D. and Prentice, R. L. (1980). *The Statistical Analysis of Failure Time Data*. New York: Wiley.

Kleinbaum, D. G., Kupper, L. L., and Morgenstern, H. (1982). *Epidemiologic Research*. Belmont: Wadsworth.

Knox, E. G. (1973). "Computer simulation of industrial hazards." *British J. Industrial Medicine*, **30**, 54–63.

Laird, N. and Olivier, D. (1981). "Covariance analysis of censored survival data using log–linear analysis techniques." *J. Amer. Statist. Assoc.*, **76**, 231–240.

Lee, A. M. and Fraumeni. J. F. (1969). "Arsenic and respiratory cancer in man: An occupational study." *J. Natl. Cancer Inst.*, **42**, 1045–1052.

Lee-Feldstein, A. (1983). "Arsenic and respiratory cancer in humans: Follow-up of copper smelter employees in Montana." *J. Natl. Cancer Inst.*, **70**, 601–610.

Lehman, E. L. (1959). *Testing Statistical Hypotheses*. New York: Wiley.

Lundin, F. E., Archer, V. E., and Wagoner, J. K. (1979). "An exposure-time–response model for lung cancer mortality in uranium miners: Effects of radiation exposure, age, and cigarette smoking." In N. E. Breslow and A. Whittemore (eds.), *Energy and Health*. Philadelphia: SIAM, 243–264.

MacMahon, B. and Pugh, T. F. (1970). *Epidemiology: Principles and Methods*. Boston: Little, Brown.

Mantel, N. and Haenszel, W. (1959). "Statistical aspects of the analysis of data from retrospective studies of disease." *J. Natl. Cancer Inst.*, **22**, 719–748.

Moolgavkar, S. H., Lustbader, E. D., and Venzon, D. J. (1984). "A geometric approach to nonlinear regression diagnostics with application to matched case-control studies." *Ann. Statist.*, **12**, 816–826.

Nelder, J. A. and Wedderburn, R. W. M. (1972). "Generalized linear models." *J. Roy. Statist. Soc. A*, **135**, 370–384.

Oakes, D. (1977). "The asymptotic information in censored survival data." *Biometrika*, **64**, 441–448.

Oakes, D. (1981). "Survival times: Aspects of partial likelihood." *Internat. Statist. Rev.*, 49, 235–264.

Pregibon, D. (1981). "Logistic regression diagnostics." *Ann. Statist.*, **9**, 705–724.

Prentice, R. L. and Self, S. (1983). "Asymptotic distribution theory for Cox-type regression models with general relative risk form." *Ann. Statist.*, **11**, 804–813.

Rao, C. R. (1965). *Linear Statistical Inference and its Applications. New York: Wiley.*

Richards, F. S. G. (1961). "A method of maximum likelihood estimation." *J. Roy. Statist. Soc. Ser. B*, **23**, 469–473.

Rothman, K. J. (1974). "Synergy and antagonism in cause-effect relationships." *Amer. J. Epidemiology*, **99**, 385–388.

Rothman, K. J. and Boice, J. D. (1979). *Epidemiologic Analysis with a Programmable Calculator*. NIH Publication 79-1649. Washington: U.S. Government Printing Office.

Roueché, B. (1967). *Annals of Epidemiology*. Boston: Little, Brown.

Schlesselman, J. J. (1982). *Case-Control Studies: Design, Conduct and Analysis*. New York: Oxford U.P.

Selikoff, I. J., Hammond, E. C., and Seidman, H. (1980). "Latency of asbestos disease among insulation workers." *Cancer*, **46**, 2736–2740.

Storer, B. E., Wacholder, S., and Breslow, N. E. (1983). "Maximum likelihood fitting of general risk models to stratified data." *Appl. Statist.*, **32**, 177–181.

Tarone, R. E. (1982). "The use of historical control information in testing for a trend in Poisson means." Biometrics, **38**, 457–462.

Thomas, D. C. (1977). "Addendum to a paper by Liddel, F. D. K., McDonald, J. C. and Thomas, D. C." *J. Roy. Statist. Soc. Ser. A*, 483–485.

Thomas, D. C. (1981). "General relative risk models for survival time and matched case-control analysis." *Biometrics*, 37, 673–686.

Thomas, D. C. (1982). "Temporal effects and interactions in cancer: Implications of carcinogenic models." In R. L. Prentice and A. S. Whittemore (eds.), *Environmental Health: Risk Assessment*, Philadelphia: SIAM, 107–121.

Thompson, R. and Baker, R. (1981). "Composite link functions in generalized linear models." *Appl. Statist.*, **30**, 125–131.

Wall, W. D. and Williams, H. L. (1970). *Longitudinal Studies and the Social Sciences*. London: Heinemann.

White, E. (1982). "A two stage design for the study of the relationship between a rare exposure and a rare disease." *Amer. J. Epidemiology*, **115**, 119–128.

Whitehead, J. (1980). "Fitting Cox's regression model to survival data using GLIM." *Appl. Statist.*, **29**, 268–275.

Whittemore, A. and Altshuler, B. (1976). "Lung cancer incidence in cigarette smokers: Further analysis of Doll and Hill's data on British physicians." *Biometrics*, 32, 805–816.

Yule, G. U. (1934). "On some points relating to vital statistics, more especially statistics of occupational mortality." *J. Roy. Statist. Soc.*, **94**, 1–84.

CHAPTER 7

Psychiatric Statistics

Morris H. DeGroot and Juan E. Mezzich

Abstract

We present a survey of the areas of psychiatric and mental-health research to which statistical theory and methodology can effectively be applied. A brief history of psychiatric statistics is presented, along with a discussion of how this branch of statistics tends to differ from other branches. We describe some important topics of current research and discuss the latest edition of the *Diagnostic and Statistical Manual of Mental Disorders* (DSM-III), the main features of which are its multiaxial approach and its use of explicit diagnostic criteria. The results of a review of the use of statistical methods in the psychiatric literature is included.

1. Introduction

Psychiatry is the branch of medicine that deals with the description, classification, diagnosis, treatment, course, and etiology of mental, emotional, and behavioral disorders in humans. By the term *psychiatric statistics* we mean the field of knowledge that pertains to the development of statistical theory and methodology appropriate for applications in psychiatry.

The field of psychiatric statistics is relevant to both psychiatric practice and psychiatric research. Furthermore, the collection of statistical data in psychiatric research can take place in a laboratory setting, a clinical setting, or an epidemiological setting. Each of these three modes of research will now be described in just a few words.

In the laboratory, psychiatric research is usually carried out by studying the biological and behavioral processes of animals under various conditions, including possibly the administration of experimental drugs. Laboratory work of this type is usually referred to as basic research.

In a clinical setting, research may be carried out on just a single patient or on a group of patients, and might deal with any or all of the aspects of psychiatry described in the first paragraph of this paper: description, classification,

Key words and phrases: classification, cluster analysis, diagnostic agreement, diagnostic criteria, DSM-III, mental health statistics, psychiatric epidemiology.

diagnosis, treatment, course, or etiology. Alternatively, a clinical trial involving a large number of subjects might be carried out in order to evaluate and compare the effects of different treatments or environments.

Finally, the epidemiological setting provides another rich area for the application of psychiatric statistics. Psychiatric epidemiology is the branch of psychiatry and epidemiology that deals with the distribution and evolution of psychiatric disorders in specified populations. In a sense, the fundamental quest of all psychiatric research is an understanding of the relationship between psychiatric disorders and the physical, biological, and social variables that determine their ontological and ecological contexts.

It is reasonable to ask whether or how psychiatric statistics differs from any other kind of statistics. In particular, why is psychiatric statistics not just a branch of biostatistics or psychometrics, the areas of statistics that deal with biomedical and behavioral issues, respectively? A glib but reasonably accurate answer to this question is that biostatistical methods are typically inadequate for handling the behavioral components of problems in psychiatric statistics, and the psychometric methods are typically inadequate for handling the biomedical components. Naturally, the field of psychiatric statistics has its roots, very deep roots, in these and other areas of statistics. However, it has become, and continues to grow as, a separate branch with an increasing need for a methodology of its own.

It is not surprising that this should be so. Traditionally, each substantive area of application of statistics has spawned new statistical methodology that is subsequently used, for better or worse, in a new substantive area until it is eventually realized that this area would be better served by its own specialized methodology, and so forth. For example, the classical statistical theory of the design of experiments and the methodology of the analysis of variance were developed, largely by R. A. Fisher, in response to the needs of agricultural research. Subsequently, these methods were used routinely in industrial experimentation and clinical medical trials. In recent years, we have seen the replacement of these methods in industrial experimentation by sequential, adaptive, and evolutionary experimentation; and in the design and analysis of clinical trials, at least in principle if not in practice, by methods of optimal stopping and survival analysis.

As we shall see later in this paper, research reported in the psychiatric literature uses very little statistical methodology beyond a few basic classical techniques. We believe that the time is ripe to bring to this important, exciting, and rapidly growing field some of the modern techniques that have been developed in other branches of statistics, suitably modified where required; and to create new statistical theory and methodology to keep pace with and spur new developments in psychiatric research.

In Section 2, we present an overview of the history of psychiatric statistics. In Section 3, some of the important areas of current research, both methodological and substantive, are mentioned. In Section 4, a description of the third edition of the *Diagnostic and Statistical Manual of Mental Disorders*

(DSM-III) is presented. In Section 5, the results of a review of the use of statistics in the psychiatric literature are described. In Section 6, we offer some concluding remarks.

2. History of Psychiatric Statistics

The following review of the historical roots of psychiatric statistics focuses on attempts to quantify central psychiatric concerns. Given that psychiatry is usually defined as the science and art of diagnosing and treating mental disorders, and given that the stage of formalization of most psychiatric treatments is still rather incipient, the content of this review mainly deals with the antecedents of various aspects of the diagnostic process, i.e. psychopathological assessment and classification. Of course there are a number of biological, psychological, and sociological disciplines connected with and of great importance for psychiatry, all of which, along with their quantitative aspects, have their own historical roots. Delineations of such roots can be found in textbooks and other sources from each discipline. On the basis of these considerations, it should be emphasized that this review does not intend to be comprehensive, and furthermore, that it reflects the authors' perspectives and interests.

Early antecedents to psychiatric statistics can be dated back to the first documented concerns about the classification of psychopathology. These include Egyptian descriptions by the physician Imhotep and date back to 3000 B.C. Ayurvedic literature from the last centuries of the pre-Christian era include the *Caraka-Samhita* and the *Susruta-Samhita*, which contain classifications of medical illnesses including mental disorders conceptualized as demonic possessions. Hippocrates (460–377 B.C.) considered mental disorders, including epilepsy, as derived from natural rather than supernatural factors and classified them according to such variables as chronicity and presence or absence of fever. Temkin (1965) emphasized that the origins of the classification of illnesses go back into a remote past, and that the beginning of the medical sciences lies somewhere along the road, not at its start.

The eighteenth century saw the inauguration of a new era in the history of classification through the work of the Swedish biologist Carolus Linnaeus and his comprehensive taxonomy of plants and animals, for which he used a binomial system. He gave tremendous impulse to the idea of a systematic examination of nature in search of stable order and patterns. He said in his *Genera Plantarum* (1737): "All the real knowledge which we possess depends on methods by which we distinguish the similar from the dissimilar.... We ought therefore by attentive and diligent observation determine the limits of the genera, since they cannot be determined *a priori*. This is the great work, the important labor, for should the genera be confused, all would be confusion." His example reinforced the teachings of his contemporary, Sydenham, on stable disease entities with fixed manifestations, and encouraged the development by Boissier de Sauvages and Cullen of monumental nosologies composed of thousands of species of diseases, organized into classes, orders, and genera.

In psychiatry, around the onset of the nineteenth century, Pinel and Esquirol inaugurated the formal study of mental statistics and emphasized clinical observation, reacting against the specific-disease-entity tradition of Sydenham. Philippe Pinel, in fact, is a towering figure in the history of psychiatry and the medical sciences, not only for the therapeutic and attitudinal era he started by freeing the mentally ill from their chains at the Bicetre and Salpetriere hospitals in Paris, but also for accumulating systematic and detailed data on each patient seen and for his attempts to analyze and categorize psychiatric symptoms. In his monumental *Traité Medico-Philosophique sur l'Aliénation Mentale*, first published in 1801, he stated that his task as a natural scientist was to observe facts and to bring some order to the chaos of the existing treatment methods by means of critical and objective (statistical) investigations.

Of no less historical importance was his disciple, Jean-Etienne Esquirol, whose clinical observations and descriptions are judged by many as being even clearer, more detailed, and more accurate than Pinel's. Esquirol was also the first psychiatrist to use statistical evidence to evaluate clinical and epidemiological questions such as the relationship between civilization and mental illness.

In 1853, a milestone in psychiatric and, in general, medical statistics took place when the International Statistical Congress requested William Farr of England and Marc d'Espine of Italy to prepare a uniform nomenclature of the causes of death applicable to all countries. After several revisions, a final version was prepared in 1891 by a committee chaired by Jacques Bertillon, chief of statistical activities of the city of Paris, and adopted by the International Statistical Institute in 1893. This International Classification of Causes of Death and its successor, The International Statistical Classification of Diseases, Injuries, and Causes of Death, have since then been revised regularly at about 10-year intervals.

In 1897, with the publication of *Le Suicide*, the sociologist Emile Durkheim made a pioneering contribution toward the use of statistical methods and the social science approach to the study of a major human and psychiatric problem. Durkheim's work on suicide remains the prototype of a systematic and persistent treatment of a subject, using as creatively as possible the data and procedures available at a given time.

In the last quarter of the nineteenth century the modern discipline of statistics was brought into being by anthropometrists, biologists, and psychologists. Such initial contributors to modern statistics as Francis Galton and Karl Pearson stem from that period. The works of both Galton and Pearson anticipated the development of factor analysis, but it was later that two psychologists brought this to fruition by their conceptualizing and measuring of human intelligence. First, Spearman (1927) formulated a two-factor theory at the turn of the twentieth century, and then Thurstone (1947) developed multiple factor analysis by building on Spearman's model in a more formal way.

In 1935, Stephenson proposed the use of *Q*-correlation coefficients computed between individuals across variables, which were useful for developing typologies through Q-factor analysis. Zubin (1938) described an interesting clustering procedure for subdividing a group into subgroups of like-structured individuals and for determining the factors that make them like-structured.

Illustrative of pioneering efforts in the quantitative study of the biology and epidemiology of neuropsychiatric disorders is the work of L. S. Penrose on mental retardation. In 1938 he completed the now classic Colchester Survey, and in 1949 he published the first edition of *The Biology of Mental Defect*. In the latter publication he quantitatively addressed the assessment and incidence of mental retardation, as well as the principles for its classification and its genetic and environmental study.

During the past two decades there has been a tremendous expansion in the development of methods related to pattern recognition and clustering (Sneath and Sokal, 1973; Hartigan, 1975). In fact, a fully fledged field of quantitative taxonomy has developed, including complete clustering techniques (both hierarchical and partition types) and intermediate methods involving graphical representation of multivariate data units. Many of these techniques have been applied to the elucidation of patient groups in psychiatry, in some cases within the framework of an evaluation of clustering methods (e.g. Bartko, Strauss, and Carpenter, 1971; Mezzich, 1978).

A major thrust in psychiatric statistics during the past two decades was given by pioneering international studies in the description of psychiatric patients. These included the U.S.–U.K. Diagnostic Project (Cooper, Kendell, Gurland, Sharpe, Copeland, and Simon, 1972), which compared the diagnostic habits of psychiatrists in these two countries, and the International Pilot Study of Schizophrenia (World Health Organization, 1973), which was addressed to the standardized identification of schizophrenia in nine countries.

These international studies have prompted the development of major methodological advances in psychiatric diagnosis, covering various phases of the diagnostic process. One of these developments, generically known as the structured clinical interview, attempts to decrease undesirable variability in the process of obtaining clinical information. One of the earliest and most widely known structured clinical interviews is the Present State Examination (Wing, Cooper, and Sartorius, 1974), which was developed in England and has been used in both of the international studies mentioned above. Several other examples of systematic evaluation procedures have been developed more recently, some of them being fully structured (e.g. the Diagnostic

Interview Schedule by Robins, Helzer, Croughan, and Ratcliff, 1981) and others semistructured (e.g. the Initial Evaluation Form by Mezzich, Dow, Rich, Costello, and Himmelhoch, 1981).

Another development aimed at reducing unreliability in the diagnostic process has been the proposal of explicit or specific diagnostic criteria. This procedure involves setting clear objective rules for assigning diagnostic categories to individuals under examination. One of the earliest and best-known sets of explicit psychiatric diagnostic criteria is that developed by Feighner, Robins, Guze, Woodruff, Winokur, and Munoz (1972). More recent examples are the so-called Research Diagnostic Criteria (Spitzer, Endicott, and Robins, 1978) and the diagnostic criteria in DSM-III (American Psychiatric Association, 1980).

Parallel to the methodological developments described above, the traditional single-label diagnostic model in psychiatry has been challenged on scientific grounds (e.g., Strauss, 1973), and, in contraposition, a multiaxial approach has been proposed. This model consists of the systematic formulation of the patient's condition, and its etiological and associated factors, in terms of several variables, aspects, or axes (such as psychopathological syndromes, concomitant physical disorders, and social functioning) which are thought to have high clinical information value and are conceptualized and rated as quasi-independent of each other. Diagnostic axes may be categorical or dimensional. In fact, most multiaxial systems reported in the literature have a mixed categorical and dimensional structure (Mezzich, 1979).

Although the impact of multiaxial diagnosis is relatively recent, its origins are not. The contrast between the multiaxial and the conventional uniaxial diagnostic system may be traced to the old nosological controversy reviewed by Kendell (1975a, p. 60) between, on the one hand, the idealized and abstract Platonic disease entity and, on the other hand, the closer-to-the-patient and therefore more "clinical" Hippocratic approach.

This contrast was revived early in this century by the argument between Kraepelin, who thoroughly endorsed the disease-entity model, and Hoche, who proposed the separation of syndrome and etiology in the diagnostic formulation. Kraepelin's position prevailed, and nothing much was heard about this issue until 1947, when Essen-Moller and Wohlfahrt suggested the amendment of the official Swedish classification of mental disorders to separate syndrome and etiology.

Increased interest in multiaxial models for psychiatric diagnosis was then prompted by various symposia on the classification of mental disorders sponsored by the World Health Organization and by the American Psychopathological Association. These symposia reviewed a number of fundamental methodological issues in diagnosis and also some germinal multiaxial ideas (e.g. Rutter, Lebovici, Eisenberg, Sneznevskij, Sadoun, Brooke, and Lin, 1969; Stengel, 1959; Zubin, 1961). More recently, in various ways building upon the diagnostic separation of syndrome and etiology put forward by Essen-Moller and Wohlfahrt (1947) and Essen-Moller (1961, 1971), a number of multiaxial models have been proposed in several parts of the world, including England (Rutter, Shaffer, and Shepherd, 1975; Rutter, Shaffer, and Sturge, 1975; Wing, 1970), Germany (Helmchen, 1975; von Cranach, 1977), Japan (Kato, 1977), Sweden (Ottosson and Perris, 1973), and the United States (American Psychiatric Association, 1980; Strauss, 1975).

3. Current Problems and Areas of Research

In this section, we shall list some of the current problems and areas of research in psychiatry in which the application of new statistical methods promises to help explicate important issues.

The development of improved diagnostic systems is of fundamental importance. This development will require both the refinement of current diagnostic categories and structural innovations. Better definitions of a case and the development of procedures for its efficient identification are relevant to the work. Obviously, methods such as cluster analysis, factor analysis, and discriminant analysis have been and will be useful for these purposes. Some pertinent references here are the works of Bartko, Strauss, and Carpenter (1971) and Mezzich and Solomon (1980) on comparative evaluations of clustering methods for psychiatric diagnosis, and of Fowler, Mezzich, Liskow, and Valkenburg (1980) on discriminant analysis of schizophrenic and affective diagnostic groups. As in many other modern statistical applications involving high-dimensional data, graphical methods promise to be useful (e.g., Mezzich and Worthington, 1978). Moreover, since the development of good diagnostic systems must be based on large numbers of subjects and large amounts of data, effective methods for the management of large-scale data files are important.

Another highly relevant methodology, which has been little used in the study of diagnostic systems, is multidimensional categorical data or contingency-table analysis. Of particular interest, however, is the controversy regarding the use of this methodology in the investigation of the social origins of depression (e.g., Brown and Harris, 1978a, b; Tennant and Bebbington, 1978; and Bebbington, 1980).

One area of high interest and controversy, in which statistical methods of grouping and classification such as cluster analysis and discriminant analysis have been extensively used, is the taxonomy of depression. A relatively large number of cluster-analytic studies of depressed patients have tended to establish an identifiable group with endogenomorphic or psychotic features, as well as one or more groups with various other forms of depression (Kendell, 1975b; Mezzich and Raab, 1983). A study by Paykel (1972) has investigated specific relationships between cluster-analytic typology and treatments of depression. An understanding of the controversy surrounding this area can be obtained from Kiloh and Garside (1963), Pilowsky, Levine, and Boulton

(1969), Kendell (1969), Kendell and Gourlay (1970), Everitt, Gourlay, and Kendell (1971), Garside and Roth (1978), and Everitt (1972, 1981).

The determination of rates of psychiatric disorders in both treated and untreated populations is an important problem of long standing. Determining the reliability and validity of psychiatric diagnosis in untreated populations involves measurement and sampling problems unique to the field of psychiatric statistics. A well-known index for assessing reliability on nominal scales is the kappa coefficient developed by Cohen (1960), which is corrected for chance agreement. More complex situations of clinical assessment have led to the development of extensions of kappa, such as those described by Kraemer (1980) and Mezzich, Kraemer, Worthington, and Coffman (1981) for problems of multiple raters formulating multiple diagnoses. For measuring reliability on dimensional scales, Bartko (1976) and Shrout and Fleiss (1979) have described correlational indices.

Longitudinal studies of the course of a psychiatric disorder are vital. In particular, these studies must focus on the relationship of the course of the disorder to the human development of the subject, environmental factors such as stressful events and social support systems, and treatment interventions. It seems clear that methods of time-series analysis, survival analysis, and stochastic control theory should give structure to these studies.

In general, we find that there has been very little stochastic modeling carried out in psychiatric research. The little that has been done is mainly in the area of modeling the diagnosis of the clinician, rather than modeling the course of a disorder or the behavior of a patient. There are two other areas of contemporary interest in psychiatric research where stochastic modeling combined with other modes of statistical analysis should be helpful. One is the elucidation of biological markers of psychiatric disorders and their interaction with psychosocial variables. The other is the study of issues peculiar to psychiatry that derive from certain life-cycle stages such as childhood and old age.

We conclude this section with brief descriptions of two further topics that illustrate different interesting and difficult aspects of psychiatric statistics. The first topic pertains to experiments with considerable data, but few subjects. The second pertains to the evaluation of the effectiveness of psychotherapies.

A characteristic of many psychiatric research studies is that they are (often necessarily) carried out with just a few subjects or even a single subject. Despite the small number of subjects, these studies typically generate a large amount of data, since each subject is usually observed intensely over a certain period of time and a large number of variables are measured.

For example, a patient suffering from depression might be observed while he sleeps during three or four nights. If EEGs are collected during these nights from four different sites on the brain, together with a detailed report of pulse rate, temperature, and various body movements, then there are enough data available to keep psychiatric statisticians fully occupied for years (Kupfer, 1981; Kupfer and Reynolds, 1983; Kupfer, Shaw, Ulrich, Coble, and Spiker, 1982; Kupfer, Spiker, Rossi, Coble, Ulrich, and Shaw, 1982).

As another example, a five-minute videotape of a psychiatric patient answering certain questions or giving his informed consent for some particular treatment contains a virtually infinite number of relevant variables that might be measured. In such studies (e.g. Roth, Lidz, Meisel, Soloff, Kaufman, Spiker, and Foster; 1982), by a careful choice of variables and appropriate statistical models, it may sometimes be possible to make good predictions about the processes involved in these subjects. But generalizations to larger groups are difficult.

The design and implementation of large-scale clinical trials to evaluate the effectiveness of various psychotherapies, by themselves and in combination with other treatment modalities, has become an important topic in research in the past few years. Some urgency has been given to this topic in the United States because both governmental medical insurance programs, such as Medicare and Medicaid, and private programs have been trying to decide which types of psychotherapy they should pay for and to what extent. Some of the key issues to be considered in the design of these trials are the choice of the particular psychotherapies to be studied (there are a sizable number of carefully developed and more or less widely used formal psychotherapy models, each having some variants), the identification of appropriate control groups (it is difficult to develop placebos for psychotherapy), and the specification of precise outcomes for each subject. Some references on these topics are Yates and Newman (1980a, b); Parloff (1980, 1982); Bergin and Lambert (1978); Smith, Glass, and Miller (1980); Imber, Pilkonis, and Glanz (1983); and Pilkonis, Imber, Lewis, and Rubinsky (1984).

The problem of determining an appropriate control group is of major importance in psychiatric research. For example, in a comparative study involving patients with some particular psychiatric disorder, the subjects in the "normal" control group cannot usually be selected as a random sample from a specified population. Typically, these subjects are volunteers who are determined to be free of the disorder being studied and who meet some other general conditions of being in good health. In sleep studies of the type described in this section, special care must be taken to allow all the subjects in the studies to become adjusted to sleeping in a new environment, to having their EEGs, temperatures, and body movements measured while they are sleeping, and in some cases also to having blood samples drawn while they are sleeping for the measurement of hormone levels. Only if this adjustment has been made can one infer that phenomena observed during the studies would also occur during normal sleep. As another example, in a longitudinal epidemiological study in which the persons in the "treatment" or "study" group have been subject to some stressful event, repeated interviews in period after period may possibly affect the responses of the control group differently from those of the treatment group. In general, the differential effects of an intervention on the treatment and control groups require careful consideration.

4. DSM-III

The first edition of the *Diagnostic and Statistical Manual of Mental Disorders* (DSM-I) appeared in 1952, and was prepared by the United States Public Health Service and the American Psychiatric Association, in response to widespread dissatisfaction with the sixth edition of the *International Classification of Diseases* (ICD-6). Its pervasive use of the term "reaction" to refer to the various diagnostic categories expressed the impact of Adolf Meyer's environmental orientation, and its frequent connotative reference to defense mechanisms reflected the predominance of psychoanalytic ideas at that time.

The second edition of the *Diagnostic and Statistical Manual of Mental Disorders* (DSM-II) was published in 1968 by the American Psychiatric Association. The preparation of DSM-II was guided by an interest in observing its compatibility with the ICD-8 catalogue of mental disorders, while defining each disorder for American use. DSM-II attempted to be more descriptive or objective in its designation of mental disorders, as reflected in the elimination of the term "reaction" from its diagnostic labels.

In 1980 the third edition of the *Diagnostic and Statistical Manual of Mental Disorder* (DSM-III), the current U.S. psychiatric diagnostic system, was published by the American Psychiatric Association. The mandate given to the development committee was to prepare a diagnostic system which was more consonant with current clinical research and experience than with the contemporary version of the International Classification of Diseases (ICD-9). Since its publication, it has attracted considerable interest, and along with a recognition of its limitations, it is being widely judged in this country and abroad as a significant step forward towards a more accurate and thorough characterization of psychiatric patients. Its most important features are the use of explicit or operational criteria for the definition of diagnostic categories and the use of a multiaxial framework. These two features, which represent some of the most important methodological developments in diagnosis in recent years, were discussed earlier in this paper.

The multiaxial system in DSM-III includes three typological axes (I, clinical psychiatric syndromes; II, personality and specific development disorders; and III, physical disorders) and two dimensional ones (IV, severity of psychosocial stressors; and V, highest level of adaptive functioning in the past year). A patient is diagnosed in a typological axis through the use of one or more qualitatively distinct categories, while in a dimensional axis he is described by the indication of his standing on an interval or rank scale.

4.1. Axis I: Clinical Psychiatric Syndromes

Axis I of DSM-III comprises the catalogue of mental disorders (other than personality and specific developmental disorders) and related conditions. Most of these disorders have explicit diagnostic criteria, one of the accomplishments of DSM-III.

4.2 Axis II: Personality and Specific Developmental Disorders

This axis tends to represent stable behavioral handicaps. However, it seems that to do justice to this conceptualization it should also include mental retardation, a shift that is being considered for a planned revision of DSM-III.

4.3 Axis III: Physical Disorders

Axis III includes any current physical disorders or conditions relevant to the understanding or management of the individual. Such conditions are catalogued in the nonmental-disorder section of the International Classification of Diseases (ICD-9-CM) (United States National Center for Health Statistics, 1978).

4.4 Axis IV: Psychosocial Stressors

Axis IV assesses psychosocial stressors judged to have been significant contributors to the development or exacerbation of the current disorder. First, specific psychosocial stressors (e.g. death of sister) are to be identified and listed in order of importance. Then, the overall stressor severity is rated, using a seven-point scale, from "1 = None" to "7 = Catastrophic."

4.5. Axis V: Highest Level of Adaptive Functioning in the Past Year

Axis V assesses an individual's highest level of adaptive functioning (for at least a few months) during the past year. Adaptive functioning is conceptualized mainly as a composite of social relations (breadth and quality of interpersonal relations with family, friends, and other people) and occupational functioning (consistency and quality of performance as worker, student, or homemaker). Use of leisure time is considered supplementally. Axis V is rated according to a seven-point scale, from "1 = Superior" to "7 = Grossly Impaired."

A format to facilitate the formulation of DSM-III diagnoses (Figure 1) and a description of its use with a large psychiatric population was recently published (Mezzich, Coffman, and Goodpastor, 1982).

Although DSM-III was developed as a national diagnostic system, its impact has been truly international. Two of its basic characteristics, the use of explicit diagnostic criteria and a multiaxial framework, are likely to be incorporated into the next edition of the International Classification of Diseases (ICD-10), to be issued late in this decade.

The definitional clarity afforded by the established diagnostic criteria has greatly increased the opportunity for data-based research. This is not only

DIAGNOSTIC SUMMARY — Codes

I. Clinical psychiatric syndromes

Main Formulation:
1. ______
2. ______
3. ______
4. ______

Alternatives to be ruled out:

II. Personality and specific developmental disorders

Main Formulation:
1. ______
2. ______
3. ______

Alternatives to be ruled out:

III. Physical disorders

Main Formulation:
1. ______
2. ______
3. ______

Alternatives to be ruled out:

IV. Psychosocial stressors

A. Ranked list:
1. ______
2. ______
3. ______
4. ______

B. Overall stressor severity:

1	2	3	4	5	6	7	0
None	Minimal	Mild	Moderate	Severe	Extreme	Catastrophic	Unspecified

V. Highest level of adaptive functioning during the past year

1	2	3	4	5	6	7	0
Superior	Very Good	Good	Fair	Poor	Very Poor	Grossly Impaired	Unspecified

Figure 1. Format for DSM-III diagnostic formulation. From Mezzich, Coffman, and Goodpastor (1982).

enhancing the investigation of the etiology and treatment of the so-defined mental disorders, but provides an empirical base for the refinement of the diagnostic criteria themselves.

The multiaxial system, with the structural flexibility that its typological and dimensional axes offer, poses a challenge to develop innovative conceptualizations and scaling arrangements in order to make the diagnostic formulation increasingly useful. New statistical methodology will also be needed to assess the reliability and the various aspects of the validity of complex diagnostic systems such as DSM-III.

5. The Use of Statistics in the Psychiatric Literature

In order to assess how many and which statistical methods are used in the current psychiatric research literature, all papers published during 1980 in three leading psychiatric journals were reviewed. These journals were the *American Journal of Psychiatry* (AJP), the *British Journal of Psychiatry* (BJP), and the *Archives of General Psychiatry* (AGP). The review covered 339 papers in the AJP, 148 in the BJP, and 110 in the AGP.

Of course, any review such as ours that is restricted to just three journals may suffer from a selection bias. Papers describing innovative statistical methodology or innovative applications of existing statistical methodology to psychiatric data can and do appear in a variety of pertinent journals. For example, the paper by Woolson, Tsuang, and Fleming (1980) describing the use of the Cox proportional-hazards model for the analysis of survival data in a psychiatric follow-up study appeared in the *Journal of Chronic Disease*. Other examples from psychology, psychometrics, and statistics journals could be cited. Nevertheless, because of the key role played by the three journals that we have selected in reflecting and influencing psychiatric research, we believe that our review is both informative and useful.

The results are presented in Table 1. Papers are classified into 16 different categories of statistical usage. A paper using more than one statistical methodology may be classified in more than one category. The percentage presented in any cell of the table is the percentage of all the different papers included in that column which were classified in the given category.

It can be seen from Table 1 that those papers that used statistical methods relied mainly on standard methods: χ^2 tests; t-tests for one and two samples; the Fisher exact test for 2×2 tables; product–moment and rank correlation coefficients; analysis of variance and covariance, and F-tests; nonparametric methods such as the sign test, the Wilcoxon signed ranks test, the Wilcoxon–Mann–Whitney test, and the Kruskal-Wallis ranks test; the kappa coefficient and other measures of inter-rater agreement and association; regression analysis; factor analysis; and discriminant analysis.

Some of the methods used, although standard, were far from elementary.

Table 1. Distribution of Papers Published in the *American Journal of Psychiatry* (AJP), *British Journal of Psychiatry* (BJP), and *Archives of General Psychiatry* (AGP) during 1980, by Categories of Statistical Usage

	AJP (1980)	BJP (1980)	AGP (1980)
Categories of statistical usage	339 papers	148 papers	110 papers
1. Expository, literature review, etc.	18 (5.3%)	6 (4.1%)	4 (3.6%)
2. No statistical data: Case reports, etc.	115 (33.9%)	12 (8.1%)	2 (1.8%)
3. Descriptive statistics only: tables, graphs, means, variances	65 (19.2%)	14 (9.5%)	11 (10.0%)
4. χ^2 and t-tests, Fisher exact test: 1 or 2 samples, contingency tables	95 (28.0%)	75 (50.7%)	66 (60.0%)
5. Product–moment correlations, rank correlations	42 (12.4%)	22 (14.9%)	30 (27.3%)
6. Analysis of variance, F-tests: 1-, 2-, and higher-way	25 (7.4%)	22 (14.9%)	32 (29.1%)
7. Nonparametric rank methods (other than rank correlations)	9 (2.7%)	17 (11.5%)	10 (9.1%)
8. Measures of association and agreement (other than correlation)	10 (2.9%)	13 (8.8%)	9 (8.2%)
9. Regression analysis: simple, multiple, polynomial, stepwise	6 (1.8%)	9 (6.1%)	10 (9.1%)
10. Discriminant and factor analyses	4 (1.2%)	6 (4.1%)	7 (6.4%)
11. Estimation: maximum likelihood, interval estimation, etc.	0 (0.0%)	3 (2.0%)	2 (1.8%)
12. Cluster analysis, classification	1 (0.3%)	0 (0.0%)	1 (0.9%)
13. Life tables, life testing, survival analysis	1 (0.3%)	0 (0.0%)	2 (1.8%)
14. Time-series analysis, spectral analysis	2 (0.6%)	0 (0.0%)	1 (0.9%)
15. Classical experimental design: Latin squares, hierarchical models	0 (0.0%)	3 (2.0%)	1 (0.9%)
16. Bayesian methods	0 (0.0%)	0 (0.0%)	1 (0.9%)

Linear models were presented that supported quite complicated covariance analyses or hierarchical structures. Furthermore, a paper occasionally appeared with a log–probit plot, an analysis using a log–linear model, a cross-validation study, a jackknife procedure, or a cluster analysis. However, there were many more missed opportunities than there were successes. For example, there were more than 30 papers reporting longitudinal studies, but only a few that used time series or survival analysis. The others fell back on the usual t-tests or covariance analyses, if they carried out any statistical analysis at all.

White (1979) reviewed the articles appearing in the *British Journal of Psychiatry* during the 12 months from July 1977 to June 1978. She found that of the 168 papers published during that period, 83 percent presented numerical results. Similarly, in our survey, we found that of the 148 papers published in that same journal during 1980, 88 percent presented numerical results. White also found that, "A total of 63 papers contained statistical errors, and at least one drew unsupportable conclusions. In many cases the errors were not considered to be severe, but they were often sufficient to raise doubts about some inferences."

We did not attempt to evaluate the papers covered by our review with regard to the presence of statistical errors. We did find, however, that in almost every paper that made use of statistical methods, the culmination of the analysis was regarded as the performance of some tests of hypotheses and the reporting of the observed p-values. Only five papers in the entire survey presented maximum-likelihood estimators, confidence intervals, or any other kind of estimation procedure besides the usual least-squares estimates in regression models. The methodology of testing hypotheses completely dominates the statistical aspects of the psychiatric research literature.

Statisticians have only themselves to blame for this deplorable situation, in which p-values are senselessly reported, with an asterisk or two indicating that they are less than 0.05 or 0.01, without any regard to the sample size or whether the "statistically significant" deviations from the null hypothesis are of any practical significance whatsoever. The statisticians' insistence that rigorous statistical analysis be applied to all substantive research has led to the unfortunate but seemingly universal view of editors that a paper is not suitable for publication unless it contains the statement "$p < 0.05$".

Psychiatric research is not alone in its dependence on hypothesis testing. Zellner (1980) found a similar situation in the economics and econometrics literature. Cornfield (1975), in his presidential address to the American Statistical Association, wrote, "Statistical theory, instruction, and practice have tended to suffer from overemphasis on hypothesis testing." The shortcomings of the usual approach to testing are well known (e.g., Zellner, 1980, or DeGroot, 1980) and have been explicitly described in the psychiatric literature by Pocock (1980). Psychiatric research must be based on the development of new methods of data analysis, statistical modeling, and the estimation of appropriate stochastic mechanisms.

6. Concluding Remarks

In this paper we have tried to emphasize the richness and diversity of the areas of psychiatric research that can benefit from appropriate statistical theory and methodology. In 1969, the *Journal of the Royal Statistical Society* published two long discussion papers with this same emphasis: Kramer (1969) described the types of data that are needed for an adequate understanding of the incidence, prevalence, and treatment of mental disorders, and Moran (1969) described the need for statistical methods in psychiatric research, including problems of psychiatric epidemiology. Of course, quantitative methods have been, and continue to be, widely used in this research. Chun, Cobb, and French (1975) list 3000 journal articles published between 1960 and 1970 dealing with the relatively narrow topic of measures of psychological assessment and mental health. Quantitative methods have been applied not only to the decision-making processes of psychiatric patients, but also to the diagnostic process of clinicians (a Bayesian approach is described by Hirschfeld, Spitzer, and Miller, 1974), and to the process of psychotherapy itself (Bellman and Smith, 1973).

Thus, as the survey in Section 5 confirmed, statistical methods are found in most psychiatric research papers. However, as the survey also revealed, there is almost total reliance on standard tests of hypotheses and the reporting of *p*-values, with only occasional use made of modern estimation methods, time-series analysis, survival analysis, exploratory data analysis, cluster analysis, discrete multivariate analysis, Bayesian methods, or stochastic modeling. As the new concepts introduced in DSM-III continue to be refined and developed, and as more longitudinal studies are carried out in psychiatric epidemiology, it will become increasingly important to use this statistical technology to create new analytic methods designed specifically for psychiatric research.

The status of psychiatric research in Britain has recently been discussed by Cranmer (1979), Mackay (1981), Peart (1979), Rawnsley (1980), and Shepherd (1981). *The Lancet* (October 3, 1981, p. 733) has echoed Peart's emphasis on the need for "bridge men" who can "bring together clinical psychiatric services and preclinical basic science and laboratory skills." In our view, psychiatric statistics is a vital bridge for psychiatry. It is the bridge that links psychiatric research with the methodology that must infuse that research. It is to be hoped that the bridge will be crowded with creative "bridge persons."

Acknowledgements

This work was carried out with the partial support of the National Science Foundation under grant DMS-8320618 and the National Institute of Mental Health under grant MH-37116. We are indebted to Evelyn J. Bromet, Gerald A. Coffman, Joseph L. Fleiss, Stanley D. Imber, David J. Kupfer, Peter A.

Lachenbruch, Loren H. Roth, Herbert Solomon, Miron L. Straf, Craig D. Turnbull, Joseph Zubin, and the referees for many helpful suggestions, and to Mark Ciancutti for carrying out the literature survey described in Section 5.

Bibliography

American Psychiatric Association (1980). *Diagnostic and Statistical Manual of Mental Disorders*, 3rd edn. Washington: Amer. Psychiatric Assoc.

Bartko, J. J. (1976). "On various intraclass correlation reliability coefficients." *Psychol. Bull.*, **83**, 762–765.

Bartko, J. J., Strauss, J. S., and Carpenter, W. T. (1971). "An evaluation of taxonomic techniques for psychiatric data." *Classification Soc. Bull.*, **2**, 2–28.

Bebbington, P. (1980). "Causal models and logical inference in epidemiological psychiatry." *British J. Psychiatry*, **136**, 317–325.

Bellman, R. and Smith, C. P. (1973). *Simulation in Human Systems: Decision-Making in Psychotherapy*. New York: Wiley.

Bergin, A. E. and Lambert, M. J. (1978). "The evaluation of therapeutic outcomes." In S. L. Garfield, and A. E. Bergin (eds.), *Handbook of Psychotherapy and Behavior Change*, 2nd edn. New York: Wiley, 139–190.

Brown, G. W. and Harris, T. (1978a). *Social Origins of Depression: A Study of Psychiatric Disorder in Women*. London: Tavistock.

Brown, G. W. and Harris, T. (1978b). "Social origins of depression: A reply." *Psychol. Medicine*, **8**, 577–588.

Chun, K.-T., Cobb, S., and French, J.R.P., Jr. (1975). *Measures for Psychological Assessment: A Guide to 3,000 Original Sources and Their Applications*. Ann Arbor, MI: Inst. for Social Research, Univ. of Michigan.

Cohen, J. (1960). "A coefficient of agreement of nominal scales." *Educ. and Psychol. Measurement*, **20**, 37–46.

Cooper, J. E., Kendell, R. E., Gurland, B. J., Sharpe, L., Copeland, J. R. M., and Simon, R. (1972). *Psychiatric Diagnosis in New York and London*. London: Oxford U. P.

Cornfield, J. (1975). "A statistician's apology." *J. Amer. Statist. Assoc.*, **70**, 7–14.

Cranmer, J. L. (1979). "Research in decline." *Bull. Roy. Coll. Psychiatrists*, **3**, 174–175.

DeGroot, M. H. (1980). "Remarks on the statistical analysis of hypotheses." In *Proceedings of the Business and Economic Statistics Section, American Statistical Association*, 204–206.

Essen-Moller, E. (1961). "On classification of mental disorders." *Acta Psychiatrica Scand.*, **37**, 119–126.

Essen-Moller, E. (1971). "Suggestions for further improvement of the international classification of mental disorders." *Psychol. Medicine.*, **1**, 308–311.

Essen-Moller, E., and Wohlfahrt, S. (1947). "Suggestions for the amendment of the official Swedish classification of mental disorders." *Acta Psychiatrica Scand. Suppl.*, **47**, 551–555.

Everitt, B. S. (1972). "Cluster analysis: A brief discussion of some of the problems." *British J. Psychiatry*, **120**, 143–145.

Everitt, B. S. (1981). "Bimodality and the nature of depression." *British J. Psychiatry*, **138**, 336–339.

Everitt, B. S., Gourlay, J., and Kendell, R. E. (1971). "An attempt at validation of traditional psychiatric syndromes by cluster analysis." *British J. Psychiatry*, **119**, 399–412.

Feigner, J. P., Robins, E., Guze, S. B., Woodruff, R. A., Winokur, G., and Munoz, R. (1972). "Diagnostic criteria for use in psychiatric research." *Arch. Gen. Psychiatry*, **26**, 57–63.

Fowler, R. C., Mezzich, J. E., Liskow, B. L., and Valkenburg, C. V. (1980). "Schizophrenia—Primary affective disorder discrimination. II. Where unclassified psychosis stands." *Arch. Gen. Psychiatry*, **37**, 815–817.

Garside, R. F. and Roth, M. (1978). "Multivariate statistical methods and problems of classification in psychiatry." *British. J. Psychiatry*, **133**, 53–67.

Hartigan, J. A. (1975). *Clustering Algorithms*. New York: Wiley.

Helmchen, H. (1975). "Schizophrenia. Diagnostic concepts in the ICD-8." In M. H. Lader, (ed.), *Studies in Schizophrenia. British J. Psychiatry*, Special Publication, No. 10, 10–18.

Hirschfeld, R., Spitzer, R. L., and Miller, R. G. (1974). "Computer diagnosis in psychiatry: A Bayes approach." *J. Nervous and Mental Disease*, **158**, 399–407.

Imber, S. D., Pilkonis, P. A., and Glanz, L. (1983). "Outcome studies in psychotherapy." In C. E. Walker (ed.), *The Handbook of Clinical Psychology: Theory, Research, and Practice*, Homewood, IL: Dow Jones–Irwin, 242–269.

Kato, M. (1977). "Multiaxial diagnosis in adult psychiatry." Paper presented at the VI World Congress of Psychiatry, Honolulu, Hawaii.

Kendell, R. E. (1969). "The continuum model of depressive illness." *Proc. Roy. Soc. Medicine*, **62**, 335–339.

Kendell, R. E. (1975a). *The Role of Diagnosis in Psychiatry*. Oxford: Blackwell.

Kendell, R. E. (1975b). "The classification of depressions: A review of contemporary confusion." *British J. Psychiatry*, **129**, 15–28.

Kendell, R. E. and Gourlay, J. (1970). "The clinical distinction between psychotic and neurotic depressions." *British J. Psychiatry*, **117**, 257–266.

Kiloh, L. G. and Garside, R. F. (1983). "The independence of neurotic depression and endogenous depression." *British J. Psychiatry*, **109**, 451–463.

Kraemer, H. C. (1980). "Extension of the kappa coefficient." *Biometrics*, **36**, 207.

Kramer, M. (1969). "Statistics of mental disorders in the United States: Current status, some urgent needs, and suggested solutions (with discussion)." *J. Roy. Statist. Soc. Ser. A*, **132**, 353–407.

Kupfer, D. J. (1981). "The application of EEG sleep in the treatment of depression." In J. Mendlewicz and H. M. van Praag (eds.) *Advances in Biological Psychiatry*, **6**. Basel, Switzerland: S. Karger, 87–93.

Kupfer, D. J. and Reynolds, C. F. (1983). "Neurophysiologic studies of depression: State of the art." In J. Angst (ed.) *The Origins of Depression: Current Concepts and Approaches*. Berlin: Springer, 235–252.

Kupfer, D. J., Shaw, D. H., Ulrich, R., Coble, P. A., and Spiker, D. G. (1982). "Application of automated REM analysis in depression." *Arch. Gen. Psychiatry*, **39**, 569–573.

Kupfer, D. J., Spiker, D. G., Rossi, A., Coble, P. A., Ulrich, R. F., and Shaw, D. H. (1982). "Recent diagnostic and treatment advances in REM sleep and depression." In P. Clayton and J. Barrett (eds.) *Treatment of Depression: Old Controversies and New Approaches*. New York: Raven Press, 31–52.

Mackay, A. V. P. (1981). "Psychiatric research." *J. Roy. Soc. Medicine*, **74**, 168–169.

Mezzich, J. E. (1978). "Evaluating clustering methods for psychiatric diagnosis." *Biol. Psychiatry*, **13**, 265–281.

Mezzich, J. E. (1979). "Patterns and issues in multiaxial psychiatric diagnosis." *Psychol. Medicine*, **9**, 125–137.

Mezzich, J. E., Coffman, G. A., and Goodpastor, S. M. (1982). "A format for DSM-III diagnostic formulation: Experience with 1,111 consecutive patients." *Amer. J. Psychiatry*, **139**, 591–596.

Mezzich, J. E., Dow, J. T., Rich. C. L., Costello, A. J., and Himmelhoch, J. M. (1981). "Developing an efficient clinical information system for a comprehensive psychiatric institute. II. Initial evaluation form." *Behavior Res. Methods and Instrumentation*, **13**, 464–478.

Mezzich, J. E., Kraemer, H. C., Worthington, D. R. L., and Coffman, G. A. (1981). "Assessment of agreement among several raters formulating multiple diagnoses." *J. Psychiatric Res.*, **16**, 29–39.

Mezzich, J. E. and Raab, E. S. (1983). *A Cross-National Assessment of Depressive Groups*. Unpublished technical report. Western Psychiatric Inst. and Clinic, Univ. of Pittsburgh School of Medicine.

Mezzich, J. E. and Solomon, H. (1980). *Taxonomy and Behavioral Science: Comparative Performance of Grouping Methods*. London: Academic.

Mezzich, J. E. and Worthington, D.R.L. (1978). "A comparison of graphical representations of multidimensional psychiatric diagnostic data." In P. Wang (ed.), *Graphical Representation of Multivariate Data*. New York: Academic.

Moran, P.A.P. (1969). "Statistical methods in psychiatric research (with discussion)." *J. Roy. Statist. Soc. Ser. A*, **132**, 484–524.

Ottosson, J. O. and Perris, C. (1973). "Multidimensional classification of mental disorders." *Psychol. Medicine*, **3**, 238–243.

Parloff, M. B. (1980). "Psychotherapy and research: An anaclitic depression." *Psychiatry*, **43**, 279–293.

Parloff, M. B. (1982). "Psychotherapy research evidence and reimbursement decisions: Bambi meets Godzilla." *Amer. J. Psychiatry*, **139**, 718–727.

Paykel, E. S. (1972). "Depressive typologies and response to amitriptyline." *British J. Psychiatry*, **120**, 147–156.

Peart, W. S. (1979). "Research in psychiatry: A view from general medicine." *Psychol. Medicine*, **9**, 205–206.

Pilkonis, P. A., Imber, S. D., Lewis, P., and Rubinsky, P. (1984). "A comparative outcome study of individual, group, and conjoint psychotherapy." *Arch. Gen. Psychiatry*, **41**, 431–437.

Pilowsky, I., Levine, S., and Boulton, D. M. (1969). "The classification of depression by numerical taxonomy." *British J. Psychiatry*, **115**, 937–945.

Pocock, S. J. (1980). "The role of statistics in medical research." *British J. Psychiatry*, **137**, 188–190.

Rawnsley, K. (1980). "Psychiatric research." *J. Roy. Soc. Medicine*, **73**, 768–769.

Robins, L. N., Helzer, J. E., Croughan, J., and Ratcliff, K. S. (1981). "National Institute of Mental Health Diagnostic Interview Schedule. Its history, characteristics, and validity." *Arch. Gen. Psychiatry*, **38**, 381–389.

Roth, L. H., Lidz, C. W., Meisel, A., Soloff, P. H., Kaufman, K., Spiker, D. G., and Foster, F. G. (1982). "Competency to decide about treatment or research. An overview of some empirical data." *Internat. J. Law and Psychiatry*, **5**, 29–50.

Rutter, M., Lebovici, S., Eisenberg, L., Sneznevskij, A. V., Sadoun, R., Brooke, E., and Lin, T. Y. (1969). "A tri-axial classification of mental disorders in childhood." *J. Child Psychol. and Psychiatry*, **10**, 41–61.

Rutter, M., Shaffer, D., and Shepherd, M. (1975). *A Multiaxial Classification of Child Psychiatric Disorders*. Geneva: World Health Organization.

Rutter, M., Shaffer, D., and Sturge, C. (1975). *A Guide to a Multiaxial Classification Scheme for Psychiatric Disorders in Childhood and Adolescence*. London: Inst. of Psychiatry.

Shepherd, M. (1981). "Psychiatric research in a medical perspective." *British Medical J.*, **282**, 961–963.

Shrout, P. E. and Fleiss, J. L. (1979). "Intraclass correlations: Uses in assessing rater reliability." *Psychol. Bull.*, **38**, 420–428.

Smith, M. L., Glass, G. V., and Miller, T. I. (1980). *The Benefits of Psychotherapy*. Baltimore: Johns Hopkins U. P.

Sneath, P.H.A. and Sokal, R. R. (1973). *Numerical Taxonomy: The Principles and Practice of Numerical Classification*. San Francisco: Freeman.

Spearman, C. (1927). *The Abilities of Man, Their Nature, and Measurement*. New York: Macmillan.

Spitzer, R. L., Endicott, J., and Robins, E. (1978). "Research diagnostic criteria." *Arch. Gen. Psychiatry*, **35**, 773–782.

Stengel, E. (1959). "Classification of mental disorders." *Bull. World Health Organization*, **21**, 601–663.

Stephenson, W. (1935). "Correlating persons instead of tests." *Character and Personality*, **4**, 17–24.

Strauss, J. S. (1973). "Diagnostic models and the nature of psychiatric disorder." *Arch. Gen. Psychiatry*, **29**, 445–449.

Strauss, J. S. (1975). "A comprehensive approach to psychiatric diagnosis." *Amer. J. Psychiatry*, **132**, 1193–1197.

Temkin, O. (1965). "The history of classification of medicial sciences." In M. Katz, J. Cole, and W. Barton (eds.), *Classification in Psychiatry and Psychopathology*. Washington: U.S. Government Printing Office.

Tennant, C. and Bebbington, P. (1978). "The social causation of depression: A critique of Brown and his colleagues." *Psychol. Medicine*, **8**, 565–575.

Thurstone, L. L. (1947). *Multiple Factor Analysis*. Chicago: Univ. of Chicago Press.

United States National Center for Health Statistics (1978). *The International Classification of Diseases, 9th Revision, Clinical Modification* (ICD-9-CM). Ann Arbor, MI: U.S. Natl. Center for Health Statistics.

von Cranach, M. (1977). "Categorical vs. multiaxial classification." Paper presented at the VI World Congress of Psychiatry, Honolulu, Hawaii.

White, S. J. (1979). "Statistical errors in papers in the British Journal of Psychiatry." *British J. Psychiatry*, **135**, 336–342.

Wing, L. (1970). "Observations on the psychiatric section of the International Classification of Diseases and the British Glossary of Mental Disorders." *Psychol. Medicine*, **1**, 79–85.

Wing, J. K., Cooper, J. E., and Sartorius, N. (1974). *The Measurement and Classification of Psychiatric Symptoms*. Cambridge: Cambridge U.P.

Woolson, R. F., Tsuang, M. T., and Fleming, J. A. (1980). "Utility of the proportional-hazards model for survival analysis of psychiatric data." *J. Chronic Disease*, **33**, 183–190.

World Health Organization (1973). *Report of the International Pilot Study of Schizophrenia*, **1**. Geneva: World Health Organization.

Yates, B. T. and Newman, F. L. (1980a). "The efficiency and cost-effectiveness of psychotherapy." In *The Implications of Cost-effectiveness Analysis of Medical Technology*. Background Paper No. 3, Office of Technology Assessment. Washington: U.S. Government Printing Office.

Yates, B. T. and Newman, F. L. (1980b). "Findings of cost-effectiveness and cost–benefit analyses of psychotherapy." In G. R. VandenBos (ed.), *Psychotherapy: Practice, Research, Policy*. Beverly Hills, CA: Sage.

Zellner, A. (1980). "Statistical analysis of hypotheses in economics and econometrics." In *Proceedings of the Business and Economic Statistics Section, American Statistical Association*, 199–203.

Zubin, J. (1938). "Socio-biologic types and methods for their isolation." *Psychiatry*, **1**, 237–247.

Zubin, J. (1961). *Field Studies in the Mental Disorders*. New York: Grune & Stratton.

CHAPTER 8

Mathematical Models for Infectious Disease Statistics

K. Dietz and D. Schenzle

Abstract

Numerous mathematical models have been developed to gain better insight into the transmission and control of infectious diseases. Yet there are many unsolved problems, partly because the models are still too simple, partly because detailed epidemiologic records are notoriously lacking. The present survey concentrates on virus infections in humans. It is shown that available data do not allow a discrimination between various plausible models for the spread of common cold in households. Similar problems of model identification arise in the analysis of age-specific sero-prevalence-data of antibodies with so-called catalytic models. From such data alone one cannot derive contact rates between different age groups, although knowledge of these rates is needed in order to evaluate the effects of mass immunization and to describe the fluctuating infection incidence patterns. A new deterministic model is presented which takes into account increased infection transmission inside schools. This provides an explanation for one- and two-year periods of recurrent measles epidemics. The paper provides an outlook to future developments in this field.

1. Introduction

The last case of naturally transmitted smallpox occurred in the town of Merka, Somalia, on October 26, 1977. In 1760 the mathematician and physician Daniel Bernoulli had developed a method to calculate the gain in life expectancy if smallpox were eliminated as a cause of death. By now about one thousand papers have contributed to mathematical epidemiology, yet smallpox eradication was achieved without the use of any of them. We agree with the assessment by Bart et al. (1983):

> Mathematical formulations have been developed to describe outbreaks of infectious diseases, to test concepts, and to provide insights into disease control and policy formulations. The resulting equations sometimes mirror the observed events, but to date have had little impact upon disease control or preventive

Key words and phrases: catalytic model, common cold, epidemiological model, hepatitis, household epidemics, infectious disease control, measles, oscillations, vaccination practice. Instead they have been used more retrospectively to reassure rather than assist in the development of policy.

The pioneers in this field have been epidemiologists with a passion for mathematics which they used to study the transmission laws of epidemics. The foundation was laid by Sir Ronald Ross, the receiver of the second Nobel prize in medicine, awarded in 1902 for his discovery that the mosquito is the vector of malaria. He derived the first threshold theorem which identifies a critical mosquito density with respect to man below which malaria cannot maintain itself (Fine, 1975a). It does not seem to be well known that he also published (Ross and Hudson, 1917) the epidemic equations which are now attributed to Kermack and McKendrick (1927) and the hyperbolic partial differential equation for age-dependent populations which is usually named after McKendrick (1926) and Von Foerster (1959). We shall later also show that the epidemic models by Soper (1929) and Reed and Frost (Fine, 1977; Frost, 1976; Sartwell, 1976) had been anticipated already in 1889 in Saint Petersburg by the physician En′ko (Moshkovskiĭ, 1950).

The models which had the greatest impact on public health programs are those of the late George Macdonald (1973), Professor of Tropical Hygiene at the University of London, and Director of the Ross Institute. He provided mathematical arguments for the global malaria eradication program of the World Health Organization (Macdonald, 1956), his quantitative conclusions for the control of the helminth disease schistosomiasis (Macdonald, 1965) stimulated long term field projects on the island of St. Lucia (Bradley, 1982), and the plan for a mass vaccination campaign against measles in West Africa was based on his simulations (Millar, 1970). Macdonald's (1957) conclusion that the control of adult mosquitos by residual insecticides is much more effective than larval control is considered by Bradley (1982) as "the single most important insight into public health from modelling."

The pioneering work by Ross and Macdonald was subsequently put on a firm basis by rigorous mathematical analysis. Lotka (1923a) examined the stability of the equilibria for Ross's models. He did not construct a new malaria model as Bradley (1982) implies. The same is true for Nåsell and Hirsch (1973) with respect to their analysis of Macdonald's (1965) schistosomiasis model. They discovered that his breakpoint concept could be interpreted as a separatrix between two stable equilibria. Nåsell (1977) and Goddard (1978) clarify how Macdonald arrived at his pronounced conclusions about the relative effect of certain control measures against schistosomiasis ("Safe water supplies are more important than latrines"). It is now clear that Macdonald's mathematical formulation of superinfection did not correspond to his biological assumption (Dietz, 1970; Fine, 1975b; Aron and May, 1982; Bailey, 1982).

Ideally the modeling is initiated on the request of the authorities who can make direct use of the results. Examples may be the large scale influenza models developed under the auspices of the Ministry of Health of the USSR

(Baroyan et al., 1977) and the bacterial disease models of the World Health Organization (Cvjetanović et al., 1978). Sometimes one finds exaggerated enthusiasm for the potentials of models: "In future, when malaria eradication is achieved in Africa, it will be due, first of all, to the computer" (Pampana, 1969). In general, however, the mathematical approach is rarely adopted in planning infectious disease control. E.g. the decision-making process which led in 1976 to the nationwide influenza vaccination program in the U.S.A., as revealed in the fascinating book by Neustadt and Fineberg (1978), did not resort to any use of the simulation models which were explicitly constructed in anticipation of a new pandemic (Elveback et al., 1975, 1976a, b).

Public health officials will pay more attention to the mathematical approach if the present trend continues to bring models into closer contact with the data. If the models can be shown to reflect reality, then their conclusions are more likely to be taken into account. But the lack of reliable data is notorious in this field. A diminishing number of diseases are notifiable. And even for those there is an unknown and probably variable degree of underreporting. Individual epidemic outbreaks are rarely fully investigated. There is always a delay until the outbreak is brought to the attention of the health authorities, and then the primary objective is control rather than the detailed collection of data. Therefore it does not surprise if some of the few well-documented epidemics have been used again and again for fitting different models. It will be seen that in some cases the data do not allow discrimination between different models. Stille and Gersten (1978) claimed that epidemic models are "tautologies or selfcontained arguments and not scientific theories which are capable of refutation by predicting experimental or observable results." In a correspondence, Koopman (1979) rightly points out that e.g. the hypothesis of a uniform contact probability of most epidemic models is subject to empirical tests, and we shall later see that it must be rejected on the basis of the available evidence.

The focus of concern of infectious disease models is the transmission of the disease agents through the population. A detailed model would describe at any time the number of parasites in each member of the host population. The term parasite is used here to comprise viruses, bacteria, protozoa, and helminths. The transmission model has to include always two basic components: (a) the course of an infection within one individual once the parasites have entered, (b) the mode of spread of parasites between individuals. There is a whole branch of mathematical biology which is concerned only with the first aspect, i.e. the interaction of antigens with antibodies and the resulting dynamical behavior (Bell et al., 1978). It does not seem to be known that McKendrick (1926) had already discussed the qualitative behavior of a system of differential equations describing the interaction of "microbes" and "antibodies." He identified the parameter ranges for which the microbes go extinct or coexist in a steady state with the antibodies and those for which the system oscillates. An attempt to link the ordinary differential equations for antigens (viruses) and antibodies for the individuals with the transmission dynamics in the whole population was made by Rvachev (1967), but this has not been followed up. Most infectious disease models neglect the number of parasites per host and reduce the whole process to a succession of two discrete states: latent and infectious. During the latent state an individual already harbors the infection but is not yet able to infect other individuals. While infectious an individual may transmit the infection to other susceptibles with a constant probability. After the infectious state the individual may become immune against further infections or return to the susceptible state. For many virus infections the sequence of states

$$\text{susceptible} \rightarrow \text{latent} \rightarrow \text{infectious} \rightarrow \text{immune}$$

represents a good approximation to reality. The main method of control consists in vaccination, i.e. the induced transfer of an individual from the susceptible into the immune state. This transfer has an effect not only on the individual in protecting him/her against an infection and its possible consequences (disease, death) but also on the community in that the infection risk to others will be reduced. It is a key problem in quantitative epidemiology to calculate the resulting incidence of infection and disease for a given rate of vaccination. The main results are concerned with the identification of the minimum vaccination coverage which ensures zero infection incidence, i.e. eradication of the virus.

In the case of bacterial and protozoan diseases one individual may be reinfected several times, but the duration and the infectivity of subsequent episodes may be reduced from the first episode. There may also be a so-called carrier state in which the individual is infectious for a prolonged period without any disease symptoms. The typical modeling approach for such diseases is the assumption of a set of compartments through which an individual passes according to certain transition rates. (Cvjetanović et al., 1978; Dietz et al., 1974). The model parameters are partly guessed, partly fitted. For most diseases in this category there is no vaccine available. The control strategies include chemotherapy and reduction of the contact rate (sanitation, quarantine, vector control, etc.). The models are therefore used to assess the relative efficiency of alternative methods of control. Chemotherapy has a benefit for the individual in reducing the duration of disease, but also for the community in reducing the risk of infection to others. On the other hand, since chemotherapy has no protective effect, the individual is again at risk to reinfection, with the result that for high contact rates the total incidence of cases may be higher with chemotherapy than without.

Anderson and May (1979) lumped viruses, bacteria, and protozoa under the term microparasites, in contrast to macroparasites, which they apply to helminths. The former have in common that they multiply in the host, whereas the latter accumulate as a consequence of repeated exposure. Since transmissibility of infection and severity of disease symptoms crucially depend on the number of helminths per host, most of the corresponding models, starting with Kostitzin (1934), take into account at least the average worm burden if not its whole distribution. The prevalence of infection may be 100% above a certain age in certain populations, yet there may be considerable differences in mean density of infection and prevalence of disease. Therefore a model which classifies individuals simply into the discrete states "susceptible" and "infectious" is inadequate. Because helminth infections and the resulting disease are usually of long duration, one cannot ignore the turnover of the human host population. Recently Hadeler and Dietz (1983) analysed a transmission model which allows age-dependent mortality of the human host in addition to differential mortality which is proportional to the number of parasites per host. There is considerable uncertainty about the mechanisms which regulate the abundance of helminths in one host. The age-specific worm densities in a stable endemic equilibrium could be explained either by age-dependent exposure or by immunity (Warren, 1973). The control methods are directed against the parasite through chemotherapy or against its transmission by contact reduction. The problem of selective chemotherapy which is concentrated on heavily infected individuals has been addressed by Anderson and May (1982a) and Dietz and Renner (1985). There have been a few preliminary attempts to fit helminth transmission models to data (Anderson, 1982; Dietz, 1982), but much more work needs to be done in this direction.

We have used the terms "epidemic" and "endemic" so far without explanation. The former relates to transient outbreaks, which are terminated when the number of infected individuals reaches zero. Much of the mathematical literature in this field is devoted to the description of various aspects of such outbreaks, e.g. the initial conditions which lead to an epidemic, the shape of the epidemic curve, the number of cases at the peak of the epidemic, the duration of the total epidemic, and the total number of cases. The practical applications are almost exclusively concerned with certain viral diseases such as measles, influenza, and the common cold. Researchers try to estimate model parameters or to test hypotheses which would provide insight into the mode of transmission. The attempts to apply mathematical control theory to single outbreaks (Wickwire, 1977) do not appear to be realistic, mainly because these models assume that the public health authorities have current information on the progress of the epidemic, which in fact is rarely available.

Much more important for the evaluation of control measures is the analysis of endemic situations, i.e. where the incidence of cases is always positive and nearly constant, at least if one takes the average over one or a few years. For these equilibrium situations one would like to know the degree of their stability. In other words, one would like to know how much control effort is needed in order to render the positive equilibrium unstable and the zero incidence equilibrium stable. For virus diseases the corresponding problem asks for the minimum vaccination coverage in order to ensure eradication or only minor outbreaks if cases are imported. For diseases transmitted by vectors (mosquitoes, flies, snails, etc.) one asks for the minimum reduction in vector density in order to reach the critical contact rate below which there is no positive equilibrium possible. Even if eradication is not the objective of a given public health program—in most cases it is not—it may be useful to know how difficult it would be to achieve it. The painful lessons of the global malaria eradication program should be heeded when discussing the present proposals for global measles eradication.

In endemic situations the age-specific prevalence of antibodies as revealed by representative cross-sectional samples of the population provides the most important information about the true incidence of a virus infection which can be used to assess the degree of underreporting. Muench (1934, 1959) introduced the term "catalytic" models for the differential equations describing the age-specific antibody prevalence in a cross-section of the population. Dietz (1975a) showed how the catalytic model can be used to estimate the proportion of susceptibles in the total population. If contact rates are homogeneous in the population, then one minus the equilibrium proportion of susceptibles is the minimum immunization coverage required to eradicate an endemic infection. We shall later show how this simple result needs to be modified for heterogeneous contact rates. It will turn out that it is impossible in principle to estimate the contact rates on the basis of incidence rates in an equilibrium situation, because one has only n equations to determine n^2 parameters, where n is the number of subgroups of the population for which one knows the incidence rates. One can narrow down the degree of uncertainty if one compares the effect of known interventions with the various model predictions. In order to construct a realistic model it is thus necessary to have data both from the baseline situation prior to intervention and during the intervention. The model may then be applied to project into the future the likely consequences of changes in the intervention. It may also be applicable to control programs in other populations provided the contact rates are similar.

When the data to be analysed consist in frequency distributions of the number of cases in households of a given size, it is obvious that a stochastic approach is appropriate. On the other hand, age-specific antibody prevalence can be described by a deterministic model. When we observe oscillations in reported incidence it is not clear whether a deterministic or a stochastic model is more appropriate. We shall later show that the two-year cycle of measles in large populations can be reproduced, with quite different assumptions, with both stochastic and deterministic models. In the context of schistosomiasis Nåsell and Hirsch (1973) have introduced the notion of a hybrid model, which is obtained by replacing certain transition rates in a stochastic model with their expectations. The resulting deterministic model can then be analysed by the methods of the qualitative theory of differential equations. Rost (1981) has shown that this approach corresponds to the mean field or Vlasov approximation to the fully stochastic model.

If the modeler attempts to be realistic, he is sooner or later forced to replace the analytic approach by simulation. For instance, Bartlett (1957) and Bailey (1967) simulated the spatial spread of epidemics. The most elaborate simulation models for virus diseases are due to Elveback et al. (1976a, b). They are extremely useful in exploring how important it is to include a certain feature into the model. But they cannot be used to optimize control strategies. Thus Longini et al. (1978) complemented the original influenza simulation model with a deterministic transmission model for which the vaccination strategy was optimized by linear programming. The design and the interpretation of simulation runs should be guided by the insights based on the analysis of simplified deterministic models. Simulation models are used in several schools for teaching medical students, who can try out different control strategies and thereby acquire a feeling for the nonlinear behavior of the system. But one learns most about the dynamics of an infectious disease if one tries oneself to construct a model.

It is impossible to cover the whole field of infectious disease modeling in this survey. We shall restrict ourselves to virus diseases. Section 2 will be concerned with the analysis of statistics related to single outbreaks in households. Section 3 gives an account of the analysis and interpretation of age-specific prevalence of immunity by the use of catalytic models. In Section 4 we review models for the description of long-term patterns in the incidence of reported cases in large populations. In the final section we make some suggestions for further work.

2. Statistics of Household Epidemics

The most complete information about the course of an epidemic is the sequence of dates when symptoms began in each case. In practice this is rarely available. Mostly the dates are grouped in weekly or biweekly intervals, or only the total number of cases at the end of the epidemic is known. For certain virus diseases like measles it is often possible to identify for each case the generation number, i.e. the number of intermediate cases which separate him from the initial case(s) who started the epidemic. Thus one can record the course of an epidemic by the number of cases in each generation. The data set which has been subject to the most attempts at mathematical analysis was collected by Brimblecombe et al. (1958) in London. All of the 45–50 families observed at any time during a two year period consisted of father, mother, two school children, and one preschool child. The 664 family epidemics of the common cold recorded were analysed for the first time by Heasman and Reid (1961). It is not mentioned how the epidemics were distributed among the families. Despite the fact that on average more than a dozen epidemics were recorded per family, the 664 epidemics are treated as independent in all analyses of this data set—a hypothesis which one could probably reject if the raw data were available.

Heasman and Reid (1961) fitted three models to the distribution of the total number of secondary cases caused by the initial case. The first one is due to Lowell J. Reed and Wade Hampton Frost. It had been used in class lectures at

Johns Hopkins University and was published posthumously (Frost, 1976). The relationship between the model and its mechanical analogue consisting of colored balls representing "susceptibles," "infectives," "immunes," and "contact neutralizers" is clarified by Fine (1977). Let S_t and C_t denote the numbers of susceptibles and cases during time interval t. If p denotes the probability that any two individuals come into contact sufficient to transmit the infection during one time period, then the Reed–Frost model determines the expected number C_{t+1} of new cases in the next generation according to the formula

$$C_{t+1} = S_t[1 - (1 - p)^{C_t}]. \tag{2.1}$$

The number S_{t+1} of susceptibles in the next generation is then given by

$$S_{t+1} = S_t - C_{t+1}. \tag{2.2}$$

The Reed–Frost model assumes that one susceptible becomes infective if he has at least one effective contact with an infective. The probability of having no contact with any of the C_t infectives is assumed to be $(1 - p)^{C_t}$. En′ko (1889) was apparently the first to deal with this problem. He argues as follows. (We change his notation to agree with the one already introduced.) Let N denote the size of the total population. Given a contact, the probability that this contact occurs with an infective is $C_t/(N - 1)$. Hence the probability of not meeting an infective is $1 - C_t/(N - 1)$. If an individual makes exactly k contacts during the interval, the probability of avoiding contact with an infective is $[1 - C_t/(N - 1)]^k$. The probability of having at least one contact with an infective is therefore $1 - [1 - C_t/(N - 1)]^k$. Hence the number C_{t+1} of infectives in the next time interval is given by the expression

$$C_{t+1} = S_t\left\{1 - \left[1 - \frac{C_t}{N - 1}\right]^k\right\}. \tag{2.3}$$

En′ko (1889) used this formula together with (2.2) to calculate the expected course of epidemics for some 40 sets of parameters, varying N, k, S_0, and C_0. He assumes a fixed number of contacts during one time interval, whereas the Reed–Frost model does not specify the distribution of the number of contacts. Let f denote the probability generating function of the number of contacts which one individual makes during one time interval:

$$f(x) = \sum_{k=0}^{\infty} p_k x^k. \tag{2.4}$$

In general, we get

$$C_{t+1} = S_t\left\{1 - f\left(1 - \frac{C_t}{N - 1}\right)\right\}. \tag{2.5}$$

If $f(x) = \exp[-K(1 - x)]$, i.e. if we assume a Poisson distribution with mean K, then

$$C_{t+1} = S_t\left\{1 - \exp\left[-\frac{KC_t}{N-1}\right]\right\}. \tag{2.6}$$

This is equivalent to the Reed–Frost model if we set

$$p = 1 - \exp\left[-\frac{K}{N-1}\right]. \tag{2.7}$$

The quantity $c = K/(N-1)$ is the average number of contacts which one individual makes with one other individual during one time interval. Assuming a Poisson distribution, Equation (2.7) gives the probability of having at least one contact with another individual. If one sets $K = r(N-1)$, i.e., the number of contacts is proportional to the number of possible persons to contact, then the probability p of any two individuals having at least one contact is independent of the population size. This assumption is implicit in the Reed–Frost model. It may be realistic for very small communities like households, but already for schools or small villages the number of contacts K probably increases less than linearly or approaches a constant asymptote.

Equation (2.5) suggests possibilities to generalize the En′ko–Reed–Frost model by choosing different distributions for the number of contacts, like the negative binomial distribution. If one takes as a special case the geometric distribution with mean K, then

$$C_{t+1} = S_t\frac{cC_t}{1 + cC_t}. \tag{2.8}$$

Figure 1 compares the course of the epidemic if one assumes a Poisson and a geometric distribution. The parameter p has been set equal to $c/(1+c)$, so that the number of the secondary cases of the initial case is the same for the two distributions.

If the average number $K = f'(1)$ of contacts of one individual in one time interval is small, Equation (2.5) may be approximated by

$$C_{t+1} = \frac{KS_tC_t}{N-1} = cS_tC_t, \tag{2.9}$$

which corresponds to the formulation by Soper (1929).

In order to get the distribution of the number of cases in any generation for the Reed–Frost model, equation (2.1) is taken as the expected value of a binomial distribution:

$$P\{C_{t+1} = i\} = \binom{S_t}{i}[1 - (1-p)^{C_t}]^i[(1-p)^{C_t}]^{S_t - i}. \tag{2.10}$$

The model by Greenwood (1931) assumes that the distribution of the number of new cases is independent of the number of present cases provided it is positive:

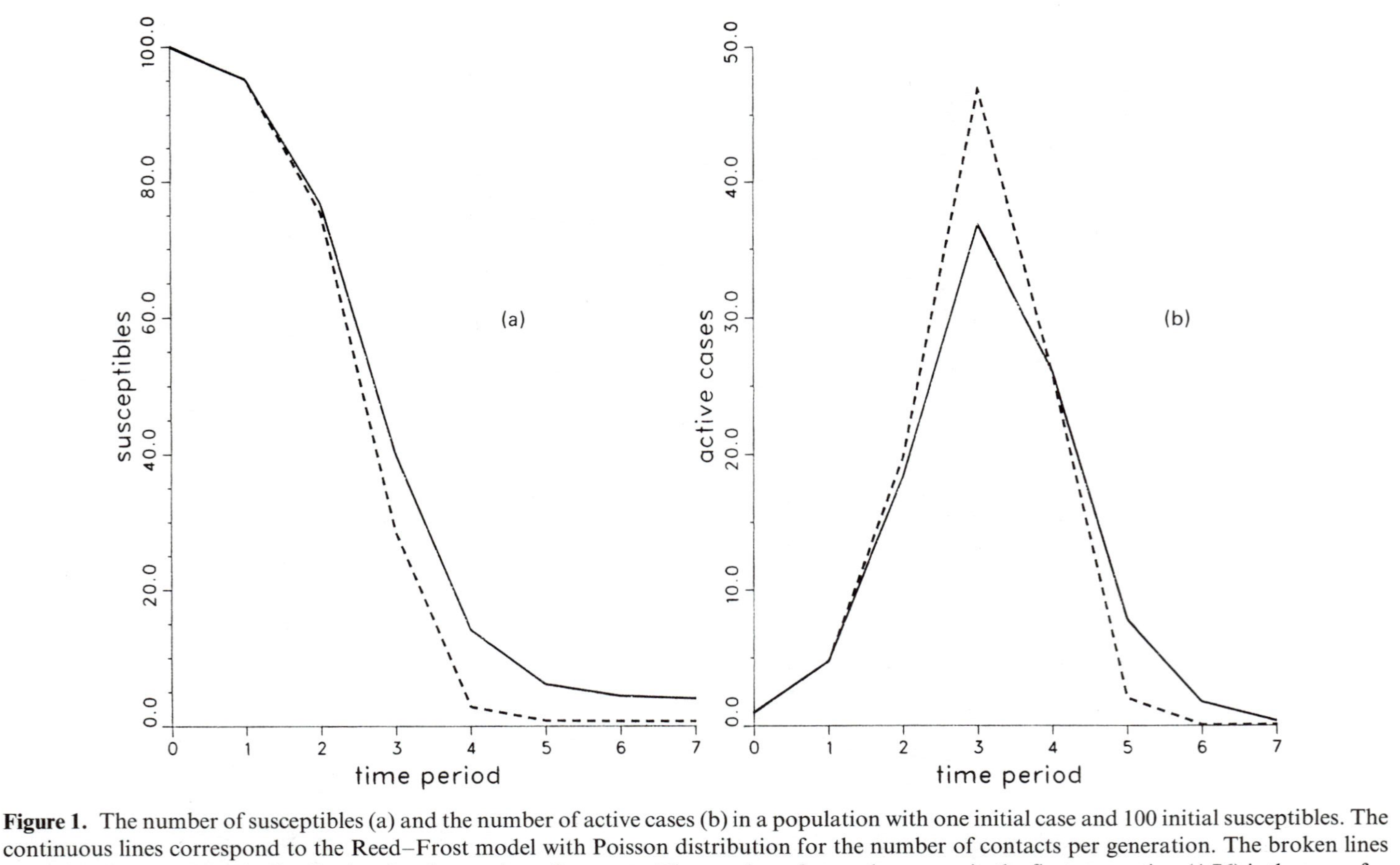

Figure 1. The number of susceptibles (a) and the number of active cases (b) in a population with one initial case and 100 initial susceptibles. The continuous lines correspond to the Reed–Frost model with Poisson distribution for the number of contacts per generation. The broken lines correspond to a geometric distribution for the number of contacts. The number of secondary cases in the first generation (4.76) is the same for both models.

$$\begin{aligned} P\{C_{t+1} = i\} &= \binom{S_t}{i} p^i (1-p)^{S_t - i} \quad &\text{for } C_t > 0, \\ P\{C_{t+1} = 0\} &= 1 \quad &\text{for } C_t = 0. \end{aligned} \tag{2.11}$$

The Greenwood assumption is usually associated with airborne spread of an infection: the risk of infection of a susceptible depends only on the presence of viruses in his environment, not on the number of infectives from which the viruses originate. Riley et al. (1978) however was able to demonstrate airborne spread of measles in an elementary school, yet the infection risk of susceptibles could be shown to depend on the number of infectives.

The third model, which Heasman and Reid (1961) fitted to the distribution of the total number of cases, is the stochastic version of the Kermack–McKendrick equations:

$$\begin{aligned} \frac{dx}{dt} &= -\beta xy, \\ \frac{dy}{dt} &= \beta xy - \gamma y. \end{aligned} \tag{2.12}$$

Here x and y denote the numbers of susceptibles and infectives, respectively, and β and γ are the contact and removal rates. These equations had already been published by Ross and Hudson (1917), but Kermack and McKendrick (1927) solved them approximately and showed that $x_0\beta/\gamma$, i.e. the initial number of secondary cases which one case could produce during his infectious period of duration γ^{-1}, had to be greater than one for an epidemic to occur. Here it is implicitly assumed that the infectious period has an exponential distribution and that there is no latent period. McKendrick (1926) had already derived the distribution of the total number of cases in a household of a given size for the stochastic analogue of (2.12) which assumes that the transitions from the state $(S_t, C_t) = (s, i)$ to the states $(s-1, i+1)$ and $(s, i-1)$ occur with the probabilities $\beta sih + o(h)$ and $\gamma ih + o(h)$ in the time interval $(t, t+h)$. It turned out that the Kermack–McKendrick model gave a better fit than the Reed–Frost model for the epidemics which occurred in families with crowded homes, i.e. with at most three rooms. Based on the distribution of the total number of cases in all epidemics combined, i.e. independent of degree of overcrowding, the Reed–Frost model just fails to be rejected at the 5% significance level, whereas the Greenwood model produces $P = 0.001$.

The next attempt to describe the distribution of the total number of cases is due to Griffiths (1973a), who fits successfully a truncated beta-binomial distribution. This model ignores completely the transmission possibilities inside the family. It simply assumes that each family member has a probability of contracting the disease independently of the other members of the family, but this probability varies between families according to a beta distribution. Since only epidemics with one initial case are recorded, the beta-binomial distribution has to be zero-truncated.

Heasman and Reid (1961) point out that it is not sufficient to consider only the total number of cases in order to discriminate between different models. They present also a table which lists the observed frequencies of each type of epidemic chain, i.e. the enumeration of the number of cases in each generation. Table 1 reproduces these observations together with six expected distributions. Heasman and Reid (1961) only fitted the Reed–Frost model. Their estimate for the contact probability $\hat{p}$ equals 0.114. The expected values were recalculated again by Becker (1981a) with slightly different results for $\hat{p} = 0.116$. In order to avoid ambiguities when quantifying goodness of fit, Schenzle (1982) calculated three P values, P_1, P_2, P_3, which result from the chi-square statistic after chains have been pooled according to the total number of cases 1, length of chain $(1, 2, 3, 4 + 5)$, and number of cases in the first generation $(0, 1, 2, 3 + 4)$, respectively. If one calculates the chi-square statistic by pooling all chains with expected frequencies less than five, one cannot reject the Reed–Frost model. The more detailed calculation of P values in Table 1 reveals that the Reed–Frost model does not account for the number of chains with length greater than three.

The Greenwood model, which assumes a constant infection probability, was formally rejected by Becker (1981a) in comparison with the Reed–Frost model. It is adequate neither for the total number of cases nor for the length of the chain. It is not surprising that it cannot be rejected with respect to the number of cases in the first generation, since the initial number of cases is one and the Reed–Frost model is equivalent with the Greenwood model.

The third model in Table 1 is due to Becker (1980), who derived explicit formulas for the individual chain probabilities in the Kermack–McKendrick model. Heasman and Reid (1961) had already found that this model described well the total number of cases of an epidemic. The corresponding P value in Table 1 of 0.569 supports this finding. But the model is inadequate according to the other two criteria.

The Kermack–McKendrick model has been shown by Becker (1980) to be a special case of a general chain binomial model where the individual probabilities that a given susceptible escapes infection by each of the infectives are independent random variables having the same beta distribution. This general model provides only a small improvement compared to the Reed–Frost model as judged by comparing the log likelihood values. The maximum value of the log likelihood function equals -816.7 when each chain frequency is fitted separately. Becker's general chain binomial model yields a value of -825.7, i.e., it can be concluded that a variability in the contact probabilities is not sufficient to explain the observed distribution of chain frequencies.

Another generalization of the classical chain binomial models has been proposed by Becker (1981a). The probabilities p_i that a given susceptible when exposed to i infectives of one generation is infected are introduced as arbitrary parameters to be estimated. For $p_i = 1 - q^i$ and $p_i = 1 - q$ we get the Reed–Frost model and the Greenwood model as special cases. One advantage of this model is the ease of obtaining the estimates of $q_i = 1 - p_i$. If m_i denotes

Table 1. Chain Models Fitted to 664 Common Cold Epidemics Recorded by Heasman and Reid (1961)

		Expected					
Chain	Observed	Reed–Frost	Greenwood	Kermack–McKendrick	Becker (1980)	Becker (1981a)	Schenzle (1982)
1-0	423	405.2	400.0	435.7	409.8	403.9	422.3
1-1-0	131	147.1	147.8	117.7	142.1	147.3	130.1
1-1-1-0	36	45.3	46.5	29.0	42.4	45.6	43.0
1-1-1-1-0	14	10.5	11.1	5.9	9.7	10.7	13.8
1-1-2-0	8	6.0	7.0	6.6	6.2	6.2	6.8
1-2-0	24	25.6	34.0	32.0	27.2	26.9	20.2
1-2-1-0	11	12.7	8.1	13.9	13.2	12.2	11.5
1-1-1-1-1	4	1.4	1.5	0.8	1.3	1.4	3.4
1-1-1-2	2	0.8	0.8	0.9	0.8	0.8	1.7
1-1-2-1	2	1.7	1.0	1.9	1.7	1.6	3.0
1-1-3	2	0.3	0.4	1.1	0.4	0.3	0.5
1-2-1-1	3	1.7	1.1	1.8	1.7	1.6	1.9
1-2-2	1	2.0	0.6	3.3	2.2	1.8	2.4
1-3-0	3	2.5	3.5	8.2	3.4	3.7}	2.3
1-3-1	0	1.1	0.5	3.6	1.5	0.0}	1.0
1-4	0	0.1	0.1	1.5	0.2	0.1	0.1
		$\hat{p} = 0.116$	$\hat{p} = 0.119$	$\hat{p} = 0.116$	$\hat{p} = 0.116$ $\hat{d} = 0.044$	$\hat{p}_1 = 0.117$ $\hat{p}_2 = 0.205$	$\hat{p}_0 = 0.107$ $\hat{a} = 0.03$
$\ln L$		−826.2	−828.1	−836.7	−825.7	−825.0	−821.2
P_1		.059	.001	.569	.100	.004	.864
P_2		.012	.003	.0002	.0028	.002	.306
P_3		.351	.176	.00008	.148	.124	.619

the total number of exposures to i infectives and x_i denotes the number of these which escape infection, then

$$\hat{q}_i = x_i/m_i. \qquad (2.13)$$

One easily verifies that $m_1 = 3397$ and $x_1 = 3000$, hence $\hat{q}_1 = 0.883$. Since the estimate of q_3 is based on only three Bernoulli trials, Becker (1981a) neglects this parameter by pooling the frequencies of the chains 1-3-0 and 1-3-1. A formal test of the Reed–Frost hypothesis $q_2 = q_1^2$ against $q_2 > q_1^2$ does not lead to its rejection.

The final model in Table 1 (Schenzle, 1982), which gives the largest P values, assumes a changing probability of escaping infection during the course of a family epidemic:

$$q = q_0 - at, \qquad (2.14)$$

where t denotes the generations of an epidemic ($t = 0, 1, 2, 3$) and a is the decrease in q per generation. The model implies that the chance of contracting an infection for a susceptible after the third generation has increased from the initial $\hat{p}_0 = 0.107$ to $\hat{p}_3 = 0.197$.

Schenzle (1982) presents three more models which assume either that the chain probabilities have a beta distribution (for the Reed–Frost and the Greenwood model) or that they assume two discrete values with probability π and $1 - \pi$ (for the Reed–Frost model). They all give satisfactory fits.

Becker (1981b) presents a further analysis of these data which is based on the estimation methods for point processes by Aalen (1978). This nonparametric approach constructs a suitable martingale and obtains an estimating equation by setting the realization of the martingale, up to a suitable stopping time, equal to its mean. Becker (1981b) thus obtains an estimate for the contact parameter and its variance. The hypothesis that the variance is zero cannot be rejected, i.e. the Reed–Frost model without variable parameters seems appropriate. But this analysis ignores the individual chain frequencies for all chains where the final number of susceptibles is greater than zero. Since the Reed–Frost model is not rejected if the chi-square statistic is based on the distribution of the total number of cases, this conclusion is not surprising. We have seen earlier that this test is not sensitive in detecting departures from the predicted distribution of chain frequencies.

A further analysis of this data set is due to Longini and Koopman (1982), who are concerned with estimating the probability of an infection from outside the household apart from the probability of an infection from other members of the household. They use only the distribution of the total number of cases and find that the probability of infection due to contact with the community is about two orders of magnitude lower than the infection probability within the household. Early on, McKendrick (1926) gave a method to estimate the ratio of external to internal infection probability. Further contributions to this problem are due to Sugiyama (1960), Kemper

(1980), Becker and Angulo (1981), Longini et al. (1982, 1984a, b), and Becker and Hopper (1983a, b).

We notice a highly desirable trend of increased concern about the statistical aspects of infectious disease modeling. It is hoped that the availability of methods for the analysis stimulates the collection of more data. From the example given it is obvious that the discrimination between models requires very detailed information. Aggregated data will inevitably lead to the rejection of the simple models. If the models are made more flexible by the introduction of more parameters, the fit can be significantly improved, but without *independent* information about the amount and the source of heterogeneity one cannot draw any firm epidemiological conclusions.

3. Age-Specific Cross-Sectional Surveys

Valuable epidemiologic information about permanently immunizing infections can be derived from the age-specific prevalence of individuals with immunity. Formerly these data were obtained from officially notified cases or by interviewing individuals for a past history of diseases, whereas more recently it has become practicable to assess the immunity status of individuals by more reliable serological methods for the detection of infection-specific antibodies. In any case, cross-sectional age-specific "spot surveys" of population immunity can be approached from three different points of view. Firstly, one may apply curve fitting techniques in order to obtain summary statistics for the purpose of description or comparison. Secondly, one may try to extract information about the intensity of past infection transmission in a population if the infection risk is not age-dependent. Finally, age-specific prevalences of immunity may be used to test hypotheses about the intensity of infection transmission between the various age groups in a population assuming a constant average infection risk.

Let $\lambda(a, t)$ denote the attack rate or force of infection for susceptibles of age a at time t. Then the proportion $S(a)$ of susceptibles in a cohort born at time t_0 decreases with age a according to the differential equation

$$\frac{dS(a)}{da} = -\lambda(a, t_0 + a)S(a), \tag{3.1}$$

assuming the initial condition $S(0) = 1$. This is the catalytic model of immunity acquisition, which describes "survival" of susceptibles under a "hazard rate" $\lambda(a, t)$. Collins (1929) made implicit use of this model, which was stated more explicitly later by Muench (1934, 1959), Wilson and Worcester (1941), Berger (1973), and Schenzle et al. (1979). In a survey carried out at a given point in time t_c the proportion $P(a)$ of previously infected individuals in a cohort of age a is

$$P(a) = 1 - \exp\left[- \int_0^a \lambda(\alpha, t_c - a + \alpha)\, d\alpha \right]. \tag{3.2}$$

Applying e.g. the maximum likelihood principle for comparing calculated values of $P(a)$ with corresponding observed prevalences of immunity, $\hat{P}(a)$, it is possible to test and identify alternative model expressions for the force of infection $\lambda(a, t)$. Often in practice due attention has to be paid to imperfect sensitivity and specifity in assessing immunity. Also real life data usually allow only the identification of simply parametrized model functions $\lambda(a, t)$, which are assumed to be independent of either time or age.

For example, in an analysis of hepatitis A prevalence data from several countries it was appropriate to concentrate on time dependence in the force of infection by putting

$$\lambda(t) = \frac{\lambda_\infty}{1 + \exp[\varphi(t - \theta)]} \tag{3.3}$$

with three adjustable parameters λ_∞, φ, and θ (Schenzle et al., 1979; Hu et al., 1984). Using this type of function, a fairly good fit to the data was obtained, thereby verifying and quantifying the impact of past sanitary improvements on the intensity of hepatitis A transmission in the sampled communities. In Germany the indigenous force of infection of hepatitis must have declined substantially during the 1950s from a value of 0.07/year to practically zero nowadays. This interpretation was confirmed by surveying the same population in 1965 and again in 1975 (Frösner et al., 1978). Using similar methods Sundaresan and Assaad (1973) were able to assess the effect of control measures against trachoma in Taiwan.

Perhaps age-specific prevalences of immunity have been most important in delineating the variable intensity of infection transmission within and between different age groups in a population. Especially for some common infections such as measles and rubella, a lot of data have been accumulated, although in the end these turn out not to be of sufficient detail and precision. Moreover the analysis and interpretation of these data is complicated by the fact that even in larger communities the infections under consideration appear in marked recurrent epidemics with interepidemic periods of one to several years (see below). Neglecting temporal changes in infection incidence may therefore lead to biased estimates of age-specific forces of infection, especially in younger age groups. Putting aside such complications, the available data display a characteristic feature of the so-called childhood infections in Europe and North America, namely the S-shaped rise in prevalence of immunity with age. This is well demonstrated in Figure 2, showing age-specific prevalences of measles immunity as observed in England and Wales, the USA, and Denmark (Fine and Clarkson, 1982a, b; Collins et al., 1942; Snyder et al., 1962; Horwitz et al., 1974). Therefore, long ago, Collins (1929) proposed to fit age-specific measles data by a "catalytic logistic curve," which is obtained essentially by assuming a linearly age-dependent force of infection:

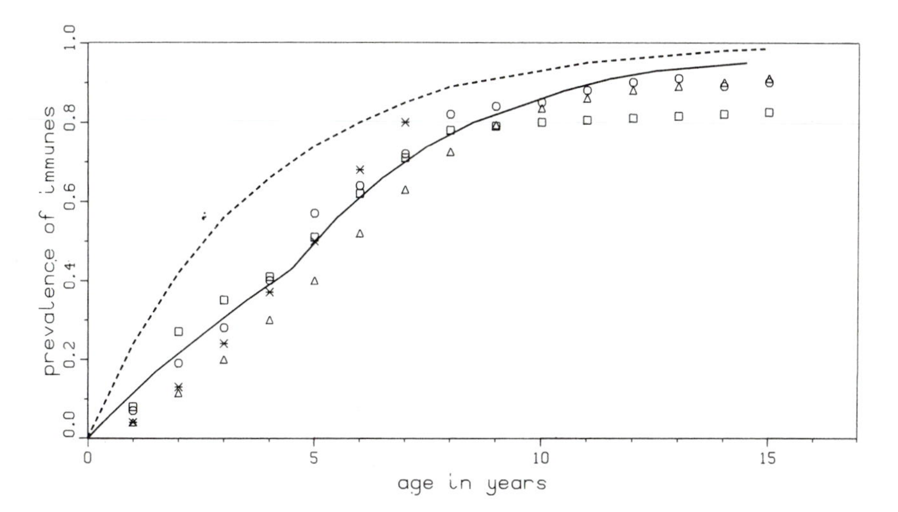

Figure 2. Observed age-specific prevalences of individuals with past history of measles infection in the U.S.A. (△, Collins et al., 1942), in Baltimore (□, Snyder et al., 1962), in England and Wales (∗, Fine and Clarkson, 1982b), and in Denmark (○, Horwitz et al., 1974). The theoretical curves are explained in Section 3.

$$\lambda(a) = b + ca. \tag{3.4}$$

The same model has been used more recently by Griffiths (1974) and by Anderson and May (1982b). The latter authors fitted measles immunity prevalences up to the age of ten by putting $b = 0.03$/year and $c = 0.057$/year^2.

Surprisingly this result is being largely ignored even in recent contributions to the modelling of the population dynamics of measles. In order to see its implications let us therefore outline a general measles model extending the approach taken earlier by Dietz (1975a). Consider a stable population in which aging is simulated by letting individuals pass through a series of "age compartments" numbered $i = 1, 2, \ldots, k$. Each of the first $k - 1$ consecutive compartments has an average sojourn time of one year, whereas the sojourn time in the final compartment is chosen to yield a total life expectancy of L years. At time t let $u_i(t)$ and $v_i(t)$ denote the proportions of individuals who are in age compartment i and susceptible or infectious, respectively. For simplicity individuals are supposed to become infectious immediately after being infected and to be removed into the class of immunes at a constant rate γ. Finally the force of infection on susceptibles in age compartment i is taken to be of the form

$$\lambda_i(t) = \sum_{j=1}^{k} \beta_{ij} v_j(t), \tag{3.5}$$

where the k^2 quantities β_{ij} denote differential contact rates. Then, choosing one year as the unit of time, the model equations read

$$\begin{aligned}
\dot{u}_1 &= \frac{1}{L} - \lambda_1(t)u_1 - u_1, \\
\dot{u}_j &= u_{j-1} - \lambda_j(t)u_j - u_j \quad \text{for } j = 2, \ldots, k-1, \\
\dot{u}_k &= u_{k-1} - \lambda_k(t)u_k - \frac{u_k}{L-k+1}, \\
\dot{v}_1 &= \lambda_1(t)u_1 - \gamma v_1 - v_1, \\
\dot{v}_j &= v_{j-1} + \lambda_j(t)u_j - \gamma v_j - v_j \quad \text{for } j = 2, \ldots, k-1, \\
\dot{v}_k &= v_{k-1} + \lambda_k(t)u_k - \gamma v_j - \frac{v_j}{L-k+1}.
\end{aligned} \tag{3.6}$$

In the special case $k = 1$ and therefore $\beta_{ij} \equiv \beta$, corresponding to uniform measles transmission in a population, the model (3.6) reduces to the model considered by Dietz (1975a). Then, in an endemic equilibrium there is a constant, age-independent force of infection $\bar{\lambda}$, and the prevalence of measles immunity increases for large γ with age as $1 - \exp(-\bar{\lambda}a)$. But this is not observed in the measles data from North America and Europe quoted above. If one tries to fit the generalized model (3.6) to the data shown in Figure 2, then two serious problems arise. Firstly, there is a lack of empirical information about the forces of infection of measles for adolescents and adults. This information is important for the estimation of the differential contact rates with and among adolescents and adults, who make up 80% of the total population. The linear function (3.4) fitted by Griffiths (1974) and Anderson and May (1982b) for children up to the age of ten is misleading for adults. Secondly, from Equation (3.5) it is obvious that a given set of endemic age-specific forces of infection may be fitted by many different matrices of differential contact rates β_{ij}. Two special choices have been considered by Cvjetanović et al. (1982) and Knolle (1983). Hence the population dynamics of measles cannot be predicted uniquely from a knowledge of age-specific forces of infection. This insight gains practical relevance in connection with the prediction of effects from mass vaccination, as we shall see in the next section.

4. Long-Term Patterns of Case Incidence

The explanation of case incidence patterns not only poses fascinating theoretical problems but also is of high relevance for the attempts to reduce incidence through mass immunization. There exist long-term records of morbidity and mortality over decades or even centuries. Thus Brownlee (1918) could apply his method of periodogram analysis to reveal various periodicities in measles mortality data from London during 1703–1918. Reliable incidence data from officially notified cases of disease do not cover such a long time period, but they suffice to recognize some characteristic trends and patterns. These vary markedly between infectious agents and communities. For example, an infection like hepatitis B may persist in small isolated populations because a proportion of infected individuals enters a long-term infectious carrier state (Black et al., 1974; Gust et al., 1979; Skinhoj et al., 1980). Hepatitis A, which is transmitted by the fecal–oral route, may be perpetuated at a fairly stable level in communities of moderate size (Wong et al., 1979; Nuti et al., 1982). Measles however has a short infectious period and requires fairly close person-to-person contact for transmission. Hence even larger communities tend to be free of infection until imported cases cause transient outbreaks, which may be separated by many years (Bartlett, 1957, 1960; Black, 1966).

We shall therefore focus our attention to one type of infection in one type of population, namely on the so-called childhood infections in large cities in Europe and North America. These infections, such as measles, mumps, and rubella, are highly contagious, directly transmitted, and permanently immunizing. Moreover they share the common feature of being endemic in large cities yet with marked oscillations. Some representative data are shown in Figure 3, which displays incidences of measles and mumps in New York City reported by Yorke and London (1973). Similar large-scale fluctuations have been observed at other places (Fine and Clarkson 1982a; Schütz, 1925; Soper, 1929).

The phenomenon of recurrent epidemics, especially the "biennial measles cycle," has long puzzled epidemiologists and mathematicians alike. A hundred years ago Hirsch (1833) pronounced that "the recurrence of the epidemics of measles at one particular place is connected neither with an unknown something (the mystical number of the Pythagoreans) nor with general constitutional vicissitudes . . . , but it depends solely on two factors, the time of importation of the morbid poison, and the number of susceptibles to it." A similar reasoning about the "accumulation of susceptibles" made Hamer (1906) believe he had explained the London measles figures. But this reasoning eventually turned out to be fallacious, and up to the present day the problem of recurrent epidemics has not been definitively settled.

It was the epidemiologist E. Martini (1921) who explicitly wrote down what still is the "standard" model for infections like measles in a homogeneous population. This model has been reformulated independently by Hethcote (1974) and by Dietz (1975a), who also established the connection with the simple catalytic model (see the previous section). Denoting by $x(t)$ and $y(t)$ the numbers of susceptibles and infectives in a population of size n, the model equations read

$$\dot{x} = \nu - \beta x y - \mu x, \tag{4.1}$$

$$\dot{y} = \beta x y - \gamma y - \mu y, \tag{4.2}$$

where ν denotes the number of births per unit of time, μ the death rate, γ the infection removal rate, and β a population- and infection-specific contact rate. Lotka (1923b) determined the equilibrium solutions of (4.1)–(4.2) and their

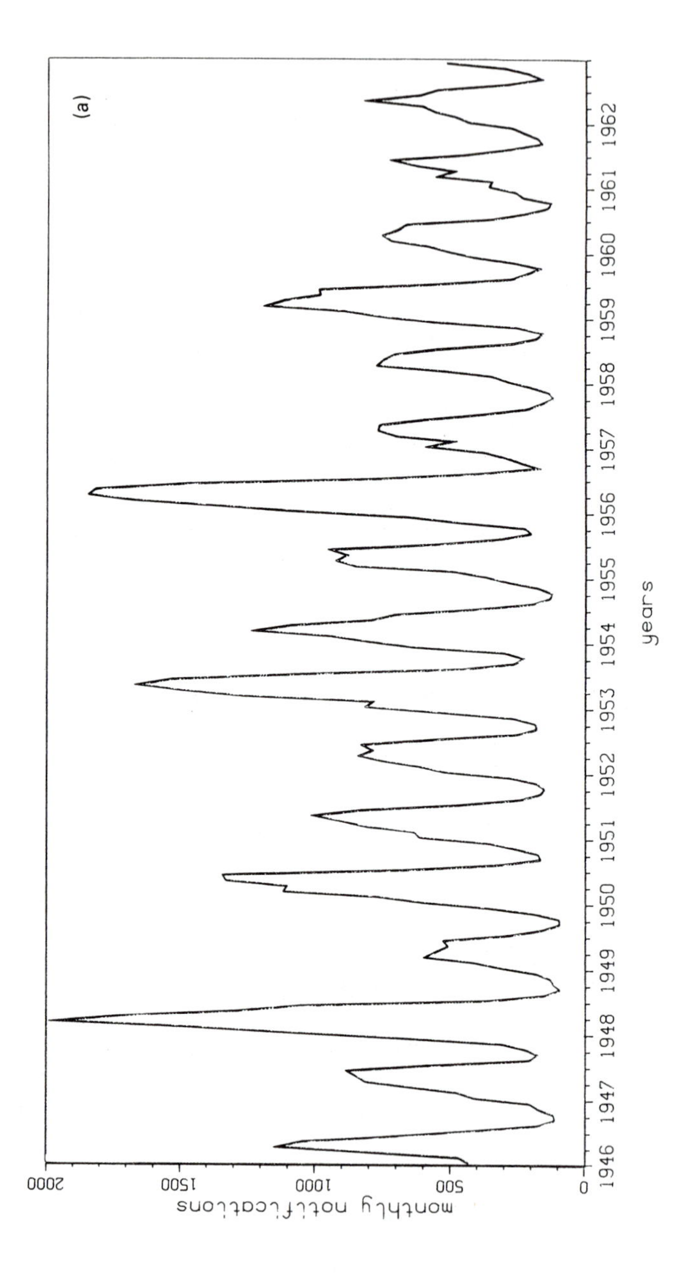
(a)
monthly notifications
0
500
1000
1500
2000
years
1946
1947
1948
1949
1950
1951
1952
1953
1954
1955
1956
1957
1958
1959
1960
1961
1962

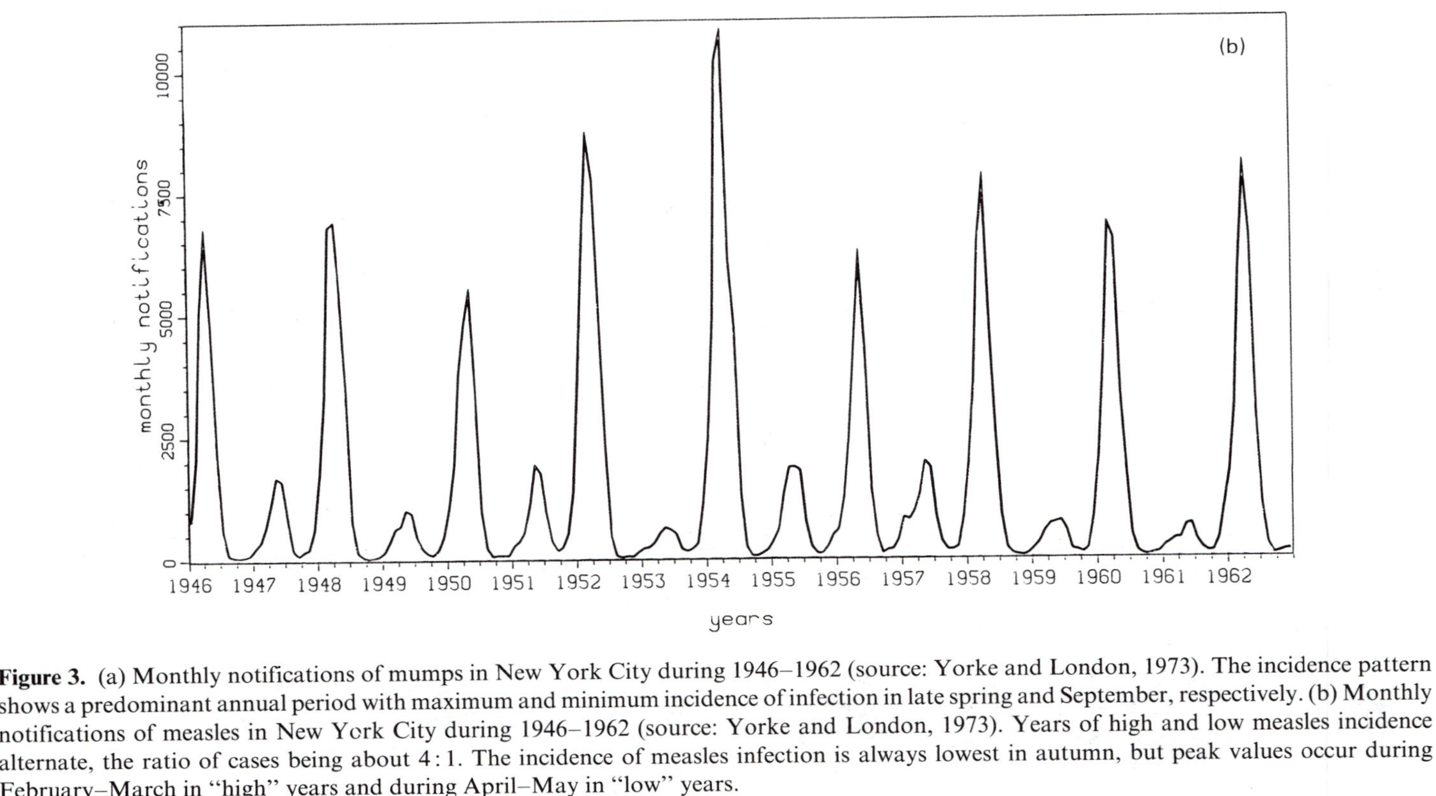

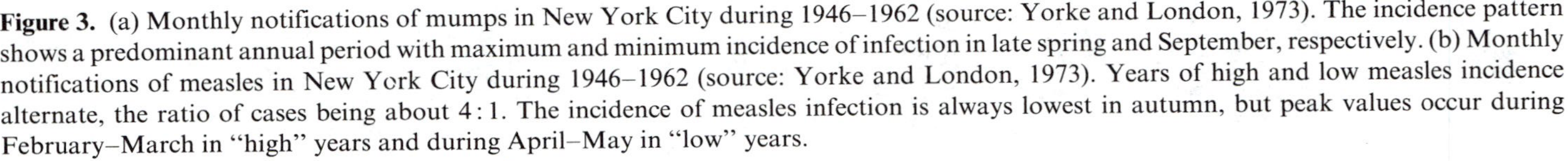

Figure 3. (a) Monthly notifications of mumps in New York City during 1946–1962 (source: Yorke and London, 1973). The incidence pattern shows a predominant annual period with maximum and minimum incidence of infection in late spring and September, respectively. (b) Monthly notifications of measles in New York City during 1946–1962 (source: Yorke and London, 1973). Years of high and low measles incidence alternate, the ratio of cases being about 4:1. The incidence of measles infection is always lowest in autumn, but peak values occur during February–March in "high" years and during April–May in "low" years.

stability properties. Apart from the trivial solution $x = v/\mu = n$, $y = 0$ there is one further stationary solution for $R = \beta n/(\gamma + \mu) > 1$:

$$\bar{x} = n/R \tag{4.3}$$

and

$$\bar{y} = \frac{v(1 - R^{-1})}{\gamma + \mu}. \tag{4.4}$$

This solution represents a stable endemic equilibrium, and hence the model (4.1)–(4.2) allows only damped oscillations, the period of which is approximately given by

$$T \approx \frac{2\pi}{(\bar{\lambda}\gamma)^{1/2}}, \tag{4.5}$$

where $\bar{\lambda} = \mu(R - 1)$ is the endemic force of infection corresponding to a mean age at infection $\bar{A} = 1/\bar{\lambda}$ (Dietz, 1975a; Anderson and May, 1982b). Therefore the Hirsch–Hamer argument does not explain recurrent epidemics under endemic conditions. [This conclusion by the way could have been reached already in 1910, when Lotka (1910) found a stable equilibrium in a chemical reaction system structurally identical to the system (4.1)–(4.2).] Soper (1929), who is commonly credited with having developed the first mathematical model for measles, does not make any reference to the work of Martini (1921) or Lotka (1923b). He too realized the damped nature of the solutions of the model (4.1)–(4.2) (with $\mu = 0$ however), and therefore he was led to consider his "point of infection law" and to investigate the discrete-time model

$$\begin{aligned} x_{t+1} &= x_t + v - \beta x_t y_t, \\ y_{t+1} &= \beta x_t y_t, \end{aligned} \tag{4.6}$$

which reduces to the second order difference equation

$$h_{t+2} + (\beta v - 2)h_{t+1} + h_t = 0. \tag{4.7}$$

This equation can be shown (see, e.g., Jordan, 1960) to yield undamped oscillations, but these are only marginally stable and their amplitude depends on the initial values. Actually Soper's (1929) model (4.6) is structurally unstable, since any reformulation including a finite infection removal rate or a death rate leads to damped solutions (Wilson and Worcester, 1945; Bailey, 1975, p. 143). This should be stated very clearly when using the discrete-time model for "illustrative" purposes, as has been done very recently by Fine and Clarkson (1982c) and by Anderson and May (1982c).

Facing the difficulties in predicting undamped oscillations, Wilson and Worcester (1945) even doubted the reality of periods, arguing that possibly "measles simply dies out and then returns." Therefore Bartlett (1956, 1957, 1960) started to analyse the stochastic version of the model (4.1)–(4.2), also putting $\mu = 0$. This has led to the concept of a critical community size necessary for the endemic persistence of an infection in a finite population. Measles, for example, appears to persist in communities of size 500,000 or larger (Bartlett 1957, 1960; Black, 1966). But even under such conditions the nonlinear mass action term apparently induces macroscopic fluctuations, the "period" of which roughly corresponds to the natural period (4.5) of the model (4.1)–(4.2). In his stochastic simulations Griffiths (1973b) took $\bar{A}$ equal to 2.6 years. This, together with a measles generation period $\Gamma = 1/\gamma$ of two weeks, yields exactly a period $T = 2$ years, which is also apparent in the numerical simulations. But the value $\bar{A} = 2.6$ years, corresponding to an endemic force of infection $\bar{\lambda} = 1/\bar{A} = 0.38$/year, hardly conforms to the data shown in Figure 2. Moreover, the stochastic model cannot explain the annual oscillations in the case of mumps. Thus it is an overstatement to say "that a stochastic investigation appears to be essential if the absence of damping in large communities ... is to receive a reasonably realistic explanation" (Bailey, 1975, p. 143). The "stochastic instability" of endemic states is an important aspect even in large populations, but it cannot be considered the proper cause of the fluctuations shown in Figure 3.

Returning to the deterministic framework, it has been suggested that one consider modifications or extensions of the model (4.1)–(4.2) in accordance with some general criteria guaranteeing sustained oscillations (Tyson and Light, 1973; Hanusse, 1972; Cronin, 1977). Thus Cunningham (1979) modified the mass action term in (4.1)–(4.2) to take the form $(xy)^r$. Then for $r > 1$ oscillations of the limit cycle type are possible, but these show little similarity with observed incidence patterns. It is also difficult to give a reasonable explanation for the exponent r. Enderle (1980) in turn considered and simulated a model, which is essentially identical with the age-compartment model (3.6) presented above. This model can produce undamped oscillations, but only for special and highly asymmetric choices of the matrix of contact rates β_{ij}. In order to simulate measles incidence patterns with realistic model parameter values, Enderle (1980) had to assume that the contact rates vary during the year.

Now it has always been known that the incidences of measles, mumps, and rubella show a strong seasonal component (Soper, 1929; Bliss and Blevins, 1959; London and Yorke, 1973; Yorke and London, 1973; Fine and Clarkson, 1982a). Therefore Lotka (1923b) pointed to a possible periodicity in the parameters of the model (4.1)–(4.2). Soper (1929) then used monthly measles figures from Glasgow in order to derive an effective annually periodic contact rate $\beta(t)$. This procedure has been repeated by London and Yorke (1973) with the New York City data of Figure 3, and also by Fine and Clarkson (1982a) with measles data from England and Wales. The resulting effective contact rates in fact vary throughout the year by ± 20–30%. Accordingly, as conjectured by Soper (1929) but denied by Bartlett (1956), the model system (4.1)–(4.2) or some variants thereof may become forced or excited into annual, biennial, or even multiannual oscillations. Subsequently this has been shown to be the case (London and Yorke, 1973; Stirzaker, 1975; Dietz 1976; Grossman et al., 1977; Grossman, 1980; Gumowski et al., 1980; Smith, 1983a, b). However, here the same problem arises as in the stochastic model: in order to obtain biennial oscillations the natural period T given in (4.5) should take a value of about two years. For infection generation periods between one and two weeks this condition requires mean ages at infection between 5.3 and 2.6 years respectively, but such low values cannot be reconciled with the data shown in Figure 2. Nevertheless, the parametric resonance approach has some appeal and epidemiological foundation—and it gives reason to think of still another mechanism causing endemic incidence fluctuation to be sustained.

Perhaps one should question Bartlett's (1956) statement that it is largely irrelevant whether seasonal changes in the contact rate "are due to atmospheric changes ... or due merely to more artificial causes such as dispersion and subsequent reassembly of school children during the summer." Atmospheric changes would indeed affect a whole population, whereas school-related events are primarily relevant to children. Moreover, after summer vacations school children are not only reassembled, but a whole cohort of mostly susceptible children enter school for the first time, thereby replacing mostly immunized children of the previously highest grade. It is not at all clear whether these events can be adequately described with a seasonally varying overall contact rate. This has been tested by simulating a modified version of the age-structured model (3.6) described above. As stated there, individuals pass continuously through the various age compartments. But now we take the first 15 compartments to represent grades, and only once every year all individuals in each grade are promoted into the next higher grade, those in the highest grade joining the residual compartment of "adults." All this happens at the beginning of each new school year, in the course of which the lowest grade is being replenished by new births. In its simplest version this model uses only two different contact rates, an overall contact rate β_1, which is increased to $\beta_1 + \beta_2$ among children in grade 7–15. Holidays and vacations are simulated by putting $\beta_2 = 0$. In the case $\beta_2 \equiv 0$ one essentially recovers the model (4.1)–(4.2).

Simulating this modified model (3.6) on a day-by-day basis yields oscillatory incidence patterns with various periods of one to six years, depending on the parameter values of β_1, β_2, and γ. Surprisingly these oscillations are obtained even without including any vacations. This demonstrates that it is the entry into school of a new grade which provides the "shove" maintaining the oscillations. However, the precise incidence pattern depends critically on how summer vacations are temporally located relative to the beginning of the school year. The pattern shown in Figure 4(a) arises if schools open on September 1 and if summer vacations begin with June 1, whereas in Figure 4(b) the school year begins on April 15 and summer vacations extend from July 1 to August 15. Both simulations include Christmas vacation from December 20 to January 9 and Easter vacation from April 1 to 14, but otherwise schools are closed only on Sundays. The pattern shown in Figure

4(a) conforms to the measles incidence pattern classified empirically as type I by Schütz (1925). It is to be found in communities where the school year begins in autumn. In Germany Schütz (1925) classified most measles incidence patterns as type II, with two peaks per year. These patterns arise because formerly school years in Germany began in April.

Thus the modified model (3.6), operating on a grade basis, yields infection incidence patterns resembling those observed for measles. At the same time it predicts the prevalence curve of measles immunity shown in Figure 2 (full line). This curve conforms to the data better than the simple catalytic curve (dashed line) predicted from homogeneous mixing models of the type (4.1)–(4.2). Note also that the performance of the model (3.6) can of course be improved further by using more than the two contact rates β_1 and $\beta_1 + \beta_2$.

Let us finally mention the relevance of the model (4.1)–(4.2) for the assessment and prediction of effects of mass vaccination. The simplest way to incorporate mass immunization into the model (4.1)–(4.2) consists in replacing the birth rate v by $v' = qv$, where q denotes the proportion of newborns still entering the pool of susceptibles. Then, because of (4.4), for $q \leq R^{-1}$ the infection cannot remain endemic and ultimately the population possesses herd immunity. For measles the prevaccination endemic proportion $\bar{x}/n$ of susceptibles is commonly estimated as about 6% (Yorke et al., 1979; Anderson and May, 1983; Hethcote, 1983). Accordingly the achievement and maintenance of herd immunity against measles requires ongoing immunization of at least 94% of each birth cohort. However, a specification of the age-structured model (4.1)–(4.2), which fits the age-specific data of Figure 2, predicts an immunization rate of only 80% to be sufficient for herd immunity against measles (Schenzle, 1985a, b). In view of current attempts to eliminate measles in U.S.A. and in other countries this statement is of practical relevance and actuality. There is thus an urgent need to investigate age-structured epidemiologic models and to test them thoroughly against empirical material.

5. Outlook

So far we have dealt with infectious disease statistics which take into account only the time dimension: the generation in an epidemic chain, the weekly or yearly notifications, the age-specific prevalence of antibodies as a reflection of infections acquired or missed in the past, and the age dependence of contact rate. This last aspect leads us naturally to the discussion of the heterogeneities in the transmission due to distance. In the section on household epidemics we mentioned several attempts to estimate the ratio of the contact rates between and within households. But for these small scale investigations there exist only very few suitable data sets with enough detail. Routine records only provide the number of cases by time and by administrative regions. The most extensive analyses of these aspects have been concerned with influenza in the USSR (Baroyan et al., 1977) and with measles in Iceland (Cliff et al., 1981). In the

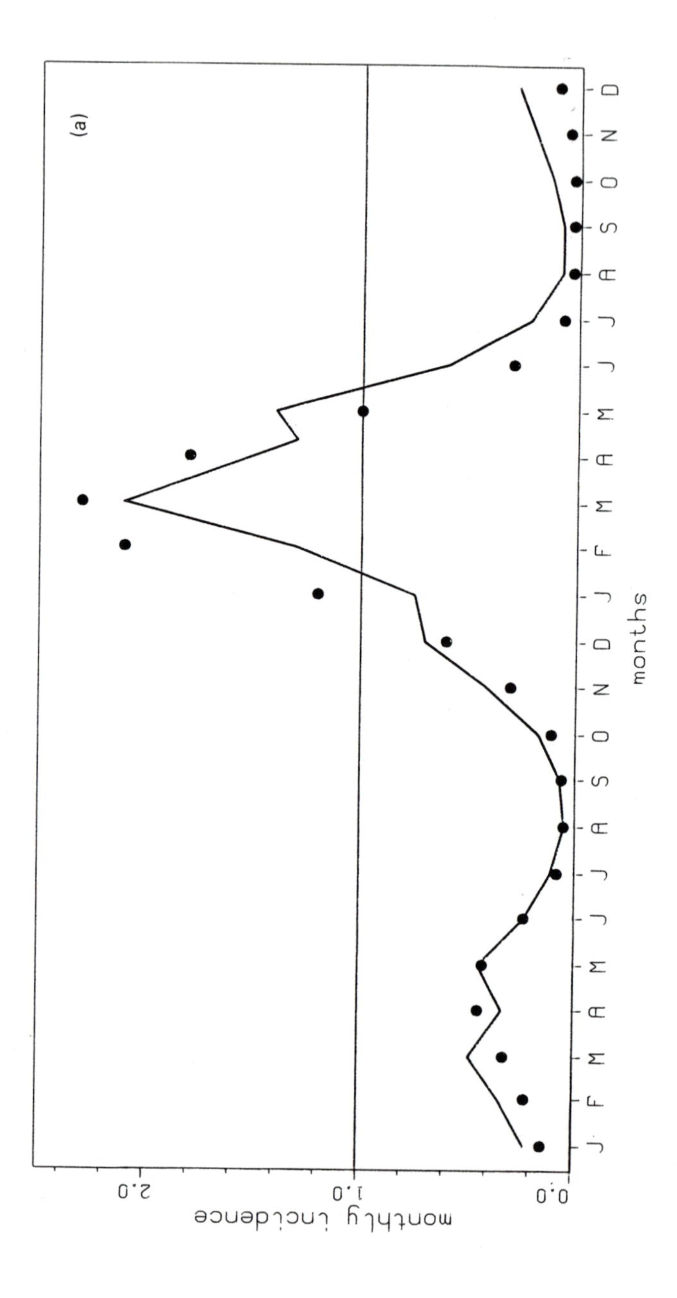
(a)
monthly incidence
0.0
1.0
2.0
months
J F M A M J J A S O N D J F M A M J J A S O N D

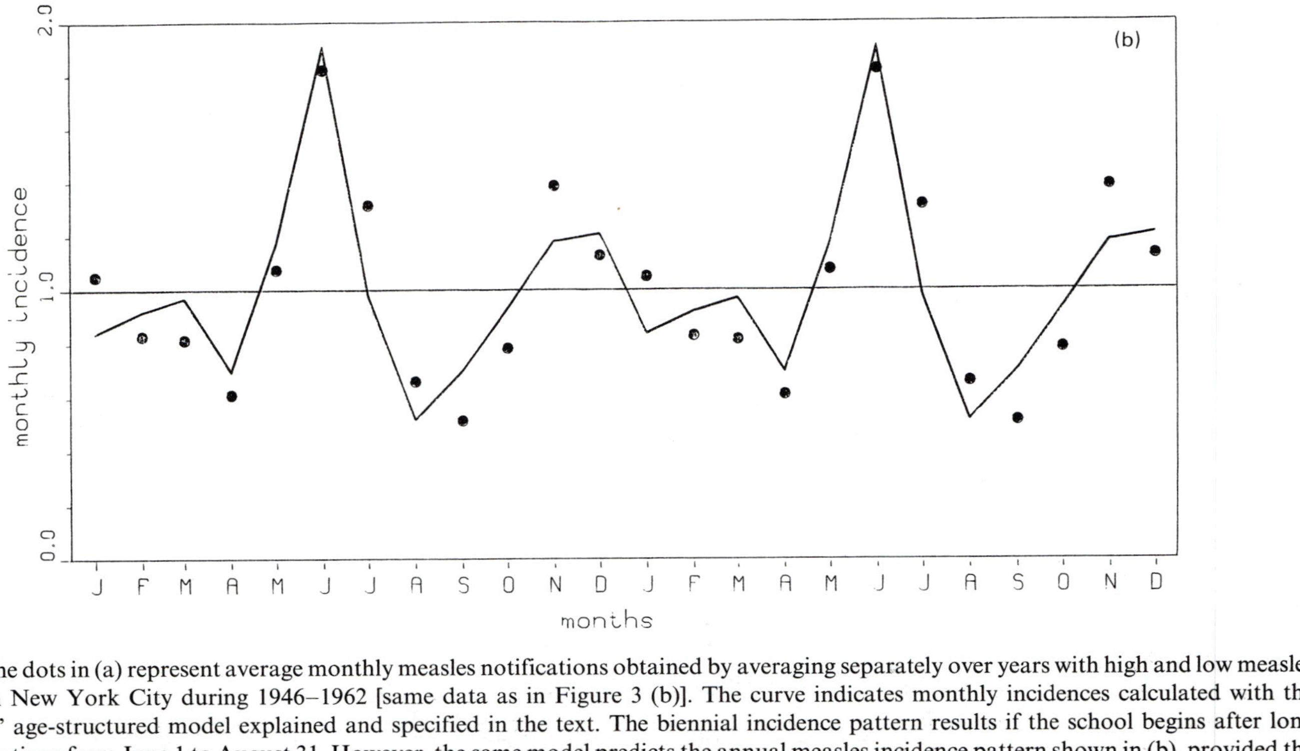

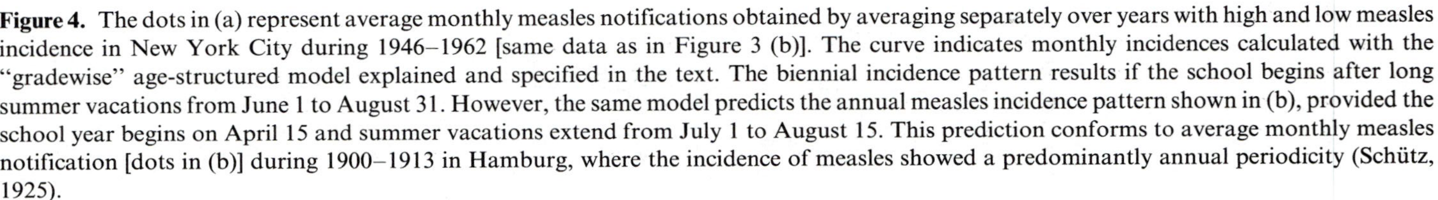
Figure 4. The dots in (a) represent average monthly measles notifications obtained by averaging separately over years with high and low measles incidence in New York City during 1946–1962 [same data as in Figure 3 (b)]. The curve indicates monthly incidences calculated with the "gradewise" age-structured model explained and specified in the text. The biennial incidence pattern results if the school begins after long summer vacations from June 1 to August 31. However, the same model predicts the annual measles incidence pattern shown in (b), provided the school year begins on April 15 and summer vacations extend from July 1 to August 15. This prediction conforms to average monthly measles notification [dots in (b)] during 1900–1913 in Hamburg, where the incidence of measles showed a predominantly annual periodicity (Schütz, 1925).

USSR influenza is recorded by day for the major towns. The model tries to predict the onset and the size of the epidemics in the years 1971–1972, 1972–1973, 1974–1975, and 1976 in the 100 largest towns given the information that the spread of influenza started in one of them. The contact rates between the towns are assumed to be proportional to the rates of public transport. The details of the results are not available in English, but Bailey (1980) and Fine (1982) provide good summaries. The measles data in Iceland consist of monthly numbers of reported cases in each of the some 60 medical districts. Between 1904 and 1975 there occurred 16 separate epidemics. The raw data are reproduced by Cliff et al. (1981) so that additional analyses with new approaches can be performed. Haggett (1982) discusses the possible relevance of this work to influenza epidemics. Both the British and the Soviet team use what Bailey (1980) calls multisite modeling, i.e. the description of the spread of epidemics in a network of subpopulations. Within each subpopulation one assumes homogeneous mixing. The contact rates between subpopulations are assumed to vary either by public transport rates or simply by geographic distance. This multisite approach to measles epidemics has also been applied to British data sets (Haggett, 1972, 1976; Cliff et al., 1975; Murray and Cliff, 1977), and Leeuwenburg et al. (1979) have published a data set from Kenya to which it could be applied.

Bailey's (1980) survey covers, in addition to these multisite models, a large number of mathematical papers which describe the spread of epidemic waves either in one- or two-dimensional space, where the space coordinates are either continuous or restricted to the lattice points. But well-defined wave fronts are only seen in the spread of certain plant diseases or wildlife diseases, such as fox rabies. In human populations they no longer occur, mainly due to air traffic. If there is a clear wavelike advance, as documented for example by Bögel et al. (1976) in the case of rabies, one may consider the prevention of further spread by ring vaccination. Thieme (1980) seems to have been the first to treat this problem mathematically. Lambinet et al. (1978) describe the spatial spread of fox rabies by a simulation model.

If there are heterogeneities in the contact rates between population subgroups, this probably has implications for the design of vaccination strategies (Becker, 1979). A uniform coverage does not appear to be optimal. On the other hand, if whole subpopulations do not participate in vaccination, one may experience epidemics even if the coverage in the total population is extremely high (CDC, 1981; Scott, 1971). The effect of nonparticipating subpopulations on the control of parasites by chemotherapy has been evaluated on the basis of Ross's malaria model by Dietz (1975b). The problems of selective vaccination (Longini et al., 1978) or chemotherapy (Anderson and May, 1982a; Dietz and Renner, 1985) have not yet received sufficient attention. The optimal strategy is likely to depend on the contact rates. Since these cannot be estimated directly, one may have to resort to proxy variables with which they are highly correlated, like age and place. For example in a study on the transmission on schistosomiasis, Rosenfield et al. (1977) proposed distance to snail-infested water as a determining factor of the contact rate. It is not clear whether heterogeneous contact rates increase or decrease the stability of endemic equilibria. The answer may depend on the strength of immunity. We conjecture that heterogeneity increases stability for helminth diseases where there is little or no immunity against reinfection. For virus diseases with lifelong immunity, homogeneous mixing seems to reduce the critical population size for the persistence of the endemic state, as shown by preliminary simulation studies by a student of ours (Brenner, 1985).

In practice there is not only heterogeneity in space but also in time. Contact rates vary by season under the direct or indirect influence of meteorological variables. This has been taken into account by the introduction of time-dependent but deterministic contact rates. It appears worthwhile to use the tools of stochastic differential equations and population processes in random environments. Becker (1977) considered a branching process with random environment in connection with the estimation of the reproduction rate of smallpox epidemics. There seems to be great potential for further developments.

A given vaccine protects only against the virus for which it induces antibodies. In order to save costs it is common practice to administer several of them simultaneously, e.g. measles, mumps, and rubella vaccine (MMR). Similarly, the provision of safe water will reduce the exposure to a whole range of waterborne diseases. The control of mosquitoes, which may transmit both malaria and filariasis in certain regions, will reduce the incidence of both diseases. These examples suggest considering the joined distribution of certain infectious diseases. If there are heterogeneities in contact rates and if several infections are transmitted via the same sort of contact, it is likely that certain individuals will acquire simultaneously more different infections than expected on the assumption of homogeneous contacts. Buck et al. (1978a, b) have provided empirical evidence for the association of multiple infections. Cohen and Singer (1979) study the joint distribution of the two malaria parasites *Plasmodium falciparum* and *Plasmodium malariae* on the basis of longitudinal data provided by the Garki project of the epidemiology of malaria in the African savannah (Molineaux and Gramiccia, 1979). They find that initially uninfected individuals tend to stay uninfected with higher frequency, and initially doubly infected individuals tend to remain doubly infected with higher frequency, than would be expected on a Markovian assumption. Stimulated by observations by Bang (1975) on the spatial distribution of certain enteroviruses in India, Dietz (1979) investigated the local stability of the four possible equilibria in a model for the interference between two viruses. The study of multiple infections is of great theoretical and practical interest, yet very little has been done in this field because of lack of suitable data.

Mathematical models in genetics and in the epidemiology of infectious diseases have long but separate histories. Only recently have the two fields joined forces to tackle the problem of co-evolution of parasite–host interaction. For the description of a single epidemic one can neglect evolutionary aspects, but not for the study of endemic equilibria, especially under long-term application of chemical control. There are not only the questions of resistance which can jeopardize the success of control programs. Basic evolutionary problems like the maintenance of sexual reproduction have been discussed in the framework of parasite transmission models. The report of the Dahlem Workshop on the Population Biology of Infectious Diseases (Anderson and May, 1982d) provides a stimulating introduction to this new field and gives key references. One of the questions is concerned with the evolution of pathogenicity of a parasite from virulence to commensalism. Some models suggest that intermediate levels of virulence may be optimal for the parasite. This interaction of genetical and epidemiological modeling is likely to be fruitful for both fields.

In this survey we have restricted ourselves primarily to models for human virus diseases because there is a clear tendency: as models become more realistic and relevant they also become less applicable to a whole class of diseases. Even within human virus diseases it is not meaningful to evaluate vaccination strategies for measles and rubella with the same models, because of different objectives in different countries: In the U.S.A. one tries to eliminate the infection; in the U.K. one only aims at the elimination of the risk of congenital rubella syndrome.

We have covered neither human diseases caused by bacteria, protozoa, and helminths, nor diseases of animals and plants (Morris, 1972; Kranz, 1974; Waggoner, 1981). It may be interesting to note that the epidemic modeling approach has also been applied to the spread of drug addiction (Egan and Robinson, 1979; Mackintosh and Steward, 1979; Hoppensteadt and Murray, 1981). Even the transmission of scientific ideas has been described by the Kermack–McKendrick equations (Goffman and Newill, 1964). The rise in the number of contributions to epidemic theory has been compared to the rise of an epidemic, which Fine (1979) exhibits graphically on the basis of Bailey's (1975) comprehensive bibliography. But epidemic theory would predict sooner or later a decline and extinction. The theory of infectious diseases will certainly not become obsolete because the diseases will be eradicated. There are good reasons to predict that smallpox will remain the single human disease for which global eradication has been successfully completed. In view of the ever increasing number of problems which are susceptible to solution, the theory of infectious diseases will stay above the threshold of extinction and will maintain a stable endemic state—possibly with oscillations.

Bibliography

Aalen, O. O. (1978). "Nonparametric inference for a family of counting processes." *Ann. Statist.*, **6**, 701–726.

Anderson, R. M. (1982). "The population dynamics and control of hookworm and roundworm infections." In R. M. Anderson (ed.), *The Population Dynamics of Infectious Diseases: Theory and Applications*. London: Chapman and Hall, 67–108.

Anderson, R. M. and May, R. M. (1979). "Population biology of infectious diseases: I." *Nature*, **280**, 361–367.

Anderson, R. M. and May, R. M. (1982a). "Population dynamics of human helminth infections: Control by chemotherapy." *Nature*, **297**, 557–563.

Anderson, R. M. and May, R. M. (1982b). "Directly transmitted infectious diseases: Control by vaccination." *Science*, **215**, 1053–1060.

Anderson, R. M. and May, R. M. (1982c). "The logic of vaccination." *New Scientist*, **96**, 410–415.

Anderson, R. M. and May, R. M. (eds.) (1982d). *Population Biology of Infectious Diseases*. Berlin: Springer.

Anderson, R. M. and May, R. M. (1983). "Vaccination against rubella and measles: Quantitative investigations of different policies." *J. Hyg. Camb.*, **90**, 259–325.

Aron, J. L. and May, R. M. (1982). "The population dynamics of malaria." In R. M. Anderson (ed.), *Population Dynamics of Infectious Diseases*, London: Chapman and Hall, 139–179.

Bailey, N. T. J. (1967). "The simulation of stochastic epidemics in two dimensions." In *Proceedings of the Fifth Berkeley Symposium on Mathematical Statistics and Probability*, **4**, 237–257.

Bailey, N. T. J. (1975). *The Mathematical Theory of Infectious Diseases and its Applications*, 2nd edn. London: Griffin.

Bailey, N. T. J. (1980). "Spatial models in the epidemiology of infectious diseases." *Lecture Notes in Biomath.*, **38**, 233–261.

Bailey, N. T. J. (1982). *The Biomathematics of Malaria*. London: Griffin.

Bang, F. B. (1975). "Epidemiological interference." *Internat. J. Epidemiol.*, **4**, 337–342.

Baroyan, O. V., Rvachev, L. A., and Ivannikov, Yu. G. (1977). *Modeling and Prediction of Influenza Epidemics in the USSR*. Moscow: N. F. Gamaleia Inst. of Epidemiology and Microbiology. (In Russian.)

Bart, K. J., Orenstein, W. A., Hinman, A. R., and Amler, R. W. (1983). "Measles and models." *Internat. J. Epidemiol.*, **12**, 263–266.

Bartlett, M. S. (1956). "Deterministic and stochastic models for recurrent epidemics." In *Proceedings of the Third Berkeley Symposium on Mathematical Statistics and Probability*, **4**, 81–109.

Bartlett, M. S. (1957). "Measles periodicity and community size." *J. Roy. Statist. Soc. Ser. A*, **120**, 48–70.

Bartlett, M. S. (1960). "The critical community size for measles in the United States." *J. Roy. Statist. Soc. Ser. A*, **123**, 37–44.

Becker, N. (1977). "Estimation for discrete time branching processes with applications to epidemics." *Biometrics*, **33**, 515–522.

Becker, N. (1979). "The uses of epidemic models." *Biometrics*, **35**, 295–305.

Becker, N. (1980). "An epidemic chain model." *Biometrics*, **36**, 249–254.

Becker, N. (1981a). "A general chain binomial model for infectious diseases." *Biometrics*, **37**, 251–258.

Becker, N. (1981b). "The infectiousness of a disease within households." *Biometrika*, **68**, 133–141.

Becker, N. and Angulo, J. (1981). "On estimating the contagiousness of disease transmitted from person to person." *Math. Biosci.*, **54**, 137–154.

Becker, N. G. and Hopper, J. L. (1983a). "Assessing the heterogeneity of disease spread through a community." *Amer. J. Epidemiol.*, **117**, 362–374.

Becker, N. G. and Hopper, J. L. (1983b). "The infectiousness of a disease in a community of households." *Biometrika*, **70**, 29–39.

Bell, G. I., Perelson, A. S., and Pimbley, G. H. Jr. (eds.) (1978). *Theoretical Immunology*. New York: Marcel Dekker.

Berger, J. (1973). "Zur Infektionskinetik bei Toxoplasmose, Röteln, Mumps und Zytomegalie." *Zbl. Bakt. Hyg., I. Abt. Orig., A*, **224**, 503–522.

Bernoulli, D. (1760). "Essai d'une nouvelle analyse de la mortalité causée par la petite vérole et des avantages de l'inoculation pour la prévenir." *Mém. Math. Phys. Acad. Roy. Sci. Paris*, 1–45.

Black, F. L. (1966). "Measles endemicity in insular populations: Critical community size and its evolutionary implication." *J. Theoret. Biol.*, **11**, 207–211.

Black, F. L., Hierholzer, W. J., De Pinheiro, F., et al. (1974). "Evidence for persistence of infectious agents in isolated human populations." *Amer. J. Epidemiol.*, **100**, 230–250.

Bliss, C. I. and Blevins, D. L. (1959). "The analysis of seasonal variation in measles." *Amer. J. Hyg.*, **70**, 328–334.

Bögel, K., Moegle, H., Knorpp, F., Arata, A., Dietz, K., and Diethelm, P. (1976). "Characteristics of the spread of a wildlife rabies epidemic in Europe." *Bull. World Health Org.*, **54**, 433–447.

Bradley, D. J. (1982). "Epidemiological models–theory and reality." In R. M. Anderson (ed.), *The Population Dynamics of Infectious Diseases: Theory and Applications*. London: Chapman and Hall, 320–333.

Brenner, H. (1985). *Simulationsstudien zu zyklisch wiederkehrenden Infektionskrankheiten in räumlich heterogenen großen Populationen am Beispiel der Masern*. Dissertation. Eberhard-Karls-Universität, Tübingen.

Brimblecombe, F. S. W., Cruickshank, R., Masters, P. L., Reid, D. D., and Stewart, G. T. (1958). "Family studies of respiratory infections." *Brit. Med. J.*, **1**, 119–128.

Brownlee, J. (1918). "An investigation into the periodicity of measles epidemics in London from 1703 to the present day by the method of the periodogram." *Philos. Trans. Roy. Soc. Lond. Ser. B*, **208**, 225–250.

Buck, A. A., Anderson, R. I., Macrae, A. A., and Fain, A. (1978a). "Epidemiology of poly-parasitism: I. Occurrence, frequency and distribution of multiple infections in rural communities in Chad, Peru, Afghanistan, and Zaire." *Tropenmed. Parasit.*, **29**, 61–70.

Buck, A. A., Anderson, R. I., Macrae, A. A., and Fain, A. (1978b). "Epidemiology of poly-parasitism: II. Types of combinations, relative frequency and associations of multiple infections." *Tropenmed. Parasit.*, **29**, 137–144.

CDC (1981). "Measles among children with religious exemptions to vaccination—Massachusetts, Ohio." *Morb. Mortal. Weekly Report.*, **30**, 550–556.

Cliff, A. D., Haggett, P., Ord, J. K., Bassett, K., and Davies, R. B. (1975). *Elements of Spatial Structure: A Quantitative Approach*. Cambridge: Cambridge U.P.

Cliff, A. D., Haggett, P., Ord, J. K., and Versey, G. R. (1981). *Spatial Diffusion: An Historical Geography of Epidemics in an Island Community*. Cambridge: Cambridge U.P.

Cohen, J. E. and Singer, B. (1979). "Malaria in Nigeria: Constrained continuous-time Markov models for discrete-time longitudinal data on human mixed-species infections." In S. A. Levin (ed.), *Lectures on Mathematics in the Life Sciences*, **12**, *Some Mathematical Questions in Biology*. Providence: Amer. Math. Soc., 69–133.

Collins, S. D. (1929). "Age incidence of the common communicable diseases of children." *Publ. Health Reps.*, **44**, 763–826.

Collins, S. D., Wheeler, R. E., and Shannon, R. D. (1942). *The Occurrence of Whoop-*

ing Cough, Chicken Pox, Mumps, Measles and German Measles in 200 000 Surveyed Families in 28 Large Cities. Special Study Series 1. Washington U.S. Public Health Service, Division of Public Health Methods.

Cronin, J. (1977). "Some mathematics of biological oscillations." *SIAM Rev.*, **19**, 100–138.

Cunningham, J. (1979). "A deterministic model for measles." *Z. Naturforsch.*, **34c**, 647–648.

Cvjetanović, B., Grab, B., and Dixon, H. (1982). "Epidemiological models of poliomyelitis and measles and their application in the planning of immunization programmes." *Bull. World Health Org.*, **60**, 405–422.

Cvjetanović, B., Grab, B., and Uemura, K. (1978). "Dynamics of acute bacterial diseases, epidemiological models and their applications in public health." *Bull. World Health Org., Suppl. No. 1*, **56**, 1–143.

Dietz, K. (1970). "Mathematical models for malaria in different ecological zones." Paper presented at Seventh International Biometric Conference, Hannover, August 16–21, 1970.

Dietz, K. (1975a). "Transmission and control of arbovirus diseases." In D. Ludwig and K. L. Cooke (eds)., *Epidemiology*. Philadelphia: SIAM, 104–121.

Dietz, K. (1975b). "Models for parasitic disease control." *Bull. Internat. Statist. Inst.*, **46**, Book 1, 531–544.

Dietz, K. (1976). "The incidence of infectious diseases under the influence of seasonal fluctuations." *Lecture Notes in Biomath.*, **11**, 1–15.

Dietz, K. (1979). "Epidemiologic interference of virus populations." *J. Math. Biol.*, **8**, 291–300.

Dietz, K. (1982). "The population dynamics of onchocerciasis." In R. M. Anderson (ed.), *The Population Dynamics of Infectious Diseases: Theory and Applications*. London: Chapman and Hall, 209–241.

Dietz, K., Molineaux, L., and Thomas (1974). "A malaria model tested in the African Savannah." *Bull. World Health Org.*, **50**, 347–357.

Dietz, K. and Renner, H. (1985). "Simulation of selective chemotherapy for the control of helminth diseases." In J. Eisenfeld and C. DeLisi (eds)., *Mathematics and Computers in Biomedical Applications*. New York: Elsevier, 287–293.

Egan, D. J. and Robinson, D. O. (1979). "Models of a heroin epidemic." *Amer. J. Psychiatry*, **136**, 1162–1167.

Elveback, L. R., Fox, J. P., and Ackerman, E. (1975). "Simulation models." *Proc. Internat. Statist. Inst.*, **46**, Book 1, 553–568.

Elveback, L. R., Fox, J. P., and Ackerman, E. (1976a). "Stochastic simulation models for two immunization problems." *SIAM Rev.*, **18**, 52–61.

Elveback, L. R., Fox, J. P., Ackerman, E., Langworthy, A., Boyd, M., and Gatewood, L. (1976b). "An influenza simulation model for immunization studies." *Amer. J. Epidemiol.*, **103**, 152–165.

Enderle, J. D. (1980). *A Stochastic Communicable Disease Model with Age-Specific States and Applications to Measles*. Ph.D. Dissertation. Rensselaer Polytechnic Inst., Troy, NY.

En'ko, P.D. (1889). "The epidemic course of some infectious diseases," *Vrac'*, **10**, 1008–1010, 1039–1042, 1061–1063. (In Russian.)

Fine, P. E. M. (1975a). "Ross's *a priori* pathometry—a perspective." *Proc. Roy. Soc. Med.*, **68**, 547–551.

Fine, P. E. M. (1975b). "Superinfection: A problem in formulating a problem." *Trop. Dis. Bull.*, **72**, 475–488.

Fine, P. E. M. (1977). "A commentary on the mechanical analogue to the Reed–Frost epidemic model." *Amer. J. Epidemiol.*, **106**, 87–100.

Fine, P. E. M. (1979). "John Brownlee and the measurement of infectiousness: An historical study in epidemic theory." *J. Roy. Statist. Soc. Ser. A*, **142**, 347–362.

Fine, P. E. M. (1982). "Applications of mathematical models to the epidemiology of influenza: A critique." In P. Selby (ed.), *Influenza Models: Prospects for Development and Use*. Lancaster: MTP Press. 15–85.

Fine, P. E. M. and Clarkson, J. A. (1982a). "Measles in England and Wales—I: An analysis of factors underlying seasonal patterns." *Internat. J. Epidemiol.*, **11**, 5–14.

Fine, P. E. M. and Clarkson, J. A. (1982b). "Measles in England and Wales—II: Impact of the measles vaccination programme on the distribution of immunity in the population." *Internat. J. Epidemiol.*, **11**, 15–25.

Fine, P. E. M. and Clarkson, J. A. (1982c). "The recurrence of whooping cough: Possible implications for assessment of vaccine efficacy." *Lancet*, **I**, 666–669.

Frösner, G., Willers, H. Müller, M., Schenzle, D., Deinhardt, F., and Höpken, W. (1978). "Decrease of incidence of hepatitis A infection in Germany." *Infection*, **6**, 259–260.

Frost, W. H. (1976). "Some conceptions of epidemics in general." *Amer. J. Epidemiol.*, **103**, 141–151.

Goddard, M. J. (1978). "On Macdonald's model for schistosomiasis." *Trans. Roy. Soc. Trop. Med. Hyg.*, **12**, 123–131. [Correction. **13**, 245.]

Goffman, W. and Newill, V. A. (1964). "Generalization of epidemic theory. An application to the transmission of ideas." *Nature*, **204**, 225–228.

Greenwood, M. (1931). "On the statistical measure of infectiousness." *J. Hyg. Camb.* **31**, 336–351.

Griffiths, D. A. (1973a). "Maximum likelihood estimation for the beta-binomial distribution and an application to the household distribution of the total number of cases of a disease." *Biometrics*, **29**, 637–648.

Griffiths, D. A. (1973b). "The effect of measles vaccination on the incidence of measles in the community." *J. Roy. Statist. Soc. Ser. A*, **136**, 441–449.

Griffiths, D. A. (1974). "A catalytic model of infection for measles." *Appl. Statist.*, **3**, 330–339.

Grossman, Z. (1980). "Oscillatory phenomena in a model of infectious diseases." *Theoret. Population Biol.*, **18**, 204–243.

Grossman, Z., Gumowski, I., and Dietz, K. (1977). "The incidence of infectious diseases under the influence of seasonal fluctuations—analytical approach." In V. Lakshmikantham (ed.), *Nonlinear Systems and Applications*. New York: Academic, 525–546.

Gumowski, I., Mira, C., and Thibault, R. (1980). "The incidence of infectious diseases under the influence of seasonal fluctuations—period models." *Lecture Notes in Medical Informatics*, **9**, 140–156.

Gust, I. D., Lehmann, N. I., and Dimitrakakis, M. (1979). "A seroepidemiologic study of infection with HAV and HBV in five Pacific islands." *Amer. J. Epidemiol.*, **110**, 237–242.

Hadeler, K. P. and Dietz, K. (1983). "Nonlinear hyperbolic partial differential equations for the dynamics of parasite populations." *Comput. Math. Appls.*, **9**, 415–430.

Haggett, P. (1972). "Contagious processes in a planar graph: An epidemiological application." In N. D. McGlashan (ed.), *Medical Geography*. London: Methuen, 307–324.

Haggett, P. (1976). "Hybridizing alternative models of an epidemic diffusion process." *Economic Geography*, **52**, 136–146.

Haggett, P. (1982). "Building geographic components into epidemiological models." In P. Selby (ed.), *Influenza Models: Prospects for Development and Use*. Lancaster: MTP Press. 203–212.

Hamer, W. H. (1906). "Epidemic disease in England." *Lancet*, **1**, 733–739.

Hanusse, M. P. (1972). "De l'existence d'un cycle limite dans l'évolution des systèmes chimiques ouverts." *C. R. Acad. Sci. Paris C*, **274**, 1245–1247.

Heasman, M. A. and Reid, D. D. (1961). "Theory and observation in family epidemics of the common cold." *Brit. J. Prev. Soc. Med.*, **15**, 12–16.

Hethcote, H. W. (1974). "Asymptotic behavior and stability in epidemic models." *Lecture Notes in Biomath.*, **2**, 83–92.

Hethcote, H. W. (1983). "Measles and rubella in the United States." *Amer. J. Epidemiol.*, **117**, 2–13.

Hirsch, A. (1883). *Handbook of Geographical and Historical Pathology*, **I**. London: New Sydenham Soc.

Hoppensteadt, F. C. and Murray, J. D. (1981). "Threshold analysis of a drug use epidemic model." *Math. Bio.*, **53**, 79–87.

Horwitz, O., Grünfeld, K., Lysgaard-Hansen, B., and Kjeldsen, K. (1974). "The epidemiology and natural history of measles in Denmark." *Amer. J. Epidemiol.*, **100**, 136–149.

Hu, M., Schenzle, D., Deinhardt, F., and Scheid, R. (1984). "Epidemiology of hepatitis A and B in the Shanghai area: Prevalence of serum markers." *Amer. J. Epidemiol*, **120**, 404–413.

Jordan, C. (1960). *Calculus of Finite Differences*, 2nd edn. New York: Chelsea.

Kermack, W. O. and McKendrick, A. G. (1927). "A contribution to the mathematical theory of epidemics." *Proc. Roy. Soc. Ser. A*, **115**, 700–721.

Kemper, J. T. (1980). "Error sources in the evaluation of secondary attack rates." *Amer. J. Epidemiol.*, **112**, 457–464.

Knolle, H. (1983). "The general age-dependent endemic with age-specific contact rate." *Biom. J.*, **25**, 469–475.

Koopman, J. S. (1979). "Models of transmission of infectious agents." *J. Inf. Diseases*, **139**, 616–617.

Kostitzin, V. A. (1934). *Symbiose, Parasitisme et Évolution (Étude mathématique)*. Paris: Hermann.

Kranz, J. (ed.) (1974). *Epidemics of Plant Diseases. Mathematical Analysis and Modelling*. Berlin: Springer.

Lambinet, D., Boisvieux, J.-F., Mallet, A., Artois, M., and Andral, L. (1978). "Modéle mathématique de la propagation d'une épizootie de rage vulpine." *Rev. Epidém. et Santé Publ.*, **26**, 9–28.

Leeuwenburg, J., Ferguson, A. G., and Omondi-Odhiambo (1979). "Machakos project studies: XIII. Spatial contagion in measles epidemics." *Trop. Geogr. Med.*, **31**, 311–320.

London, W. P. and Yorke, J. A. (1973). "Recurrent outbreaks of measles, chicken-pox and mumps. I. Seasonal variation in contact rates." *Amer. J. Epidemiol.*, **98**, 453–468.

Longini, I. M., Jr., Ackerman, E., and Elveback, L. R. (1978). "An optimization model for influenza A epidemics." *Math. Biosci.*, **38**, 141–157.

Longini, I. M., Jr. and Koopman, J. S. (1982). "Household and community transmission parameters from final distributions of infections in households." *Biometrics*, **38**, 115–126.

Longini, I. M., Jr., Koopman, J. S., Monto, A. S., and Fox, J. P. (1982). "Estimating household and community transmission parameters for influenza." *Amer. J. Epidemiol.*, **115**, 736–751.

Longini, I. M., Jr., Monto, A. S., and Koopman, J. S. (1984a). "Statistical procedures for estimating the community probability of illness in family studies: Rhinovirus and influenza." *Internat. J. Epidemiol.*, **13**, 99-106.

Longini, I. M., Jr., Seaholm, S. K., Ackerman, E., Koopman, J. S., and Monto, A. S. (1984b). "Simulation studies of influenza epidemics: Assessment of parameter estimation and sensitivity." *Internat. J. Epidemiol.*, **13**, 496–501.

Lotka, A. J. (1910). "Contribution to the theory of periodic reactions." *J. Phys. Chem.*, **14**, 271.

Lotka, A. (1923a). "Contributions to the analysis of malaria epidemiology." *Amer. J. Hyg.*, **3**, Suppl. 1, 1–121.

Lotka, A. J. (1923b). "Martini's equations for the epidemiology of immunising diseases." *Nature*, **111**, 633–634.

Macdonald, G. (1956). "Theory of the eradication of malaria." *Bull. World Health Org.*, **15**, 369–387.

Macdonald, G. (1957). *The Epidemiology and Control of Malaria*. London: Oxford U.P.

Macdonald, G. (1965). "The dynamics of helminth infections with special reference to schistosomes." *Trans. Roy. Soc. Trop. Med. Hyg.*, **59**, 489–506.

Macdonald, G. (1973). *Dynamics of Tropical Disease* (Collected papers; L. J. Bruce-Chwatt and V. J. Glanville, eds.). London: Oxford U.P.

McKendrick, A. G. (1926). "Applications of mathematics to medical problems." *Proc. Edinburgh Math. Soc.*, **44**, 98–130.

Mackintosh, D. R. and Stewart, G. T. (1979). "A mathematical model of a heroin epidemic: Implications for control policies." *J. Epidemiol. Comm. Health*, **33**, 299–304.

Martini, E. (1921). *Berechnungen und Beobachtungen zur Epidemiologie und Bekämpfung der Malaria*. Hamburg: W. Gente.

Millar, J. R. (1970). "Theoretical and practical problems in measles control." Center for Disease Control. *Smallpox Eradication Program Reports*, **4**, 165–176.

Molineaux, L. and Gramiccia, G. (1979). *The Garki Project: Research on the Epidemiology of Human Malaria in the Sudan Savannah of West Africa*. Geneva: World Health Organization.

Morris, R. S. (1972). "The use of computer modelling in studying the epidemiology and control of animal disease." In A. Madsen and P. Willeberg (eds.), *Proceedings of the International Summer School on Computers and Research in Animal Nutrition and Veterinary Medicine*. Copenhagen: Frederiksberg Bogtrykkeri, 435–463.

Moshkovskĭ, Sh. D. (1950). *Basic Laws of Malaria Epidemiology*. Moscow: Izdat. AMN SSSR. (In Russian.)

Muench, H. (1934). "Derivation of rates from summation data by the catalytic curve." *JASA*, **29**, 25–38.

Muench, H. (1959). *Catalytic Models in Epidemiology*. Cambridge, MA: Harvard U.P.

Murray, G. D. and Cliff, A. D. (1977). "A stochastic model for measles epidemics in a multi-region setting." *Trans. Inst. Brit. Geographers, N.S.*, **2**, 158–174.

Nåsell, I. (1977). "On transmission and control of schistosomiasis, with comments on Macdonald's model." *Theoret. Population Biol.*, **12**, 335–365.

Nåsell, I. and Hirsch, W. M. (1973). "The transmission dynamics of schistosomiasis." *Comm. Pure Appl. Math.*, **26**, 395–453.

Neustadt, R. E. and Fineberg, H. V. (1978). *The Swine Flu Affair: Decision-Making on a Slippery Disease*. Washington: U.S. Dept. HEW.

Nuti, M., Ferrari, M. J. D., Franco, E., Taliani, G., and De Bac, C. (1982). "Seroepidemiology of infection with hepatitis A virus and hepatitis B virus in the Seychelles." *Amer. J. Epidemiol.*, **116**, 161–167.

Pampana, E. J. (1969). *A Textbook of Malaria Eradication*, 2nd edn. London: Oxford U.P.

Riley, E. C., Murphy, G., and Riley, R. L. (1978). "Airborne spread of measles in a suburban elementary school." *Amer. J. Epidemiol.*, **107**, 421–432.

Rosenfield, P. L., Smith, R. A., and Wolman, M. G. (1977). "Development and verification of a schistosomiasis transmission model." *Amer. J. Trop. Med. Hyg.*, **26**, 505–516.

Ross, R. and Hudson, H. P. (1917). "An application of the theory of probabilities to the study of *a priori* pathometry—part III." *Proc. Roy. Soc. Ser. A*, **93**, 225–240.

Rost, H. (1981). "On the method of hybrid model approximation." In *Proceedings of the 6th Conference on Probability Theory, Brasov* 1979. Editura Acad. Rep. Soc. Romania, 185–194.

Rvachev, L. A. (1967). "A model for the connection between processes in the organism and the structure of epidemics." *Kibernetika*, **3**, 75–78. (In Russian.)

Sartwell, P. E. (1976). "Memoir on the Reed–Frost epidemic theory." *Amer. J. Epidemiol.*, **103**, 138–140.

Schenzle, D. (1982). "Problems in drawing epidemiological inferences by fitting epidemic chain models to lumped data." *Biometrics*, **38**, 843–847.

Schenzle, D. (1985a). "Control of virus transmission in age-structured populations." In *Proceedings of an International Conference on Mathematics in Biology and Medicine, Bari, July 18–22, 1983. Lecture Notes in Biomathematics* (in press).

Schenzle, D. (1985b). "An age-structured model of pre- and post-vaccination measles transmission." *IMA J. Math. Appl. Med. Biol.* (in press).

Schenzle, D., Dietz, K., and Frösner, G. (1979). "Hepatitis A antibodies in European countries II. Mathematical analysis of cross-sectional surveys." *Amer. J. Epidemiol.*, **110**, 70–76.

Schütz, F. (1925). *Die Epidemiologie der Masern*. Jena: G. Fischer.

Scott, H. D. (1971). "The elusiveness of measles eradication: Insights gained from three years of intensive surveillance in Rhode Island." *Amer. J. Epidemiol.*, **94**, 37–42.

Skinhoj, P., Mathiesen, L. R., and Cohn, J. (1980). Persistence of viral hepatitis A and B in an isolated caucasian population. *Amer. J. Epidemiol.*, **112**, 144–148.

Smith, H. L. (1983a). "Subharmonic bifurcation in an *S-I-R* epidemic model." *J. Math. Biol.*, **17**, 163–177.

Smith, H. L. (1983b). "Multiple stable subharmonics for a periodic epidemic model." *J. Math. Biol.*, **17**, 179–190.

Snyder, M. J., McCrumb, F. R., Bigbee, T., Schluederberg, A. E., and Togo, Y. (1962). "Observations on the seroepidemiology of measles." *Amer. J. Dis. Child.*, **103**, 250–251.

Soper, H. E. (1929). "Interpretation of periodicity in disease prevalence." *J. Roy. Statist. Soc.*, **92**, 34–73.

Stille, W. T. and Gersten, J. C. (1978). "Tautology in epidemic models." *J. Inf. Diseases*, **138**, 99–101.

Stirzaker, D. R. (1975). "A perturbation method for the stochastic recurrent epidemic." *J. Inst. Math.* Appl., **15**, 135–160.

Sugiyama, H. (1960). "Some statistical contributions to the health sciences." *Osaka City Medical J.*, **6**, 141–158.

Sundaresan, T. K. and Assaad, F. A. (1973). "The use of simple epidemiological models in the evaluation of disease control programmes: A case study of trachoma." *Bull. World Health Org.*, **48**, 709–714.

Thieme, H. R. (1980). "Some mathematical considerations of how to stop the spatial spread of a rabies epidemic." *Lecture Notes in Biomath.*, **38**, 310–319.

Tyson, J. J. and Light, J. C. (1973). "Properties of two-component bimolecular and trimolecular chemical reaction systems." *J. Chem. Phys.*, **59**, 4164–4172.

Von Foerster, H. (1959). Some remarks on changing populations." In F. Stohlman, Jr. (ed.), *The Kinetics of Cellular Proliferation.* New York: Grune and Stratton, 382–407.

Waggoner, P. E. (1981). "Models of plant disease." *BioScience*, **31**, 315–319.

Warren, K. S. (1973). "Regulation of the prevalence and intensity of schistosomiasis in man: Immunology or ecology?" *J. Inf. Diseases*, **127**, 595–609.

Wickwire, K. (1977). "Mathematical models for the control of pests and infectious diseases; a survey." *Theoret. Population Biol.*, **11**, 182–238.

Wilson, E. B. and Worcester, J. (1941). "Contact with measles." *Proc. Nat. Acad. Sci. U.S.A.*, **27**, 7–13.

Wilson, E. B. and Worcester, J. (1945). "Damping of epidemic waves." *Proc. Nat. Acad. Sci. U.S.A.*, **31**, 294–298.

Wong, D. C., Purcell, R. H. and Rosen, L. (1979). "Prevalence of antibody to hepatitis A and hepatitis B viruses in selected populations of the South Pacific." *Amer. J. Epidemiol.*, **110**, 227–236.

Yorke, J. A. and London, W, P. (1973). "Recurrent outbreaks of measles, chickenpox and mumps. II. Systematic differences in contact rates and stochastic effects." *Amer. J. Epidemiol.*, **98**, 469–482.

Yorke, J. A., Nathanson, N., Pianigiani, G. and Martin, J. (1979). "Seasonality and the requirements for perpetuation and eradication of viruses in populations." *Amer. J. Epidemiol.*, **109**, 103–123.

CHAPTER 9

Evolutionary Origins of Statisticians and Statistics

J. Durbin

Abstract

Statistics is basically mathematical, so the first question considered is why the human species can do mathematics as well as it can. Since the genetic composition of mankind cannot have changed much in the past ten or twenty thousand years, we have to explain man's capacity for mathematics in terms of traits evolved for the needs of the hunter–gatherers of prehistory. An explanation is given in terms of the survival advantages of logical reasoning and symbolic thinking. The second question considered is why the mathematics that humans do in their minds works as well as it does when it is applied in the real world. It is suggested that the most likely explanation is that, at least to a good approximation, the world has the simple mathematical structure that it appears to have, and that man's mind evolved by natural selection to mirror this structure. Since the relation between the mathematical models used in statistics and external reality is not essentially different from the relation between mathematics and reality in other branches of the natural sciences, it is not surprising that statistical theory works as well as it does in the real world. An attempt is made to apply Darwinian thinking to the cultural evolution of statistics. While the success of this is debatable, it is suggested that looking at the development of statistics from an evolutionary perspective helps to improve our understanding of some of the factors involved, at least to a modest degree.

1. Introduction

The celebration of the Centenary of the International Statistical Institute challenges us to reexamine the fundamentals of our subject. Among those fundamentals is the philosophical basis for statistical work. When in 1982 the Editors invited me to contribute this paper, they suggested that I should "provide something which shows that science is done by people." Now 1982 was the year in which the centenary of Darwin's death was being commemorated, and like many others I was greatly interested in the vigorous public discussions that were taking place at the time, among biologists and non-biologists alike, about developments in evolutionary biology and the impact of the Darwinian revolution on mankind's view of itself and of the world. In view of the near coincidence of the ISI and Darwin centenaries it seemed to

Key words and phrases: cultural evolution, Darwin, foundations of mathematics, genes, natural selection.

me that an appropriate response to the Editors' invitation would be for me to explore the extent to which Darwinian thinking, together with relevant recent developments in evolutionary biology, can help towards the formulation of a more acceptable philosophy for statistics.

In this paper I shall examine just two aspects of what is obviously a very large subject. First, I shall consider how, in the process of biological evolution, mankind came to acquire the intellectual capacity to formulate mathematical theories of statistical analysis which work in practice in the real world. Secondly, I shall consider the cultural evolution of human knowledge and explore how far a Darwinian interpretation of this cultural evolution helps to improve our understanding of the development of science in general and of statistical science in particular.

I have elected to bring out these issues for discussion now because I believe that it will be useful for them to be debated within the statistical profession. In my view the time has come for the establishment of a better philosophical base for statistics than exists at present. Conventional philosophy seems inadequate to the task, since it appears to start by taking man's intellectual capacity as given. My belief is that a satisfactory system of philosophy cannot be constructed without first having an acceptable hypothesis about where man's capacity for effective rational thought comes from. There is no evidence that the philosophical profession is engaged in modifying its theoretical structures in the light of recent discoveries in evolutionary biology, even though it is already clear that the ultimate philosophical implications of these discoveries will be profound. Yet the message of work in molecular biology over the past thirty years is that the essential structure of the inheritance mechanism is far simpler than might previously have been expected. Under the circumstances, it seems to me both right and proper for the members of the statistical profession to debate among themselves the philosophical implications for their subject of recent developments in evolutionary biology without waiting to be told what to think by philosophers and biologists.

It is unnecessary for me to elaborate here on the essential characteristics of statistical science. These characteristics arise from variations in observations of real phenomena. Statistical analysis is based on mathematical models derived ultimately from probability theory. The relation between these models and reality is not essentially different from the relation between mathematical models and reality in other branches of science such as physics. Thus the task of explaining man's capacity to construct theoretical models for statistics and why such models work in the real world is essentially the task of explaining how man acquired the capacity to do mathematics and why the mathematics that he does in his mind works as well as it does in the outside world.

The tentative conclusions that emerge from the analysis of this paper are as follows. The human mind and brain evolved as a control mechanism for man's behaviour. The special mental capacities that man developed relative to other animals were in language, for communication, in logical thinking, for analysis, and in simulation, for rapid manipulation of symbols of linguistic, logical, and physical entities. Mankind's ability to do mathematics arose partly because of the human mind's capacity to think logically and partly because of its ability to manipulate symbols of physical entities. The simplest hypothesis that can explain why the mathematics that human beings do in their minds works so beautifully when it is applied to the real world is that, to a very good approximation, the part of the physical universe close to man in space and time has a simple physical and logical structure similar to that which modern science perceives it to have. In other words, the real world as seen by man is basically simple and there are no demons "out there" who arbitrarily and unpredictably interfere with the regularities observed by science. Once this hypothesis is accepted, it seems to me natural that the logic that eventually evolved in man's mind should mirror, to the extent needed for his survival, the logical structure of the external world. Since statistical theory is essentially mathematics and mathematics is essentially logic, it does not seem as surprising as might have appeared at first sight that the analysis we create in our minds works so well when we apply it to the real world.

The basic evolutionary model that will be used in this paper is described in the next section. In Section 3 this model is employed to study the biological evolution of mankind's capacity to do mathematics and hence to do statistics. Later on in this section we consider the relation between mathematical reasoning in the human mind and events in the real world from the standpoint of evolutionary thinking. In Section 4 we consider the evolution of human culture from a Darwinian point of view and explore the extent to which our understanding of the development of science in general and statistical science in particular is enhanced by regarding them in these terms.

This paper is a cross-disciplinary venture, and in writing it I have had to step outside my own area of expertise into other specialists' territories. I am conscious of the risk of having committed errors and infelicities in other disciplines, though I have tried to reduce these as much as I can. My overriding objective has not been to make pronouncements about other subjects but to provoke discussion on philosophical matters among my fellow statisticians. If I can succeed in this I shall regard the risk as having been well worth taking.

2. The Evolutionary Model

My object in this section is to describe a model for the evolution of the human species which will serve as a basis for the reasoning later in the paper. I shall assume that the reader is generally familiar with the basic ideas of Darwin's theory of natural selection and Mendelian genetics, but in order to make the article accessible to a wide readership I shall not assume familiarity with details of modern molecular biology. I shall give only a bare outline of the facts and omit inessential qualifications. I ask the indulgence of those whose knowledge of the matter is greater than my own.

Each human cell, other than egg and sperm cells, contains 23 pairs of chromosomes, one member of each pair from the father and the other from the mother. The functional part of each chromosome, and this is the only part that concerns us, consists of a single molecule of the nucleic acid DNA arranged in the form of a double helix. Each of the two complementary strands of the helix consists of a long string of nucleotides. There are only four different nucleotides in DNA. A gene consists of a set of nucleotide pairs in a compact segment of the DNA molecule and determining a specific characteristic of the organism. This is not a precise definition of a gene, but there does not seem to be general agreement on the matter, and this seems the appropriate definition for our purpose. The arbitrariness in the concept of a gene is discussed by Dawkins (1982, pp. 85, 86). The situation is further complicated by the occurrence of split genes and gene clusters; see, for example, Bodmer (1981, 1983). The reader who prefers precision may adopt Watson's (1977, p. 190) definition, summed up by the phrase "one gene—one polypeptide chain," without materially affecting the argument. The essential function of the DNA molecules is to provide the information needed for the development of the organism.

Egg and sperm cells differ from other cells in that they have 23 single chromosomes instead of 23 pairs. At the moment of conception these combine to give 23 pairs in the fertilised egg. Each of the single chromosomes in an egg cell is made up by combining segments of DNA chosen from the corresponding chromosomes of the grandfather and grandmother; similarly for each sperm cell. It appears that the points at which the DNA molecules are broken for recombination into a single DNA molecule correspond exactly and that these points are chosen by a probabilistic mechanism. The situation is slightly different for the sex-determining X and Y chromosomes, but for brevity, since the details are not essential for our purposes, I shall not discuss them further. All the genetic information required for the future development of a human being is contained in the 46 DNA molecules present in the fertilised egg immediately after conception.

Evolution occurs as a result of changes in DNA molecules called mutations. These can occur in a variety of ways of which I shall mention three. First, radiation from outer space or elsewhere can damage components of DNA in any cell. The cell has a repair mechanism, but this can make mistakes, thus giving rise to a change in DNA information. A second type of change can occur on cell division. When the cell divides, the two strands of DNA separate and each provides the basis for the construction of a new double helix; mistakes can occur in this process. Thirdly, segments of DNA can detach themselves and become inserted in new positions along the molecule. All these mechanisms for mutation are effectively probabilistic in their operation.

The important point about the structure we have described is its relative simplicity. A gene can be thought of as a sentence written in a language possessing an alphabet containing only four letters, each sentence containing an instruction about a particular characteristic of the individual. Moreover, the way the instructions are initially carried out is also simple, since the information coded in the nucleotide sequence of DNA essentially relates only to the construction of proteins, and each protein is composed of a string of elements of which there are only twenty different kinds.

My contention is that the simplicity of this structure assists us substantially in our thinking about the evolutionary process. Had the essentials of the inheritence mechanism consisted of a highly complex biochemical process which the layman could not be expected to comprehend, it would be much more difficult for him to become involved in discussions of the implications of biological evolution. I shall not comment at all on how this basic structure came itself to be evolved; the interested reader is referred to Orgel (1973). Further details about the inheritance mechanism can be found in any modern elementary book on molecular biology or genetics, such as Watson (1977) or Bodmer and Cavalli-Sforza (1976). We end this brief discussion of DNA structure by quoting the prophetic final two sentences of Bodmer (1981):

> The whole DNA sequence will eventually be known, and also, but even more eventually, its meaning will be understood. This knowledge will have profound implications for all aspects of human activities and endeavors and surely will, in the long run, contribute positively to the betterment of our society.

We now consider how natural selection works in the light of this mechanism. It is well known that genetic mutations are the basic source of evolutionary change. If a mutation changes a gene into a form more favourable to the individual's chance of survival to parenthood, then the chance of survival of his genes is increased. Thus an important consequence of selection is the survival of favourable genes in the DNA of the species. However, natural selection is not the only means by which species evolve. For example, Bodmer (1983) states, "There are basically three types of causes for the propagation of a new variant in a population, namely (1) random genetic drift, if the variant is adaptively neutral; (2) the classical Darwinian process of natural selection; and (3) non-Mendelian processes which lead to preferential segregation of one or other allele in a heterozygote." Nevertheless natural selection is the most potent of these so we shall concentrate attention on it in this paper.

We now consider what the basic unit of selection is, that is, what is the entity for whose benefit adaptations by natural selection take place. So far as Darwin's own views are concerned, Alexander (1979, p. 24) states, "One has to conclude from extensive reading of Darwin that he was uncertain about the identity of units of selection." He goes on to say,

> For almost one hundred years the question of the potency of selection at different levels was not even clearly posed, and its resolution has proved so complex that many aspects remain to be analyzed.... Nevertheless a general consensus among evolutionary biologists is clear; adaptiveness is not appropriately assumed at any higher level of organization than is necessary to explain the trait in question. This means that genes and their simplest groupings—about which Darwin had no knowledge—are the usual units of selection.

Alexander (1979) and Dawkins (1976, 1982) call such a gene or group of genes a *genetic replicator*. From this point of view an individual organism or cooperative group of organisms is just a vehicle for the transmission of genes from generation to generation. The outcome of natural selection is then just to enhance the survival probabilities of the genetic replicators.

I find this proposition not completely convincing in the particular form in which Alexander and Dawkins present it. It seems clear that selection must take place in all important respects at the individual level, since (minor qualifications apart) the individual is the smallest unit that survives and has progeny. Thus one cannot speak of a genetic replicator as a unit of *selection*. A further objection is that the concept of the gene itself as a single unit has recently tended to become less clearcut. On the other hand there is no doubt that the most important consequence of selection for the evolution of a species is the preservation in its DNA of favourable genes. We are therefore led to the proposition that, so far as the evolution of the human species is concerned, the main unit on which selection takes place is the individual, while the main outcome of selection is the survival of favourable genes in the DNA.

We are now able to formulate our basic evolutionary model. This consists of the chromosome inheritance mechanism, random genetic mutations and natural selection leading to the survival of favourable genes in the DNA of the species. We shall need to bear this model in mind when considering the evolution of the human capacity to do mathematics in the next section.

Three readers of the first draft of this paper independently referred to the Eldredge–Gould (1972) theory of punctuated equilibria and suggested that this was in conflict with neo-Darwinism as summarised above. This theory suggests that evolution occurs in sharp jumps separated by long periods of stability. While the facts in the fossil record are not in dispute, there does not appear to by any real justification for the notion that there is a serious conflict. For example, a distinguished progenitor of the theory, Ernst Mayr, says (1982, p. 1130), "The claim that evolutionary stasis, as postulated by some adherents of the theory of punctuated equilibria, is in conflict with either neo-Darwinism or the evolutionary synthesis is quite without foundation." The claim that there is a conflict with neo-Darwinism is refuted in detail by Charlesworth et al. (1982).

An interesting collection of recent papers on evolutionary topics, together with a commentary intended "to help non-specialists to find their way through papers which are rather technical," has recently been published in paperback by J. Maynard Smith (1982).

3. Biological Evolution of Mankind's Ability to Do Mathematics

Statistical theory is essentially mathematical, so the capacity to construct and exploit statistical theory is effectively the capacity to do mathematics. In this section we shall consider how mankind came to develop this capacity during the evolutionary process. Using the evolutionary model of the last section, this means that we shall regard the ultimate mathematical capacity of each human being as entirely determined by the sequences of nucleotide pairs in the 46 DNA molecules in the fertilised egg immediately after conception. I am not, of course, suggesting here that there is a specific gene for mathematics. What I am saying is that the potential capacity for symbolism and logical analysis that is needed for mathematics must be present at conception in the information structure of DNA in a form capable of subsequent development. Of course, the extent to which this potential is realised in any individual depends on the interaction between the individual's genes and his environment. This means that only a tiny proportion of mankind's mathematical potential is ever realised, since on a worldwide basis it is only rarely that environmental circumstances exist that will permit the realisation of a particular individual's mathematical potential.

Having recognized that an individual's potential capacity for mathematical work must be present in his DNA, we must next solve the mystery of how it got there. On the basis of our evolutionary model, and minor qualifications apart, the only acceptable solution is that random mutations occurred and those that led to better adaptation to the environment became incorporated into the DNA of the species. The problem about accepting this as an explanation of the origin of the capacity for mathematics is that the human species cannot have changed very much in its genetic makeup in the past ten or twenty thousand years. On the other hand agriculture did not begin until about ten thousand years ago. It follows that the genetic makeup of mankind must have evolved to satisfy the survival needs of hunter–gatherer peoples in the days of prehistory. We therefore have to explain, by means of evolutionary arguments, how the powerful intellectual capacity that mankind exploits so successfully today came to be evolved for the quite different needs of hunter-gatherer peoples many thousands of years ago.

There is an even deeper mystery to be explained. Every scientist knows that the mathematics that is done in the human mind works extremely well in applications to the real world, provided, of course, that appropriate models are being correctly used. It is as though there is almost perfect harmony between logical operations in the mind and physical events in the world. But why should this be so? Philosophers and others have pondered on this question throughout the ages without, as far as I know, discovering any convincing answers. My contention is that by approaching the question from the standpoint of evolutionary biology one can explain the mystery more or less satisfactorily; I shall attempt to do this later in the section.

I shall approach these questions from the standpoint of scientific realism, that is I shall assume firstly that our perceptions of the world arise from real phenomena outside ourselves, and secondly that our perceptions are in harmony with those aspects of the phenomena that impinge upon us in a sufficiently important way. In this respect I agree with the ethologist Konrad Lorenz, who stated (1977, p. 7):

> I have only modest hopes of understanding the meaning or the ultimate values of our world, but I am unshakeably convinced that all the information conveyed to us by our cognitive apparatus corresponds to actual realities. This attitude rests on the realization that our cognitive apparatus is itself an objective reality which has acquired its present form through contact with an adaptation to equally real things in the outer world. What we experience is indeed a real image of reality—albeit an extremely simple one, only just sufficing for our own practical purposes; we have developed "organs" only for those aspects of reality of which, in the interest of survival, it was imperative for our species to take account, so that selection pressure produced this particular cognitive apparatus.

I shall start out with the proposition that the mind and brain together constitute the control mechanism which directs the behaviour of the individual. In more primitive organisms this mechanism operates according to a simple optimality criterion, namely the maximisation of the probabilities of the replication and survival of the individual's DNA, but in mankind, as in some other species, the behaviour is complicated by many factors including cultural factors. The relation between mind and brain can be thought of as partially analogous to the relation between the software and hardware of modern information-processing equipment. Information flows into the brain and is processed in it according to directions determined by the mind, part of the end product being instructions from the brain to the body.

The concept of "mind" is not a well-defined one. I am using the word "mind" in what I regard as its everyday sense and hope the reader will not quibble about this.

Let us now consider the distinguishing characteristics of the human mind and how these came to be evolved. The special mental capacities that mankind developed relative to other animals were in language, for communication; in logical thinking, for analysis; in memory, for information storage; and in simulation, for planning and organisation. There are some surprising features about these attributes. The first is that the gap between man's proficiency in these respects and that of the animal species nearest to him is far greater than might have been expected. Secondly, the evolution of mankind from its immediate ancestors was extremely rapid relative to the usual evolutionary time scale. Thirdly, the level attained by mankind in each of the four capacities I have mentioned seems enormously higher than would be thought necessary to sustain the hunter–gatherer way of life that mankind practiced ten or twenty thousand years or so ago.

Darwin himself mentioned two partial explanations of these phenomena, namely warfare between groups of humans and their immediate predecessors, which placed a premium on planning and organisation and which often resulted in the extermination of the unsuccessful, and sexual selection, with both sexes choosing mates. In both cases the selection process can operate independently of competition from other species and environmental pressures, and its impact can be very rapid. A third hypothesis, that of gene–culture coevolution, has recently been suggested by a number of biologists, including Monod (1971), Lumsden and Wilson (1981, 1983), and Dobzhansky and Boesiger (1983). This is that cultural innovations can have an effect on selection analogous in some respects to that of DNA mutations and thus give rise to genetic changes.

For a definition of culture we shall follow Cavalli-Sforza and Feldman (1981) and adopt a truncated form of Webster's dictionary definition: "the total pattern of human behavior and its products embodied in thought, speech, actions and artifacts, and dependent upon man's capacity for learning and transmitting knowledge to succeeding generations." Some useful qualifications to this definition, which we have not the space to discuss here, are made by Cavalli-Sforza and Feldman. It should be emphasised that in this section we are concerned only with the effect of cultural development on the information present in mankind's genes. We leave aside until the next section autonomous cultural evolution unrelated to genetic changes in man.

In order to accelerate our discussion of the gene–culture coevolution hypothesis let us take it for granted that an important reason for the evolutionary success of man's immediate ancestors was the development of cooperative behaviour. This would obviously be facilitated by the use of primitive forms of language, tools, and organisation, all of which, in the sense in which I am using the term, would normally be thought of as cultural phenomena. As time progressed cultural innovations would occur for a variety of reasons unconnected with mutations in DNA. Some of these would be advantageous for survival to those individuals with a predisposition to exploit them effectively. Insofar as any such predisposition has a genetic origin, the survival probabilities of the corresponding genes would be enhanced. Thus the genes that tend to survive are not only those resulting from favourable mutations that give a direct adaptive advantage, but any genes that predispose individuals to the effective exploitation of cultural change. What I intend to do now is use this reasoning to throw light on the development of man's innate capacity for language and for logical thought.

I appreciate that the hardline mutationist could claim that there is nothing essentially new here, since any genes that responded favourably to cultural innovations could be regarded as arising from earlier mutations, while the cultural innovations could be thought of as part of a changing environment. Nevertheless, I think that the idea of treating cultural innovations as having an effect on genetic development analogous to, but distinct from, mutations in DNA is an illuminating one, particularly when, as here, our objective is to explain the development of man's cultural prowess.

I shall now attempt to outline a plausible explanation of how man came to develop his genetic capacity to do mathematics. Foremost among the component elements of this is linguistic ability, and in introducing this I can hardly do better than quote Monod (1971, pp. 126–129):

> As soon as a system of symbolic communication came into being, the individuals, or rather the groups best able to use it, acquired an advantage over others incomparably greater than any that a similar superiority of intelligence would have conferred on a species without language. We see too that the selective pres-

> sure engendered by speech was bound to steer the evolution of the central nervous system in the direction of a special kind of intelligence: the kind most able to exploit this particular, specific performance with its immense possibilities.
>
> This hypothesis would be little more than attractive and reasonable if it were not justified by certain linguistic evidence being compiled today. The study of children's acquisition of language irresistably suggests that this astonishing process is by its very nature profoundly different from the orderly apprenticeship of a system of formal rules. The child learns no rules, and he does not try to imitate adult speech.... This process is, it seems, universal and its chronology the same for all tongues.

Monod reviews some relevant evidence from Lenneberg (1967) and elsewhere, and concludes that the universal capacity that children have for the rapid acquisition of language over the first few years of life must have evolved by gene–culture coevolution. Monod quotes in support the claims of Chomsky and his followers that all human languages have essentially the same underlying structure, suggesting that this structure must be innate and characteristic of the species. Chomsky's (1980) book is particularly relevant here. Monod concludes "the linguistic capacity revealed in the course of the brain's epigenetic development is today part of 'human nature,' itself defined within the genome in the radically different language of the genetic code."

An analysis of how far linguistic capacity is innate is presented from the standpoint of psycholinguistics by Aitchison (1983). She comes to the following conclusion:

> We may summarize the contents of Chapters 1 to 7 by saying that we found strong evidence in favour of the suggestion that a human's ability to talk is innate. However, when it came to discovering exactly what is innate, we ran into difficulties. Chomsky's assertion that children are endowed with highly structured acquisition device which incorporates a set of formal linguistic universals seems over-optimistic. His proposals are not borne out by the evidence. We preferred instead, the suggestion that children have an inbuilt ability for processing linguistic data—an ability which seems to be partly separate from general cognitive abilities.

However, I am left with the impression that the agreement between Monod and Aitchison is greater than their differences.

Man's capacity for language requires the possession of considerable ability in symbolic manipulation relative to the lower animals. But what was the mechanism by which this evolved to its present advanced intellectual level? The answer given by Monod is simulation, that is, the ability to conjure up in the mind and analyse at great speed images of reality. In (1971, p. 145) he says: "It is the powerful development and intensive use of the simulative function that, in my view, characterizes the unique properties of man's brain." Later, he continues "If we are correct in considering that thought is based on an underlying process of subjective simulation, we must assume that the high development of this faculty in man is the outcome of an evolution during which natural selection tested the efficacy of the process, its survival value. The very practical terms of this testing have been the success of the concrete action counselled and prepared for by imaginary experimentation."

It has to be said that we have now moved out of the realm of empirical observation into that of speculative reasoning, but let me affirm that I find Monod's reasoning convincing. His proposition is that man's development of the capacity for symbolic thought led to, or was at least accompanied by, the development of the capacity to simulate, in the mind, representations of external phenomena. Symbolic manipulation of these images enabled man to forecast, plan, invent, learn, and organise. Those individuals or groups who could perform these operations better had higher survival probabilities under environmental pressure, warfare, sexual selection, gene–culture coevolution, etc. Thus the propensity towards successful intellectual activity became part of the genetic information in man's DNA, the test of success being consistency of the outcome of the activity with concrete reality.

Although this is a fascinating theory, nobody can prove whether the picture it presents of man's intellectual development is correct, since parts of it are necessarily speculative. However, I find the theory plausible enough to make it the basis of the remainder of this section. As always, the answer to the sceptic is to challenge him to produce a more acceptable model.

Our next task is to explain how this model for the evolution of man's intellect can account for the development of the mathematical ability needed to do statistics. It is clear that if we go back ten thousand years or so our ancestors did not do any mathematical work as such, but on the other hand their genetic capacity for mathematical work must have been comparable with our own, since there cannot have been much genetic change since that time. We must, therefore, explain the mathematical ability that we have today in terms of symbolic manipulations developed by man long ago for other purposes. With this in view we rule out concepts remote from everyday experience, such as those arising in quantum mechanics and relativity theory, and direct our attention only to mental operations that were evolved for thinking about phenomena which were of the same order of magnitude as man himself and which impinged on his everyday life. We first consider the evolution of the underlying intellectual capacity to do mathematics as it exists today and then consider the step by step historical development of mathematics to see if different qualities might have been needed for man to achieve that.

There are many basic attributes and activities of the human mind in addition to language that are related to different branches of mathematics, such as counting, manipulation of geometrical images, recognition of symmetries, construction of dynamic maps of moving bodies, and probabilistic assessment. It would be interesting to speculate on the evolutionary origins of mental skills such as these and to trace the connections with various branches of modern mathematics. However, this would be too big a task to be undertaken in this paper, so I shall concentrate on the particular aspect of the matter that interests me most. This is the problem of devising a convincing explanation of how the human mind developed the capacity to construct modern pure mathematics at the highest levels of rigour and abstraction, using only intellectual equipment that was evolved for the needs of the hunter–

gatherers of prehistory. Though the task seems ambitious I am fortified by the conviction that there *must be* a rational explanation. In making the attempt I shall not consider at all the actual historical evolution of mathematics or the social or environmental forces that led to this historical evolution, though these are obviously extremely important in understanding how mathematics actually developed. What I shall do instead is base my treatment on work done during the present century which shows that the whole of present-day mathematics can be rigorously constructed from a few simply stated propositions concerning logic and set theory. We shall find that it is not difficult to relate these propositions to the linguistic and logical abilities that were developed in the evolution of early man. I am not claiming that the particular axiomatisation of mathematics that I use is unique; only that it is convenient and adequate for our purpose.

In the development of the ideas developed in the next few paragraphs I have received considerable help generously given by my colleague Ken Binmore. I will base my treatment on Binmore's (1980) undergraduate text, but the material is widely available elsewhere, e.g. in the first two chapters of Cohen (1966).

Binmore begins with an elementary introduction to predicate logic. From our point of view this can be regarded as formalising the manner in which the words "and," "or," "if," and "not" are used in ordinary language (provided the quantifiers "for any" and "there exists" are thought of as generalisations of the conjunctions "and" and "or" to the case where many statements must be linked). Therefore, none of the logical rules is inaccessible to someone who is capable of constructing grammatically correct sentences and using them clearly in normal discourse.

In addition to logic we require some axioms about the properties of sets. Various formalisations are available, but the most widely used system is Zermelo–Fraenkel set theory. In this theory, everything is regarded as a set and only two primitive connectives, namely equality($=$) and the membership relation ($\in$), are admitted. The number of axioms on which the theory is based can vary, but only six are strictly necessary. Apart from "bookkeeping" aspects, the axioms state the rules by means of which one is allowed to form a set from the sets already available.

Binmore gives (1980, pp. 26, 27) an informal introduction to the axioms. He begins by explaining the ideas behind the first four of them and points out that "the nature of these assumptions is such that it is natural to take them completely for granted without a second thought." On p. 27 he points out that two further axioms, those of infinity and of choice, are needed, and goes on to say, "It is a remarkable fact that the whole of mathematics can be based on the principles of logic and these six simple assumptions about the properties of sets." With this background Binmore builds up informally the real number system. A rigorous derivation of the cardinal system and some of its properties from logic and set theory has been given by Abian (1965).

What are the implications of these theories as far as the evolution of the capacity of the human mind to do mathematics is concerned? The answer is that, in principle, the ability to do mathematics does not require basic mental skills whose acquisition by the hunter–gatherer of prehistory would be difficult to justify on evolutionary grounds. As far as logic is concerned, the usages of ordinary language are essentially adequate provided they are employed systematically and precisely. For set theory, leaving aside difficulties about the axiom of infinity and the axiom of choice that I shall return to in a moment, little more is required than that the mind be capable of recognising similarities between distinct objects and registering this similarity symbolically in the mental simulator. The survival value of such a facility is evident.

Now we come to the difficulties. The axiom of infinity asserts the existence of infinite sets, and one might therefore ask how or why the hunter–gatherer gained his insight into the infinite during the evolutionary process. The axiom of choice, which states that given any collection of nonempty sets there is a way of choosing an element from each of the sets, is even less intuitive. The controversies that have raged for many years about these axioms, or rather the concepts underlying them, have been colourfully described by Kline (1980, particularly Chapter 9).

It might be thought at first that the nonintuitive nature of these two axioms casts doubt on the thesis propounded here about the evolutionary origins of mathematics, but Professor Binmore has pointed out to me that this is not so; in fact the opposite is the case. All that is required is first that the human mind manipulates finite sets of symbols in its simulator according to the rules of set theory, and secondly that no constraint evolved in the mind *forbidding* the concept of an infinite set and its manipulation in an analogous way. It is hard to think of any survival factor during the evolution of early man that would give rise to any such constraint. Infinite sets and the axiom of choice then emerged as cultural artifacts quite recently and have probably survived, insofar as they have survived, purely for Darwinian reasons because they have been found convenient. They need not have any counterpart whatever in man's perception of reality. Looking at the matter from the evolutionary standpoint in fact seems to help one understand better the doubts that many mathematicians have about these and other nonintuitive concepts.

In claiming that the whole of mathematics could in principle be deduced from set theory and logic I am not, of course, suggesting that anyone should use, or indeed has ever used, this approach in practice to obtain new mathematical results. There are invariably more expeditious ways of arriving at specific mathematical propositions.

My purpose in the previous few paragraphs has been to attempt to explain how the human intellect has the capacity to do present-day mathematics. However, it could be argued that it would be more relevant to discuss the chronological development of mathematics and to consider the mental equipment needed to take the successive steps in this development. Obviously, nobody in earlier centuries had heard of predicate logic and Zermelo–Fraenkel set theory, and certainly the way mathematics developed historically had nothing to do with the constructions we have been discussing. Strictly speaking, questions of cultural development are dealt with in the next section, but let me answer the point briefly here. An excellent account of the cultural evolution of mathematics from earliest times has been given by Wilder (1968). Study of this book suggests that the qualities needed for the development of mathematics step by step throughout the ages are exactly the same as those discussed above, namely the capacity for logical thought and the ability to conceptualise external phenomena. It was obviously also important that man evolved the capacity to construct accurate geometrical images in his mind of physical objects, as well as accurate dynamic maps of moving objects such as animals running and missiles in flight.

We next consider how to explain why the mathematics that we do in our minds works so well when we apply it to the real world. Ever since John Stuart Mill, attempts have been made by philosophers to argue that mathematics is essentially empirical in the sense that it is part of science. A review of some relatively recent writings in this area has been given by Lakatos (1976). Pure mathematicians have been rather cool toward these ideas, since they are well aware that the direct comparison of their theorems with events in the world is never part of their professional activity. Nevertheless the question is an intriguing and important one.

My belief is that the key to the resolution of the mystery of why mathematics works in the real world is to be found in evolutionary biology. The analysis made earlier in the paper led to the conclusion that mankind's capacity for mathematics arose through adaptation by means of natural selection. The previous few paragraphs showed that the functioning of our reasoning processes is explicable in terms of a simple mental structure based on logic and set theory. These reasoning processes would not have been consolidated in our minds by natural selection unless they had been found effective in practice. What do these considerations tell us about the relation between mathematics and the world?

Let us confine our attention to the part of the world that is near to man's experience in space and time. Thus we rule out cosmology and quantum mechanics. We begin with the hypothesis that, to a very good approximation, at least on the surface, the world has the simple physical and logical structure that science perceives it to have. This is, of course, nothing more than the working hypothesis that the scientist employs in his everyday work. The inherent plausibility of this hypothesis is reinforced by the reflection that if there were any *other* forces or influences in the world that impinged on the physical experiences of mankind and other creatures to a substantial extent, and if these forces or influences were to exhibit regularities in space or time, then the flexibility of the DNA inheritance mechanism and the adaptive power of natural selection are such that it seems almost certain that at least one organism would have evolved to occupy some appropriate ecological niche by exploiting these regularities. No such organism has been found. Moreover, no convincing evidence of extrasensory forces that have a substantial physical influence on mankind has ever been produced. We shall therefore accept the hypothesis as valid. Granted this, it is beyond question that, to a very good approximation, the world works in a way that is consistent with the laws of mathematics. The justification of this claim is that, when predictions are made based on the mathematical models of science, the predictions are found to be correct to an appropriate order of approximation.

Let us therefore consider a situation in which the world has a simple mathematical structure and living creatures are evolving. These creatures are capable of almost limitless complexity because of the immense amount of information that can be carried by DNA in the chromosomes. One can understand how the brains and central nervous systems of animals evolved by enhancing the animals' survival probabilities. One can also understand the adaptive advantage to animals of the ability to manipulate in the mind symbols of external phenomena. Since these external phenomena work according to the laws of logic and mathematics, at least to a good approximation, it does not seem so surprising that during evolution man's intellectual processes should have developed in a way that is consistent with the behaviour of such phenomena in the external world. We conclude that the most plausible explanation of the harmony between mathematical calculations in the mind and external reality is first that the real world conforms to mathematical laws and secondly that man's mind evolved by natural selection to mirror this structure. Perhaps I may be permitted an excursion into the obvious in order to point out that if the world had not been so simple, the evolution of living organisms would have been very different.

It is interesting to consider briefly what light the evolutionary approach throws on the difficulties that mathematicians have experienced during this century in formulating an acceptable logical basis for their subject. One of the most distinguished workers in the foundations of mathematics, Paul Cohen, states on the first page of his (1966) book:

> In the 19th Century, the objections regarding the use of convergent series and real numbers were met by Cauchy, Dedekind, Cantor and others only to be met by more profound criticisms from later mathematicians such as Brouwer, Poincaré and Weyl. The controversy which followed resulted in the formation of various schools of thought concerning the foundations. It is safe to say that no attitude has been completely successful in answering the fundamental questions, but rather that the difficulties seem to be inherent in the very nature of mathematics.

A fascinating nontechnical discussion of problems in the foundations of mathematics has been given by Kline. He states (1980, p. 276), "The efforts to eliminate possible contradictions and establish the consistency of the mathematical structures have thus far failed.... The recent research on foundations has broken through frontiers only to encounter a wilderness."

I submit that by approaching the origins of man's mathematical capacity from the standpoint of evolutionary biology we are able to contemplate these problems with greater equanimity. We can persuade ourselves to accept mathematical reasoning as mirroring the structure of the behaviour of phenomena locally in our part of the universe in space and time. To have imagined the possibility of constructing a consistent mathematical system in which all true propositions could be deduced from a few simple axioms turns out to have been just a dream. To have suggested that the mathematics that we create in our minds is a purely formal system with an absolute validity that holds not only for our own universe but for all possible universes was a delusion. Instead let us marvel at the success of our hunter–gatherer ancestors in evolving an intellectual instrument which, from a practical point of view, mirrors so effectively the local behaviour of the universe in our own neighbourhood.

The last chapter of Kline (1980) is devoted to this very question of relation between mathematics and the real world, but he does not consider it from an evolutionary standpoint. He quotes statements on the matter by twenty distinguished mathematicians, scientists, and philosophers, but none of them looks at the problem from a biological perspective. Kline concludes by leaving the correspondence between mathematics and reality as an unexplained mystery. Since Darwin's *Origin of Species* was published in 1859, it seems to me extraordinary that neither Kline nor any of the distinguished people he quotes suggest an explanation in terms of Darwinian evolution.

There are of course many other aspects to be explained, for example, the variability in mankind of mathematical ability. Why should people differ as much as they do? The general question of biological diversity is discussed by Shorrocks (1978); we shall not consider it further here. Explanations of general attributes needed to do mathematics and science such as curiosity, energy, imagination and so on are also outside the scope of this paper.

Discussions about human heredity and its interaction with the environment have constantly been bedevilled by politics, those on the left tending to emphasise the importance of environmental factors and those on the right the importance of heredity. I would like to express the hope that the present paper will be accepted by readers as entirely non-political.

During the past decade much of the heredity–environment controversy has centered on human sociobiology. As a serious academic discipline, sociobiology has been established for many years. It may be defined as the systematic study of the biological basis of social phenomena in organisms of all kinds. Almost all serious research in the field has been concerned with species other than man; a good idea of the range of problems studied in the field can be obtained from King's College Sociobiology Group (1982). An impression has been created that the name sociobiology was invented quite recently by E. O. Wilson. However, as P. P. G. Bateson pointed out in the preface to the book just cited, the name has been in use at least since the 1940s.

The current controversy was initiated by the inclusion in Wilson's (1975) book of a chapter entitled "From Sociobiology to Sociology" in which Wilson attempted to apply sociobiological arguments to human behaviour. As Bateson said in the preface just cited, "... we should, perhaps have been quicker to realize how much opinion would be polarized by the recent attempts to inject a particular brand of biology into the social sciences. The lacerations resulting from the ensuing ideological conflict have not yet healed, and in many places 'Sociobiology' is either a battle cry or a term of abuse."

One of the leading critics of Wilson's approach to human sociobiology is his Harvard colleague, the distinguished evolutionary biologist and paleontologist, S. J. Gould. Having mentioned Wilson's book, let me in the interests of impartiality refer the reader to Gould's criticisms in the last two chapters of his 1977 book and the last chapter of his 1981 book.

My own views on human sociobiology are expressed by the word "scepticism." I believe that the sensible statistician should evaluate the claims by sociobiologists about human nature with the same scepticism that he would employ to evaluate claims about genetic differences in intelligence between races or between the sexes. I bring up these points here because I fear that my arguments about the genetic origins of the capacity for mathematics might not be properly considered in some circles because of a mistaken suspicion that they are connected with unsubstantiated claims of human sociobiology. In fact my paper up to this point has been concerned only with the genetic origins of intellectual capacity and not with human behavior. As far as I am aware, none of the biologically qualified critics of sociobiology disputes the validity of considering the study of human *capacities* in various directions from a genetic standpoint. Their objections have only been about genetic explanations of *behaviour*. My next section is indeed about behaviour, but does not seek to explain it in genetic terms. Thus I think it would be unjustified for my arguments to be dismissed merely by erroneously attaching the label "sociobiological" to them.

4. Cultural Evolution and Statistics

In this section we assume that the genetic make-up of man has been reasonably stable over the past few thousand years and consider how his culture has evolved over this period. We shall start with culture in general, then go on to the evolution of science, and finish by considering the special position of statistics.

The basic proposition I shall consider, albeit in a tentative spirit, is that cultural evolution is essentially Darwinian in the sense that it evolves by a process analogous to natural selection. Taking as a special case the world of ideas, the proposition asserts that alternative ideas compete, that good ideas eventually succeed in competition with inferior ideas, and that these then become incorporated into human culture. The same concept applies to physical artifacts, techniques of organisation, and so on. Cultural innovations such as new ideas or artifacts can be regarded as analogous to mutations in biological evolution, and the transmission of cultural items from one individual or group to another can be regarded as analogous to the transmission of genes from parents to children. There is one very important difference in that the stock of recorded knowledge, artifacts, works of art, etc. has an existence of its own outside the minds of individuals; there is no analogue of this in the gene stock in biological evolution.

For readers new to this concept an easy access to the essentials of the argument is the final chapter of Dawkins (1976). Dawkins introduces the name *meme* for what he regards as the basic unit of selection in cultural evolution and goes on to develop the analogy with biological evolution. Dawkins makes further detailed comparisons between memes and genes and replies to some critics of the meme concept on pp. 109–112 of (1982).

A heroic effort to develop a mathematical theory of cultural evolution has been made by Cavalli-Sforza and Feldman (1981). They introduce their cultural unit in the following way (p. 15): "Let us call this selection *cultural*, and define it on the basis of the rate or probability that a given innovation, skill, type, trait or specific cultural activity or object—all of which we shall call, for brevity, *traits*—will be accepted in a given time unit by an individual representative of the population." Traits can be discrete or continuous. In their preface Cavalli-Sforza and Feldman claim, "what emerges from the theoretical analysis is the idea that the same frame of thought can be used for generating explanations of such diverse phenomena as linguistics, epidemics, social values and customs, and the diffusion of innovations" but it is an open question how far they have succeeded.

Another major effort to develop a mathematical theory of cultural transmission has been made by Lumsden and Wilson (1981). Their main interest is in cultural and genetic coevolution, but their theory is relevant to the study of cultural development alone. They call the unit of selection the *culturgen* and discuss (1981, pp. 7, 26–30) its definition in detail and compare it with those of many other workers. The Lumsden–Wilson concept is a general one and covers a wide variety of types of cultural unit. A useful review of the literature on coevolution and cultural evolution is given (1981, pp. 256–264). A non-technical presentation of part of their work is given in Lumsden and Wilson (1983).

Finally, I should mention that Alexander gives a detailed discussion (1979, pp. 73–82) of the similarities and differences between cultural and biological evolution and provides many references.

Before turning to the evolution of science and statistics I would like to make some general remarks about cultural evolution. First, I should make the obvious point that not all cultural change occurs as an accumulation of small increments—revolutions can occur for such reasons as military conquest, religious conversion, and political upheaval. For changes that take place incrementally I can certainly see some value in the biological analogy, regarding cultural innovations as mutations that become incorporated into human culture as the result of some sort of selection process. However, in biological evolution the evolutionary model can be formulated with clarity and its operation understood with relative ease. The criterion for successful selection is just survival to reproduce offspring. The Darwinian model for biological evolution thus not only is credible but also has considerable explanatory power. The same holds, in my view, for the gene–culture coevolutionary model, at least when expressed in general terms, since once again there is a clearcut criterion for selection success, namely gene survival. In the case of pure cultural evolution, however, the situation is immensely more complex. There is first the problem of defining the selection unit, and secondly that of formulating acceptable models for cultural transmission. Then there is the further question of who determines whether one particular cultural innovation should be accepted into the cultural heritage rather than another and why. Another relevant factor is the environmental and social pressure acting on the people concerned, which can be both powerful and complex. Then there is the point that in adopting an innovation the decision-takers have to take account of the costs of the innovation as well as the benefits it confers. Moreover, there are such influences as fashion and tradition and the belief that there are deep underlying cultural trends in society that influence mankind's destiny. Finally, it often appears that cultural evolution has a greater similarity to Lamarckian inheritance of acquired characteristics than to Darwinian natural selection. We are not therefore in a situation where there is an established general theory of cultural evolution which can be applied to specific cases. Rather, we should think of it as an illuminating paradigm which can be applied with advantage to specific areas provided that appropriate attention is given to special factors applying to those areas.

While not overlooking these difficulties, I would like to propose the following tentative hypothesis for the behaviour of individuals and groups when considering whether to adopt innovations in science and statistics. We assume that such individuals and groups take into account the following criteria:

(1) Subjective factors such as curiosity, understanding, enjoyment, aesthetic considerations and so on.
(2) Objective benefits such as material wealth and power, both over the environment and over people.
(3) Costs, both material and nonmaterial.
(4) Environmental and social opportunities and pressures.

In considering the extent to which cultural changes are Darwinian, we ourselves will have to bear in mind possible underlying non-Darwinian trends in environment and culture.

Now I turn to the evolution of science and immediately encounter a difficulty. It would be convenient for this paper if a consensus existed among philosophers of science about what science is, how it developed, and how it is done. However, the amount of disagreement is large, which seems strange when one considers that within any particular branch of science the scientists concerned usually agree quite closely. To make my point I refer the reader to the Proceedings of part of an International Colloquium in the Philosophy of Science edited by Lakatos and Musgrave (1970). The origin of this part of the

Colloquium was this. In 1962 T. S. Kuhn published an eloquent and persuasive book called *The Structure of Scientific Revolutions* which challenged some existing theories in the philosophy of science and had a significant impact. A number of my philosopher colleagues at the London School of Economics, notably Sir Karl Popper and Imre Lakatos, disagreed with some aspects of Kuhn's ideas and organized a session at the Colloquium so that the matter could be debated. There are nine papers from this session in the Proceedings, the first and last by Kuhn. It is clear from a perusal of the papers that the disagreement is substantial. To establish this I will quote just one passage from Kuhn's concluding paper *Reflections on my Critics* (Lakatos and Musgrave, 1970, p. 232):

> One especially interesting aspect of this volume is, then, that it provides a developed example of a minor culture clash, of the severe communication difficulties which characterize such clashes, and of the linguistic techniques deployed in the attempt to end them. Read as an example, it would be an object for study and analysis, providing concrete information concerning a type of developmental episode about which we know very little. For some readers, I suspect, the recurrent failure of these essays to intersect on intellectual issues will provide the book's greatest interest.

What is the layman to do in a situation in which there is no consensus among specialists to which he can refer? The subject is too complex for me to attempt to chart my own route. It seemed to me that the only sensible course was to select a single authority and base my exposition on his writings. For this purpose I have chosen Sir Karl Popper. Although I admire Popper and even his critics agree that he is a leading authority in the field, it is not necessary for anyone to assume that I accept his views in detail, nor will I attempt to persuade the reader into an acceptance of Popper's system of thought. It is sufficient for me to claim that Popper's philosophy of science represents a sufficiently close approximation to the views and needs of statisticians for my purpose.

For those who have not read Popper but would like to sample his writings I commend the recent paperback collection edited with an introduction by Miller (1983). Also of great interest are a collection of Popper's essays on the evolution of knowledge (1979) and his autobiography (1976), both in paperback. To those who prefer Kuhn to Popper let me suggest that the differences between them is not as great as they might believe, and let me quote in support the following statement by Kuhn on p. 798 of his article in Schilpp (1966), "On almost all occasions when we turn explicitly to the same problems, Sir Karl's view of science and my own are very nearly identical."

Let us briefly consider what science is about, recognising that for reasons of space all our statements will have to be crudely oversimplified. We statisticians regard science as made up of observations and models. We use models to explain the behaviour of real phenomena, recognizing that at best a model is only an approximation to reality. The credibility of a model is determined by how closely it fits the observations. Given two models which fit the data equally well and are equal in other relevant respects, we prefer the simpler one.

The Darwinian theory of the evolution of science is a development of the following basic idea. Given two competing models for the same phenomena, the one which survives is the one with the greatest explanatory power. It is likely that this idea is quite old, probably going back to the nineteenth century; indeed, Popper seems to imply this on p. 67 of (1979). However, the earliest explicit reference known to me is Popper's *Logik der Forschung*, published in German in 1934. The English translation (1959) states on p. 108 the following (Popper uses "theory" for "model"):

> It may now be possible for us to answer the question: How and why do we accept one theory in preference to others?
>
> The preference is certainly not due to anything like an experiential justification of the statements composing the theory; it is not due to a local reduction of the theory to experience. We choose the theory which best holds its own in competition with other theories; the one which, by natural selection, proves itself the fittest to survive. This will be the one which not only has hitherto stood up to the severest tests, but the one which is also testable in the most rigorous way. A theory is a tool which we test by applying it, and which we judge as to its fitness by the results of its applications.

Popper developed this idea further later; see, for example, Popper (1976, 1979) and Miller (1983). There are two main respects in which Popper's treatment seems to me to be deficient from the standpoint of this paper. The first is that, like most philosophers, Popper proceeds by taking the human capacity for thought as it now is for granted. As indicated in the previous section, my view is that if we want to understand in a realistic way why we think what we think we ought to consider where our capacity for thought came from. Admittedly, such an exercise is highly speculative, but is not much of philosophy speculative? The second point is that the analogy between new ideas in science and biological mutations seems to me weaker than Popper suggests. Good ideas in science do not occur randomly like mutations but are often purposive developments by scientists of earlier, less good ideas. Thus they are sometimes more Lamarckian than Darwinian in essence.

Apart from these two points, and considering the development of science purely from the standpoint of epistomology, my inclination is to accept Popper's Darwinian theory of the evolution of scientific knowledge in its main essentials. The reader may feel that there is some inconsistency between this statement and the scepticism I expressed a few paragraphs ago in relation to cultural evolution in general, but in this is not so. The reason is that Popper uses a clear criterion for the selection mechanism which determines which of two competing theories survives, namely explanatory power. If the reader turns back to the four criteria I listed for cultural selection in general he will see that Popper's treatment falls into only the first of these. He is concerned only with our understanding of the world and not at all with such questions as costs and benefits or technological or environmental feasibility. Thus Popper's scheme is much closer to biological evolution, which also has a simple selection mechanism.

From the standpoint of this paper a treatment which regarded statistics only as part of pure science would be quite unacceptable, since statistical applications extend into an immensely wide range of human activities: technology, industry, government, policy analysis, the decision and management sciences, and information science, to mention just a few. Nevertheless there is a sense in which all practising statisticians are essentially scientists, since all are concerned with the collection, analysis, organisation, and presentation of observational data, together with the fitting and testing of models intended to represent the behaviour of the phenomena under study. The essential difference from pure science is that by far the greater part of this activity is aimed directly at applications in the real world rather than towards achieving an understanding of the universe. Because of this, we need to pay serious attention to items (2), (3), and (4) in my list of selection criteria as well as item (1).

It would obviously not be practicable for me in this article to attempt to review the whole range of statistics and its historical development. A general review of the field is given in the *International Encyclopedia of Statistics* (two volumes) edited by Kruskal and Tanur (1978). This contains a series of articles ranging over the whole of statistics by an international panel of experts. I commend particularly the article entitled "The history of statistical method" by Sir Maurice Kendall, which provides a compact summary of the history of statistics and provide references to further historical sources. I should also perhaps mention Westergaard's (1932) book, which reviews the development of statistics from the sixteenth to the end of the nineteenth century, mainly in the social and economic fields.

What I shall attempt to do here is give a brief indication of the extent to which Darwinian thinking might help us to understand the development of statistics. From this point of view it is convenient to divide statistical work into three categories:

(a) statistics as a pure science,
(b) statistics applied to other sciences,
(c) statistics intended for direct application in the real world.

I hope that no one will regard it as invidious if for the sake of brevity I subsume pure and applied probability under the word "statistics." The word "science" is used here in a very general sense and is intended to include the technological and social sciences.

The essential ingredient which characterises statistical work is variation among observations. It is common knowledge that the basic impetus for the initial development of all important branches of statistics came from applications, either to other sciences or directly to the real world. This applies even to the purest part of the subject, mathematical probability, which originated in all important respects in the study of games of chance in the sixteenth and seventeenth centuries. Nevertheless, underlying each such branch resides a core of pure theory which may be classified under category (a). Since each of these theories is a branch of pure science, it is reasonable to treat it as Darwinian under Popper's description of the evolution of knowledge. Within each core of theory, innovations become accepted as significant contributions to the extent that they are perceived by the statistical community as enhancing understanding of the subject to a substantial degree in competition with other theoretical proposals. I want to emphasise here that it is statisticians themselves who make the selection and the criterion they use is explanatory power.

It is sometimes suggested that probability models of real phenomena are intrinsically different from other mathematical models in that they have a different relationship to the phenomena they seek to explain or describe, but I do not accept this. I believe that the relationship is not essentially different from that in other mathematical sciences such as physics. In both cases the purpose of the model is to approximate the behaviour of observations of real phenomena as closely as possible. In neither case will it be possible to achieve a "true model" since it seems unlikely that humans will ever achieve a knowledge of ultimate reality. Thus it seems to me futile to agonise about whether random phenomena really exist in some ultimate sense "out there" in the universe. One thinks here of Einstein's remark that God does not throw dice. It is sufficient to agree that many phenomena appear to behave randomly and can be modelled to a sufficiently close approximation by mathematical models derived from probability. There is of course sometimes a difficulty in checking the fit of the models to observations because of the intrinsic variability of the observations, but that is another matter. I am referring here to the use of objective probability models for natural phenomena. Probability models can, of course, also be used for subjective modelling of human uncertainty, for example in decision analysis, providing the axioms of probability theory are satisfied sufficiently closely.

The position under category (b) is slightly different. Here it is not statisticians who decide ultimately what statistical innovations are incorporated into scientific practice, but the scientists who use statistics in their work. In deciding whether to adopt a new statistical technique a scientist has to balance the cost and effort of using it against its value in his research. In taking his decision, environmental factors such as computing capability and the availability of expert statistical advice, or at least adequate information, play an important part. Ultimately, the value of the scientist's research is judged by his peers according to its explanatory power in the area of science concerned. Thus the process by which statistical innovations are accepted in other sciences can be regarded as Darwinian, but the selection process is complicated, both because costs as well as benefits have to be considered, and because environmental factors are important.

Category (c) is somewhat similar. In the private sector, costs, benefits, and environmental constraints have to be evaluated by managers who are considering whether or not to adopt a statistical innovation, the ultimate criterion of selection being, in theory at least, the estimated net contribution to the profits of the organization concerned. In the public sector, the responsible officials behave similarly except that the ultimate selection criterion they use is their perception of the net public benefit.

In the last few paragraphs we have seen how Darwinian-type selection of statistical innovations can operate locally in particular situations, and this can be an important factor in statistical development. However, these considerations alone do not explain the tremendous growth throughout the world in the use of statistics over the past century. To understand other aspects of this growth we must look on the one hand towards changes in the environment within which statistical work is done, and on the other hand to external causal forces which have led to an increase in the demand for statistical analysis. Environmental changes of importance include changes in computing technology and the spread of statistical education. I think it is debatable whether Darwinian thinking can contribute more than marginally to the understanding of these environmental developments. I would myself be inclined to consider them from a relatively straightforward point of view as topics in technological and educational history.

As for the causal forces underlying the worldwide increase in demand for statistics, these are obviously related to the scientific and informational needs of modern societies, particularly the advanced industrial societies. Explaining how these societies developed is a matter for social, economic, and cultural historians, and for me to offer comments would take me beyond my competence, except to venture the opinion that evolutionary theory will play a bigger part in the deliberations of these historians in the future than in the past.

It is obvious from the qualifications that I have made from time to time that I regard my attempt at constructing a theory of statistical evolution on Darwinian lines as only partially successful. Nevertheless I believe that the analysis can help to clarify the examination of some of the factors that are operating when the subject is developing. In this spirit let me close the paper by making a modest attempt to anticipate by means of evolutionary thinking some of the developments that might occur in statistics during the next few years. The basic approach is first to try to estimate the changes in the environment that can be expected over the period and then to try to identify the places where the changing environment presents opportunities for the introduction of successful innovations by individual statisticians and groups of statisticians.

It is clear that the greatest changes in the environment for statistics over the next few years will arise because of developments in computing. I am not thinking here just of statistical computing, but also more generally of deep changes in the nature of western society, especially in the way work is done, arising from developments in automation and information technology. Following earlier movements from agriculture into manufacturing industry, a movement from manufacturing into service industries, accompanied by high unemployment in some countries, is now taking place at a rapid rate. These upheavals will have drastic consequences for the nature of our society, but I cannot comment on these here except to point to the steep rise in the demand that can be expected for people with skills in information processing and data analysis in the most general sense of these terms. One can envisage countless opportunities for the modelling of systems of all kinds on computers, with analysis partly by mathematical methods and partly in some cases by simulation. The management of data streams in real time will become of great importance. In all the sciences—natural, technological, social, and management—mathematical models of growing complexity and of great diversity will be increasingly employed. These will need to be fitted to data, assessed for performance, and then used for analysis. In the experimental sciences it will be possible to analyse data sets of immensely greater complexity than has previously been envisaged. Sometimes this will be done in real time on dedicated computers in the laboratory as the data are generated. It will be increasingly possible to perform experiments on mathematical models in the computer instead of on reality itself in the laboratory.

Within the educational system there will be an increasing emphasis on quantitative methods generally, accompanied, perhaps, by a declining emphasis on traditional statistical inference. Courses will deal with mathematical modelling, computer science, problem solving, and techniques currently treated under the name operational research, as well as updated versions of classical statistics and probability. Students in all fields will be much better informed about the value of sound quantitative analysis than in the past.

In statistics itself there will be better packages with improved user friendliness and interactive facilities, easier data handling, vastly improved graphics, and a steadily increasing consultative element based initially on expert systems, and, in the remoter future, on fifth generation computers. More routine statistical analysis will be done by amateur statisticians from other fields who may have an interest in the subject and may have attended courses on it but who cannot pretend to have an expert knowledge.

How will specialist statisticians respond to this changing environment and the opportunities it offers them to introduce successful innovations and thus make contributions of lasting value to the culture of our society? This will depend to a substantial degree on the human as well as the intellectual qualities of the individual statisticians themselves, such as their vision, imagination, energy, and courage. They will need to recognise that statistics is a subject that feeds on developments in other fields, so they will need to be outward-looking and willing to make efforts to learn about other subjects as well as their own.

The opportunities for imaginative and important contributions by innovative statisticians will be substantial. Following my biological analogy, if statisticians do not move into the ecological niches created by the changing environment, the world will move ahead without them. Other, more creative and energetic mutants will occupy these ecological niches instead. The outcome may not be as good, but in the long run nobody will realise that. Society will doubtless just assume, as it usually does, that what occurred was inevitable and that whatever happened was for the best—a supposition that every individual knows from his personal experience to be fallacious.

Acknowledgements

Because this paper is interdisciplinary as well as unorthodox, I was well aware while writing it of the possibility of error, so I thought it prudent to circulate the first draft fairly widely and invite comment. I am grateful to the following whose comments have helped me to improve the paper substantially, not only by removing errors but also in more positive ways: J. Aitchison, D. J. Bartholemew, K. G. Binmore, W. F. Bodmer, Sir John Boreham, L. L. Cavalli-Sforza, H. E. Daniels, R. Dawkins, R. M. Durbin, A. W. F. Edwards, S. E. Fienberg, D. J. Finney, P. R. Fisk, E. A. Gellner, J. Hajnal, C. Howson, J. G. Jenkins, J. F. C. Kingman, M. Kline, W. H. Kruskal, J. Maynard Smith, H. Shieham, C. S. Smith, M. D. Steuer, G. S. Watson, J. Worrall, P. Urbach. Responsibility for any errors remaining is my own.

Bibliography

Abian, A. (1965). *The Theory of Sets and Transfinite Arithmetic*. Philadelphia: Saunders.

Aitchison, J. (1983). *The Articulate Mammal: An Introduction to Psycholinguistics*, (paperback 2nd edn.) London: Hutchison

Alexander, R. D. (1979). *Darwinism and Human Affairs*. Seattle: Univ. Washington Press. (1980, London: Pitman.)

Binmore, K. G. (1980). *The Foundations of Analysis: A Straightforward Introduction. Book 1, Logic, Sets and Numbers*. Cambridge: University Press.

Bodmer W. F. (1981). "Gene clusters, genome organization and complex phenotypes. When the sequence is known, what will it mean?" *Amer. J. Hum. Genet.*, **33**, 664–682.

Bodmer, W. F. (1983). "Gene clusters and genome evolution."In D. S. Bendall (ed.), *Evolution from Molecules to Men*. Cambridge: University Press, 197–208.

Bodmer, W. F. and Cavalli-Sforza, L. L. (1976). *Genetics, Evolution and Man*. San Francisco: Freeman.

Cavalli-Sforza, L. L. and Feldman, M. W. (1981). *Cultural Transmission and Evolution: A Quantitative Approach*. Princeton: University Press.

Charlesworth, B., Lande, R., and Slatkin, M. (1982). "A neo-Darwinian comment on macroevolution." *Evolution*, **36**, 474–498.

Chomsky, N. (1980). *Rules and Representations*. New York: Columbia U. P.

Cohen, P. (1966). *Set Theory and the Continuum Hypothesis*. New York: Benjamin.

Dawkins, R. (1976). *The Selfish Gene*. Oxford: University Press. (Paperback, 1978, London: Granada.)

Dawkins, R. (1982). *The Extended Phenotype: The Gene as a Unit of Selection*. Oxford: Freeman.

Dobzhansky, T. and Boesiger, E. (1983). *Human Culture: A Moment in Evolution*. New York: Columbia U. P.

Eldredge, N. and Gould, S. J. (1972). "Punctuated equilibria: An alternative to phyletic gradualism." In T. J. M. Schopf (ed.) *Models in Paleobiology*. San Franciso: Freeman Cooper, 82–115.

Gould, S. J. (1977). *Ever Since Darwin* (paperback). New York: Norton.

Gould, S. J. (1981). *The Mismeasure of Man* (paperback). New York: Norton.

King's College Sociobiology Group (1982). *Current Problems in Sociobiology. Cambridge U. P.*

Kline, M. (1980). *Mathematics: The Loss of Certainty*. Oxford: University Press. (Paperback).

Kruskal, W. H. and Tanur, J. M. (eds.) (1978). *The International Encyclopedia of Statistics*. New York: Free Press.

Lakatos, I. (1976). A renaissance of empiricism in the recent philosophy of mathematics. *Brit. J. Philos. Sci.*, **27**, 201–223.

Lakatos, I. and Musgrave, A. (1970). *Criticism and the Growth of Knowledge* (paperback). Cambridge: University Press.

Lenneberg, E. (1967). *Biological Foundations of Language*. New York: Wiley.

Lorenz, K. (1977). *Behind the Mirror: A Search for a Natural History of Human Knowledge* (paperback). London: Methuen.

Lumsden, C. J. and Wilson, E. O. (1981). *Genes, Mind and Culture: The Coevolutionary Process*. Cambridge: Harvard U. P.

Lumsden, C. J. and Wilson, E. O. (1983). *Promethean Fire: Reflections on the Origin of Mind*. Cambridge: Harvard U. P.

Maynard Smith, J. (1982). *Evolution Now: A Century after Darwin* (paperback). London: Nature and Macmillan.

Mayr, E. (1982). "Speciation and macroevolution." *Evolution*, **36**, 1119–1132.

Miller, D. (1983). *A Pocket Popper*. London: Fontana Paperbacks.

Monod, J. (1971). *Chance and Necessity*. New York:, Knopf. (Translated from *Le Hazard et la Necessité*, 1970, Paris: Editions du Seuil. English ed., 1972, London: Collins. Paperbacks, 1974, London: Fontana; and 1977, London: Fount.)

Orgel, L. E. (1973). *The Origins of Life: Molecules and Natural Selection*. New York: Wiley.

Popper, K. R. (1959). *The Logic of Scientific Discovery*. London: Hutchinson. (Revised paperback, 1980.)

Popper, K. R. (1976). *Unended Quest: An Intellectual Autobiography*. London: Fontana (Paperback).

Popper, K. R. (1979). *Objective Knowledge: An Evolutionary Approach* (revised paperback ed.) Oxford: Clarendon.

Schilpp, P. A. (ed.) (1966). *The Philosophy of Karl R. Popper*. La Salle: Open Court.

Shorrocks, B. (1978). *The Genesis of Diversity* (paperback). London: Hodder and Stoughton (Paperback).

Watson, J. D. (1977). *Molecular Biology of the Gene*, 3rd ed. Menlo Park: Benjamin.

Westergaard, H. (1932). *Contributions to the History of Statistics*. London: King.

Wilder, R. L. (1968). *Evolution of Mathematical Concepts*. New York: Wiley. (Paperback, 1974, London: Transworld Student Library.)

Wilson, E. O. (1975). *Sociobiology: The New Synthesis* Cambridge, MA: Harvard U. P.

CHAPTER 10

Statistical Computing: What's Past Is Prologue

William F. Eddy and James E. Gentle

Abstract

We sketch an outline of statistical computation, an arena of increasing importance in statistics. Its roots are in the numerical analysis of statistical techniques, and its leaves are nearly everywhere that statisticians go. After an introduction which covers the short history of statistical computing, we discuss in some detail the various areas of statistics which have been affected by computers. The most important need of statisticians interested in using computers is software that works; we consider the possible sources of software and discuss the validity of computer programs. The currently faster-growing area in statistical computation is that of graphics; we briefly discuss the past, present, and future of statistical graphics. One of the most important statistical uses of computing power is to run simulations, and we discuss the Monte Carlo method and the problems of generating random numbers. Finally we address the revolution that is occuring with the advent of inexpensive microcomputers. There are obviously a myriad of benefits, but we mention a few cautions.

1. Introduction

According to Fisher (1925), "Statistics may be regarded as (i) the study of populations, (ii) the study of variation, (iii) the study of methods of the reduction of data." At the time that sentence was written the size and complexity of statistical analyses was limited by human computational ability. The advent of programmable digital computers has changed the picture substantially.

The work of a practicing statistician is focused on data. Statistical activities encompass planning, collecting, arranging, storing, displaying, summarizing, modeling, and predicting. Modern digital computers can provide substantial assistance in the execution of all these activities. Hence the close relationship of the fields of statistics and computing is no surprise. Mathematical statisticians who may be concerned with data, only in an indirect way, sometimes perform extensive computations in Monte Carlo simulations to evaluate statistical procedures. Those whose primary concern is teaching statistics use computers to illustrate certain concepts, to produce special data sets for student exercises, and even to engage in a dialog with the student to introduce

Key words and phrases: graphics, microcomputers, Monte Carlo, random numbers, software.

or review statistical concepts. In summary, even those statisticians who are not directly involved with applications are finding ever more uses for computers.

1.1. Early Developments

Because of the close association of statistics and computing, statisticians have been major users of the latest developments in computing machinery and major users of media for storage of data in a form accessible by machines. Although Charles Babbage produced plans for an "analytical engine" in the early 1830s, there were few practical mechanical aids for computation until later in that century. The first significant development in media for machine data entry and storage was the punched card, introduced in 1890 by Herman Hollerith of the U.S. Census Bureau. The punched card as a device to store information had evolved from the early wooden cards devised by the Frenchman J. M. Jacquard in 1805 to control the pattern woven into cloth by a loom. Hollerith also invented machines to read and interpret the cards, to accumulate counts, and to sort the cards. The use of Hollerith cards and the tabulating machines allowed the processing of the 1890 Census to be completed in two years, whereas the manual processing of the 1880 Census had required ten years. Punched cards continued to be the most commonly used medium for data and program storage until the mid 1970s, when magnetic tape and disk storage became widespread.

Familiarity with computing techniques on the mechanical calculators of the 1920s was of such importance to statisticians of that time that H. A. Wallace (later a Vice President of the U.S.A.) and George W. Snedecor wrote a booklet *Correlation and Machine Calculation* and conducted classes at Iowa State College in the use of the machines. Likewise, at Rothamsted Experimental Station the computing room was an important part of that center of early activity in statistics. In the early 1920s, R. A. Fisher spent many hours there with the "Millionaire," a fairly sophisticated mechanical calculator (Box, 1978).

Desk calculators were commonplace in the offices of statisticians from the 1920s until ready access to digital computers became available to many in the 1960s. With the advent of the von Neumann stored-program computers in the 1940s and 1950s, statisticians were quick to make use of them for statistical data processing and for computation of tables of probabilities, moments, and other numerical quantities characterizing statistical distributions.

There have been two general lines of development in the use of computers by statisticians. One has been the use for tabulations and basic summary statistics ("data processing") and the other has been the development of numerical algorithms. The early more substantive work in the field of statistical computing was generally led by numerical analysts, or by statisticians with a strong interest in numerical mathematics. Nonnumeric techniques for database management, for displays, and for exploratory and distribution-free methods have recently come to be very important, and much interesting work is currently under way in these areas.

1.2. Statistical Computing as a Subdiscipline

Beginning about 1966, a series of events occurred which served to define the nascent subdiscipline of statistical computing. J. A. Nelder and B. E. Cooper organized a meeting on "Statistical Programming" held in December of that year at the Atlas Computer Laboratory in Chilton, United Kingdom. This meeting consisted of a few invited presentations and a number of informal discussions, generally concentrating on issues relating to programming languages and program development. The papers presented at the meeting and transcriptions of the discussions were published in *Applied Statistics*, **16**, No. 2 in 1967. An important outcome of the meeting was that a Working Party on Statistical Computing was appointed, with Nelder as chair. In the next few years, the Working Party published guidelines for program development and began administering a new section on Statistical Algorithms in *Applied Statistics*. The algorithms section publishes descriptions of programs and the actual code itself, in either FORTRAN or ALGOL.

Nancy Mann and Arnold Goodman organized a conference "Computer Science and Statistics: A Symposium on the Interface," held February 1, 1967, in Santa Monica, California. This symposium, which was chaired by Nancy Mann, was devoted to consideration of a number of topics in statistical computing as well as to topics in statistical evaluation of computer systems. A series of Symposia on the Interface has been held almost yearly since 1967 at various locations in the United States and Canada. The proceedings of these conferences, which have been published since the fourth Interface in 1970, provide a record of developments in statistical computing and the kinds of issues that have concerned workers in the field.

Also in 1967 the first textbook devoted to statistical computing, *Statistical Computations on a Digital Computer*, by W. J. Hemmerle, appeared. At the same time regular courses in statistical computing began to be offered in departments of statistics at several universities.

In 1968 the American Statistical Association formed a Committee on Computers in Statistics. The main activity of this committee was organizing sessions at the annual meetings of the ASA. In 1972 the Statistical Computing Section of the ASA was formed, with W. J. Dixon as chair. One of the more visible activities of this Section is in organizing sessions at the ASA annual meetings. The proceedings of the sessions organized by the Statistical Computing Section have been published since 1975.

The Conference on Statistical Computation held at the University of Wisconsin in 1969 brought together an international group to discuss changes in the teaching and practice of statistics wrought by computers. The proceedings

Table 1. Major Periodicials in Statistical Computing by Inaugural Year

Year	Periodical
1970	*Computer Science and Statistics: Proceedings of the Symposium on the Interface*
1972	*Journal of Statistical Computation and Simulation*
1974	COMPSTAT: *Proceedings in Computational Statistics*
1974	*Statistical Software Newsletter*
1975	*Communications in Statistics, Part B*
1975	*Proceedings of the Statistical Computing Section of the American Statistical Association*
1980	*SIAM Journal on Scientific and Statistical Computing*
1983	*Computational Statistics and Data Analysis*

of this conference, *Statistical Computation*, edited by Roy Milton and John Nelder, contain important early articles on topics such as data structures, statistical languages and systems, and numerical analysis with applications in statistics.

The first of an important biennial conference was convened in Vienna, Austria, in 1974. The COMPSTAT conferences, which were held in Berlin in 1976, in Leiden in 1978, in Edinburgh in 1980, in Toulouse in 1982, and in Prague in 1984, have drawn attendees from all over the world. The proceedings of all of the COMPSTAT conferences have been published.

In 1977 the International Association for Statistical Computing was formed as a section of the International Statistical Institute. The European Regional Section of this association has been a sponsor of the COMPSTAT conferences and, since 1982, has cooperated with the Institut für Medizinische Informatik und Systemforschung of the Gesellschaft fur Strahlen- und Umweltforschung (GSF) in publishing the Statistical Software Newsletter. This newsletter, which was begun by GSF in 1974, carries articles on statistical computing, news about developments in software, and book reviews.

In general, the literature of statistical computing has been dispersed through a number of journals, in applied statistics, in computer science, and in numerical analysis. The proceedings of conferences mentioned above are timely sources of information on developments in the field. The major sources are summarized in Table 1.

2. Areas of Impact

2.1. Analysis of Data

The most important influence of computing on statistics has, naturally, been in the analysis of data. There has been a marked increase in speed of computation and, hence, in the size of data sets analyzed. Also, there has been an attendant increase in the complexity of the models that are contemplated and the procedures that are utilized. The most significant changes, however, have occurred not in the routine statistical analysis of data but in the myriad other tasks that computers can and do perform.

Before the advent of digital computers the size of a data set that could actually be analyzed was extremely limited. For example, consider a simple least-squares linear regression analysis. A problem with one thousand observations on ten independent variables would require *weeks* just to accumulate the cross-products matrix on a mechanical calculator. The same matrix can be accumulated in *seconds* on a modern digital computer, once the data has been converted to computer-readable form. This is roughly a millionfold speedup. A natural consequence is that data sets can be analyzed that are considerably larger than in the past. Regression problems with one million (10^6) observations are not common but can be now handled routinely; one billion (10^9) observations is certainly conceivable. Regression problems with thousands of variables are also extremely rare but can be handled; tens of thousands of variables are possible.

Additionally, digital computers have made it easier to generate or collect very large data sets. Often a device capable of converting analog to digital (A–D) information is used as a primary collection means. For example, there are many such devices aboard satellites and space vehicles. The information collected is then transmitted (perhaps after some preliminary processing) via telemetry back to earth. The resulting data sets often contain 10^{10} or 10^{12} bits of information.

The first step in the analysis of a data set, once it is in computer-readable form, is cleaning and editing. Computers make it possible not only to verify easily that individual variables have all their values within the appropriate range, but also to verify that the joint behavior of several variables is within the appropriate range. Furthermore, various imputation procedures to fill in missing data can be easily implemented.

The most widely known of these imputation procedures is the "hot-deck" method, which could not be implemented at all without high-speed computation. Assume that individuals completing a questionnaire are divided into categories. A stored value for each category is determined, based on past data. During the processing of new data each individual is assigned to a category. If a new questionnaire has all items complete, then the new responses replace those stored for that category. If the questionnaire has a missing item, then that item is replaced with the corresponding stored item in the hot deck. The hot-deck procedure is probably the most widely used imputation method because it is so easy to implement; however, it relies on some important assumptions concerning the missing data and should be used with some caution (see, e.g., Bailar and Bailar, 1978; Panel on Incomplete Data, 1983).

One extremely important effect of the increase in speed of computation is the increase in the complexity of models which are fitted to data. It is extremely difficult to quantify this increase in complexity; but there are certainly many procedures in use today which would not be contemplated without a computer available to do the calculations. One of the more widely used such procedures is the bootstrap. Efron (1979) gives an interesting discussion of the bootstrap as well as other subsampling and resampling methods, such as the jackknife; these procedures may require orders of magnitude more computations than are required by traditional techniques.

2.2. Display of Data

Probably the most widespread nonnumeric statistical use of computers is for the generation of pictorial displays. Such displays range from the ordinary scatter plot (conceived and used long before computers), to projections of multivariate data into specially chosen coordinates, to semigraphical and nongraphical displays.

Many software packages for statistical analysis now provide extensive graphics capabilities. In addition, special-purpose graphics packages have been developed that allow special transformations and projections of the data. Additional dimensions to the display can be variously added by perspective, motion, color, and sound. Most of these packages are in their infancy; statisticians can expect a rapid increase in the availability of such packages as the price of computing and of graphical display devices decreases.

We will comment further on developments in computer graphics in a later section.

2.3. Testing Hypotheses

One area of data analysis that was introduced long before computers but that was infeasible without them is randomization inference. The fundamental idea is due to Fisher (1935) in the context of hypothesis testing. Early work was devoted to showing that various parametric testing procedures were good approximations to the permutation procedure. There has been a recent resurgence of interest in randomization inference, spurred, no doubt, by the increasing availability of the needed computing power (see, e.g. Pagano and Tritchler, 1983).

The idea behind permutation tests is fairly straightforward. Suppose there are two sets of data $X_1, \ldots, X_m$ and $Y_1, \ldots, Y_n$. Under the hypothesis that the two sets of data come from the same distribution, the labeling of each observation as an X or a Y is arbitrary. Consider all $(n + m)!/(n!m!)$ possible assignments of labels to the observations, and for each possible assignment compute $\bar{X} - \bar{Y}$. Then the level of significance of $\bar{X} - \bar{Y}$ for the observed assignment of labels is just the fraction of those values of $\bar{X} - \bar{Y}$ computed for the other assignments which were more extreme.

G. A. Barnard (1963) suggested use of "Monte Carlo tests" for testing hypotheses for which the test statistic does not have a known distribution. The idea is to simulate samples from a population satisfying the null hypothesis, compute the test statistic from each sample, and compare the computed value from the actual sample with the quantiles of the test statistic from the simulated samples. Besag and Diggle (1977) use such a test in investigating the spatial pattern of 65 pine trees growing on a square plot of ground. To test the null hypothesis of a uniform distribution over the plot, they use the distances between all pairs of trees and form a statistic similar to a chi-square goodness-of-fit statistic for distances between two random points in the square. The statistic obviously does not have a chi-square distribution even under the null hypothesis, because the distances are correlated. They simulate 99 samples of 65 points uniformly distributed over the square, and for each sample they compute the chi-square-like statistic. They do not reject the null hypothesis, since the statistic computed from the actual sample was near the median of the statistics from the simulated samples. The statistic would have been significant at the 0.01 level if it had been from a chi-square distribution.

2.4. Development of Theory

There are several areas not associated with data analysis where computers can and will have a substantial impact on statistics. In theoretical research, the most important of these areas is in mathematical experimentation, including but not limited to Monte Carlo studies. An area that has not yet been exploited by statisticians is the use of computers for symbolic calculation.

One of the earliest uses of digital computers was for random sampling experiments. Such experiments were anticipated by von Neumann (1945) when he wrote: "It [the computer] will certainly open up a new approach to mathematical statistics; the approach by computed experiments. . . ." The Monte Carlo method as it has become known is simple, elegant, and impossible without a computer. There are two general classes of Monte Carlo problems. The first is the use of computers to emulate a stochastic phenomenon; an early example was a study of neutron diffusion described briefly in Hammersley and Handscomb (1964). The second is the random sampling experiment to evaluate integrals; in its simplest form this is done by uniform sampling: the proportion of points falling within the desired region (appropriately scaled) is an estimate of the integral.

The extent of the role of Monte Carlo studies in the development of statistical theory can be seen in the form and content of research articles in leading statistical journals. Table 2 shows, for two journals (the *Journal of the American Statistical Association* and *Biometrika*), the total number of articles published in selected years and the number of articles in those journals containing reports of Monte Carlo studies.

Table 2. Articles Reporting Monte Carlo Results

Year	Journal	Total articles	Articles reporting Monte Carlo studies
1973[a]	*JASA*	170	32
	Biometrika	90	20
1980	*JASA*	132	48
	Biometrika	111	30
1983	*JASA*	111	28
	Biometrika	92	32

[a]Reported by Hoaglin and Andrews (1975).

A typical one of these research articles could be outlined as follows:

(i) Statement of the problem and discussion of the standard methods.
(ii) Description of the new approach and results.
(iii) Mathematical derivation of properties of the new approach (this is often limited to asymptotic results).
(iv) Report of the Monte Carlo study.

The Monte Carlo study is commonly used to answer the question, "How well does it (the estimator, test statistic, etc.) perform?" We will discuss Monte Carlo studies further in a later section. There are, of course, other areas of mathematical experimentation on a computer (see, e.g., Grenander, 1982).

Computers have also affected the direction of the development of statistical theory, by giving practical interest to the study of computation-bound techniques, such as multivariate analyses, randomization, and permutation tests, as well as the Monte Carlo tests referred to in the section on analysis of data above.

One area of computation which has not yet been used very extensively by statisticians is symbolic computation. There are now a variety of systems, the best-known being MACSYMA, which are able to perform the sort of basic manipulations we all learn in school (see, for example, Gong, 1983). They can operate on symbolic expressions (i.e., formulas) and can integrate, differentiate, rearrange orders of summation, reduce fractions to lowest terms, etc. While these systems do not yet seem able to easily handle asymptotic formulas, they should have considerable value in the derivation of finite-sample distributions and other similar tasks.

2.5. Teaching of Statistics

If the measure is amount of computer time, students are almost certainly the largest consumers of statistical computation. The trend will almost certainly continue as price reductions increase availability. Of course, the major use of computing for teaching statistics is the use of statistical packages to perform the routine numerical calculations. In the future this will change as computers assist or take over student–teacher communication and the assigning, collecting, and grading of homework and examinations.

2.6. Information Storage and Retrieval

The current uses of computers involve storage of data, computations on data, displays, and report writing. Computers are ideally suited to storing and retrieving any kind of information, not just data to be analyzed, but information on how to analyze data and where to find additional information on a particular topic in the statistical literature. The research statistician can search a database, using a few selected keywords, to find articles and abstracts of articles on the subject of interest. Currently the bibliographic information for all entries in *Mathematical Reviews*, from 1959 to the present, and the reviews themselves, from mid 1979 to the present, are available through several commercial online database services. Much of the statistical literature is covered by *Mathematical Reviews*. The bibliographic files of *Current Index to Statistics* (a joint publication of the American Statistical Association and the Institute of Mathematical Statistics) will also soon be made available online. Complete texts of articles and books will also be available more widely, and searching for topics will be performed by computers.

Another way in which a computer's information base will be utilized will be in the analysis of data. In a trivial example of such a process, the analyst tells the computer that he wants to do an analysis and the computer displays a "menu" listing the possibilities. Each choice on the menu leads to submenus, and the analyst follows a decision tree to the analysis that may be appropriate. In a more sophisticated environment a computer may perform some computations and use summary statistics to aid it in suggesting an approach for the analyst. These "expert systems" are currently in their infancy, but have an important future role in the activities of the data analyst (see, e.g., Chambers, 1981). Some statisticians have expressed concern about the expert systems usurping the role of the statistical consultant. A more important concern, however, is with the current use of statistical analysis programs that give the user no indication of whether or not the use is correct.

3. Software for Statistical Computing

Software for statistical applications has evolved from stand-alone programs designed for a specific type of analysis to general-purpose systems which combine data-management and statistical-analysis capabilities. The earlier programs usually required that numeric codes be given to indicate options for the analysis (if the program allowed options) and that the data be entered in a specified format. The newer systems, which by the early 1970s were beginning to be widely distributed, provided simple commands for data-management capabilities as well as for multiple analyses in a single run. Schucany et al. (1972) provide a classification scheme and a survey of the statistical software extant in the early 1970s. The explosive growth in software during the decade of the 1970s can be observed by comparing the list of software given by Schucany et al. with the list in the survey of Francis (1981). The developments of those years were in more extensive use of graphics; interactive systems; the ability to perform complex analyses, especially of linear models, using a simple, but general, command syntax; and the ability to handle a wide range of specialized analyses. Perhaps the most significant development of statistical software during that period, however, was the wide distribution achieved by the major software packages and the increasing reliance of applied statisticians on commercial software. The abilities implemented in the major statistical software and the nature of the implementation often dictate the type of analysis performed. (Of course, this is not a new phenomenon; statisticians have always had to choose an analysis based on the tools, computational or otherwise, at their disposal.)

Software development has generally been a cottage industry. There is often a lack of accountability by developers; testing and validation are difficult. Problems with programs may not be identified until after several years of use, if ever. To compound the problem, many programs are developed on an ad hoc basis but then spread into general use. These programs may perform very well on the problems for which they were intended, but for other problems they may give incorrect results. Since these more general problems were not contemplated in the original design, however, the program may make no input validity checks so as to warn of inappropriate use.

The software for statistical analysis relies to a large extent on the libraries of subprogram modules that have been developed. Many of these subprograms have been distributed through the algorithms sections of *Applied Statistics*, *ACM Transactions on Mathematical Software*, and *Communications in Statistics, Part B.*

As the complexity and the number of the software systems has increased, the need has arisen for objective evaluation and comparison of these systems by persons who are knowledgeable in the area and who will devote the necessary time to the task. A committee for this purpose was established by the Statistical Computing Section of the American Statistical Association in 1974. This committee has since issued periodic reports of its activities and sponsored sessions on software evaluation at various meetings. Another committee of the Statistical Computing Section is concerned with the basic algorithms for statistical computation, and has worked on a classification scheme and index of available algorithms.

4. Graphics

Funkhauser (1938), in a survey of the development of techniques for statistical graphics, referred to the period 1860–1880 as the "golden age" of graphical techniques. While it is true that the innovations of that period, such as nomography and log–log paper, were important, the true golden age of graphics has just begun, and it is computers that have precipitated it. The speed of computers, the ability of computers to drive display devices, and the technological advances producing these peripheral devices, have truly changed the role of graphics in the everyday life of a statistician.

Prior to the advent of computer-generated graphics, and even when only rudimentary computer graphics capabilities were available, pictorial displays were generally limited to two types: rough, preliminary scratches, intended often to be seen only by the person producing them, and laboriously produced graphs and charts, intended more to summarize the results of the analysis than to be used to analyze. In the absence of computer facilities, the applied statistician's dictum "Plot the data!" entailed considerable time and effort. The problem was (and is?) that not enough plots were examined: plots of various transformations, plots of various adjusted variables, and plots of various intermediate and supplemental diagnostic statistics. Computers have added little to the simplest forms of scatter plots, but, of course, they make numerical calculation so fast and easy that many more such plots can be created. Also, the statistician can consider new variables to plot. One such example is the use of partial residual plots in regression analysis; these would not have been contemplated without a computer to do both the arithmetic and the requisite plotting.

A variety of graphical packages have recently been developed to aid in the exploration and study of large multivariate data sets. These packages typically provide the ability to rotate, rescale, and relocate coordinate systems. The data or a subset of them are then projected on a two-dimensional display. The algorithms for these kinds of plots are interesting and have captured the attention of many computer scientists working in algorithm development and analysis. Very significant hardware developments have also occurred recently: (1) increased resolution and improved quality of the display devices, such as bit-map graphics terminals; (2) graphic input devices, such as the "mouse"; and (3) continuing increase in the speed of the hardware, without which some graphics applications would not be practical.

Computer graphics has led to development of new tools for understanding data, such as the various pictographs for representation of multivariate data. The earliest of these is probably the faces of Chernoff (1973); some of the more recent are the trees and castles of Kleiner and Hartigan (1981). None of these displays is really practical without sufficient computational power to generate and display the pictures.

The 1977 Sheffield Conference of Graphical Methods in Statistics, attended by over three hundred participants, marked a growing awareness among statisticians of the potential for computer graphics in data analysis (see Cox, 1978, and other papers in the same issue of *Applied Statistics*). Three years later, in 1980, a second Sheffield Conference was convened with a broader theme: Looking at Multivariate Data. The importance of graphics in understanding multivariate data is underscored by the number of the papers at this more general conference that dealt with graphical displays (see Barnett, 1981, for the published papers). In the United States, the National Computer Graphics Association (NCGA) sponsors annual conferences surveying advances in the field and allowing vendors of hardware and software to display their wares. In 1984, for the first time, the American Statistical Association was a cosponsor of the NCGA conference, Computer Graphics '84, held in Anaheim, California. The ASA has a Committee on Statistical Graphics that surveys developments in the area and sponsors sessions at meetings of the ASA. Another indication of the increase of interest in statistical graphics was the appearance in 1983 of three books devoted to the subject (Chambers et al., 1983; Schmid, 1983; Tufte, 1983.)

Currently, the major statistical journals have some difficulty in publishing high-quality graphics. Most will not accept displays using colors and the reproduction of black-and-white graphics is often not good. The first paper with color displays in a statistical journal was Fienberg (1979), published in the *American Statistician*. Since then, that journal has published more displays in color, but the cost remains a deterrent. The additional production costs for one article using color in the November 1984 issue were over 15 percent of the total cost of an issue without color. Just as computer typesetting has reduced composition costs of ordinary text, computer production of displays, especially those with color, will bring down costs for use of better graphics in statistical journals. The December 1984 issue of the *Journal of the American Statistical Association* has an article with color graphics, and we believe color will be used routinely in the future in statistical journals, especially in articles on data analysis.

One of the most significant developments we emphasize throughout this paper is the decrease in the cost of computing. This will have a particularly profound effect on the increased use of computer graphics because of the high utilization of computer resources in producing graphics.

5. Random Numbers and Monte Carlo

As can be seen in the proportions of journal articles reporting Monte Carlo studies, these experiments continue to be an important use of computers for statisticians. Both the number of Monte Carlo studies and the size of the experiments have increased as the cost of computing has decreased. An example of the increased size of Monte Carlo experimentation is the preliminary study of a permutation statistic conducted by one of us recently. It involved approximately 10^6 random observations and a total of 10^{11} arithmetic operations; the results suggested the statistic was not useful and were discarded.

Monte Carlo experimentation is quite similar to random sampling experiments (perhaps deceptively so). Consider, for example, the problem of determining the "best" estimator of the mean in a mixture of two normal distributions with common mean and variances differing only by a scale constant. The candidate estimators include the sample mean, the sample median, the trimmed mean, and the Winsorized mean. Since they are all unbiased, the natural criterion is small variance. The Monte Carlo method involves generating samples from the mixed distribution, computing each of the statistics for each sample, and assessing the relative variances by comparing the sample variances. An objective in any statistical experiment is to reduce the effect of noise on the inference to be made. In this example, we wish to estimate the variance of the ith estimator, $V(T_i)$, and we desire our variance estimator itself to have small variance. What we are really interested in, however, is $V(T_i) - V(T_j)$, for any two of the estimators. Since the estimators are unbiased,

$$\begin{aligned} V(T_i) - V(T_j) &= E(T_i^2) - E(T_j^2) \\ &= E(T_i^2 - T_j^2). \end{aligned}$$

Since the estimators are positively correlated, the sample estimate of $E(T_i^2 - T_j^2)$ has smaller variance than the variance of the difference of the estimates of $V(T_i)$ and $V(T_j)$. This example illustrates the simplicity of Monte Carlo experimentation; but it also makes the point that statisticians must also do what they tell experimenters in the other sciences to do: plan experiments to improve their efficiency. The designs for Monte Carlo experimentation can take advantage of many of the standard methods, such as blocking, stratified sampling, etc.

The problem of generating samples on a computer for Monte Carlo studies has been extensively investigated. The term "pseudorandom numbers" is used for the artificially generated numbers comprising such samples. Since transformations from uniform random variables to random variables with other interesting distributions are particularly simple, the basic approach is to generate a sample that appears to have come from a uniform distribution and then transform that sample to a sample that appears to have come from a distribution of interest. There are several methods to generate samples that appear to have come from a uniform distribution. A simple one that is widely used and allows several variations is the multiplicative congruential generator. Beginning with a randomly chosen d_0, it is defined recursively by

$$d_i = ad_{i-1} (\text{mod } m).$$

Depending on the fixed values of a and m, the d's will appear to have a discrete uniform distribution over the integers $1, 2, \ldots, m - 1$. These numbers are then scaled into the interval $[0, 1]$ on division by m.

Since the number of computations in Monte Carlo studies may be extremely large, much emphasis has been placed on making random-number generators very fast. Various shortcuts (among other reasons) have resulted in some very poor random-number generators. The user of the Monte Carlo method must exercise care in selecting a generator. The undue concern over speed that leads to bad generators is seen to be misplaced when one considers that in most studies, the generation of the random numbers represents only a small fraction of the total computational time.

The Monte Carlo studies reported in the statistical literature need to be recognized by editors and referees for what they are, and high standards of scientific experimentation and reporting must be required of them. See Hoaglin and Andrews (1975) for further discussion of this point.

6. Impact of Microcomputers

The most significant phenomenon of the late 1970s and early 1980s has been the development and widespread distribution of small, inexpensive computers, variously called desktop computers, personal computers, or microcomputers. The low initial cost of these machines has allowed them to be purchased by individuals or purchased in quantities by institutions and industry. Microcomputers are ideally suited to the needs of statisticians working in developing countries. Not only because of their low initial cost, but also because of their less stringent environmental requirements, microcomputers have brought computing capabilities to many statisticians who formerly had limited access to computers. Instead of occupying large, specially designed rooms with sensitive temperature and humidity controls, microcomputers need only some table space in an office and essentially the environmental controls necessary for human comfort.

The low operating cost has also had a profound effect on the way statisticians carry on their work. A research statistician, for example, who is investigating alternative methods of inference may set up a Monte Carlo experiment and turn the machine on to run overnight or over the weekend. The size of the sample may even approach the period of many random-number generators in use currently. In addition, distribution-free procedures or other techniques requiring time-consuming permutations may be tried, since the computer time is essentially free. Although the microcomputer may run more slowly than the large mainframe by several orders of magnitude, the cost of use may be essentially zero.

The growth in the number of users of computers brought about by the microcomputers has been accompanied by changes in the user interface. The word most often used to describe the microcomputer systems, at least by their developers, is "user friendly." The language of communication with computers more nearly resembles that of the researcher's native tongue than that of the microcode of computers. Software developers attempt to provide human-language messages when error conditions arise. Instead of messages referring to some error code or to the contents of some register in the machine, modern software provides messages referring to the user's input, using variables or expressions from the user's own statements. While this trend toward "friendliness" has been going on for some time on the mainframe computers, the proliferation of personal computers has accelerated the changes in attitudes of both users and developers of software.

The improvement in attitudes of software writers toward the user interface has not always been accompanied by concern for correct and stable numerical computations. Although, as mentioned above, the early leaders in statistical computing were numerically oriented, many of the recent initiates to computing using home computers are not knowledgeable in numerical analysis. The result is that much of the software written for microcomputers and distributed informally is not of high quality. This is also the case for much of the system software, such as compilers, that is currently available for the smaller computers. Libraries of basic functions, such as logarithm and exponential, often contain numerically unstable programs. One can hope that this situation will be corrected over time as good software drives out the bad, but until it is, statistical software packages for microcomputers will be difficult to develop.

The full potential of microcomputers cannot be realized without the ability to transfer files from one to another and to mainframe computers, possibly at remote sites, through a network that allows one to use his own workstation for routine computing and editing of local files, but to have access to a more extensive set of software or to the larger storage devices on mainframe computers. Researchers using different workstations can also share files through the network.

7. Future Trends

The past of statistical computing has been heavily concentrated on numerical analytic solutions to particular problems of interest to statisticians. What are the challenges of the future?

Obviously the cost of computing, either the capital cost or the unit cost, is going to continue to decrease. The result will be an extremely rapid increase in the number and variety of machines over the next 10 years. Simultaneously, there will be a rapid increase in the number of available statistical packages. Unfortunately, in the immediate future, there will be a great deal of bad software. One can only hope that there is not a Gresham's law for statistical software; indeed, one can hope that as the number of machines, programs, and people "doing" statistics increases the bad software will eventually be driven from the market place by the good software. How can we assure the quality of statistical software that may be used by those who are not knowledgeable about statistics?

One of the reasons for the decline in cost of computing is the mass production of hardware. This will allow more manufacturers to produce different computer models just as they do with other mass-produced items. With this ability the possibility arises of machines which are specifically designed to perform statistical calculations. How can we both encourage and guide the development of specialized machines for statistical computation?

In the past all (or at least nearly all) statistical programs have been written in the FORTRAN programming language. At first this was because FORTRAN was the best possible choice for scientific and numerical programming. More recently it has been because of the widespread availability of the language. There are now a wide variety of languages, each with its own adherents and proponents, that might be better choices for writing statistical programs. How are we are going to persuade statisticians to consider the possibilities? And how are we going to influence the designers of future languages to incorporate features that are of particular use to statisticians?

The proliferation of small inexpensive computers encourages their use in the acquisition of data. The greater speed with which they will operate means that data sets will increase greatly in size. Fortunately the increase in speed means that larger data sets can be analyzed. But the ease of data acquisition will encourage the collection, analysis, and preservation of useless data. How can we encourage the collection of data that have some current (or future) value and discourage the same for worthless data?

There is a related problem that arises from the improvement in intercomputer communication: privacy. This matters to us not only as private individuals but as ethical professional statisticians. How can we forestall the invasion of individual privacy that occurs as databases are amalgamated and as the controlling organizations become more powerful? We perceive this as the most pressing social impact of the proliferation of computers and feel that statisticians can have much to say about its resolution.

Acknowledgement

The work of one of us (W.F.E.) was supported in part by the Office of Naval Research under contract N00014-84-K-0588.

Bibliography

Bailar, J. C. and Bailar, B. A. (1978). "Comparison of two procedures for imputing missing survey values." In *Proceedings of the Survey Research Methods Section of the American Statistical Association.*

Barnard, G. A. (1963). "Discussion of Professor Bartlett's paper." *J. Roy. Statist. Soc., Ser. B.*, **25**, 294.

Barnett, V. (1981). *Interpreting Multivariate Data.* New York: Wiley.

Besag, J., and P. J. Diggle (1977). "Simple Monte Carlo tests for spatial pattern." *Appl. Statist.,* **26**, 327–333.

Boardman, T. J. (1982). "The future of statistical computing on desktop computers." *Amer. Statist.*, **36**, 49–58.

Box, J. F. (1978). *R. A. Fisher. The Life of a Scientist*. New York: Wiley.

Chambers, J. M. (1981). "Some thoughts on expert software." In W. F. Eddy (ed.) *Computer Science and Statistics: Proceedings of the 13th Symposium on the Interface*. New York: Springer, 36–40.

Chambers, J. M., Cleveland, W. S., Kleiner, B., and Tukey, P. A. (1983). *Graphical Methods for Data Analysis*. Belmont, CA: Wadsworth.

Chernoff, H. (1973). "The use of faces to represent points in k-dimensional space graphically." *J. Amer. Statist. Assoc.*, **68**, 361–368.

Cox, D. R. (1978). "Some remarks on the role in statistics of graphical methods." *Appl. Statist.*, **27**, 4–9.

Efron, B. (1979). "Computers and the theory of statistics: Thinking the unthinkable." *SIAM Rev.*, **21**, 460–480.

Fienberg, S. E. (1979). "Graphical methods in statistics." *Amer. Statist.*, **33**, 165–178.

Fisher, R. A. (1925). *Statistical Methods for Research Workers*. Edinburgh: Oliver & Boyd. (14th edn., 1970.)

Fisher, R. A. (1935). *The Design of Experiments*. London: Oliver & Boyd.

Francis, I. (1981). *Statistical Software. A Comparative Review*. Amsterdam: North Holland.

Funkhauser, H. G. (1938). "Historical development of the graphical representation of statistical data." *Osiris*, **3**, 269–404.

Gong, G. (1983). "Letting MACSYMA help." In J. E. Gentle (ed.), *Computer Science and Statistics: Proceedings of the 15th Symposium on the Interface*. Amsterdam: North-Holland, 237–244.

Grenander, U. (1982). *Mathematical Experiments on the Computer*. New York: Academic.

Hammersley, J. M. and D. C. Handscomb (1964). *Monte Carlo Methods*. London: Methuen.

Hoaglin, D. C., and D. F. Andrews (1975). 'The reporting of computation-based results in statistics." *Amer. Statist.*, **29**, 122–126.

Kleiner, B. and J. A. Hartigan (1981). "Representing points in many dimensions by trees and castles " (with discussion). *J. Amer. Statist. Assoc.*, **76**, 260–276.

Pagano, M. and D. Tritchler (1983). "On obtaining permutation distributions in polynomial time." *J. Amer. Statist. Assoc.*, **78**, 435–440.

Panel on Incomplete Data, Committee on National Statistics (1983). *Incomplete Data in Sample Surveys. Vol. 1, Report and Case Studies* (W. G. Madow, H. Nisselson, and I. Olkin, eds.); *Vol. 2, Theory and Bibliographies* (W. G. Madow, I. Olkin, and D. Rubin, eds.); *Vol. 3, Proceedings of the Symposium* (W. G. Madow and I. Olkin, eds.). New York: Academic.

Schmid, C. F. (1983). *Statistical Graphics. Design Principles and Practices*. New York: Wiley-Interscience.

Schucany, W. R., Minton, P. D., and B. S. Shannon, Jr. (1972). "A survey of statistical packages." *Computing Surveys*, **4**, 65–79.

Tufte, E. R. (1983). *The Visual Display of Quantitative Information*. Cheshire, CT: Graphics Press.

von Neumann, J. (1945). *Memorandum on the Program of the High-Speed Computer*. Princeton: Institute for Advanced Study.

CHAPTER 11

Conditional Econometric Modeling: An Application to New House Prices in the United Kingdom

Neil R. Ericsson and David F. Hendry

Abstract

The statistical formulation of the econometric model is viewed as a sequence of marginalizing and conditioning operations which reduce the parametrization to managable dimensions. Such operations entail that the "error" is a derived rather than an autonomous process, suggesting designing the model to satisfy data-based and theory criteria. The relevant concepts are explained and applied to data modeling of UK new house prices in the framework of an economic theory-model of house builders. The econometric model is compared with univariate time-series models and tested against a range of alternatives.

1. Introduction

The feature which distinguishes econometric modeling from time-series analysis is the integral role of economic theory in orienting the former. At one extreme, a univariate time-series model is inherently mechanistic and has little or no need for subject-matter knowledge. Often, the procedure for choosing a model can be automated so as to satisfy appropriate criteria, such as minimizing a residual variance adjusted for degrees of freedom. Even a bivariate model needs little more than common sense in selecting a relevant covarying series. At the other extreme, prior to data analysis a formal intertemporal optimization model can be developed for the behavior of rational economic agents who fully account for all relevant costs and available information. The data evidence is then used to calibrate the unknown parameters of the theoretical model. Any required data transformations are derived from the theory (e.g., moving averages might represent "permanent" components, or residuals from auxiliary regressions might act as "transitory" or "surprise" effects).

When formulated as such, the "data-driven" and "theory-driven" ap-

Key words and phrases: conditional models, diagnostic testing, dynamics, evaluation criteria, exogeneity, expectations, house prices, information sets, marginalizing, time series.

proaches to modeling have been viewed as competitive rather than complementary [see, for example, Naylor et al. (1972) and Granger and Newbold (1977)]. Confrontations between the rival strategies in terms of forecasting accuracy have not generally been kind to supporters of "theory-driven" modeling (see Nelson, 1972), although that is neither surprising nor definitive in view of the choice of a mean-squared-error criterion; moreover, such results depend on the length of the forecast horizon (see the discussion in Kmenta and Ramsey, 1981).

Each extreme also has severe drawbacks. The first approach is open to such difficulties as spurious correlations (witness Coen, Gomme, and Kendall, 1969) and often yields forecasts outside the estimation sample which are poorer than the data-based within-sample fit. "Theory-driven" models tend to manifest symptoms of dynamic misspecification (e.g., residual autocorrelation afflicts many equations in large-scale econometric models) and often fit poorly due to being excessively restricted. Since, in principle, econometric specifications nest time-series formulations [see Zellner and Palm (1974) and Prothero and Wallis (1976)], the complementarity of information from data and theory bears stressing and argues for an integrated approach.

In practice, a complete spectrum of views exists concerning the "best" combinations of theory and data modeling, and most practitioners blend both elements to produce a mixture often labeled "iterative model building." The statistical properties of such mixed approaches have proved hard to analyse, especially since the initial theory may be revised in the light of any anomalous data evidence. However, some Monte Carlo evidence is available (see Kiviet, 1981, 1982) highlighting the difficulties of selecting models from noisy data. Moreover, pretest theory indicates severe inferential problems in simplifying models by using sample evidence (see Judge and Bock, 1978). Nevertheless, little empirical research in economics commences from fully prespecified models which adequately represent all salient features of the data. Consequently, by default, many important aspects of most models have to be selected from the observed sample, including the choices of alternative potential explanatory variables, lag reaction profiles, functional forms, error properties, seasonal variations, and even the evolution over time of parameters of interest.

In the present approach, the data analysis is strongly guided by prior economic theory. The theory suggests the form that the model *class* should have in order to satisfy a number of reasonable properties likely to obtain in any static-equilibrium state of the world. The functional form is specified to ensure invariance to a range of transformations, and the length of the longest lag in the maintained model is preassigned a value such that we would be surprised if even longer lags were needed to make the model adequately characterize the data. Such an approach produces a general maintained model, which is usually heavily overparametrized. Reduction of the general model by data-based sequential simplifications in the light of the prior theory yields a parsimonious summary which aims to be both data-coherent and theory-consistent, with interpretable parameters corresponding to nearly orthogonal variables [see Trivedi (1984) for a general discussion of this strategy for model selection].

At this stage, the model has been *designed* to satisfy a range of statistical and economic criteria. Since the criteria may conflict, some of the "art" apparent in modeling remains, perhaps necessitating appropriate compromises between tractability, coherency, and credibility. Moreover, no unique path for simplification exists, so the final model may vary with the investigator. However, by sharing a common initial model and subjecting selected simplifications to testing on later data and against rival models, the strategy has some inbuilt protection against choosing poor or nonconstant representations. It is important to stress that the prototypes of the model presented below were first developed in 1978 and have altered rather little since 1980 despite several later tests. Section 2 presents a more extensive discussion of the empirical econometric modeling methodology to establish terminology and exposit the main concepts.

The topic of our application naturally determines the formulation of the theoretical model. Casual observation suggests that a vast complex of social, economic, and demographic factors influence house prices. Here we are concerned with modeling the average price of newly completed private dwellings in the United Kingdom (from 1959–1982), denoted by *Pn*, and shown in Figure 1. A separate model of the price at which the existing stock of housing is transacted (denoted by *Ph*) is presented in Hendry (1984). Heuristically, we consider the joint density of (*Pn*, *Ph*) as being factored into a conditional density for *Pn*, given *Ph*, and a marginal density for *Ph*, with our models for *Pn* and *Ph* corresponding to the latter two densities. The economic justifications for the resulting parameters of the conditional density being of interest are presented in Section 3 in the context of a theoretical model of the decisions of the construction industry.[1] Section 3 also briefly considers alternative approaches to modeling *Pn* based on construction costs and on (marginal) models derived from the supply of and demand for new dwellings.

In fact, the general formulation of earlier models of new house prices has been of an interaction between supply and demand, with prices implicitly determined by the level which equates supply and demand for new housing. The validity of this market-clearing paradigm is not obvious in the United Kingdom, where the volume of new housing is small compared to the total volume of transactions in existing houses. Also, there is clear evidence of large changes over time in the stock of completed but unsold houses, indicative of a nonclearing market. That aspect therefore requires appraisal, so the available data are described in Section 3.

Section 4 reports the various empirical results obtained and discusses the

[1] Note that few agents buy a weighted average of new and second-hand housing, hence our desire to model both variables rather than an overall "house price index." In any case, the relative price of new to existing dwellings is crucial to the construction industry.

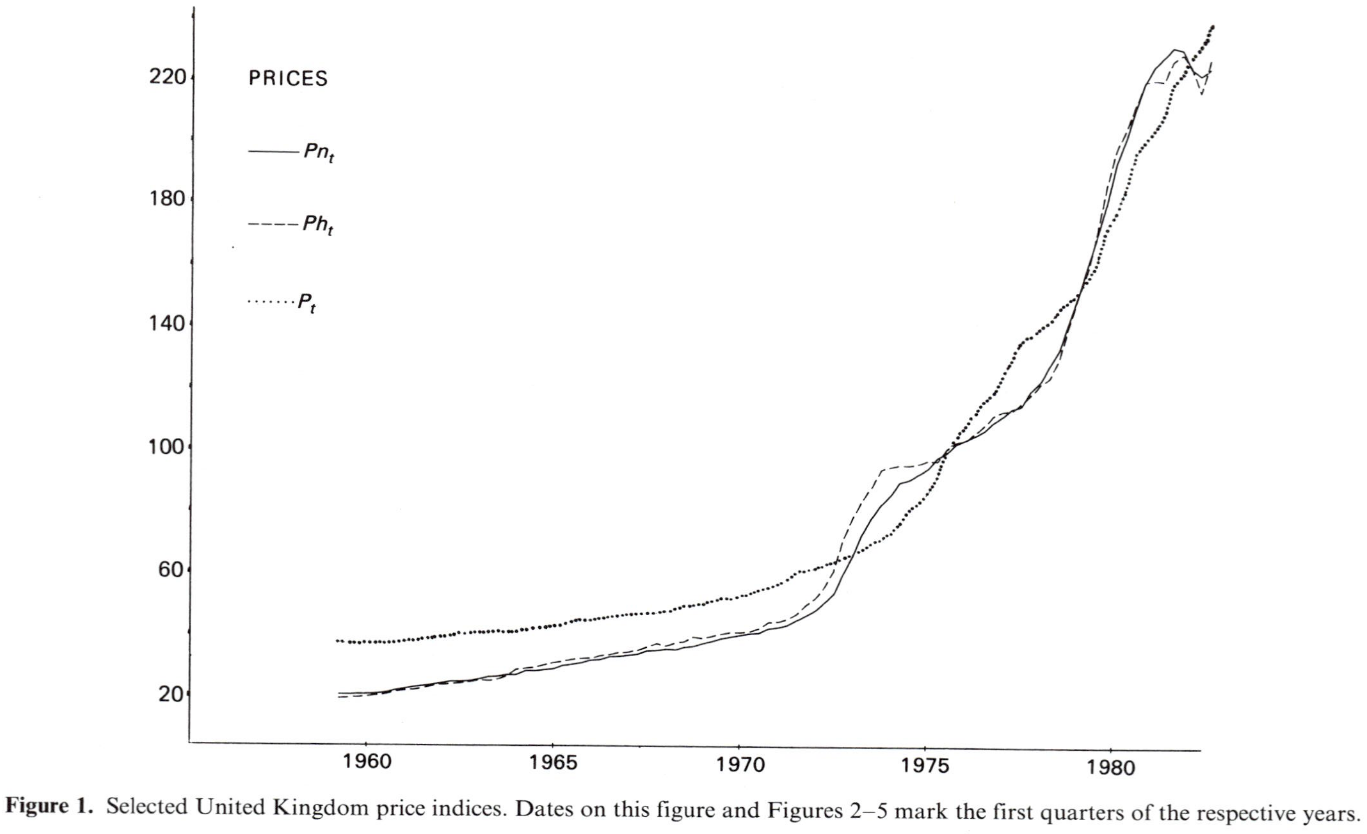

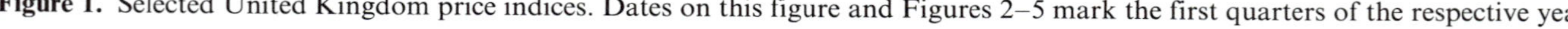

Figure 1. Selected United Kingdom price indices. Dates on this figure and Figures 2–5 mark the first quarters of the respective years.

light they throw on understanding the determination of housing prices. Section 5 briefly concludes the study.

2. Econometric Modeling

In this section, we discuss relevant aspects of our statistical approach for modeling new house prices: see Hendry and Richard (1982, 1983) for more detailed discussion and bibliographical information. Modeling is viewed here as an attempt to characterize data properties in simple parametric relationships which remain reasonably constant over time, account for the findings of preexisting models, and are interpretable in the light of the subject matter. The observed data $(\mathbf{w}_1 \cdots \mathbf{w}_T)$ are regarded as a realization from an unknown dynamic economic *mechanism* represented by the joint density function:

$$D(\mathbf{w}_1 \cdots \mathbf{w}_T \mid \mathbf{W}_0; \boldsymbol{\psi}), \tag{2.1}$$

where T is the number of observations on $\mathbf{w}_t$, $\mathbf{W}_0$ denotes the initial conditions, and $\boldsymbol{\psi}$ is the relevant parametrization. $D(\cdot)$ is a function of great complexity and high dimensionality, summarizing myriads of disparate transactions by economic agents and involving relatively heterogeneous commodities and prices, as well as different locations and time periods (aggregated here to quarters of a year). Limitations in data and knowledge preclude estimating the complete mechanism.

A *model* for a vector of observable variables $\{\mathbf{x}_t\}$ can be conceptualized as arising by first transforming $\mathbf{w}_t$ so that $\mathbf{x}_t$ is a subvector, then (implicitly) marginalizing the joint density $D(\cdot)$ with respect to all variables in $\mathbf{w}_t$ other than $\mathbf{x}_t$ (i.e., with respect to those variables not considered in the analysis). That produces the reduced density $F(\mathbf{x}_1 \cdots \mathbf{x}_T \mid \mathbf{X}_0; \boldsymbol{\theta})$, where $\boldsymbol{\theta}$ is the induced function of $\boldsymbol{\psi}$. Next, one sequentially conditions each $\mathbf{x}_t$ on past observables to yield:

$$F(\mathbf{X}_T^1 \mid \mathbf{X}_0; \boldsymbol{\theta}) = \prod_{t=1}^{T} F(\mathbf{x}_t \mid \mathbf{X}_{t-1}; \boldsymbol{\theta}), \qquad \boldsymbol{\theta} \in \boldsymbol{\Theta}, \tag{2.2}$$

where $\mathbf{X}_j^i = (\mathbf{x}_i \cdots \mathbf{x}_j)$ and $\mathbf{X}_j = (\mathbf{X}_0 \ \mathbf{X}_j^1)$ for $T \geq j \geq i \geq 1$. The usefulness of (2.2) depends on the actual irrelevance of the variables excluded from $D(\cdot)$, on the suitability of the parametrization $\boldsymbol{\theta}$ (which may include "transients" relevant to subperiods only) and on the adequacy of the assumed form of $F(\cdot \mid \boldsymbol{\theta})$. Aspects of those conditions are open to direct testing against the observed data. Although the choice of functional form is of considerable importance, it depends intimately on the nature of the problem. Thus, for this general analysis, we assume that the time series $\mathbf{x}_t$ has been appropriately transformed so that only linear models need to be considered (so $\mathbf{x}_t$ may involve logarithms, ratios, etc. of the original variables). Finally, $\mathbf{x}_t$ is partitioned into $(\mathbf{y}_t', \mathbf{z}_t')'$, where $\mathbf{y}_t$ is to be explained conditional on $\mathbf{z}_t$, corresponding to the claimed factorization:

$$F(\mathbf{x}_t \mid \cdot; \boldsymbol{\theta}) = F(\mathbf{y}_t \mid \mathbf{z}_t, \cdot; \boldsymbol{\phi}_1)\, F(\mathbf{z}_t \mid \cdot; \boldsymbol{\phi}_2), \tag{2.3}$$

where $\boldsymbol{\phi}' = \mathbf{g}(\boldsymbol{\theta})' = (\boldsymbol{\phi}_1' : \boldsymbol{\phi}_2') \in \boldsymbol{\Phi}_1 \times \boldsymbol{\Phi}_2$, and all the parameters of interest can be obtained from $\boldsymbol{\phi}_1$ alone. If such conditions are fulfilled, then $\mathbf{z}_t$ is said to be weakly exogenous for $\boldsymbol{\phi}_1$ and only the conditional model $F(\mathbf{y}_t \mid \mathbf{z}_t, \cdot; \boldsymbol{\phi}_1)$ needs to be analysed, greatly simplifying the modeling exercise if there are many variables in $\mathbf{z}_t$. This formulation is discussed more fully in Engle et al. (1983) and Florens and Mouchart (1980), and builds on the work of Koopmans (1950) and Barndorff-Nielsen (1978).

Adding the further assumption that the maximum lag length of dependence in (2.2) is fixed at $l < T$ periods, then the conditional linear model can be written as

$$\mathbf{y}_t = \mathbf{B}_0 \mathbf{z}_t + \sum_{i=1}^{l} \mathbf{B}_i \mathbf{x}_{t-i} + \boldsymbol{\varepsilon}_t \qquad t = 1, \ldots, T, \tag{2.4}$$

ignoring transients to simplify notation. That provides the general maintained model and is estimable by a variety of methods. Rather clearly, however, many drastic *a priori* assumptions have been made in order to formalize (2.4), and such assumptions need not be valid empirically. Consequently, we now consider how to evaluate such models.

Any postulated model can be evaluated by comparing its claimed properties with its actual behavior. As formulated, (2.4) entails restrictions relative to six different sources of information, which are summarized as:

(A) the history of the $\{\mathbf{x}_t\}$ process, denoted by $\mathbf{X}_{t-1}$ [namely, only $\mathbf{X}_{t-1}^{t-l}$ is relevant if (2.4) is valid];
(B) the current value of $\mathbf{x}_t$ (namely, it is valid to condition $\mathbf{y}_t$ on $\mathbf{z}_t$);
(C) the future of the $\{\mathbf{x}_t\}$ process (namely, the parameters remain constant on $\mathbf{X}_T^{t+1}$);
(D) the subject-matter theory [so that (2.4) is consistent with the available theory];
(E) the structure of the measurement system (e.g., definitional constraints must not be violated); and
(F) rival models (which should not contain additional information relevant to explaining $\{\mathbf{y}_t\}$).

We now consider empirical model selection criteria derived from each of those information sources.

(A): A crucial aspect of evaluating the empirical validity of (2.4) concerns the properties of $\{\boldsymbol{\varepsilon}_t\}$. If the assumptions underlying (2.4) are a good approximation, then $\boldsymbol{\varepsilon}_t \approx \mathbf{y}_t - E(\mathbf{y}_t \mid \mathbf{I}_t)$ where $\mathbf{I}_t \equiv (\mathbf{z}_t, \mathbf{X}_{t-1})$. Thus $\{\boldsymbol{\varepsilon}_t\}$ is a *derived* (rather than an autonomous) process, which by construction is uncorrelated with $\mathbf{I}_t$ and hence is an *innovation* relative to $\mathbf{I}_t$. One set of tests of (2.4) seeks to evaluate the extent to which the calculated residuals are consistent with $\{\boldsymbol{\varepsilon}_t\}$ being such an innovation process.

Several particular hypotheses can be investigated as follows. Firstly, defining white noise by the second-order property that, for $E(\boldsymbol{\varepsilon}_t) = \mathbf{0}$, $E(\boldsymbol{\varepsilon}_t \boldsymbol{\varepsilon}'_{t-k}) = \mathbf{0}$ for all $k \neq 0$ (i.e., $\boldsymbol{\varepsilon}_t$ is unpredictable from its own past alone), one could test for residual autocorrelation. For example, suppose an investigator postulated a model with a maximal lag length l^*. If l^* were less than l and/or the elements of $\{\mathbf{B}_i, i = 0, \ldots, l^*\}$ were inappropriately restricted, then the residuals might manifest serial correlation. Hence, *criterion (i)* of model adequacy is that the residual process (i.e., that which is left unexplained after modeling is ended) should be empirical *white noise* (see Granger, 1983). Note that an autocorrelated error can be "explained" in part (e.g., by Box–Jenkins methods). Also, $\{\boldsymbol{\varepsilon}_t\}$ need not be homoscedastic in (2.4), so that inferences may have to allow for potential heteroscedasticity. Fortunately, heteroscedastic-consistent covariance matrices can be constructed with ease (see White, 1980; Domowitz and White, 1982; Messer and White, 1984), and a variety of tests for residual heteroscedasticity is available.

Next, the assertion that $\boldsymbol{\varepsilon}_t$ is an innovation relative to $\{\mathbf{z}_t, \mathbf{X}_{t-1}^{t-l}\}$ entails that lags longer than l are redundant in (2.4) and that selecting $l^* < l$ is invalid. Thus, the residuals in (2.4) should have the smallest generalized variance (adjusted for degrees of freedom) in this class of constructed error processes. That property is called *parsimonious variance dominance* and provides *criterion (ii)*. If a model did not have white-noise residuals, it could be variance-dominated by a corresponding model which also "mopped up" the residual autocorrelation parsimoniously. Thus, (i) is a necessary condition for (ii), but is not sufficient, emphasizing that white-noise residuals are a minimal requirement for model adequacy, whether or not modified to account for parsimony (e.g., see Schwarz, 1978). Models derived by sequentially simplifying unrestricted representations such as (2.4) tend to have innovation errors. Conversely, the previously noted drawback of "theory-driven" modeling (that the associated errors are not innovations) is easily understood if the theory is not sufficiently general to posit the "correct" value l of l^* *a priori*.

(B): The validity of the assertion of *weak exogeneity* is *criterion (iii)*. Unfortunately, weak exogeneity *per se* is not easily tested in a class of models like (2.4); and to do so may require modeling $\{\mathbf{z}_t\}$, thereby defeating the main purpose of the conditioning assumption. However, if the data generation process of $\{\mathbf{z}_t\}$ does not stay constant over the sample, yet $\boldsymbol{\phi}_1$ is constant in (2.3), then this enhances the credibility of the weak exogeneity assertions underlying (2.3). When $\boldsymbol{\phi}_1$ is invariant to changes in $\boldsymbol{\phi}_2$ and (2.3) is valid, then $\mathbf{z}_t$ is said to be super exogenous for $\boldsymbol{\phi}_1$.

(C): *Parameter constancy* (after duly incorporating all relevant transients) provides *criterion (iv)*. The formulation in (2.4) explicitly defines certain parameters ($\mathbf{B}_0 \cdots \mathbf{B}_l$), changes in which would invalidate the model. It seems natural to seek models with constant parameters, whatever the purpose of the modeling exercise, and to test assertions of constancy as a check on the usefulness of the model.

Summarizing, (i) + (ii), (iii), and (iv) respectively relate to the validity of assumptions concerning *lagged*, *contemporaneous*, and *leading* data relative to any given observation at time t. In econometrics, (i)–(iv) are reasonably conventional criteria for selection and evaluation of models. A model which satisfies (i)–(iv) will be useful for forecasting $\mathbf{y}_t$ if $\mathbf{I}_t$ is available when forecasts are made; however, if $\mathbf{z}_t$ is contemporaneous with $\mathbf{y}_t$, then $\mathbf{I}_t$ will rarely be known at time t-1. Also, $\mathbf{I}_t$ need not be a "good" information set for explaining $\mathbf{y}_t$, nor need the $\{\mathbf{B}_i\}$ in (2.4) bear sensible economic interpretations or be constants across different states of the world. Consequently, while these data criteria are necessary, they are not sufficient to justify a given model for inference, forecasting, or policy analysis. Indeed, three further criteria are of equal importance.

(D): The first of these, *criterion (v)*, is *theory consistency*, which is also standard in econometrics and requires that an empirical model should reproduce the theory from which it is ostensibly derived under the hypothetical conditions relevant to that theory. That may sound weak, but some published equations violate (v), and finding a model *form* which is theory-consistent in several different but relevant hypothetical states of the world can be nontrivial.

(E): Next, *data admissibility*, *criterion (vi)*, entails that a model should be unable to predict data values which violate definitional constraints. For example, that prices are always nonnegative or that houses not started cannot be completed are data requirements and so should be satisfied automatically (i.e., with probability one) by the model.[2] Clearly, (vi) is closely related to the choice of functional form.

(F): Finally, and perhaps most importantly, *encompassing* [labeled *criterion (vii)*] requires that any model $F(\cdot)$ claimed to adequately represent the data generation process $D(\cdot)$ should be able to account for the results obtained by other models of that process. That follows because if one knew the mechanism generating all the data (as in a Monte Carlo study, say), which here would be $D(\mathbf{w}_1 \cdots \mathbf{w}_T \mid \mathbf{W}_0; \psi)$, then by formal reductions equivalent to those which produced $F(\cdot)$, one could deduce what parameter values should be found in other models of the mechanism (at least in large samples). Consequently, if the selected model is claimed to characterize the data process adequately, it too should satisfy that requirement and allow the results of rival models to be derived. Should the estimated parameters of rival models which are in fact obtained differ significantly from those derived using the selected model, then that would contradict the assertion that the selected model adequately described the data generation process. Thus, encompassing requires that any model $F(\cdot)$ of the mechanism generating $\{\mathbf{x}_t\}$ should mimic that property of $D(\cdot)$ and be able to account for the empirical results reported by rival models of $\{\mathbf{x}_t\}$.

Before concluding this section, we consider the implications of this concept

[2] Typical models of completions have them determined as a distributed lag on starts. Even if the weights on the lags sum to unity, any (e.g.) positive stochastic disturbance to the equation implies houses completed which were not started.

and its relation to testing nonnested hypotheses, which can be seen most easily for two rival linear models:

$$H_1: E(y_t \mid \mathbf{z}_{1t}) = \mathbf{z}'_{1t}\boldsymbol{\delta}_1 \tag{2.5}$$

and

$$H_2: E(y_t \mid \mathbf{z}_{2t}) = \mathbf{z}'_{2t}\boldsymbol{\delta}_2, \tag{2.6}$$

where each hypothesis separately asserts $v_{it} \equiv y_t - \mathbf{z}'_{it}\boldsymbol{\delta}_i \sim IN(0, \sigma_{ii})$. In (2.5) and (2.6), $\boldsymbol{\delta}_1$ and $\boldsymbol{\delta}_2$ are $k_1 \times 1$ and $k_2 \times 1$ vectors of unknown parameters, and $\mathbf{z}_{1t}$ and $\mathbf{z}_{2t}$ (generic symbols for sets of regressors) have (at least) some variables which are not in common. For simplicity, we assume they have none in common. Formally, the joint density of y_t, $\mathbf{z}_{1t}$, and $\mathbf{z}_{2t}$ can be factorized as $F(y_t \mid \mathbf{z}_{1t}, \mathbf{z}_{2t}, \cdot)\, F(\mathbf{z}_{1t} \mid \mathbf{z}_{2t}, \cdot)\, F(\mathbf{z}_{2t}, \cdot)$. Note that, given the joint density, both (2.5) and (2.6) must be *derived* representations; hence, while separate, they are also interrelated. Here, (2.5) entails the conditional irrelevance of $\mathbf{z}_{2t}$ in explaining y_t given $\mathbf{z}_{1t}$. Under joint normality, $\mathbf{z}_{1t}$ and $\mathbf{z}_{2t}$ can be linked using:

$$\mathbf{z}_{1t} = \boldsymbol{\Pi}\mathbf{z}_{2t} + \boldsymbol{\zeta}_{1t}, \tag{2.7}$$

where $\boldsymbol{\Pi}$ is defined by $E(\mathbf{z}_{1t} \mid \mathbf{z}_{2t}) = \boldsymbol{\Pi}\mathbf{z}_{2t}$ [so $E(\mathbf{z}_{2t}\boldsymbol{\zeta}'_{1t}) = \mathbf{0}$], and (again for expositional simplicity) we assume $E(\boldsymbol{\zeta}_{1t}\boldsymbol{\zeta}'_{1t}) = \boldsymbol{\Omega}$. From (2.5) and (2.7),

$$y_t = \boldsymbol{\delta}'_1\boldsymbol{\Pi}\mathbf{z}_{2t} + (v_{1t} + \boldsymbol{\delta}'_1\boldsymbol{\zeta}_{1t}) = \boldsymbol{\delta}'_2\mathbf{z}_{2t} + v_{2t}. \tag{2.8}$$

Consequently, (2.8) is what the model (2.5) predicts the model (2.6) should find, so that if (2.5) is to encompass (2.6), it must be the case that

$$H_a: \boldsymbol{\delta}_2 = \boldsymbol{\Pi}'\boldsymbol{\delta}_1 \tag{2.9a}$$

and

$$H_b: \sigma_{22} = \sigma_{11} + \boldsymbol{\delta}'_1\boldsymbol{\Omega}\boldsymbol{\delta}_1. \tag{2.9b}$$

The hypothesis in (2.9a) is called parameter encompassing, and that in (2.9b) variance encompassing, where a least-squares notion of encompassing is being employed. In passing, there seems little point in testing (2.5) against (2.6) or *vice versa* unless both models do satisfy their claimed formulations, which first requires evaluating both on criteria (i)–(iv) at least. If so, then (in large samples) the nonnegative definiteness of $\boldsymbol{\Omega}$ entails that H_b cannot hold unless $\sigma_{11} < \sigma_{22}$, i.e., H_1 variance-dominates H_2.[3] Thus, variance dominance is necessary, but not sufficient,for variance encompassing. That in turn entails that encompassing is asymmetric in the present context: if (2.5) encompasses (2.6), the converse is false. Also, as H_a is sufficient for H_b, it is readily established that least-squares encompassing is transitive. Neither of (2.5) or (2.6) may encompass the other, in which case a more general model is

[3] Formally, variance dominance refers to the underlying (and unknown) error variances. In practice, we often say a model variance-dominates another if the *estimated residual* variance of the former is smaller than that of the latter.

necessary. Thus, encompassing defines a partial ordering over models, an ordering related to that based on goodness-of-fit; however, encompassing is more demanding. It is also consistent with the concept of a progressive research strategy (e.g., see Lakatos, 1974), since an encompassing model is a kind of "sufficient representative" of previous empirical findings.

More generally, an encompassing strategy suggests trying to anticipate problems in rival models of which their proponents may be unaware. For example, (2.5) may correctly predict that $\{v_{2t}\}$ is not white noise, or that $\boldsymbol{\delta}_2$ is not constant over $t = 1, \ldots, T$ (e.g., if $\boldsymbol{\delta}_1$ is constant but $\boldsymbol{\Pi}$ varies as t does, then (2.5) predicts that $\boldsymbol{\delta}_2$ should vary with t). Corroborating such phenomena adds credibility to the claim that the successful model reasonably represents the data process, whereas disconfirmation clarifies that it does not.

For the specific class of linear models, the propositions in (2.9) are testable by a large range of tests. Of these, perhaps the best known belong to the class of one-degree-of-freedom tests proposed by Cox (1961, 1962) using a modified likelihood-ratio statistic and implemented by Pesaran (1974) for models like (2.5) versus (2.6). That class seems to test H_b, which is necessary but not sufficient for H_a when $k_2 > 1$. [See the discussion following the survey by MacKinnon (1983).] Mizon and Richard (1983) and Mizon (1984) present equations for generating a very large class of tests of either H_a or H_b, or other functions of parameters for which encompassing is deemed relevant. Under H_1, $E(y_t | \mathbf{z}_{1t}, \mathbf{z}_{2t}) = \mathbf{z}'_{1t}\boldsymbol{\delta}_1$, so that $E(v_{1t} | \mathbf{z}_{2t}) = 0$. Consequently, if $\boldsymbol{\delta}_1$, $\boldsymbol{\delta}_2$, and $\boldsymbol{\Pi}$ are separately estimated under their own assumptions and used to construct an estimate of $\boldsymbol{\alpha} = \boldsymbol{\delta}_2 - \boldsymbol{\Pi}'\boldsymbol{\delta}_1$ (so H_a becomes $\boldsymbol{\alpha} = \mathbf{0}$), then a minor transformation of the Wald test of $\boldsymbol{\alpha} = \mathbf{0}$ yields the conventional F-test on the marginal significance of adding (the nonredundant elements of) $\mathbf{z}_{2t}$ to (2.5) [see Atkinson (1970), Dastoor (1983)]. It is unsurprising that in the present linear context there should be no sharp dichotomy between nested and non-nested approaches to testing (2.5) against (2.6). However, the union of (2.5) and (2.6) must always encompass both, so *parsimonious encompassing* is essential to avoid vacuous formulations. For example, if adding $\mathbf{z}_{2t}$ to (2.5) produces an insignificant improvement in fit, then (2.5) parsimoniously encompasses the model embedding (2.5) and (2.6). This aspect of simplicity, therefore, remains important in establishing credible models. For an empirical attempt at encompassing a range of disparate models using an embedding strategy, see Davidson et al. (1978) and the follow-up in Davidson and Hendry (1981); conversely, Bean (1981) investigates encompassing using the Cox approach. Note, however, from (2.9b), that only one direction of testing is really worthwhile and that all models in Bean's study that are variance-dominated are rejected.

Given the best available theoretical formulation of any problem, it seems sensible to *design* models to satisfy (i)–(vii) as far as possible, recognizing the possibility of conflict between criteria for any limited class of models under consideration. In particular, data admissibility is remarkably difficult to achieve in practice without simply asserting that errors are drawn from truncated distributions with bounds which conveniently vary over time; and theory consistency and variance dominance also may clash, especially for parsimonious formulations. How a compromise is achieved must depend on the objectives of the analysis (e.g., forecasting, policy advice, testing economic theories, etc.) as well as on creative insights which effect a resolution.

Since (i)–(vii) are to be satisfied by appropriate choice of the model given the data, relevant "test statistics" are little more than selection criteria, since "large" values on such tests would have induced a redesigned model. Genuine tests of a data-based formulation then occur only if new data, new forms of tests, or new rival models accrue. Such an approach is similar in spirit to the data-based aspect of Box and Jenkins's (1976) methods for univariate time-series modeling, but emphasizes the need to estimate the most general model under consideration to establish the innovation variance. Moreover, existing empirical models and available subject-matter theory play a larger role, while being subjected to a critical examination for their data coherency on (i)–(vii).

Indeed, econometric analysis always has involved a close blend of economic theory and statistical method (e.g., see Schumpeter, 1933). That economic analysis should be used is unsurprising, but the role of statistics has proved more problematical in terms of a complete integration of the economic and statistical aspects of model formulation, even though Haavelmo (1944) stressed the necessity of carefully formulating the statistical basis of an economic theory-model. He also showed the dangers of simply "adding on" disturbance terms to otherwise deterministic equations and asserting convenient properties to justify (say) least-squares estimation.

Conversely, massive difficulties confront any purely data-based method, since the interdependence of economic variables entails a vast array of potential relationships for characterizing their behavior. That aspect is discussed more fully in Hendry et al. (1984), but the theory-model in Section 3 highlights the existence of many derived equations from a small set of "autonomous" relationships. Since economic systems are far from being constant, and the coefficients of derived equations may alter when any of the underlying parameters or data correlations change, it is important to identify models which have reasonably constant parameters and which remain interpretable when some change occurs. That puts a premium on good theory.

While our paper remains far from resolving these fundamental issues, it seeks to link the two aspects by using considerations from both Sections 2 and 3 in formulating the empirical equations of Section 4.

3. The Economic Theory-Model

As noted above, it is important to distinguish autonomous from derived relationships in a nonconstant world; yet in practice it is exceptionally difficult to do so. The main basis for any asserted status of an equation must be its correspondence (or otherwise) to a theoretical relationship. Thus, to guide the data analysis, we present a suggestive, if somewhat simplistic, theoretical model which highlights what dependencies between variables might be anticipated.

Most house builders are small in terms of the fractions of the markets they supply (housing) and from which they demand inputs (labor, capital, land, materials and fuel). In the longer run, competitive forces might be expected to operate in such conditions so that only normal profits are earned (the going rate of return on capital), with builders who fail to minimize costs eventually being eliminated. Consequently, despite its artificiality, insight can be gained by analysing the decision processes of a single builder who produces homogeneous units, faces given costs, uses best practices, and seeks to optimize his expected long-run return. Those assumptions would allow one to formulate an "optimal-control" model yielding linear, intertemporal decision rules which maximize the expected value of the postulated objective function conditional on costs and demand, by using the certainty-equivalence principle (e.g., see Theil, 1964, pp. 52ff.). An analytical solution can be obtained only by postulating known and constant stochastic processes for the uncontrolled variables. Such an assumption in effect removes the uncertainty from the problem, and will be interpreted here as narrowing the applicability of the resulting theory to an equilibrium world: that is, one which is stationary, essentially certain, and devoid of problems like evolving seasonality, adverse weather, changes in legislation or tastes, and so on. Nevertheless, the resulting equations help to constrain the equilibrium solutions of the empirical model as well as to indicate relevant variables and parametrizations of interest.

Builders, as location-specific suppliers of new dwellings, have some element of monopolistic power and can influence sales somewhat by (say) advertising. In the medium term, they can determine the volume or the price of their new construction (or possibly some combination thereof); usually, their supply schedule reflects a willingness to supply more houses with higher profitability of construction. Conversely, final purchasers demand more housing as its price falls relative to that of (e.g.) goods and services, and choose between new and second-hand units on the basis of their costs (other factors being constant). In a schematic formulation which deliberately abstracts from dynamics, completions of new houses in period t (denoted by C_t) are produced from a stock of uncompleted dwellings (U_{t-1}), with variations in the rate of completions depending on changes in new house prices (Pn_t) and in construction costs (CC_t). We use a log–linear representation

$$c_t^s = \beta_0 + \beta_1 u_{t-1} - \beta_2 cc_t + \beta_3 pn_t \qquad (\beta_i \geq 0, \quad i = 1,2,3), \tag{3.1}$$

where lowercase variables denote logarithms of the corresponding capitalized variables and C^s denotes the planned supply of completions. Letting S_t denote starts of new dwellings, then

$$U_t \equiv U_{t-1} + S_t - C_t. \tag{3.2}$$

Thus, U_t is the (end-of-period) integral of past starts less completions, and stock-flow ratios (e.g., U_t/C_t) are crude measures of the average lag between starting and ending construction. Since adjustment costs (hiring and firing workers, paying overtime or idletime, etc.) suggest that change is costly and costs are incurred by maintaining the inventory U_{t-1}, a builder minimizing costs should aim for a constant rate of production.[4] Thus, $\beta_1 \approx 1$ seems likely in (3.1), corresponding to $C = KU$ in equilibrium (K a positive constant).

Since (3.1) is conditional on the preexisting stock of work in progress, the role of cc_t and pn_t is to alter the mean lag around $\exp(-\beta_0)$, with the main impact of changes in long-run profitability being via the level of u_{t-1}. Thus, it seems reasonable to expect $\beta_2 \approx \beta_3$ with both being relatively small.

On the demand side, population, income, interest rates, and the relative price of new to second-hand housing are the main determinants of purchases of completions. Again we use a log–linear equation

$$c_t^d = \gamma_0 + \gamma_1(y - n)_t + \gamma_2 n_t - \gamma_3(pn - ph)_t - \gamma_4 R_t, \tag{3.3}$$

where C_t^d is the demand for completions, N_t is the total number of families in the relevant geographical region,[5] Y_t is total real personal disposable income, and R_t is the interest rate. It is unclear how significantly demographic factors should influence the relative price of new to existing housing (since much of their effect will be reflected in the conditioning variable Ph). So, to a first approximation, we assume $\gamma_1 \approx \gamma_2 \approx 1$; hence n_t can be dropped, leaving y_t to capture both scale changes (e.g., via population size) and changes in real personal disposable income per capita. Since this abstract analysis assumes homogeneous housing units,[6] a very large value of γ_3 might be anticipated, reflecting a willingness to switch freely between otherwise identical new and almost new dwellings, depending on their respective prices.

The overall demand for housing relates to the national stock, H_t; and as C_t is a small fraction of that stock, the average second-hand house price Ph_t is determined primarily by the demand for housing in relation to the preexisting stock, H_{t-1}. Given Ph_t, (3.3) then determines the demand for new houses which is confronted with a supply of C_t^s dwellings. In general, C_t^d and C_t^s will not be equal, and builders will either experience unsatisfied demand or end up holding unsold houses. Either way, they must adjust by changing output or price. However, if γ_3 is large, disequilibrium will persist until Pn is fully adjusted to Ph. Simultaneously, H_t must be altering, given the equation

$$H_t = (1 - \delta_t)H_{t-1} + C_t + O_t \tag{3.4}$$

where δ_t is the rate of destruction of houses and O_t is other net sources of housing supply (e.g., from the governmental or rental sectors). The whole

[4] Note that we have abstracted from weather effects, etc.

[5] Families already housed may nevertheless wish to switch to a new house.

[6] In practice, it would be desirable to allow for changes in their attributes and composition.

stock-flow system evolves until an appropriate combination of stock and price results, with flows in balance.

If we consider a static equilibrium defined by $C_t^d = C_t^s$ and all change ceasing, then, from (3.4),

$$c = \ln \delta + h, \tag{3.5}$$

using log–linear equations where $\delta_t = \delta$ and $O_t = 0$ for simplicity. Further, we assume that the function for the total demand for housing can be written as

$$h = \lambda_0 + \lambda_1 y - \lambda_2(ph - p) - \lambda_3 R - \boldsymbol{\lambda}_4' \mathbf{z}, \tag{3.6}$$

and the function for the volume of work in progress is

$$u = \kappa_0 + \kappa_1(pn - cc) + \boldsymbol{\kappa}_2' \mathbf{z}, \tag{3.7}$$

where $\mathbf{z}$ denotes other exogenous influences [such as technology in (3.7)] and P is the overall price level of goods and services. Together with (3.1) and (3.3), we obtain five equations to determine the equilibrium values of c, h, u, pn, and ph, given y, δ, p, R, cc, and $\mathbf{z}$. Consequently, in such an equilibrium pn is determined by cc and the factors affecting profitability. Since the system evolves so long as a disequilibrium persists, pn_t will reflect current and past values of construction costs. Nevertheless, "explaining" pn_t by conditioning on $\{cc_{t-i}, i = 0, \ldots, l\}$ alone would not necessarily produce a useful model.[7]

One way to see such an argument is to consider (3.1) and (3.3) being equated instantaneously by Pn_t adjusting so that $C_t^s = C_t^d$. Then

$$pn_t = (\gamma_3 + \beta_3)^{-1}[(\gamma_0 - \beta_0) + \gamma_3 ph_t + \beta_2 cc_t + \gamma_1 y_t - \beta_1 u_{t-1} - \gamma_4 R_t]. \tag{3.8}$$

For large values of γ_3, we have $pn_t \approx ph_t$ and the influence of cc_t becomes small.[8] A generalization of (3.8) is the basis for the empirical model presented below, which turns out to yield estimates consistent with the view that γ_3 is indeed large and $\gamma_1 \approx 1$. However, there is evidence that $c_t^s \neq c_t^d$ in general, using data on a series for the stock of unsold completions, US_t (which was collected only up until 1978). From the identity

$$US_t \equiv US_{t-1} + C_t - Sl_t, \tag{3.9}$$

where Sl denotes sales of completions, then $\Delta_1 us_t$ reflects changes in net demand relative to the outstanding stock.[9] As shown in Figure 2, US_t has fluctuated substantially. A model for US_t is developed at the end of Section 4.

As a consequence of these considerations, our model class was required to reproduce (3.8) only under equilibrium assumptions, but otherwise was determined empirically by commencing from an unrestricted autoregressive-

[7] In particular, being a derived relationship, its parameters need not be very constant.

[8] Appendix A sketches an alternative theoretical derivation for a profit-maximizing builder which yields a solution similar to that in (3.8).

[9] Defining the lag operator L as $Lx_t = x_{t-1}$, then we let $\Delta_1 x_t = (1 - L)x_t$. More generally, $\Delta_j^i x_t = (1 - L^j)^i x_t$. If i and/or j is undefined, it is taken to be unity.

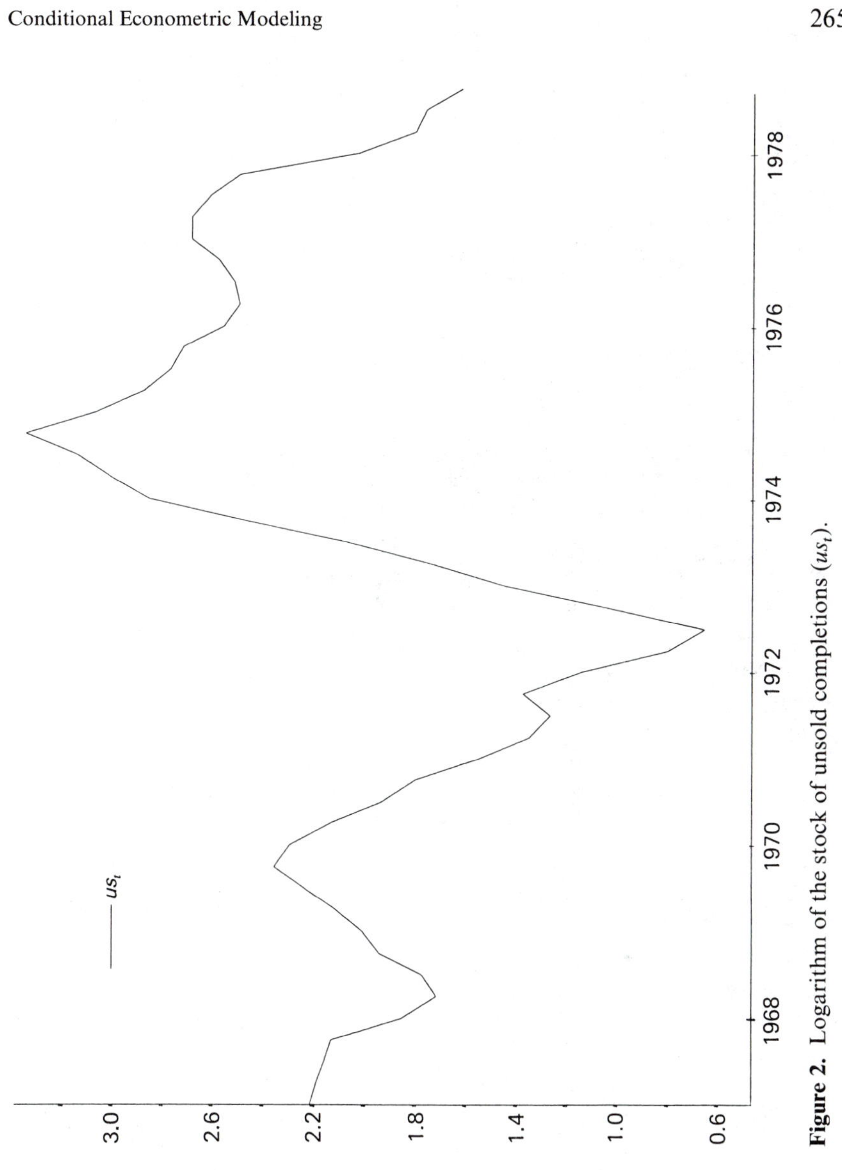

Figure 2. Logarithm of the stock of unsold completions (us_t).

Table 1. A Comparison of $\hat{\sigma}$ for Different Models of $\Delta_1 pn_t$

	Univariate autoregression Eq. (4.1)	Bivariate model Eq. (4.2)	Econometric model Eq. (4.4)
$\hat{\sigma}$	1.54%	1.30%	0.94%
k	3	6	12

distributed lag equation in which all variables entered with up to four lags. Additive seasonal intercepts were included, since none of the data series was seasonally adjusted, but these did not prove significant. The log–linear specification was retained because it ensured positive predictions of prices, yielded parameters which were elasticities and so could in principle be constant over time, and allowed freedom to switch between any of a number of sensible alternatives for the dependent variable [such as pn_t, $(pn - ph)_t$, $(pn - cc)_t$, $(pn - p)_t$, $(pn - p - y)_t$, and changes in all of these] without altering the specification of the model. Moreover, a constant *percentage* residual standard error also seemed a reasonable requirement.

The basic difficulties inherent in econometric modeling have now been introduced, and prior to empirical implementation it is worth considering: why bother? A straightforward answer is that anything less than a properly specified, autonomous relationship may "break down" whenever the statistical properties of any of the actually relevant variables alter. In practice, such events occur with monotonous regularity. Seen from this perspective, time-series models are simply very special cases of econometric equations in which all (for univariate models) or most (for "transfer function" models) covariates are ignored. Consequently, they too should regularly "break down" or mispredict; and, in practice, they do indeed do so. However, tests of predictive failure of time-series models may have low power because the models themselves fit poorly, in which case large mispredictions are needed to obtain "significant" outcomes. Econometric equations also often badly mispredict, revealing their inadequacy; but they remain susceptible to progressive improvement using the approach discussed in Section 2.

Nevertheless, much of the reduction in the error variance may derive from only a few additional factors, as the residual standard deviations $\hat{\sigma}$ in Table 1 illustrate.[10] There, k is the total number of regressors in each model, the sample size is 94 [1959(i)–1982(ii)], and the mean and standard deviation of $\{\Delta_1 pn_t\}$ (the regressand in every case) are 2.60% and 2.37% respectively. As can be seen, three variables effect a 35% reduction in $\hat{\sigma}$ over the unconditional standard deviation, whereas it takes twelve to effect a 60% decrease.

A stronger justification than goodness-of-fit is needed for the larger size of the econometric specification. The two natural arguments are the resulting

[10] For comparison with Tables 1 and 2 and equation (4.4), Ericsson (1978) obtains $\hat{\sigma} = 1.14\%$ [1958(i)–1974(iv)], and Hendry (1980, p. 31), $\hat{\sigma} = 0.93\%$ [1958(iv)–1976(iii)].

understanding of how the market functions (of obvious importance for predicting the complicated indirect effects flowing from changing government policies) and the feedback to improve economic analysis of markets in general (so that theories can commence from corroborated models, rather than from *a priori* assertions). The following results should help in judging the realism of such justifications.

4. Empirical Findings

As an illustration of the methodology described in Sections 2 and 3, we model the determinants of new house prices and present evidence for market disequilibria using the series on unsold completions. Before doing so, we first develop simple time-series models of Pn: such models establish a useful baseline against which to evaluate our econometric model of Pn. Further, those models will allow insight into the role of economic theory in econometric modeling.

The historical quarterly time series for the rates of change of pn and $pn - p$ (i.e., nominal and real new house prices respectively) are shown in Figures 3 and 4. The series are highly volatile, and two large "booms" in 1971–1974 and 1977–1979 are evident. Also, $\Delta_1 pn_t$ has tended to be positive (in almost every quarter), whereas $\Delta_1(pn - p)_t$ often has been negative. Precise data definitions are recorded in Appendix B. Next, the relative price $(pn - ph)_t$ is shown in Figure 5 and reveals substantial swings: generally, $pn - ph$ falls (rises) when ph is rising (falling) most rapidly, suggestive of pn adjusting to lagged (and possibly current) ph.

A fifth-order autoregression for pn suggested the following simplified model:

$$\widehat{\Delta_1 pn_t} = \underset{(.06)}{0.50}\, \Delta_2 pn_{t-1} - \underset{(.10)}{0.23}\, \Delta_1 pn_{t-3} + \underset{(.003)}{0.006}, \tag{4.1}$$

$$T = 94, \qquad R^2 = 0.59, \qquad \hat{\sigma} = 1.54\%,$$

$$\eta_1(6{,}85) = 0.7, \qquad \eta_2(6{,}85) = 0.1, \qquad \eta_3(2{,}89) = 1.0, \qquad \eta_5(4{,}87) = 0.6,$$

$$\xi_6(1) = 0.08, \qquad \xi_7(2) = 27.0.$$

In (4.1), coefficient standard errors are shown in parentheses, T denotes the sample size, R^2 is the squared multiple correlation coefficient, and $\hat{\sigma}$ is the residual standard deviation. The $\eta_i(\cdot)$ are test statistics labeled as far as possible to correspond to the order of the criteria in Section 2, and the figures in parentheses are their degrees of freedom. All statistics except $\eta_2(\cdot)$ are viewed as Lagrange-multiplier or efficient score statistics (see Rao, 1948). Under the relevant null, η_i is distributed in large samples as a central F; similarly, $\xi_i(j)$ is asymptotically $\chi^2(j, 0)$. The statistics are

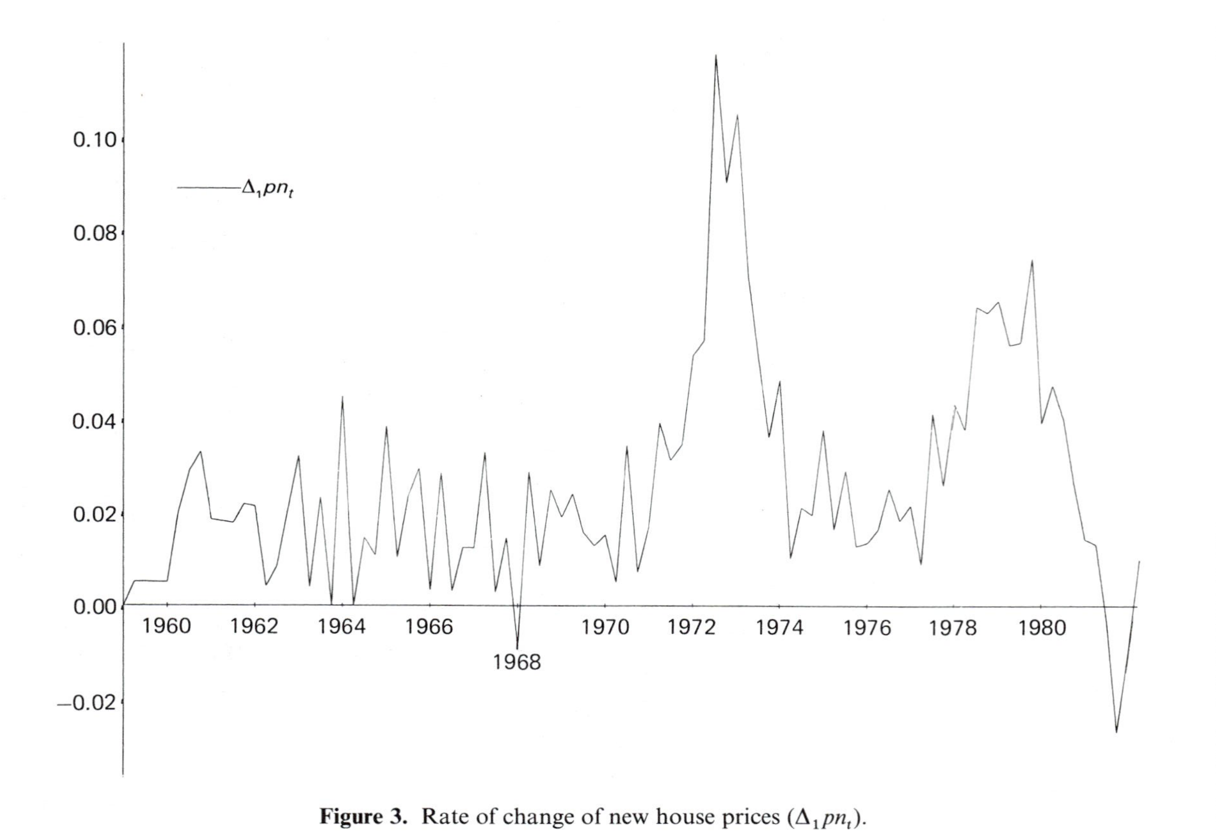

Figure 3. Rate of change of new house prices ($\Delta_1 pn_t$).

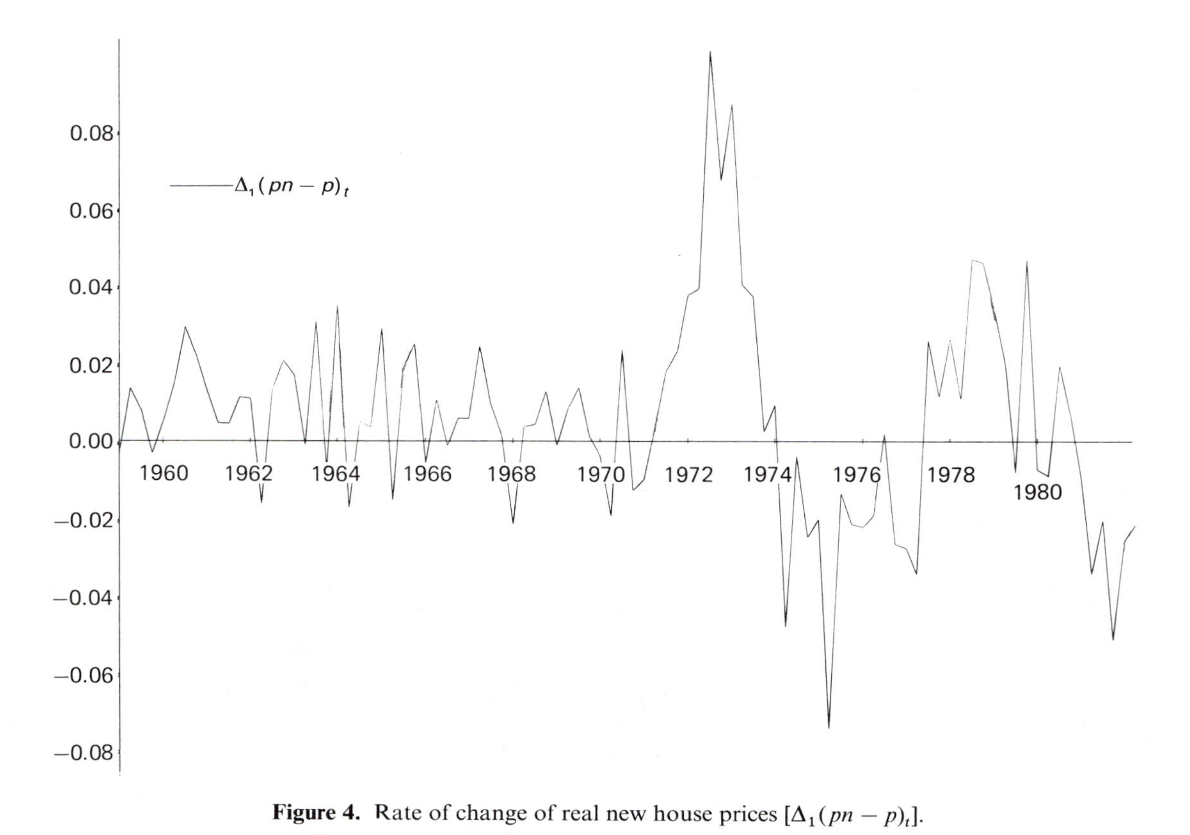

Figure 4. Rate of change of real new house prices $[\Delta_1(pn - p)_t]$.

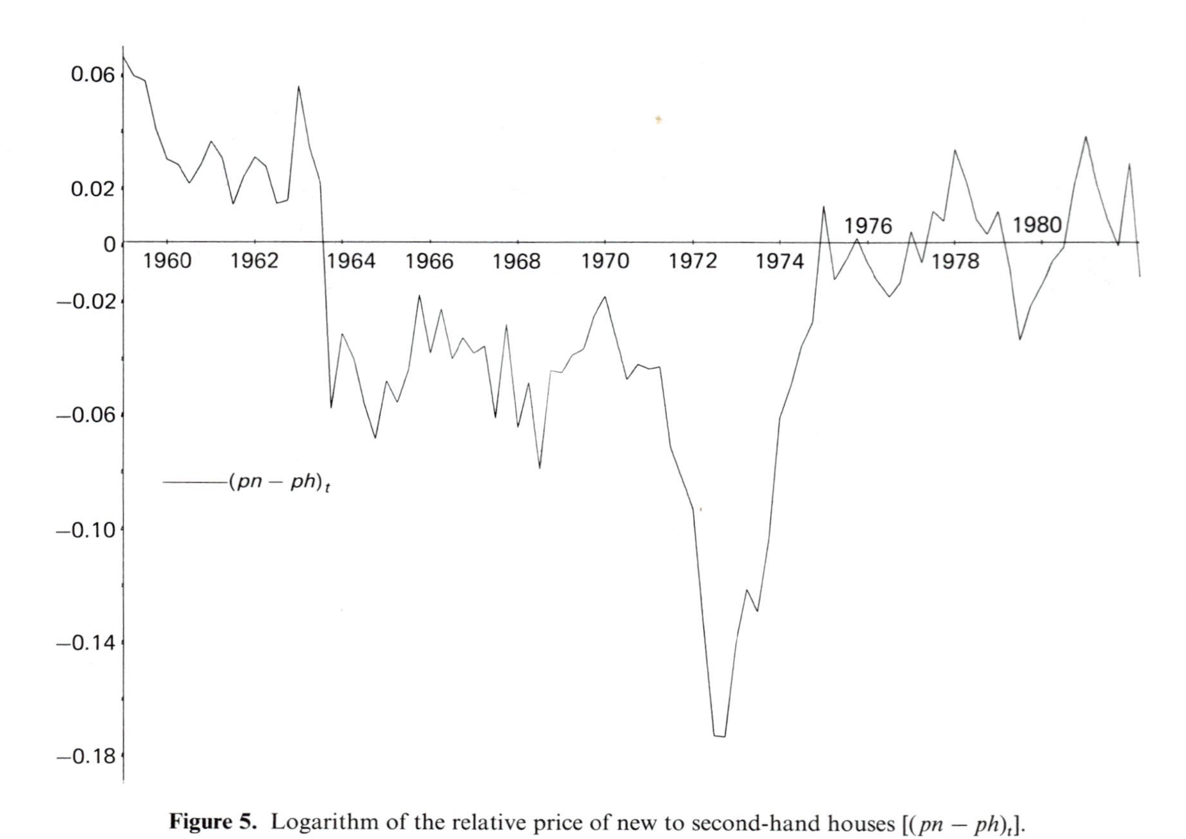

Figure 5. Logarithm of the relative price of new to second-hand houses $[(pn - ph)_t]$.

$\eta_1(\cdot) =$ Lagrange-multiplier statistic for testing against residual autocorrelation (see Godfrey, 1978; Harvey, 1981, p. 173),

$\eta_2(\cdot) =$ Wald statistic for testing against the relevant unrestricted maintained model [e.g., in (4.1), a fifth-order autoregression with seasonally shifting intercepts],

$\eta_3(\cdot) =$ Chow's (1960, pp. 594–595) statistic for testing parameter constancy,

$\eta_5(\cdot) =$ White's (1980) statistic for testing against residual heteroscedasticity,

$\xi_6(\cdot) =$ Engle's (1982) ARCH statistic (i.e., for testing against first-order autoregressive conditional heteroscedasticity),

$\xi_7(\cdot) =$ Jarque and Bera's (1980) statistic for testing against nonnormality in the residuals (based on skewness and excess kurtosis).

From $\eta_1(\cdot)$, (4.1) has white-noise residuals, and $\eta_2(\cdot)$ confirms it is an acceptable simplification of the autoregression. The parameters are not significantly nonconstant over the last two observations (despite measurement problems noted below) or indeed over several longer test samples.[11] There is no evidence of residual heteroscedasticity, but the residuals are highly nonnormal, reflecting the marked failure of (4.1) to predict the large changes in *pn* observed during the boom periods.

An indirect check on the usefulness of the theoretical model of Section 3 is that lagged values of *ph* should be informative about present *pn* since it is assumed that the market for the stock of houses nearly clears each period, whereas that for the flow adjusts more slowly while builders adapt to disequilibria. Thus, a bivariate model of *pn* on lagged values of *pn* and *ph* should perform better than (4.1). Simplifying from a bivariate model with up to fifth-order lags on *pn* and *ph* yielded the following equation:

$$\begin{aligned}\widehat{\Delta_1 pn_t} &= \underset{[.08]}{0.55}\,\Delta_1 ph_{t-1} + \underset{[.07]}{0.29}\,\Delta_1 pn_{t-2} - \underset{[.004]}{0.004}\\ &\quad + \underset{[.004]}{0.011}Q_{1t} + \underset{[.004]}{0.011}Q_{2t} + \underset{[.005]}{0.008}Q_{3t},\end{aligned} \tag{4.2}$$

$$T = 94, \qquad R^2 = 0.72, \qquad \hat{\sigma} = 1.30\%,$$

$$\eta_1(6,82) = 1.1, \qquad \eta_2(8,80) = 0.5, \qquad \eta_3(12,76) = 2.2, \qquad \eta_5(13,75) = 3.9,$$

$$\xi_6(1) = 0.01, \qquad \xi_7(2) = 5.4.$$

$[\cdot]$ denotes White's (1980) heteroscedasticity-consistent estimate of the standard error. Manifestly, $\Delta_1 ph_{t-1}$ is a highly significant predictor of $\Delta_1 pn_t$: thus, (4.2) is as usable for one-step-ahead predictions as (4.1), but has a significantly smaller residual variance. Although (4.1) and (4.2) are not

[11] It should be noted that $\eta_3(\cdot)$ will usually reflect changes in σ as well as in the regression coefficients.

nested, (4.1) is a special case of the unrestricted version of (4.2) which the latter parsimoniously encompasses. Directly testing the significance of lags of *ph* in the unrestricted version of (4.2) yields $\eta_2(5,80) = 8.3$, so that the white-noise residual of (4.1) is far from being an innovation on the joint information set generated by (pn, ph). Conversely, the errors on (4.2) are accepted as being an innovation process on that information set. However, $\eta_5(\cdot)$ reveals residual heteroscedasticity in (4.2), and $\eta_3(\cdot)$ indicates the possibility of nonconstant parameters.

That last problem is probably due primarily to measurement errors. During late 1981, commercial banks began to rapidly expand their loans for house purchases in competition with building societies.[12] Banks lent mainly against more expensive dwellings, and in larger than average loans. Thus, they attracted a distinctly biased sample of house purchasers. However, the data series are based on returns for the average prices of houses sold with a mortgage from a building society. Consequently, the series were distorted for a period until building societies attracted back a representative selection of house purchasers. For second-hand house prices, the main biases appear to have been in 1981(iv) and 1982(i) and (ii) (the termination of our data period). Here, the model is conditional on lagged *ph*; and as the relative distortions between the price series for new and existing house purchases is not known, it is difficult to assert any precise pattern for the residuals consequent on the measurement problem. For example, if the bias in observing $\Delta_1 ph_t$ as $\widetilde{\Delta_1 ph_t}$ is d_t and that for $\Delta_1 pn_t$ as $\widetilde{\Delta_1 pn_t}$ is e_t, then

$$\widetilde{\Delta_1 pn_t} = 0.55\,\widetilde{\Delta_1 ph_{t-1}} + 0.3\,\widetilde{\Delta_1 pn_{t-2}} + \{e_t - 0.3e_{t-2} - 0.55d_{t-1}\} + \varepsilon_t, \quad (4.3)$$

where the innovation error is denoted by ε_t and we assume (4.2) holds for the correctly measured series. (Seasonal factors are ignored for simplicity.) The restriction that $e_t = d_t$ with the latter measured as -2.1%, -2.1%, and $+4.2\%$ in the three relevant quarters (from Hendry, 1984) could be rejected. This finding is consistent with the observed predictions from (4.1), which suggest a smaller but more prolonged distortion, but is also interpretable as evidence against the hypothesis that the large value of $\eta_3(\cdot)$ here (or the corresponding test statistic for the *ph* model) is mainly due to mismeasurement. Until later data become available to clarify the issue, some doubt must remain concerning how distorted the series are over the last four observations. Below, however, we will continue to act as if the hypothesis were valid and, from the patterns of the residuals in (4.1) and (4.2), construct a dummy variable D with the values $(1\ \ 1\ \ 0\ \ -1\ \ -1)$ from 1981(iii) to 1982(iii).[13]

[12] These last are friendly (non-profit-making) societies whose primary function is to act as financial intermediaries between savers and potential house owners seeking mortgages. For econometric analyses of their behavior, see Hendry and Anderson (1977) and Anderson and Hendry (1984).

[13] The last observation lies outside our data period and hence entails a testable prediction of the model.

Table 2. An Unrestricted Model for $\Delta_1 pn_t$[a]

Variable	$j = 0$	1	2	3	4
$\Delta_1 pn_{t-j}$	-1	.05 (.16)	.25 (.16)	.03 (.15)	.13 (.11)
$\Delta_1 ph_{t-j}$	.35 (.09)	.05 (.13)	$-.06$ (.13)	.05 (.12)	$-.07$ (.12)
$\Delta_1 u_{t-j}$	—	.00 (.07)	$-.19$ (.08)	.01 (.08)	$-.11$ (.08)
$\Delta_1 y_{t-j}$	.05 (.08)	.21 (.14)	.21 (.12)	.14 (.12)	.05 (.10)
$\Delta_1 (cc - p)_{t-j}$	.13 (.11)	.11 (.12)	.13 (.11)	.05 (.11)	$-.03$ (.10)
$\Delta_1 p_{t-j}$	.06 (.18)	.20 (.20)	.11 (.19)	.11 (.18)	.04 (.17)
Q_{jt}	.46(1.21)	.003(.008)	.001(.010)	.001(.008)	—
ph_{t-j}	—	$-.09$ (.07)	—	—	—
$(u - y)_{t-j}$	—	$-.01$ (.02)	—	—	—
y_{t-j}	—	$-.06$ (.13)	—	—	—
$(cc - p)_{t-j}$	—	—	—	—	.18 (.10)
p_{t-j}	—	—	—	—	.10 (.05)
$(pn - ph)_{t-j}$	—	$-.48$ (.15)	—	—	—

[a] $T = 94$, $R^2 = 0.87$, $\hat{\sigma} = 1.094\%$, $\eta_3(11,45) = 1.8$, $\xi_1(12) = 9.7$, $\xi_6(1) = 0.00$. $\{Q_{jt}, j = 0, \ldots, 3\}$ denote a constant and three seasonal shift dummy variables. $\xi_1(12)$ is Box and Pierce's (1970) statistic based on the residual correlogram.

The unrestricted fourth-order autoregressive-distributed lag representation with (3.8) as its static-equilibrium solution is shown in Table 2, in a reparametrization intended to aid interpretability of this highly overparametrized equation.[14] Unsurprisingly, few of the regressors have "*t*-statistics" in excess of 2, although that should not be interpreted as entailing their irrelevance, given the generally high correlations between successive lags of economic variables. Notwithstanding their standard errors, many of the coefficients are negligibly small. However, three of the variables in levels omitted from both (4.1) and (4.2) have large coefficients, highlighting the role of the "equilibrating" mechanisms postulated in Section 3.

To an economist, both (4.1) and (4.2) have fatal flaws as claimed autonomous relationships, even though neither is an unacceptable data description on (i)–(iv) of Section 2. Concerning (4.1), the rate of inflation of new house prices is modeled as independent of inflation of goods and services; and if $\Delta_1 pn_t$ became constant at μ, the model would restrict μ to the single value 2.6% (the sample mean). Both implications are implausible, although only prolonged changes from the sample behavior would reveal the empirical inadequacy of the model. Similar difficulties afflict (4.2) even though $\Delta_1 ph_{t-1}$ is used: for example, under a constant growth rate μ^* in second-hand house prices (and averaging over the seasonals), $\mu \approx 0.8\mu^* + 0.005$. Thus, the prices Pn and Ph would diverge indefinitely unless $\mu^* \approx 2.6\%$.

The significance of $(pn - ph)_{t-1}$ in Table 2 can now be seen in perspective: any divergence of the house prices alters the conditional growth rate of pn

[14] Computer-program limitations precluded additional lags or the inclusion of further variables, so R_t and its lags were omitted.

relative to ph so as to bring the relative price (Pn/Ph) into line [see Granger and Weiss (1983) for a discussion on the relationship between time-series models, error-correction models like the one in (4.4), and the existence of long-run relationships between variables]. Consistent with (3.8), the long-run relative price varies with real construction costs. That evidence is sufficiently favorable to the theory in Section 3 to merit parsimonious modeling by forming interpretable functions of the regressors in Table 2 and marginalizing with respect to all the other potential variables. En route, a distributed lag in R_{t-j} ($j = 1, \ldots, 4$) also was allowed for, and that is reflected in the finally selected specification.

Since it illustrates the relative roles of all the potential explanatory factors, the following equation was selected as being the most interesting for present purposes:

$$\begin{aligned}\widehat{\Delta_1 pn_t} = {} & \underset{[.06]}{0.67}\, A_4(\Delta_1 ph_t) + \underset{[.05]}{0.19}\, \Delta_1^2 pn_{t-2} + \underset{[.04]}{0.11}\, \Delta_3(cc - p)_t \\ & - \underset{[.03]}{0.17}\, \Delta_1 u_{t-2} + \underset{[.04]}{0.09}\, \Delta_2 y_{t-1} - \underset{[.5]}{1.9}\, D^\circ \\ & - \underset{[.06]}{0.26}(pn - ph)_{t-1} - \underset{[.015]}{0.027}(ph - p)_{t-1} + \underset{[.03]}{0.13}(cc - p)_{t-4} \\ & - \underset{[.012]}{0.023}(u - y)_{t-1} - \underset{[.13]}{0.23}\{R(1 - \tau)\}_{t-1} - \underset{[.050]}{0.077},\end{aligned} \tag{4.4}$$

$$T = 94, \qquad R^2 = 0.86, \qquad \hat{\sigma} = 0.94\%,$$

$$\eta_1(6{,}76) = 0.9, \qquad \eta_3(2{,}80) = 0.9, \qquad \xi_6(1) = 0.2, \qquad \xi_7(2) = 1.5,$$

where $A_4(x_t) = 0.1\sum_{i=0}^{3} (4 - i)x_{t-i}$, a normalized linearly declining distributed lag; τ is the standard tax rate, so $R(1 - \tau)$ is the after-tax interest rate; and $D^\circ = D/100$, so that its coefficient represents 1.9%. The first six regressors represent disequilibrium or growth factors, all of which vanish in a static state; and the last six represent levels which persist in the equilibrium solution. Within those sets, variables are organized by influences from house prices, costs, net demand or supply, and other factors.

To analyse (4.4), we first derive its static solution by setting all growth rates equal to zero:

$$\begin{aligned} pn - ph = {} & -\underset{(.05)}{0.11}(ph - p) + \underset{(.09)}{0.52}(cc - p) \\ & -\underset{(.05)}{0.09}(u - y) - \underset{(.70)}{0.89}R(1 - \tau) - \underset{(.18)}{0.30}\,. \end{aligned} \tag{4.5}$$

The quoted standard errors are asymptotic approximations for nonlinear functions of estimated parameters, based on the covariance matrix of the estimates in (4.4). The coefficients in (4.5) can be interpreted in the light of

(3.8) (assuming $\gamma_1 = \beta_1 = 1$). That entails $\beta_2 = 5.8$, $\beta_3 = 1.2$, $\gamma_3 = 9.9$. The last two of those three coefficients are of the anticipated size and sign, but the first is so much larger than expected as to be somewhat implausible (suggesting that supply is very elastic in response to changes in real costs, rather than in response to profitability). However, in (4.4) the coefficients of $(ph - p)_{t-1}$ and $(cc - p)_{t-4}$ would have to be equal in magnitude and opposite in sign for β_2 to equal β_3 when (4.5) is interpreted as (3.8). For comparison, the static solution implied by the unrestricted model in Table 2 is

$$pn - ph = -0.19(ph - p) + 0.38(cc - p) - 0.02(u - y) - 0.13y + 0.02p. \tag{4.6}$$

Given the uncertainty inherent in the unrestricted model, the two derived equilibria are acceptably similar.

Of course, direct estimation of (3.8) (i.e., omitting all dynamics) is not overly enlightening; but, for completeness, we record such results.[15]

$$\begin{aligned}\widehat{(pn - ph)}_t = &- 0.35ph_t - 0.02u_t + 0.23y_t + 0.42(cc - p)_t + 0.32p_t \\ &+ 0.82\{R(1 - \tau)\}_t - 2.09 \qquad (4.7)\\ T = 94 \quad &\hat{\sigma} = 1.75\% \quad \eta_1(6,78) = 9.6.\end{aligned}$$

While (4.7) does not even fit as well as (4.1) (which phenomenon does not entail that equilibrium economic theory is vacuous), a coherent pattern of estimates emerges across the empirical counterparts of (3.8), with the last coming close to implying that $\beta_2 \approx \beta_3$ (but $\gamma_1 \neq 1$).

The test statistics of (4.4) suggest white-noise errors [with (4.4) parsimoniously dominating the model in Table 2 in terms of the value of $\hat{\sigma}$], constant parameters (rejected if D° is omitted), and approximately normal errors. Thus, the equation offers a reasonable data description and is consistent with the economic analysis of Section 3. Figure 6 shows the track of $\widehat{\Delta_1 pn_t}$ from (4.4) against $\Delta_1 pn_t$.

Next, we consider the dynamics of (4.4). The reaction of pn_t to a change in ph_t is quite rapid initially (e.g., Pn_t would change by over 0.6% by the end of six months in response to a 1% change in Ph_t), followed by an oscillatory convergence. For construction costs, however, a much slower adjustment pattern is observed, the main feedback in (4.4) being lagged by one year. Such lag responses are also consistent with the theory model of Section 3 and show the need to model *Pn* given *Ph* and solve for the long-run role of construction costs rather than to model *Pn* directly, given *CC* (omitting *Ph*). Note that, because Table 2 includes the "construction cost" hypothesis as a special case, the present approach automatically encompasses that rival model

[15] To be comparable with (4.5), p is included and the interest rate is after tax. The substantial residual autocorrelation in (4.7) precludes calculation of sensible estimates of standard errors. Equation (4.7) also includes Q_i, $i = 1,2,3$.

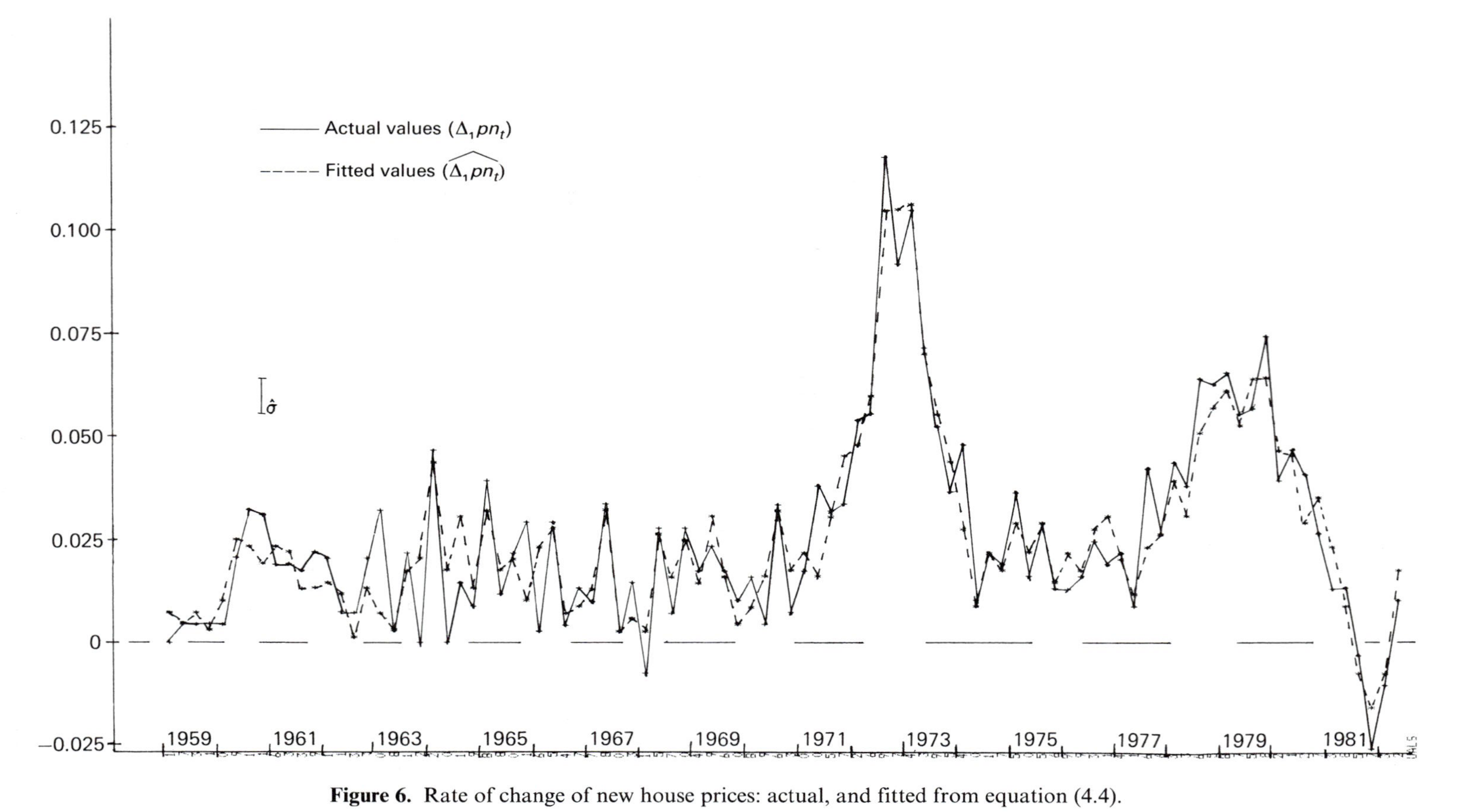

Figure 6. Rate of change of new house prices: actual, and fitted from equation (4.4).

(although Appendix A should clarify that several different interpretations are possible). The dynamic impacts of changes in u and y have appropriate signs and seem more important quantitatively than their equilibrium impacts. The interest-rate coefficient is small and not well determined.

The above interpretations of the individual coefficients of (4.4) rest upon an implicit assumption of relatively orthogonal regressors. Table 3 reports the matrix of correlations for the whole sample. The figures in brackets are the partial correlations from estimating (4.4), and it is noteworthy that five of these have the opposite sign to the corresponding simple correlation, highlighting the difficulty of interpreting simple correlations directly when in a multivariate context. Of the regressor intercorrelations, 36 are smaller than 0.5 in absolute value and only two are larger than 0.75. Since both involve the term $\{R(1-\tau)\}_{t-1}$, and since that variable plays a small role in the model, it would seem sensible to further simplify the model by omitting interest rates altogether. Doing so, $\hat{\sigma}$ increases to 0.95% and $(u-y)_{t-1}$ ceases to be significant, whereas $(ph-p)_{t-1}$ becomes better determined, with most of the remaining coefficients being unaltered.

Since the data behave very differently before and after 1971, and since the sample is large enough to produce sensible estimates from each half, we tested the model by fitting it separately to the two subsamples. That is a demanding test in the present context, since "success" (i.e., nonrejection of parameter constancy) would imply an ability to track turbulent data from estimates based on a quiescent period. Conversely, rejection would imply a need to revise the specification, though without clarifying precisely how. In the event, the subsample estimates of σ (in models without interest rates) were 0.80% and 0.89% respectively, against the whole-period figure of 0.95%. An F-test of constancy across subsamples yielded $\eta(10,73)=3.1$, thus rejecting the null. Most of the estimates were in fact fairly similar between the subperiods and the whole sample, but those for $(u-y)_{t-1}$ changed sign, as did that for $(cc-p)_{t-4}$ in the first subperiod. Otherwise, the second-period estimates were similar to those for the whole period, with rather more rapid adjustment, consistent with the need for builders to respond more rapidly to large disequilibria and/or substantial changes.

The last stage of the analysis is to examine the evidence for market disequilibria using the short series on unsold completions (US_t) for 1967–1978. As before, we first record the unrestricted log–linear representation in which us_t is explained by up to two lags of y, r, $pn-p$, and $ph-pn$ (see Table 4). There, $\sum$ denotes the sum over j of coefficient estimates for a given variable, from which a derived long-run solution analogous to equation (4.6) above can be derived. The mean and standard deviation of us_t and $\Delta_1\ us_t$ are (2.1, 63%) and (0, 21%) respectively; so, even if $\Delta_1 us_t$ were the dependent variable in Table 4, R^2 would still exceed 80% despite the large value of $\hat{\sigma}$. All of the effects of r, y, $pn-p$, and $ph-pn$ are sensibly signed, the last of these yielding a static relative price elasticity in excess of 4.5 (though some care is required in interpreting these magnitudes, as us_t is conditioned on pn_t).

Table 3. Data Correlation Matrix for the Variables in (4.4)[a]

		$\Delta_1 pn_t$	1	2	3	4	5	6	7	8	9
1.	$A_4(\Delta_1 ph_t)$	0.85 (0.77)									
2.	$\Delta_1^2 pn_{t-2}$	0.31 (0.37)	0.19								
3.	$\Delta_3(cc-p)_t$	0.51 (0.29)	0.51	0.05							
4.	$\Delta_1 u_{t-2}$	0.06 (−0.47)	0.21	0.09	0.28						
5.	$\Delta_2 y_{t-1}$	0.31 (0.25)	0.32	0.09	0.37	0.62					
6.	$(pn-ph)_{t-1}$	−0.55 (−0.43)	−0.52	−0.13	−0.51	−0.14	−0.15				
7.	$(ph-p)_{t-1}$	0.41 (−0.17)	0.36	−0.06	0.37	−0.19	−0.06	−0.51			
8.	$(cc-p)_{t-4}$	−0.03 (0.43)	−0.09	−0.12	−0.18	−0.21	−0.16	0.36	0.54		
9.	$(u-y)_{t-1}$	0.05 (−0.20)	0.03	0.03	0.19	0.05	−0.08	−0.61	0.15	−0.38	
10.	$\{R(1-\tau)\}_{t-1}$	0.11 (−0.15)	0.06	−0.14	0.03	−0.28	−0.19	0.03	0.78	0.84	−0.30

[a] Figures in brackets are the partial correlations from estimating (4.4).

Table 4. An Unrestricted Model for us_t[a]

Variable	$j = 0$	1	2	$\sum$
us_{t-j}	−1 (—)	0.74 (0.21)	−0.15 (0.16)	−0.41
r_{t-j}	0.49 (0.15)	−0.11 (0.22)	0.45 (0.18)	0.83
y_{t-j}	−0.68 (1.13)	−1.97 (0.93)	−0.33 (1.23)	−2.98
$(pn - p)_{t-j}$	1.11 (1.65)	1.43 (2.50)	−0.84 (1.41)	1.70
$(ph - pn)_{t-j}$	−1.84 (1.37)	0.11 (1.70)	−0.16 (1.92)	−1.89
$Q_{j+1,t}$	−0.13 (0.08)	−0.14 (0.06)	−0.08 (0.08)	—
Constant	32.5 (10.8)	—	—	—

[a] $T = 45$, $R^2 = 0.988$, $\hat{\sigma} = 9.0\%$, $\eta_1(6,21) = 1.96$, $\eta_3(6,21) = 2.25$, $\xi_6(1) = 2.4$.

A simplified representation of the model in Table 4 is given below:

$$\begin{aligned}\widehat{\Delta_1 us_t} = {} & \underset{[.08]}{0.51}\, \Delta_2 r_t + \underset{[.77]}{1.34}\, \Delta_1 (pn - p)_{t-1} + \underset{[.09]}{0.94} r_{t-2} - \underset{[.30]}{3.08} y_{t-1} \\ & + \underset{[.69]}{1.43}(pn - ph)_t + \underset{[.22]}{1.65}(pn - p)_t - \underset{[.07]}{0.37} us_{t-1} + \underset{[3.1]}{33.4} \\ & - \underset{[.04]}{0.09} Q_{1t} - \underset{[.04]}{0.15} Q_{2t} - \underset{[.05]}{0.08} Q_{3t}, \\ & T = 45, \quad R^2 = 0.87, \quad \hat{\sigma} = 8.6\%, \\ & \eta_1(6,28) = 1.6, \quad \eta_2(7,27) = 0.5, \quad \eta_3(6,28) = 1.6, \\ & \xi_6(1) = 1.9, \quad \xi_7(2) = 0.3. \end{aligned} \tag{4.8}$$

All of the test criteria are acceptable, but, although both $\eta_1(\cdot)$ and $\eta_3(\cdot)$ are smaller than in Table 4, neither is greatly favorable to the model. Unsold completions seem to be extremely sensitive to all of the demand factors and to adjust fairly rapidly, but still to reveal that disequilibria persist for around about three quarters to a year on average.

5. Conclusion

A complete statistical analysis of the model-building procedures applied above is bound to indicate that the finally selected model is subject to considerable uncertainty in its specification. Most Monte Carlo studies hold the model specification fixed as the sample size varies and still yield large uncertainty regions. Moreover, in dynamic equations, nominal and actual test sizes can depart radically for relevant sample sizes, and test powers often are unimpressive. When the equation formulation is itself data-based, models can have but a tentative status. In Monte Carlo terms, large variability seems likely to arise from the selection process.

Some protection against "spurious" estimates is provided by having a predefined maintained hypothesis embodying subject-matter knowledge; Sections 2 and 3 above discussed the principles underlying the model of Table 2. However, the reparametrized equation (4.4) which summarizes the salient features of that table reflects a larger element of judgment; the reported coefficient standard errors are much smaller in (4.4) than in Table 2, primarily because of the reduced collinearity, but also partly because the imposed restrictions are acceptable through being apparent in the unrestricted model. Different samples and/or investigators could produce simplifications different from (4.4). The alternative of not basing the selection of the model upon the data is even less appealing.

Partial protection is offered by testing the selected model for its ability to encompass rival hypotheses as well as by checking parameter constancy on later data. Models which are encompassing and remain constant over time are useful tools for later applications. The present equation for new house prices fits substantially better than any preexisting models of house prices [e.g., see the estimates discussed in Nellis and Longbottom (1981)]. However, it is an explanation conditional on contemporaneous second-hand house prices Ph_t: marginalizing with respect to Ph_t would increase the residual standard deviation from $\hat{\sigma}$ to $\tilde{\sigma} \equiv (\hat{\sigma}^2 + \hat{a}^2\hat{\omega}^2)^{1/2}$, where $\hat{a}$ is the estimated coefficient of Ph_t in (4.4) (i.e., 0.26) and $\hat{\omega}^2$ is the estimated error variance of the model for Ph_t [e.g., $\hat{\omega} \approx 1.4\%$ in Hendry (1984)]. Thus, $\tilde{\sigma} \approx 1.02\%$ is implied, which remains smaller than in existing models which do not include Ph_t in modeling Pn_t. Nevertheless, we intend to conduct direct tests between the various marginal models of pn in due course, and hope to account for the subsample variation in $\hat{\sigma}$.

A mixed outcome, in which some developments have been implemented while others remain to be carried out, is fairly typical of empirical econometric research. Viewed as part of a research strategy in which anomalies point towards new research areas, remaining problems become a future stimulus rather than a major drawback. They are also a caution to the limitations of a model rather than a definitive rejection, and a sign that stringent evaluation criteria are being demanded rather than that econometric modeling is not worth undertaking.

Appendix A. A Simple Theoretical Model of the Market for New Housing

Consider a builder constructing dwellings subject to a Cobb–Douglas production function of the form

$$C = K_0 M^{\alpha} L^{\beta}, \qquad \alpha + \beta \leq 1, \tag{A.1}$$

where α, β, and K_0 are positive constants; M denotes direct inputs (workers and materials, with a price index per unit dwelling given by CC); and L denotes land (with a price per plot of Pl). Ignoring any fixed costs, profits are given by

$$\pi = Pn \cdot C - CC \cdot M - Pl \cdot L; \tag{A.2}$$

and the objective is to maximize π subject to (A.1), where CC, Pl, and Ph are taken as given (i.e., the builder is small relative to the whole market). The assumption that $\alpha + \beta \leq 1$ is made to rule out increasing returns as the scale of production grows. In practice, there are elements of local monopolistic power, so the demand function facing the builder is postulated to be

$$C = K_1 \cdot (Pn/Ph)^{\gamma}, \qquad \gamma < 0, \tag{A.3}$$

where K_1 is a positive constant.

The algebra of maximizing π subject to (A.1) and (A.3) is tedious but well-known (for an excellent introduction, see Smith, 1982) and yields the solution that *pn* is a weighted average of *ph*, *cc*, and *pl*, with the individual weights being dependent on α, β, and γ, and where (e.g.) doubling all nominal prices would leave decisions about quantities unaltered. In the important special case that $\alpha + \beta = 1$ (so constant returns prevail), *pn* only depends on *cc* and *pl* with weights α and $1 - \alpha$. However, *Ph* would then proxy *Pl*, since any increase in land prices would be reflected in increased prices for existing housing. Thus, an alternative representation would be $pn = \lambda cc + (1 - \lambda)\, ph$ (+ demand factors), which is closely similar to (3.8). Note that the functional form of the model is log–linear, given the postulated production and demand functions. Also, if (A.3) is "inverted" to express *Pn*/*Ph* as a function of *C*, the latter can be eliminated from the schedule for the supply of new housing, as in the main text. Conversely, since *u* in (3.8) is endogenous to the construction sector, a formulation dependent only on "outside" influences would necessitate substituting in its determinants.

Appendix B. Data Definitions

Variable[a]	Definition	Source[b]
C	Private housing completions (GB)	E.T.
CC	Cost of new construction index (1975 = 100)	M.D.S., H.C.S.
D	Dummy variable for 1981(iii)–1982(iii)	See text
P	General index of retail prices (1975 = 100)	E.T.
Ph	Index of prices of comparable dwellings (second-hand houses) on which transactions were completed	B.S.A.
Pn	Average price at completion of new dwellings on which new building-society mortgages were accepted (1975 = 100)	E.T., B.S.A.
R	Bank of England's minimum lending rate to the market (the base rate from 1981 (iii)); or mortgage rate	F.S.
S	Private housing starts (GB)	E.T.
τ	Standard rate of income tax	A.A.S.
U	Uncompleted houses (GB): constructed from the identity $U_t \equiv U_{t-1} + S_t - C_t$, with several benchmark surveys as checks	
US	Stock of unsold completions	H.C.S.
Y	Real personal disposable income (1975 prices)	E.T.

[a] All variables are quarterly, seasonally unadjusted.

[b] A.A.S., *Annual Abstract of Statistics*, H.M.S.O.; B.S.A., Building Societies Association *Bulletins* and *Compendium of Statistics*; E.T., *Economic Trends*, Annual Supplements, 1980–1984, H.M.S.O.; F.S., *Financial Statistics*, H.M.S.O.; H.C.S., *Housing and Construction Statistics*, H.M.S.O.; M.D.S., *Monthly Digest of Statistics*, H.M.S.O.

Acknowledgements

This chapter is based on work initially undertaken in Ericsson (1978) and extended for the U.K. Department of the Environment in Hendry (1980). We are indebted to Frank Srba for invaluable help in carrying out the preliminary analyses; to Jon Faust for his excellent research assistance; and to Anthony Atkinson, Julia Campos, John Muellbauer, Andrew Rose, and two anonymous referees for their helpful comments. Recent research has been supported by grants from the Economic and Social Research Council to the MIME programme at the London School of Economics, and by E.S.R.C. grants HR8789 and B00 220012 to Nuffield College. We are grateful for the financial assistance from the E.S.R.C., although the views expressed in this paper are solely the responsibility of the authors and should not be interpreted as reflecting those of the E.S.R.C., the Board of Governors of the Federal Reserve System, the Federal Reserve Banks, or other members of their staffs.

Bibliography

Anderson, G. J. and Hendry, D. F. (1984). "An econometric model of United Kingdom building societies." *Oxford Bull. Econom. and Statist.*, **46**, 185–210.

Atkinson, A. C. (1970). "A method for discriminating between models" (with discussion) *J. Roy. Statist. Soc., Ser. B*, **32**, 323–353.

Barndorff-Nielsen, O. (1978). *Information and Exponential Families in Statistical Theory*. New York: Wiley.

Bean, C. R. (1981). "An econometric model of manufacturing investment in the U. K." *Economic J.*, **91**, 106–121.

Box, G. E. P. and Jenkins, G. M. (1976). *Time Series Analysis: Forecasting and Control*, revised ed. San Francisco: Holden-Day.

Box, G. E. P. and Pierce, D. A. (1970). "Distribution of residual autocorrelations in autoregressive–integrated moving average time series models." *J. Amer. Statist. Assoc.*, **65**, 1509–1526.

Chow, G. C. (1960). "Tests of equality between sets of coefficients in two linear regressions." *Econometrica*, **28**, 591–605.

Coen, P. G., Gomme, E. D., and Kendall, M. G. (1969). "Lagged relationships in economic forecasting" (with discussion) *J. Roy. Statist. Soc., Ser. A*, **132**, No. 2, 133–163.

Cox, D. R. (1961). "Tests of separate families of hypotheses." In J. Neyman (ed.), *Proceedings of the Fourth Berkeley Symposium on Mathematical Statistics and Probability*, **1**. Berkeley: Univ. of California Press, 105–123.

Cox, D. R. (1962). "Further results on tests of separate families of hypotheses." *J. Roy. Statist. Soc., Ser. B*, **24**, 406–424.

Dastoor, N. K. (1983). "Some aspects of testing non-nested hypotheses." *J. Econometrics*, **21**, 213–228.

Davidson, J. E. H., Hendry, D. F., Srba, F., and Yeo, S. (1978). "Econometric modelling of the aggregate time-series relationship between consumers' expenditure and income in the United Kingdom." *Economic J.*, **88**, 661–692.

Davidson, J. E. H. and Hendry, D. F. (1981). "Interpreting econometric evidence: the behaviour of consumers' expenditure in the U. K. " *European Economic Rev.*, **16**, 177–198.

Domowitz, I. and White, H. (1982). "Misspecified models with dependent observations." *J. Econometrics*, **20**, 35–58.

Engle, R. F. (1982). "Autoregressive conditional heteroscedasticity with estimates of the variance of United Kingdom inflations." *Econometrica*, **50**, 987–1007.

Engle, R. F., Hendry, D. F., and Richard, J. -F. (1983). "Exogeneity." *Econometrica*, **51**, No. 2, 277–304.

Ericsson, N. R. (1978). *Modelling the Market for Owner-Occupied Housing in the United Kingdom: An Exercise in Econometric Analysis*. M.Sc. Thesis. London School of Economics.

Florens, J.-P. and Mouchart, M. (1980). "Initial and sequential reduction of Bayesian experiments." CORE discussion paper 8015. Louvain-la-Neuve, Belgium: Univ. Catholique de Louvain.

Godfrey, L. G. (1978). "Testing against general autoregressive and moving average error models when the regressors include lagged dependent variables." *Econometrica*, **46**, 1293–1301.

Granger, C. W. J. (1983). "Forecasting white noise" (with discussion). In A. Zellner (ed.), *Applied Time Series Analysis of Economic Data*. Washington: U.S. Bureau of the Census, 308–326.

Granger, C. W. J. and Newbold, P. (1977). "The time series approach to econometric model building." In C. A. Sims (ed.), *New Methods in Business Cycle Research*. Minneapolis: Federal Reserve Bank of Minneapolis, 7–21.

Granger, C. W. J. and Weiss, A. A. (1983). "Time series analysis of error-correction models." In S. Karlin, T. Amemiya, and L. A. Goodman (eds.), *Studies in Econometrics, Time Series, and Multivariate Statistics*. New York: Academic, 255–278.

Haavelmo, T. (1944). "The probability approach in econometrics." *Econometrica*, **12**, suppl., i–viii, 1–118.

Harvey, A. C. (1981). *The Econometric Analysis of Time Series*. Oxford: Philip Allan.

Hendry, D. F. (1980). *An Econometric Model of the UK Housing Market*. London: Economists Advisory Group.

Hendry, D. F. (1984). "Econometric modelling of house prices in the United Kingdom." In D. F. Hendry and K. F. Wallis (eds.), *Econometrics and Quantitative Economics*. Oxford: Basil Blackwell, Chapter 8, 211–252.

Hendry, D. F. and Anderson, G. J. (1977). "Testing dynamic specification in small simultaneous systems: An application to a model of building society behavior in the United Kingdom." In M. D. Intriligator (ed.), *Frontiers in Quantitative Economics*, Vol. IIIA. Amsterdam: North-Holland, Chapter 8c, 361–383.

Hendry, D. F., Pagan, A. R., and Sargan, J. D. (1984). "Dynamic specification." In Z. Griliches and M. D. Intriligator (eds.), *Handbook of Econometrics*, Vol. II. Amsterdam: North-Holland, Chapter 18, 1023–1100.

Hendry, D. F. and Richard, J.-F. (1982). "On the formulation of empirical models in dynamic econometrics." *J. Econometrics*, **20**, 3–33.

Hendry, D. F. and Richard, J.-F. (1983). "The econometric analysis of economic time series" (with discussion). *Internat. Statist. Rev.*, **51**, 111–163.

Jarque, C. M. and Bera, A. K. (1980). "Efficient tests for normality, homoscedasticity and serial independence of regression residuals." *Economics Lett.*, **6**, 255–259.

Judge, G. G. and Bock, M. E. (1978). *The Statistical Implications of Pre-test and Stein-Rule Estimators in Econometrics.* Amsterdam: North-Holland.

Kiviet, J. F. (1981). "On the rigour of some specification tests for modelling dynamic relationships." Paper presented at the Amsterdam Conference of the Econometric Society.

Kiviet, J. F. (1982). "Size, power and interdependence of tests in sequential procedures for modelling dynamic relationships." Discussion paper, Univ. of Amsterdam.

Kmenta, J. and Ramsey, J. B. (eds.) (1981). *Large-Scale Macro-econometric Models.* Amsterdam: North-Holland.

Koopmans, T. C. (1950). "When is an equation system complete for statistical purposes?" In T. C. Koopmans (ed.), *Statistical Inference in Dynamic Economic Models.* Cowles Commission Monograph 10. New York: Wiley, Chapter 17, 393–409.

Lakatos, I. (1970). "Falsification and the methodology of scientific research programmes." In I. Lakatos and A. Musgrave (eds.), *Criticism and the Growth of Knowledge.* Cambridge: Cambridge U. P., 91–195.

MacKinnon, J. G. (1983). "Model specification tests against non-nested alternatives" (with discussion). *Econometric Rev.*, **2**, 85–158.

Messer, K. and White, H. (1984). "A note on computing the heteroskedasticity-consistent covariance matrix using instrumental variables techniques." *Oxford Bull. Econom. and Statist.*, **46**, 181–184.

Mizon, G. E. (1984). "The encompassing approach in econometrics." In D. F. Hendry and K. F. Wallis (eds.), *Econometrics and Quantitative Economics.* Oxford: Basil Blackwell, Chapter 6, 135–172.

Mizon, G. E. and Richard, J. -F. (1983). "The encompassing principle and its application to testing non-nested hypotheses." CORE discussion paper 8330. Louvain-la-Neuve, Belgium: Univ. Catholique de Louvain, forthcoming, *Econometrica.*

Naylor, T. H., Seaks, T. G., and Wichern, D. W. (1972). "Box–Jenkins methods: an alternative to econometric models." *Internat. Statist. Rev.*, **40**, 123–137.

Nellis, J. G. and Longbottom, J. A. (1981). "An empirical analysis of the determination of house prices in the United Kingdom." *Urban Stud.*, **18**, 9–21.

Nelson, C. R. (1972). "The prediction performance of the FRB-MIT-PENN model of the U.S. economy." *Amer. Econom. Rev.*, **62**, 902–917.

Pesaran, M. H. (1974). "On the general problem of model selection." *Rev. Econom. Stud.*, **41**, 153–171.

Prothero, D. L. and Wallis, K. F. (1976). "Modelling macroeconomic time series" (with discussion). *J. Roy. Statist. Soc., Ser. A*, **139**, 468–500.

Rao, C. R. (1948). "Large sample tests of statistical hypotheses concerning several parameters with applications to problems of estimation." *Pro. Cambridge Philos. Soc.: Math. and Phys. Sci.*, **44**, 50–57.

Schumpeter, J. (1933). "The common sense of econometrics." *Econometrica*, **1**, 5–12.

Schwarz, G. (1978). "Estimating the dimension of a model." *Ann. Statist.*, **6**, 461–464.

Smith, A. (1982). *A Mathematical Introduction to Economics.* Oxford: Basil Blackwell.

Theil, H. (1964). *Optimal Decision Rules for Government and Industry.* Amsterdam: North-Holland.

Trivedi, P. K. (1984). "Uncertain prior information and distributed lag analysis." In D. F. Hendry and K. F. Wallis (eds.), *Econometrics and Quantitative Economics.* Oxford: Basil Blackwell, Chapter 7, 173–210.

White, H. (1980). "A heteroskedasticity-consistent covariance matrix estimator and a direct test for heteroskedasticity." *Econometrica*, **48**, No. 4, 817–838.

Zellner, A. and Palm, F. (1974). "Time series analysis and simultaneous equation econometric models." *J. Econometrics*, **2**, 17–54.

CHAPTER 12

Large-Scale Social Experimentation in the United States

Stephen E. Fienberg, Burton Singer, and Judith M. Tanur

Abstract

Randomized controlled experimentation has a long history. But the transition from statistical theory to statistical practice remains a difficult one, especially when the experiments are large in scale. In this paper, we review several major randomized social experiments carried out in the United States beginning in the early 1970s. We describe briefly the policy settings out of which these experiments grew, and we indicate how this type of experimentation fits into the historical development of randomized experimentation more generally. We give detailed design features for five different sets of social experiments, and we briefly summarize some of the analyses of the experimental data. We end the paper with a discussion of analytical strategies for social experiments, and possible alternatives to large-scale experimentation, including combining the results of several experiments, and evolutionary designs. Throughout, we place special emphasis on the possible roles of models and of social and economic theory in both the design and the analysis of experiments, and we describe several open research problems requiring the attention of mathematical statisticians and economists.

1. Introduction

When partisans advocate a policy change, they often believe that the implementation of their plan will produce obviously positive results. Their opponents are likely to contend that either negative results or no change at all will be effected. Typically the truth is somewhere in between—small gains are more likely than large gains or no gains at all, and sometimes the gains are concentrated in a portion of the target population (Gilbert, Light, and Mosteller, 1975). But such small or localized gains are often worth finding because they can aggregate to large savings or impressive overall improvement in well-being. For example, in the United States there is a debate about whether the possible loss of Social Security-related benefits inhibits individuals who have been certified as disabled from attempting to return to work. One calculation

Key words and phrases: allocation models, combining results, evolutionary designs, factorial designs, finite selection model, optimal designs, stochastic-process models.

suggests that a rise in annual terminations from disability rolls from the current 3% to only 4% for a one-year cohort of newly awarded beneficiaries would result in the saving of some $200 million for the Social Security Trust Fund over the life of that cohort. Effecting such a change is the goal of the treatment in a proposed Work Incentive Experiment. Detecting such potentially valuable small gains calls for careful investigation; evidence of their existence can easily be lost unless very sensitive tools are used to gauge the effect of policy changes. The most sensitive such tool is the large-scale randomized social experiment, in which the act of randomization creates the foundation both for precise analysis of outcomes and for direct causal evaluation of potential policy changes.

The notion of relying upon a randomized controlled experiment to inform public policy debate gained currency in the United States about twenty years ago. In the early 1960s, the political zeal accompanying President Lyndon Johnson's "War on Poverty" collided with the realization that the knowledge base about the target population was so weak that it was virtually impossible to formulate and *defend* an effective antipoverty strategy. One of the principal foci of discussion was the concept of a negative income tax as a substitute for existing welfare programs. The essential idea was that a floor was to be placed under income, and all persons or families would be entitled to this amount from the government if they had no earnings or other income. Families or persons with positive incomes would receive subsidies until their total income (including subsidies) reached a breakeven point. Subsidies for persons with positive incomes would be reduced by a "tax rate" such that for every dollar earned the subsidy would be reduced by an amount less than one dollar.

The notion of a negative income tax met with considerable resistance, primarily because many people believed that the guarantee of a minimum level of living would provide an irresistible inducement for a significant number of persons, especially able-bodied males, to reduce their hours of work or to stop work entirely. Although economists had attempted to develop estimates of the effect of a negative income tax on work effort, the results had been decidedly inconclusive (Cain and Watts, 1973). There was, therefore, a live policy issue at hand that could not be satisfactorily resolved on the basis of available data: Are cash allowances whose net benefits decline as work income increases an inducement for recipients to reduce work significantly? The suggestion to resolve the debate by controlled experimentation was put forth in 1966 (see e.g. Kershaw and Fair 1976, p. XV), and led directly to the first large-scale controlled experiment in the U.S. which used a social program as a treatment: the New Jersey Income Maintenance Experiment.

The New Jersey Experiment and the other income maintenance experiments of that first generation (Seattle/Denver, Gary, Rural) produced as offspring other U.S. socioeconomic experiments—among them the Housing Allowance Demand Experiment, the National Health Insurance Experiment, and a series of Time-Of-Use (TOU) Electricity Pricing Experiments. In this paper, we focus on these studies in an attempt (i) to characterize the domain of large-scale social experiments, and (ii) to describe open problems that must be addressed if we are to develop a deeper understanding of the process of experimentation as a vehicle for measuring behavioral responses to social programs.

Randomized social experimentation has a long history, predating by several decades the New Jersey Income Maintenance Experiment. Lest the reader lose sight of this fact, we begin in Section 2 with some history that will help put the large-scale U.S. socioeconomic experiments discussed here into perspective. We also note some features of these experiments that distinguish them both from earlier social experiments and from earlier experiments in the agricultural, industrial, and medical sciences. We do not discuss ethical issues in any detail, because these are discussed articulately and at length in Gilbert, Light, and Mosteller (1975, Section VI). Basically, our position coincides with theirs: "We need to decide whether fooling around with people's lives in a catch-as-catch-can way that helps society little is a good way to treat our citizens, and we need to understand that in current practice this is the principal alternative to randomized controlled field trials."

In Section 3 we give an outline of the background and the relevant policy questions for the four major sets of social experiments—income maintenance, housing allowance, health insurance, and peak-load pricing—as well as for a yet-to-be-implemented experiment on work incentives and Social Security disability payments. We also summarize some of the results from the completed experiments. We then turn in Section 4 to a discussion of design features of the U.S. social experiments, and how they relate to the statistical literature on design of experiments. Ferber and Hirsch (1979, 1982) also describe some of these and related materials, in a less technical form.

Much of the analysis work for these large-scale experiments is still not publicly available. Nevertheless, we perceive a set of issues regarding analytical strategies, which we describe in Section 5.

As a scientific aid to policy formation, is there an alternative to the use of a single large-scale experiment? We address this issue in Section 6. Two technical appendices discuss a research agenda of mathematical statistics and microeconomic issues which must be resolved if our suggested alternatives to past experiments are to be implemented.

It is important to note that a social experiment is unlikely to come close to exactly duplicating a likely policy. The traditional "model-free" use of experimental results can thus be viewed as misleading unless the proposed policy is exactly implemented by the experiment. Thus to make sense of the results of any such experiment we require some form of model for self-selection and behavioral adjustment to interpolate or, more likely, to extrapolate to a menu of policy alternatives (Heckman, 1978). We consider the adaptive approach suggested in Section 6 as an alternative (but not necessarily a model-free one) to the Heckman position that an experiment is useful only as a *supplement* to detailed, less costly, nonexperimental evidence.

In attempting to relate the design, execution, and findings of these large-scale social experiments to the policy issues they were intended to address, we encounter a paradox. Behavioral reactions to policy changes such as the implementation of a negative income tax or a housing allowance program are not expected to be immediate or even short term, and thus the experiments we shall describe were in the field for extended periods of time, often several years. Moreover, the complexities of the data collection (and measurement) and analyses added to the delays before even preliminary results were in hand. In the meantime, however, political priorities, administrations, and budgets all changed, so that the old policy debates had been replaced by new ones (see the warning about this problem in Rivlin, 1974). The questions the social experiments had been designed to answer were no longer the burning issues of the day. Thus we cannot write the ideally logical review that would illustrate how the results of the experiments informed or influenced policy debates. Rather we need to rely on judgements (i) that experimental findings expand the knowledge base of the social sciences and thus are inherently valuable, and (ii) that policy questions may recur in similar forms, so that either specific findings or an accumulation of findings may be of use in the future when political priorities, administrations, and budgets change yet again.

Given this random nonstationary political environment combined with similar nonstationarities imposed on the experimental settings as such, we are confronted with the question of whether some form of adaptive designs could be more effective in producing useful comparisons for policy purposes. Although we address this issue in somewhat more detail in Section 6, the basic question must, for the present, go unanswered, and thus it represents an important research challenge.

In this paper we do not deal with a variety of other reasonably large-scale social experiments, primarily due to limitations of space and the lack of full sets of published materials. For example, the U.S. Department of Defense has sponsored and continues to sponsor experiments on manpower planning and recruitment (see e.g. Haggstrom, 1976, and Haggstrom et al., 1981), and there are several experiments in the criminal-justice area (for one of the better examples see Rossi, Berk, and Lenihan, 1980, and the discussion of their work in Cronbach, 1982).

2. Some History

Randomized controlled experiments originated in agriculture (e.g. for a systematic review see Fisher, 1935). The original experiments described by Fisher and his Rothamsted colleagues were small in scale, often involving the subdivision of individual fields into plots. Yet one of the earliest examples of the implementation of Fisher's ideas in the U.S., taking place in Minnesota, involved the comparison of the yields of ten varieties of barley in a randomized block experiment, carried out at six locations in two successive years (Immer, Hayes, and Powers, 1934). Thus concern for aspects of large-scale experimention at multiple sites dates back almost to the beginning of the statistical literature on randomized experiments.

With the adoption of statistical experiments in areas other than agriculture, the methodology underwent transformation and elaboration. As Cox (1958) points out, applications to manufacturing processes faced problems caused by the much larger number of factors affecting the processes. Fractional factorial designs were originated to allow the simultaneous manipulation of a multiplicity of factors (Finney, 1945), and found many applications in large-scale industrial experimentation. At the same time, because manufacturing processes can produce output quickly (without the necessity of waiting for crops or livestock to mature), industrial applications provided the opportunity for developing evolutionary designs (Box and Wilson, 1951).

The adoption of randomized controlled experiments for medical–surgical research also brought about elaboration of the methodology. Now the experimental material became people, rather than crops or chemicals, and people with specific diseases in specific population groups. Not only did these new experiments need to provide mechanisms for informed consent and double blindness, but also, in order to gather sufficient subjects suffering from a relatively rare or expensive-to-treat disorder, the experiments were—of necessity—large scale, involving patients at several (or many) hospitals or clinics, typically in different cities. Note that the role of multiple sites here is different from that in the Minnesota agricultural experiment described earlier. There, six different sites were used because of a concern for the generalizability of the experimental findings to a broad spectrum of preexisting conditions (e.g. rainfall and soil), whereas in the medical context the shift to multiple sites is often necessitated by the lack of sufficient subjects at a single site. As a consequence the multiplicity of sites in medical experiments increases administrative complexity and increases the possibility of variation in the application of the treatments. Such increased variation is both a liability and an advantage. It is liability if it is so great that its effects swamp those of the treatments, but if the treatment effects are strong enough to withstand inter-site variation, the investigator in such a multi-site medical experiment (like the experimenter in a multi-site agricultural experiment) has greater warrant to generalize findings to other treatment sites than would an investigator working at a single site.

Concomitant features leading to the complexity of many medical experiments include the multiplicity of variables of interest (both dependent and independent) and the length of time required to observe certain kinds of outcomes (e.g. see the discussion in Meier, 1975). At any rate, several multi-site cooperative randomized clinical trials were launched during the 1960s (e.g. see the report on the University Group Diabetes Program, 1970), and some of the lessons learned from them should have (but seem not to have) influenced the planning of the social experiments of the 1970s. Perhaps the most important such lesson is the need for centralized planning and coordination of a multi-site experiment to ensure comparability of findings across sites and minimize variation in designs and data collection details. Indeed, we believe that the 15 TOU electricity pricing experiments sponsored by the Department of Energy—see e.g. Hendricks and Koenker (1979) and Aigner (1982)—could have been substantially improved if lessons from multi-site medical interventions had played a role in the planning process. We discuss this issue further in Section 4.

A major difference between multi-site medical experiments and the social experiments described in this paper relates to the utility of their findings. Even if a medical trial involving new therapies extends over many years, the diseases which inspired the trial are likely to continue to exist when the results of the experiment are in. As we noted in the introduction, the same is not necessarily true of long-term social experiments, where the findings may be of only limited policy interest by the time they are available. In addition, when randomized experimentation is adopted for use in social experiments, the administrative problems caused by the multiplicity of sites or the large scale of medical experiments are exacerbated. The treatment given is often more complicated than a medical or surgical protocol and must often be applied over a much longer time period. Sometimes, as in the income-maintenance experiments, treatment delivery involves setting up an entire social welfare agency. In other cases, crucial data are collected by people for whom data collection is not a primary activity—as in the Los Angeles peak-load pricing experiment, where linesmen and meter readers carried out interviews (Berry, 1979). Then, careful administrative coordination between the "action team" and the "research team" is necessary to ensure treatment integrity (e.g. Archibald and Newhouse, 1980). Further, outcomes of the experiment are often more difficult to measure when social behavior is being examined than when one is evaluating the response to a medical treatment. Use of electricity by time of day, decrease in work effort, and use of medical facilities are all examples of conceptually complicated responses that had to be measured in the social experiments discussed below.

In this brief review of the evolution and spread of randomized controlled experiments, we have focused on those developments that necessitated a move towards large-scale experimentation. Parallelling these developments has been the expanded use of randomized experiments on a smaller scale to evaluate social and sociomedical programs. The need for randomization to control for bias in such settings was recognized over 50 years ago by Gosset in his description of the famous Lanarkshire milk "experiment" (Student, 1931). Gilbert, Light, and Mosteller (1975), Boruch (1975), and Riecken et al. (1974) document many of the more recent social experiments (from the 1950s and 1960s). The experiments of the 1970s which we describe here differ from those of the preceding decades in scale, complexity, and cost. These differences lead to interesting design considerations.

3. Examples of Large-Scale Social Experiments—Some Policy Questions

3.1. Income-Maintenance Experiments

In the mid 1960s, as we noted above, much of the resistance to legislation implementing a negative-income-tax program of any kind was based on the belief that a program of cash transfers from the federal government to the working poor would simply act as a work disincentive. Thus, proponents of negative-income-tax programs felt it as important to assemble the strongest possible evidence to counter such suspicions. Because there was no clear-cut empirical evidence in either direction, a series of experiments was initiated in New Jersey and portions of Pennsylvania; in Seattle, Washington and Denver, Colorado; in Gary, Indiana; and in a limited number of rural communities, each with the primary goal of assessing whether a variety of income guarantees (support levels) and accompanying tax rates would act as work disincentives in low-income populations of the same type that would subsequently be eligible for a national income-transfer program if it were to be implemented. Table 1 summarizes the basic treatment structure of these income-maintenance experiments. (Each ran for 3 years, with the exception of Seattle–Denver, where 75% of the families were enrolled for 3 years, 25% for 5 years, and 100 families for 20 years.)

There are a variety of design issues in the negative-income-tax experiments which we will not discuss in detail. For example, in the New Jersey experiment, which was carried out in Trenton and two other sites, the investigators substantially overestimated the numbers of eligible white families. To achieve ethnic balance, the investigators added Scranton, Pennsylvania to the study.

Table 1. Treatments in Income-Maintenance Experiments[a]

	NJ–PA					Rural					Gary			Seattle–Denver		
	125	100	75	50		125	100	75	50		115	90		150	125	100
													80[b]	×	×	×
70		×	×		70	×		×					70	×	×	×
													70[b]		×	×
										60	×	×				
50	×	×	×	×	50	×	×	×	×				50	×	×	×
										40	×	×				
30			×	×	30		×	×								

Source: Lauwagie, B. (1977). Unpublished tabulation prepared for Social Science Research Council Conference on Social Experimentation, New Orleans, La., Feb., 1978.

[a] Columns show support levels as percentage of poverty level; rows show tax rates.

[b] Declining with increasing income.

Not only did this partially confound site with ethnic background, but it raised the issue of possibly different labor markets and their impact on outcomes.

In any social experiment, it is a difficult and subtle process to define and measure a response variable in such a way that the measurement is both a reliable and a valid indicator of the concept being examined. These issues became salient early on in the income-maintenance experiments, where the concept was labor supply. There are at least four measures of the amount of labor supplied by a household: labor-force participation, employment, hours of work, and total earnings. On the surface it would seem that earnings might furnish the best summary measure of experimental effects on labor supply. Experimental families, however, filled out an income form every four weeks, an activity not required of control families. Thus there could have been a differential learning process in which experimental families grasped more quickly than control families that what was to be furnished was earnings before taxes and other deductions, not take home pay. This could cause spurious differentials in earnings in favor of the experimental group, especially during the early part of the experiment. Indeed, in New Jersey the experimental group actually did show higher earnings than the control group. Hours of work per week was thus a better measure of labor-supply response. Table 2 gives a rough summary of that indicator averaged over treatment groups.

The point estimates suggest that, overall, there was a small but consistent—across experiments—reduction in hours of work in response to the various

Table 2. Labor–Supply Results. Hours worked per week: mean differences between experimental and control groups.

	Husbands		Wives		Female heads	
	Difference	% difference[a]	Difference	% difference[a]	Difference	% difference[a]
New Jersey:						
White	−1.9	−5.6	−1.4	−30.6	N/A[b]	N/A
Black	0.7	2.3	0.1	2.2	N/A	N/A
Hispanic	−0.2	−0.7	−1.9	−55.4	N/A	N/A
Rural (nonfarm):						
NC blacks	−2.9	−8.0	−5.2	−31.3	N/A[c]	N/A
NC whites	−2.1	−5.6	−2.2	−21.5	N/A	N/A
IA whites	−0.5	−1.2	−1.2	−20.3	N/A	N/A
Seattle–Denver:						
All races	−1.8	−5.3	−2.1	−14.6	−2.6	−11.9
Gary:						
Black	−1.0	−2.9	0.1	1.0	−1.9	−25.9

Source: Fisk and Roth (1980, p. 37).

[a] Using control group as base.

[b] No female-headed families enrolled in New Jersey experiment.

[c] Too few enrolled in Rural experiment for analysis.

guarantee levels and tax rates. This general trend was consistent with the *a priori* predictions of economic theories of labor-supply response to the financial incentives. Substantial deviation from this average response occurred for selected population subgroups and treatment combinations in some of the experiments. For example, blacks and Spanish-Americans in the Seattle and Denver experiments had substantially greater reductions in hours of work than whites in the same treatment groups. Heads of households enrolled in the five-year Seattle–Denver program had substantially larger labor-supply responses than those enrolled in the three-year program, and females reduced their relative participation more than males. However, the *overall small response* suggested that many of the *a priori* fears of opponents of negative-income-tax programs were unfounded, at least in the locations where the experiments were carried out.

Borjas and Heckman (1978) argue that an analysis of labor-supply estimates, using a few reasonable criteria derived from economic theory, substantially narrows the range of variability of estimates for policy purposes. Specifically, they focus on prime-age males, and their approach involves formally adjusting for various biases of the sort we describe briefly in Section 4.1 below. Their argument is that, without a model for interpreting and extrapolation, figures such as those in Table 2 have little or no direct meaning for policy.

There has been considerable criticism to the effect that the income-maintenance experiments did not end up addressing the public-policy issues of greatest interest. For example, Rossi and Lyall (1976) and Cronbach (1982) claim that Congress wanted to know the probable cost of a national program, and the design was not appropriate for this purpose. Their view is that the regression model design (which we describe below in Section 4.1) diverted the New Jersey–Pennsylvania experiment (and its successors) from its proper function. Cronbach (1982, p. 260) also notes that "family stability, a variable that the original study learned about only incidentally, loomed large in later policy discussion." There was a somewhat sharper focus on issues of family stability in the Seattle–Denver program, primarily as a result of the efforts and analyses of Hannan, Tuma, and Groeneveld (1976, 1977). For an extensive critique and useful appraisal of the income-maintenance experiments, see Neuberg (1985).

3.2. Housing-Allowance Experiments

Prior to the Housing and Urban Development HUD Act of 1965, no U.S. housing policy or support program for low-income families had incorporated the notion of direct cash transfers to a designated population of eligibles for the express purpose of helping them pay the costs necessary for living in the housing of their choice. Since the passage of the Housing Act of 1937, however, there had been repeated debate in congressional circles on the topic of such housing allowances, as opposed, for example, to government sponsored construction of low-income housing projects.

Critics of housing allowances had argued that giving low-income families cash earmarked for housing would cause a substantial rise in general housing prices because an inflationary competition for acceptable dwellings would be generated as persons living in substandard housing would qualify for assistance and seek better accommodations. Many also felt that landlords would simply raise the rents of program recipients, thereby eliminating any advantages the subsidies might provide.

Persons favoring housing allowances often argued that such a program would be less expensive than government assistance for newly constructed housing. They also felt that a program of direct cash transfers would be more equitable, since those receiving payments would live in units comparable to those of families just above the income limits for eligibility in the program. New low-income housing, on the other hand, might be especially desirable but would not be available to slightly higher-income families. Further issues concerned the advisability of fostering neighborhood segregation by income; the construction of low-income housing often has this effect.

Since there was no empirical basis of projecting behavioral responses to a housing-allowance program as recently as 1970, the U.S. Congress authorized 20 million dollars for an Experimental Housing Allowance program in the 1970 HUD Act. The principal issues to be addressed were: (i) What do families actually do with housing allowances? (ii) How are housing markets affected by allowances? and (iii) How and at what costs might a housing-allowance program be administered?

Because of the complexity of these questions, it was decided early on to address them, respectively, in three separate studies: a demand experiment, a supply experiment, and an administrative-agency experiment. Of these, only the demand experiment was a controlled randomized study. The supply "experiment" involved the *observation* of the response of the full housing market at two sites (Brown County, Wisconsin and St. Joseph County, Indiana) to advertised availability of housing allowances for all low-income families. There was no randomization of individual households into special plans, since the aim of the supply study was to assess the collective response of an entire site to a special stimulus. Cost constraints dictated concentration on two rather small sites (250,000 population in 1970), but the sites were chosen according to a quasi-experimental design (Campbell and Stanley, 1963) that contrasted a tight housing market in a racially homogeneous area with a loose housing market in a racially segregated area—a primarily white suburban section surrounding a heavily black central city.

Similarly, the administrative-agency "experiment" was a performance evaluation of eight agencies as potential administrators of a housing-allowance program. Each agency was required to advertise availability of the program to eligible households, select and screen applicants, certify income, etc. Our emphasis here and in subsequent sections will be on the demand experiment;

however, the reader should consult Bradbury and Downs (1981) for a comprehensive review of the full range of housing-allowance experiments.

The treatment structure of the demand experiment had two components: one of two payment formulas and a housing requirement. One payment formula involved a housing-gap specification, and the other was a percent-of-rent formula. The housing gap G was defined as the difference between the estimated annual rent required for a family of particular size in a specified location and some percentage of the family's income (that percentage was set at either 15%, 25%, or 35%, in line with traditional budget prescriptions for family spending for housing). In conjuction with the housing-gap formula, basic housing quality standards were set: one was a minimum set of physical standards, two others used minimum rents as surrogates for physical quality, and a fourth had no standard. In all there were 12 housing-gap treatment groups. The other payment formula directly allowed 20%, 30%, 40%, 50%, or 60% of the families' actual rent. Thus we have a total of 17 possible treatment groups and a control group. The demand experiment was conducted over a three-year period at two sites: Pittsburgh, Pennsylvania along with surrounding Allegheny County, and Phoenix, Arizona along with Maricopa County. At each site roughly 1240 families were enrolled in one of the experimental plans and another 400 to 500 families were assigned to the control group.

In the debates about housing-allowance programs prior to the HUD-sponsored experiments, a key variable in projecting costs was the proportion of eligible families that would respond to advertising and self-select to enroll in the program. The demand experiment provided the first empirical evidence on this point, and a major finding was that eligible families were remarkably apathetic about the program—of those enrolled at the start, only half participated fully enough to receive assistance payments. Furthermore, at both sites fewer than 20% of all households were eligible to be candidates for the program. Thus contrary to many *a priori* expectations, one learns that low-income families do not flock to participate in a cash-allowance program even when they are contacted individually; and they do not want to change their housing even when offered a subsidy of as much as $100 per month.

Of families that *did* participate in the program and *did not* have payments tied to housing standards, approximately $\frac{2}{3}$ lived in substandard housing. Thus to ensure that recipient households live in acceptable housing, the experimental results suggest that payments must be tied to housing standards. Among participants in the experiments, preprogram rent burdens averaged roughly 40% of gross income. If housing allowances are counted strictly as payments to reduce housing expenditures rather than also as additional income, they typically reduce rent burdens to about 25% of income. But few enrollees were anxious to increase their housing expenditures beyond the amount needed to meet program requirements. An important policy implication of the finding is that it seems pointless to force people to live in standard rather than substandard housing units if they really don't want to. Thus if there are to be further subsidies for low income populations, one can reasonably ask whether they should simply be expanded income maintenance programs rather than specific housing-allowance programs. Since most money given for housing allowances is used for nonhousing purposes, this would both reduce administrative costs and increase the freedom of recipients by eliminating constraints on the manner in which they choose to spend their money.

3.3. A Health-Insurance Study

In the U.S., there is sharp disagreement about specific plans for health insurance and very little sound empirical evidence about the effects and costs of alternative programs. With the central purpose of providing reliable information to aid in the development of national health insurance legislation, the RAND Health Insurance Study (HIS) was initiated in 1972. This investigation "was specifically designed to estimate how various insurance plans affect the demand for medical care and the health status of individuals" (Newhouse, 1974). The major treatment variables related to the levels of coinsurance for different types of medical care and to the maximum percentage of income spent for medical care. By coinsurance is meant the percentage of the cost of services that are to be paid by the insured person. The experimental plans were of five principal types: (i) free care, (ii) 25% coinsurance, (iii) 50% coinsurance, (iv) 95% coinsurance, (v) 95% coinsurance to $150 per year per individual with no coinsurance above that amount. By further refining these categories the researchers used 11 treatment plans along with two control groups. For full details see Newhouse et al. (1979).

The experiment was conducted at six sites and involved a total of 2750 families (most for 3 years, some for 5 years). An important feature of the HIS, representing a considerable improvement over the income-maintenance experimental designs, was direct measurement of possible Hawthorne effects (i.e. effects that are the result of experimentation itself and not effects attributable to the specific treatment; see Mosteller, 1978, p. 210) and an assessment of whether the frequency with which people filled out health forms affected their utilization of health services. In particular, one control group was given no treatment and selected by random assignment from a total sample of eligible families, while another was simply enrolled in a group health program. Then a random part of one of the control groups was asked to fill out forms on a continuing basis while the other part received no such request.

A variety of initial summary analyses are described in Newhouse et al. (1981). For purposes of formulating legislation on health insurance with coinsurance features, it would be critically important to be able to anticipate sources of large expenditures. The exploratory analyses described by Duan et al. (1983) already indicate that distinctions should be made between subpopulations that are nonspenders, ambulatory spenders with no inpatient utilization, and spenders with inpatient utilization. The last group contains a relatively small number of individuals, but many of these people contribute to the far-right tail of the distribution of the logarithm of positive expenditures. Thus careful assessment of the nature of a long-tailed distribution will play a central role in the predictions of costs of various insurance plans. Another important aspect of the analyses to date is the strong interaction between the age of an individual and the insurance plan. Adults respond to plans by changing their conditional probability of utilizing inpatient services, whereas children do not. These suggestions from early analyses of HIS data should be treated with great caution, since full analyses and a final report lie several years in the future. Nevertheless, the reader should consult Duan et al. (1983) for an exemplary illustration of analytical strategies to detect and support conclusions from social experiments.

3.4. Time-of-Use Electricity Pricing Experiments

Beginning in 1975 the Department of Energy (DOE) initiated a series of 15 time-of-use (TOU) rate experiments involving time-of-day and seasonally varying prices for electricity. The main goal of the experiments was to determine whether TOU pricing would produce sufficient alteration in power use patterns of residential customers so that power companies would require lower generating capacity and thereby could use more cost-effective generating equipment.

In the U.S., electricity is commonly billed under a declining block rate structure. Larger consumers (e.g. businesses) pay less per kilowatt hour (kWh) than smaller consumers (e.g. households), and for any consumer the marginal cost per kWh declines as usage increases. Although other countries use peak load pricing schemes, such schemes had not been tried in the U.S. prior to the DOE experiments. If the charge for electricity is highest when demand is greatest, demand should smooth out. Then it becomes cheaper for a utility to supply electricity, because the less efficient production plants do not have to be put into operation to meet peak loads and the need to build new plants decreases. The purpose of the experiment thus was to investigate whether the smoothing out of demand could be accomplished by manipulation of prices for electricity over the 24-hour day and from season to season, and whether such smoothing could be accomplished without unduly increasing consumers' expenditures for electricity. A secondary purpose was to generate accurate data on level of load (under both conventional and peak-load rate structures) to be used for reliable load forecasting and potentially for comparisons across times and locations (Manning, Mitchell, and Acton, 1979, p.136; Acton, Manning, and Mitchell, 1978).

An important requirement for assessing a TOU price response is wide price variation in the design. Despite the fact that the TOU experiments had a common sponsor, the designs were not coordinated across sites in order to ensure substantial price variation and comparability. In our view, this is regrettable. The current consensus judgments on all 15 TOU experiments—see e.g. U.S. Department of Energy (1978), Hendricks and Koenker (1979), and Aigner (1982)—are that there are six studies which have substantial price variation and overall high-quality designs. These are: Arizona, Wisconsin, Los Angeles Department of Water and Power, Southern California Edison Co., Oklahoma, and North Carolina Power and Light. Even within this list, however, there is considerable variation in the basic designs, thereby making cross-site comparison somewhat difficult—but see, in this connection, Aigner and Leamer (1984).

In order to illustrate the structure of TOU experiments we describe the Los Angeles Department of Water and Power (LADWP) and South California Edison (SCE) studies in some detail.

The LADWP experiment began in mid 1976, with some 1200 households enrolled voluntarily for 30 months; data collection was completed in mid 1979. There were two subexperiments, one manipulating prices by time of day, the other by season of the year. Key treatment variables (some of which only apply to the time-of-day experiment) were the differential between peak price and off-peak price (peak prices were from \$.05 to \$.13 per kWh; off-peak prices were \$.01 to \$.02), number of days in the week with peak periods (5 vs. 7), length of peak period (3, 9, or 12 hours), and time of the peak period. From this large number of possible treatments, design points were chosen and sample sizes allotted to them using a finite selection model and an allocation model subject to subjective weighting for policy importance and to cost constraints. The outcome variable was the change in electric usage by time of use (measured by specially installed meters). Among concomitant variables measured were weather patterns and household characteristics such as appliances owned, income, prior consumption level, and family members at home by time of day. These variables were measured from prior Department of Water and Power records, 1970 census tract figures, and face-to-face interviews at the beginning, middle, and end of the experiment.

The LADWP experiment was able to answer some policy questions, and furnish information that policy makers could use to guide some decisions. For seasonal peak pricing the net welfare gains were very small—on the order of a few cents per month. Permanent household differences accounted for 90% of the variation in consumption; the response attributable to weather variation was minimal. The researchers concluded that seasonal peak pricing offered no gains over conventional pricing or time-of-day variations (Lillard and Acton, 1980). On the other hand, time-of-day pricing was found to be useful—but only for a particular part of the population. Metering costs were prohibitive unless the household consumption was very high (over 1100 kWh/month) or the household had a swimming pool. Note again the phenomenon that a small gain, or one applying to a small part of the population, can be impressive when aggregated; the affected subpopulation accounts for only 5% to 10% of Los Angeles households, but for as much as 20% of residential electricity consumption. Policymakers are presented with two options: mandatory time-of-day rates could be prescribed for specific residential subclasses, or rate structures could be redesigned in such a way that time-of-day rates become attractive for just those who realize net welfare gains for the system. This latter option requires structuring rates in such a way as to encourage consumers to shift to time-of-day pricing only if they also change their consumption patterns; any other structuring would permit consumers to decrease their bills without changing usage patterns (Acton and Mitchell, 1979; Acton et al., 1982; Manning and Acton, 1980).

Complementary to the LADWP study was the Southern California Edison (SCE) experiment, where 480 residential customers faced time-differentiated rates for electricity for 24 months beginning in March, 1978. Unlike the LADWP experiment, where participation was voluntary, the SCE study had mandatory participation (with a participation incentive), revenue-neutral rates (i.e. a rate schedule designed to balance power-company costs and revenues), and peak/off-peak price ratios that ranged from 3 : 1 to 9 : 1 for a 10-hour peak-period definition. The sample households were spread over three temperature zones, five consumption levels, and four peak/off-peak price ratios. There was a control group of 120 customers facing standard rates; however, 20 of these controls received the same educational material as those in the experimental groups. This pure education treatment was designed to isolate the effects of these materials apart from the TOU rates. Allocation of customers to treatment groups proceeded as in a standard factorial design, but with an overrepresentation of higher-consumption customers relative to the population served by SCE.

In an interesting analysis of the SCE data, which included estimation of elasticities of substitution, Lillard and Aigner (1984) determined that under conditions present in the experiment there would be a revenue gain by SCE (i.e., customers would pay more) from peak-load pricing unless total consumption were reduced by more than 10%. Nevertheless, there are revenue-neutral TOU rate schedules that also result in improvements in customer welfare. When translated into dollar amounts, all of the welfare gains are sufficient to offset the additional metering costs associated with a move to mandatory TOU rates.

3.5. A Social Security Work Incentive Experiment

For some years the Social Security Administration has been planning what, if carried out, will be the largest social experiment yet. Can recipients of Social Security disability benefits be encouraged to go back to work if they are given more time to make the attempt without losing either their income benefits or their entitlement to medical benefits? The Congressional interest in this experiment, expressed explicitly in the enabling legislation, is the saving of monies. Such savings will arise if work attempts are sustained and cash benefits terminated.

The Congressional mandate for the Work Incentive Experiment—the operationalization of policy questions—has several influences on the design of the proposed experiment. First, the need to respond to direct Congressional questions about the effects of the various treatments (benefit/earnings offset, extension of trial work period, extension of Medicare eligibility after return to work, shortening of waiting period for eligibility for Medicare, and some combinations of these) suggests that analyses, no matter how complicated, must be reported in a manner that is understandable by those not technically trained in statistics. In addition, the lack of any *a priori* defensible model for the response of beneficiaries to the experimental manipulations dictated a return to a more traditional orthogonal experimental design (instead of designs based on econometric or regression models such as those used in the other experiments described above). Each treatment group represents a program alternative with short-run Social Security costs greater than those of the existing program. Thus the existence of long-run savings, not the short-run labor-supply response, is the ultimate outcome of interest. This will place considerable pressure on the analysis of the experimental outcomes, because the investigators cannot afford to wait to measure long-run savings directly.

Next, the Congressional mandate specified that the experimental results must reflect possible national implementation—hence a national sample is called for. Luckily such a sample is feasible if it uses the existing national structure of district Social Security offices, and this links to a separate decision to the effect that the treatment should be applied face-to-face by an official of the Social Security Administration. The approximately 1300 such Social Security offices are to be organized into 120 primary sampling units, or clusters, by beneficiary load. Some 15–20 of these (mostly in metropolitan areas) will be chosen for inclusion with certainty; the remainder will be stratified by region, and clusters will be chosen by probability methods for a total of 40–50 clusters. New beneficiaries will themselves be stratified by age and diagnosis, with planned overrepresentation of young people likely to try to return to work. During the year-long enrollment period, a complete replicate of the experiment (4 beneficiary strata by 10 experimental groups) will be chosen within each cluster. Thus the plan is to analyze the results both as an experiment and as a complex survey in order to make population estimates based on finite population sampling.

4. Detailed Design Features: Allocation

Because of the enormously high costs of the large-scale social experiments described in Section 3, from the outset strenuous efforts were made to optimize their design in order to obtain maximum information for the invested resources. Given a large set of possible experimental conditions, this design problem becomes one of choosing the treatment combinations and the families to be assigned to them in the way that will be most informative for policy and that does not violate budget constraints. The choice of treatment combinations was based in each instance (except for the Work Incentive Experiment) on both economic theory and on practical political considerations, and we have summarized this information in Section 3. Here we focus on the allocation problem and how it was addressed in each experiment. For a general introduction to criteria for optimal design allocation see Atkinson (1985, in this volume).

4.1. Negative-Income-Tax Experiments

The design of the negative income tax experiments was based on an assumed linear model for the labor-supply responses. The basic approach was initiated by Conlisk and Watts (1969) for the New Jersey Income Maintenance Experiment and was adopted for use in all of the subsequent negative-income-tax experiments. The regression model used was of the form

$$\mathbf{y} = \mathbf{x}\boldsymbol{\beta} + \mathbf{e}, \quad \text{where } E(\mathbf{e}) = \mathbf{0} \text{ and } \mathrm{Var}(\mathbf{e}) = \sigma^2\mathbf{I}. \tag{4.1}$$

The design problem is to choose in some optimal fashion the rows of $\mathbf{x}$, that is, the treatment combinations to be used and the number and types of experimental units to assign to each. It was assumed that the goal of the experiment is to obtain an accurate estimate of $\mathbf{P}\boldsymbol{\beta}$ (a linear combination of the elements of $\boldsymbol{\beta}$), and that the relative policy importance of the elements of $\mathbf{P}\boldsymbol{\beta}$ could be expressed as a diagonal matrix of weights, $\mathbf{W}$. If c_i is the cost of one observation at the ith design point (i.e. at regressor row $\mathbf{x}_i$), then the budget constraint can be expressed as

$$\sum_{i=1}^{m} c_i n_i \leq C, \tag{4.2}$$

where n_i is the number of observations at the ith design point and C is the total cost. Then the design problem was to find $n_1, \ldots, n_m$ which minimizes

$$\mathrm{trace}[\mathbf{P}'\mathbf{W}\mathbf{P}(\sum_{i=1}^{m} n_i \mathbf{x}_i'\mathbf{x}_i)^{-1}]. \tag{4.3}$$

(e.g. see Conlisk and Watts, 1979). Here all possible linear combinations of $\boldsymbol{\beta}$ are viewed as potentially interesting. This is a version of what is referred to in the optimal-design literature as A-optimality (e.g. see the discussion in Aigner, 1979 and in the paper by Atkinson, 1985, in this volume).

The Conlisk–Watts model embodies certain weaknesses. Not only is it very sensitive to the specification of the response function, but when viewed as a means of stratified sampling for allocation of experimental units to conditions, it has the following additional drawbacks (Morris, 1979): (i) it can use only a limited number of stratification variables, (ii) it requires that potential stratifying variables that are continuous be categorized, (iii) because it operates as if the population of experimental units were infinite, it can allocate more specific kinds of units to treatments than are actually available, and (iv) it focuses on the estimation of the parameters of the response function and not on differences in response among various negative-income-tax programs (Keeley and Robins, 1978).

For the New Jersey experiment the sample was initially divided into three strata according to "normal" income, and then the Conlisk–Watts model was used with the full set of possible payment-by-tax-rate design points illustrated in Table 1 plus a 9th control point, i.e. (0, 0) level. Special allowance was made for attrition of sample families at low payment levels, and higher error variances at high payment levels (Conlisk and Watts, 1979), after the original allocation placed more than $\frac{2}{3}$ of the families in the control condition and very few in conditions with low tax rates. When combined with the stratification of families by pre-experimental "normal" income, even these alterations in the allocation resulted in fewer low-income families being allocated to expensive treatment combinations (because the cost of such treatments for very poor people would be very high). As a consequence, analyses exploring the interactions between preexperimental income and experimental treatment on labor-force response were extremely difficult to carry out.

As Hausman (1981) points out, these and other problems all resulted because family income is intertwined with labor-supply response, i.e. the response variable in the key regression model of expression (4.1). But family income is also related to (i) the definition of the target population or sampling frame (which in New Jersey limited participation to those families with relatively low family incomes), (ii) treatment assignment (with lower-income families being assigned mostly to low subsidy plans, as we just noted above), and (iii) possibly nonrandom attrition. Standard analyses which do not take these linkages into account may lead to badly biased results. Unfortunately all of the negative-income-tax experiments and the housing-allowance demand experiment were afflicted by these problems, even though some of the problems were well understood quite early on (e.g. see Cain and Watts, 1973, Chapter 9).

While the subsequent negative-income-tax experiments all used variants of the Conlisk–Watts allocation model, a radical change was made in the Seattle–Denver experiments. Rather than rely on the quadratic response function with interactions used in the New Jersey experiments, for which the resulting assignment is very sensitive to misspecification, the Seattle–Denver experiments used an ANOVA model with some ad hoc interaction terms (Conlisk and Kurz, 1972; Keeley and Robins, 1978).

4.2. The Housing-Allowance Demand Experiment

As the description of Section 3 suggests, the housing-allowance demand experiment in Pittsburgh and Phoenix was complex, with 17 different treatment conditions plus a control group. Randomization in the demand experiment occurred in the first stage of a multistep selection process leading to the final allocation of families to experimental treatments. The full process began with a screening interview of approximately 40,000 households at each site. Then those families that appeared to be eligible for an allowance program were randomly assigned to either the set of 17 treatment plans or the control group. Those allocated to the treatment group were then randomly allocated to one of the 17 treatment plans. The approximately 4000 assigned families at each site were interviewed in depth to obtain demographic, economic, and personal preference characteristics. On the basis of data from these interviews, eligible families were offered enrollment in the housing-allowance program. At each site roughly 1240 families were enrolled in one of the experimental plans, and another 400 to 500 families were assigned to the control group.

Despite a regression-model specification that was planned for the analysis phase, those involved in the design of the experiment at Abt Associates were unwilling to make the sharp delineation of both the objective and response functions required for the Conlisk–Watts optimal allocation scheme. Thus an attempt was made to achieve a balanced allocation among the treatment groups as they would exist *at the analysis stage*. What this means is that a smaller allocation was made to those treatments with a higher payout than to those treatments with lower payouts and more stringent restrictions, because the investigators anticipated higher attrition among the latter families. This scheme implicitly assumes a randomness model for the attrition process both within and between treatments that we find highly questionable.

In the housing-allowance experiment there was a longitudinal followup to assess housing turnover, but as in the negative-income-tax experiments, this longitudinal feature of the anticipated data analysis was not incorporated into the planning when alternative experimental designs were considered. Of course, when assessing housing turnover rates we would be dealing with counting processes rather than continuous response variables, and the standard optimal-allocation results would not really be appropriate. Designs analogous to that for the pure-birth-process example, which we present in Appendix B, are needed for such longitudinal studies.

To summarize, the key feature in the design of the housing-allowance experiment was the pulling back from the Conlisk–Watts optimal-allocation approach, and an attempt to implement a more traditional balanced design (in the analysis phase after allowing for attrition).

4.3. The Health Insurance Study

The design of the Health Insurance Study (HIS) is probably best viewed as a reaction to the deficiencies of the Conlisk–Watts allocation scheme coupled with a desire to use information on a large number of covariates. This led to the development by Morris (1979) of the finite selection model (FSM), so named because it explicitly makes assignments to treatments by selecting subjects from the finite population available for assignment. This model deals directly with objections (i), (ii), and (iii) to the Conlisk–Watts model listed in Section 4.1 above.

The FSM was actually used in two separate modes in the HIS. First it was used to do the *allocation* of numbers of subjects to the eleven treatment plans and two control groups. As such, this is a modification of Conlisk–Watts to assure that the number of families in a given stratum assigned to achieve the optimum does not exceed the actual number available. The major gains from the FSM come in its second phase, where it was actually used in selecting subjects from a finite pool to fill the allocation plan in such a way as to achieve "balance" on a set of 24 covariates such as wage and nonwage income, family size, health status, insurance, etc. The FSM used in this second phase can be summarized as follows.

We have pre-experimental observations on N families as entries in an $N \times k$ matrix $\mathbf{X}_N$, $k - 1$ columns for the covariates ($k - 1 = 24$ in the HIS) and one for fitting the constant term in the model

$$\mathbf{Y} = \mathbf{X}_N \boldsymbol{\beta} + \mathbf{e}, \tag{4.4}$$

with usual assumptions about $\mathbf{e}$. In the Conlisk–Watts allocation we assign families at random to the treatment groups within strata. In the FSM we improve upon this random allocation, by allowing the treatment groups to take turns in selecting a "cost-effective" family from the remaining pool of unselected families. At each stage the cost-effective choice is determined by minimizing a trace criterion based on the linear model (4.4) similar to that which would be associated with a Conlisk–Watts-like allocation scheme were it able to handle a large number of covariates. For further details on the choice criterion, and a discussion of selection order and other refinements of the basic methods, see Morris (1979). Using the FSM in the HIS actually led to a 25% improvement over random allocation as measured by a weighted sum of variances of the coefficients in the regression model (4.4), as well as variances of other interesting linear combinations of coefficients. Furthermore, the achievement of balance of covariates across treatments is a fundamentally important aspect of the FSM.

In summary, we note that the FSM used for design allocation in the Health Insurance Study overcame some of the criticisms of the Conlisk–Watts approach used in the negative-income-tax experiments. The approach still depended on the proper specification of a complex linear model, and the possibilities for bias were many.

4.4. Time-of-Use Electricity Pricing Experiments

4.4.1. *Los Angeles Department of Water and Power (LADWP) Design.* The sample was drawn in several stages beginning with the frame of approximately 960,000 residential customers of LADWP as of late 1975. Primary sampling areas with a total of 280,000 households were selected to represent a diversity of income and ethnic groups as well as one census tract from each of the 15 city council districts in Los Angeles. The number of households available for further sampling was reduced at this point by excluding 78,000 customers who had lived at their current address for less than one year.

The remaining 202,000 households were stratified into four groups according to bimonthly power consumption in kWh. Then a sample of 6184 customers was randomly selected by usage strata and weather zones—coastal, valley, and civic center—with high-usage customers having a higher probability of remaining in the sample than low-usage customers. Finally, participating households were selected from the sample of 6184 customers by utilizing a regression model with the FSM—as in the Health Insurance Experiment—to ensure that each TOU rate group was balanced across such covariates as average monthly consumption, household income, and the presence of electric air conditioning.

Roughly one customer out of six—1,093 out of 6,184—was placed on one of 40 experimental rate schedules. In addition, there was a control group of 175 customers who simply continued on the existing declining block rate schedule. The experimental rate schedules consisted of 34 nonseasonal TOU rates, 2 nonseasonal flat rates, and 4 seasonal flat rates. (Three unique flat rates are represented in each season and in various combinations to produce the four seasonal and two nonseasonal flat-rate schedules.) The TOU and flat rates were designed to permit calculation of price elasticities of demand by time of use, both within and across groups.

4.4.2. *Southern California Edison (SCE) Design.* As indicated previously, the SCE design was a factorial experiment focusing primarily on comparisons of peak with off-peak power consumption. Thus it differed from the LADWP design in that the sample allocation was not optimized for estimation within an *a priori* specified model incorporating covariates. In addition, participation in the SCE experiment was intended to be mandatory, though the California Public Utilities Commission eventually allowed exclusion of unwilling participants. In fact, there were very few refusals, possibly as a result of lack of publicity for the exemption procedure at the time of initial customer contact. Thus any behavioral interpretations of response to the SCE rate schedules must be further contrasted with responses in the LADWP experiment, where participation was voluntary. It should also be noted that voluntary participation in LADWP seems inappropriate for the evaluation of a program that was intended to be mandatory. (Of course, there are ethical question connected with any sort of mandatory participation, some of which are discussed in Gilbert, Light, and Mosteller, 1975, Section VI.)

In contrast to coordinated multi-site experimentation in medical-intervention trials, the decidedly different designs and participation conditions just between LADWP and SCE indicate that there is no analogue of the site-to-site comparability of large-scale medical studies in the TOU experiments. Furthermore, SCE and LADWP are only two out of 15 TOU experiments, each of which operated as an autonomous study. Of particular significance is the fact that, unlike SCE and LADWP, many of the experiments had only one set of TOU prices, thereby limiting references to a single statistical comparison of control-group and experimental household responses. Here the Department of Energy goal of generalizability of conclusions is placed in serious jeopardy simply as a result of the lack of national coordination of the TOU experiments.

5. Analytical Strategies

5.1. Descriptive Models vs. Specifications Embedded in Formal Theories

A common feature of many analyses of the experiments described in Section 3 is that the primary dependent variables—e.g. labor supply as measured by hours of work per week in the income-maintenance experiments—were viewed as being related to a vector of independent variables, **x**, via systems of equations which were *not* grounded in formal economic theories. While economically plausible arguments, frequently quite detailed, were put forth in support of particular variables as important determinants of the dependent variables, the economic theory stopped influencing the analytic framework at this point. Standard statistical models, incorporating the designated independent variables, were actually fitted to data from the experiments, and interpretation of the influence of these variables was then based on parameter estimates and variability assessments within these ad hoc (relative to any economic theory) models. In our opinion this practice leaves open the question of economic and behavioral interpretations of the estimated models, a particularly troublesome situation in that it is precisely such interpretations which are of central importance for explanations of *why* particular interventions do or do not achieve desired goals.

An exception to this practice is Lillard and Aigner's (1984) estimation of elasticities of substitution in the Southern California Edison (SCE) time-of-use (TOU) pricing experiment. They assumed *a priori* that households behave *as if* they maximize a household utility function. In turn, they assumed that the utility function is homothetic (a form of scale invariance) and both separable and homogeneous of degree 1 in the two variables, peak and off-peak kilowatt hours of power consumed. The homothetic separability assumption has important interpretive consequences in that it is a necessary and sufficient condition for the decomposition of the household consumption decision into two stages: (i) the allocation of total consumption expenditures among electricity and other items, (ii) within the electricity expenditures, the allocation of consumption by time of day. Although homotheticity and separability are, in principle, assumptions capable of empirical examination, the SCE data do not allow for such testing. Nevertheless these assumptions were maintained in the analysis. This inadequacy of the SCE data is a consequence of the fact that the experiment was not designed with estimation of elasticities of substitution and testing of functional forms of utility functions as primary objectives.

In addition to an assumed homothetic and separable direct utility function, a constant-elasticity functional form was assumed in the indirect utility function. This leads to a representation of the logarithm of the ratio of peak to off-peak power consumption as a linear function of the logarithm of the ratio of peak to off-peak prices. The intercept and slope in this linear relationship are household and time dependent. Furthermore, the negative of the slope can be interpreted as the elasticity of substitution between peak and off-peak kilowatt hours. The intercept and slope were then decomposed into a linear combination of weather variables, dummy variables for central or room air conditioning, unobserved variables identified with "unmeasured household characteristics that affect the household's responsiveness to temperature changes over time," and interaction terms among these variables. Parameter estimation was facilitated by introducing disturbance terms in the linear equation relating power consumption and prices and in the linear decomposition of the coefficients. The disturbance terms and unmeasured variables were assumed to be jointly normally distributed. This distributional assumption is not grounded in economic theory; its primary role is to facilitate estimation of structural parameters such as the elasticity of substitution and correlations relating appliance choice to time-of-day power consumption.

Among the findings based on the modeling framework outlined above and the SCE data are:

1. TOU pricing offers greater value for the largest customers, especially those consuming 8880 kWh or more per year. These customers have elasticity anywhere from 2 to 4 times the elasticities for other groups in the sample. Typically, but not always, the elasticity is greatest for those without air conditioning in cool climates and those with central air conditioning in hot climates.
2. Correlations relating appliance choice to TOU consumption—a relation involving unobserved variables—are jointly and individually insignificant. This suggests that a simpler model specification than the multilevel system described above can be estimated, with appliance terms taken as given, without introducing bias in the coefficient estimates due to the endogeneity of air-conditioner ownership and the fact that both it and TOU consumption are dependent upon temperature level, household responsiveness to temperature variation, and other household characteristics.

In addition to the fact that these conclusions are based on a homothetic, separable direct utility function with corresponding indirect utility containing a constant-elasticity functional form, strong untested distributional assumptions were imposed on the unobserved variables and residuals. In particular, joint normality of residuals and unmeasured household characteristics which influence household responsiveness to temperature changes over time was an assumption made to simplify maximization of a likelihood and testing for correlations between residuals (when viewed as random variables) and unmeasured household characteristics. Normality as such has nothing in particular to do with response to electricity pricing plans. Thus a question of dependence of elasticity estimates on the normality assumption is raised, suggesting an alternative strategy and a currently unresolved research problem. In particular, if an elasticity of substitution within a homogeneous degree 1 utility function is the primary focus of interest, then we may view other parameters in the model specification for power consumption as incidental parameters. Then we require an estimation algorithm for the elasticity that is insensitive to a wide range of functional forms that can be assumed for distributions of residuals and unmeasured variables. Such methods have been put forth in the literature on discrete choice processes in Cosslett (1983), for example, but have thus far not been widely used in empirical analyses (an exception is the work of Gallant and Koenker, 1983). The estimation of elasticities in TOU pricing experiments would seem to be an ideal site for applying such technology, particularly since economic theory does not currently provide sharp restrictions on distributions of residuals and unmeasured variables.

5.2. Longitudinal Analyses

Although all of the major social experiments had longitudinal data-collection components, none of them based their designs on dynamic models of a form that subsequently appeared in analyses. In particular, economic theory guided the specification of counting-process models in analyses of the demand for episodes of medical treatment in the Health Insurance Study (HIS) (Keeler and Rolph, 1982). Important conclusions from these analyses were: (i) people do not appear to change the timing of medical purchases to reduce costs; (ii) the cost of episodes was the same on all experimental plans, thereby greatly simplifying the future simulation and prediction of medical expenses.

The initial analyses of the HIS estimated the effects of the plan on annual expenses. However, an interest in assessing the effects of insurance with scopes of coverage, bases for copayment, or deductibles differing from those in the experiment itself requires that we know how decisions to buy medical services during the year are actually made. Such decisions are analyzed in terms of episodes that contain all of the spending associated with a given bout of illness, a chronic condition, or an elective medical procedure.

Individuals were assumed to experience a Poisson-distributed number of episodes in a year with expected value λ_i for person i given by a multiplicative regression equation with

$$\ln \lambda_i = \mathbf{X}_i'\boldsymbol{\beta} + \varepsilon_i \tag{5.1}$$

where $\mathbf{X}_i$ is a vector of attributes—including the insurance plan—and $\exp(\varepsilon_i)$ was assumed to be gamma-distributed. Thus the population-level description of the episode counts is a negative binomial distribution. This kind of specification is supported by tests of the count data that indicate:

1. Episodes occur "randomly" over time.
2. The mean number of outpatient episodes on the free plan is 4.98 with a standard deviation of 4.26—clearly indicative of a non-Poisson model. A homogeneous Poisson distribution with this mean should have standard deviation $= 4.98^{1/2} = 2.23$ rather than 4.26. A plausible explanation for this discrepancy from a Poisson model is that the observations come from a heterogeneous population with each individual having his/her own Poisson rate.

In addition to the conclusions (i) and (ii) mentioned above on timing and costs, analysis based on the mixed Poisson specification suggests that although price affects the number of episodes experienced by participants, it has little effect on the cost of each episode. Furthermore, episode rates differ most between full coverage and other plans as a group, with less difference between low and high levels of coinsurance. Health status, age, sex, and doctor visits the year before the experiment are the most important determinants of episode rates.

These results from a longitudinal analysis could have been maximally defensible for a given budget of observation if the data collection had been optimized for estimation and hypothesis testing within the class of mixtures of Poisson processes. That optimal designs for counting processes can make a genuine difference in subsequent inferences is clarified by the discussion in Appendix B on optimal designs for pure birth processes. It should also be emphasized that longitudinal analyses from nonoptimal data were carried out in the Seattle and Denver income-maintenance experiments (SIME/DIME) in studies of marital dissolution by Hannan, Tuma, and Groeneveld (1976, 1977). As in the longitudinal analyses of the HIS, fine-grained analyses would have been far more defensible if marital dissolution as part of a dynamic process had guided the SIME/DIME designs.

5.3. Qualitative Distinctions Between Statistical Models and Models Derived from Economic Theory

Our emphasis on the desirability of grounding models utilized in social experiments in formal behavioral theory is based on much more than the predictive uses of the models emphasized above. In particular, assessments of the influence of observed covariates on a dependent variable (such as the number of episodes of medical treatment in a year or the duration of a spell of unemployment) are sensitive to the functional forms of the models. Thus conclusions about the influence of marital status on the duration of a spell of unemployment in a standard statistical model can—depending upon your *belief* in a class of models—be viewed as an artifact of the model. With a specification that is derived from economic theory and that provides an equally good numerical description of observed durations, the conclusions can be the opposite of those arrived at through the use of a statistical model. (See the related discussion in the paper by Ericsson and Hendry, 1985, in this volume.)

To clarify how such reversals of conclusion can arise, we consider an experiment designed to see whether it is possible to reduce the durations of spells of unemployment in a particular population. Suppose that the treatments are in the form of assistance in the job search process so that different treatment levels give rise to different reductions in the cost of search. A conventional statistical model of the distribution of the length of a spell of unemployment imposes a multiplicative structure such that

$$\log \operatorname{Prob}(\text{duration of spell} > t) = -H(t)U(\mathbf{x}), \tag{5.2}$$

where $H(t)$ is the cumulative number of terminations in a population during a time interval of length t, and $U(\mathbf{x})$ is a nonnegative function of observed covariates including the cost of search. A standard specification is

$$U(\mathbf{x}) = \exp(\mathbf{x}'\boldsymbol{\beta}); \tag{5.3}$$

e.g., see Lancaster and Nickell (1980).

The specification in (5.2) and (5.3) should be contrasted with one from an optimizing model—described in more detail in Appendix A—and given by

$$\log \operatorname{Prob}(\text{duration of spell} > t) = \lambda t[1 - F(rV)], \tag{5.4}$$

where

$$\begin{aligned} \lambda &= \text{rate of confrontation of an individual with job offers,} \\ F(w) &= \text{distribution of wage offers,} \\ r &= \text{rate of interest,} \\ V &= \text{value of search.} \end{aligned}$$

The value of search V is functionally dependent on the cost of search via the relationship

$$c + rV = \frac{\lambda}{r}\int_{rV}^{\infty} (w - rV)\, dF(w), \qquad V > 0, \tag{5.5}$$

where c is the cost of search per unit time. The essential point here is that c is one of the covariates in the statistical model and influences the duration of a spell in an algebraically different form from its implicit appearance in a specification derived from economic theory. Furthermore c may itself be dependent on marital status. Thus, we may make qualitatively different inferences about the impact of marital status, depending upon which model is utilized in an analysis. The central message of these alternative specifications is that a multiplicative structure like that in the statistical model is virtually never a consequence of economic theories of search during unemployment. It is this distinction that should force us to exercise considerable caution in the interpretation of analyses of data in social experiments.

Of course, as we have already noted, economic theory is often assumption laden. For example, the assumption that individuals participating in a social experiment are as dedicated to expected-utility maximization as are many econometricians should be of as much concern to us as the differences between the conclusions of such optimizing models and statistical models. The *satisficing* approach to utility functions (e.g. see Simon, 1957, pp. 241–260) puts less demands on individuals as actors, but even it may not be an adequate assumption (e.g. see the discussion of this point in Pottinger, 1983).

6. Alternatives to Large-Scale Experiments

We have seen that large-scale social experiments are costly and time consuming. The social cost may well have to be borne as solutions are sought for long-standing and severe social problems (Gilbert, Light, and Mosteller, 1975). Nonetheless, it is legitimate to inquire if there are alternatives to the kinds of large-scale social experiments described in this paper.

At the outset, we note that we do not see observational or nonrandomized trials as substitutes for randomized experiments. The evidence is rather clear that nonrandomized trials give little information about cause-and-effect relations. Thus, while such trials are sometimes necessary in situations where randomization is impossible, we believe they are to be avoided whenever alternatives exist. The kind of alternatives we do have in mind are combinations or series of smaller experiments as substitutes for single large-scale studies.

6.1. Combining the Results of Several Experiments

Combining results from many experiments has a long history in agriculture (e.g. see Cochran 1937, 1954), in bioassay (Bliss, 1952), and in high-precision measurement in physics (Cohen, Crowe, and Dumond, 1957). The existing procedures for combining results depend on the lack of interactions between treatment and experimental site, and whether the observations can be considered to measure the same quantity from site to site and are of equal precision. As interaction increases and precisions become less equal, the required methodology—much of which remains to be developed—becomes more complicated. In addition, most existing methods assume that the experiments combined are a random sample of all experiments on which we seek information—a rather tricky assumption at best. Recent uses of this kind of meta-analysis in education, psychology, and medicine are reviewed in Hedges and Olkin (1981) and Glass, McGaw, and Smith (1981). An extended example with commentary is presented by Rosenthal and Rubin (1978).

A new development in the context of social experiments is the adaptation of multilevel analyses by Aigner and Leamer (1984) to support transferability of findings from one site to another in the TOU pricing experiments. Their methods are based on a random-coefficient linear regression model. The idea is that each of the utility companies under consideration is assumed to have a separate set of parameters in the linear function. If there were no relationship between these parameter sets, then each utility would be forced to estimate its own model separately, and utilities not involved in experiments could not infer the responsiveness of their customers to TOU pricing from the experience of others. On the other hand, if parameters were identical across utilities, this would suggest that data from all utilities could be pooled to form a single estimate of the effect of TOU pricing that would apply equally well to utilities involved in experiments and those which did not conduct such studies.

In the random-coefficient model, the degree of similarity in corresponding parameters is determined by the variance—across utilities—of each coefficient. If this variance is negligible, we say that data transfer is maximal and that conclusions are generalizable from one utility to another *and* to new sites. If the variance is very large, then coefficients are viewed as dissimilar and data transferability is not justified. In Aigner and Leamer's analysis, different coefficients exhibit different degrees of transferability, the degrees determined by relative magnitudes of coefficient variances across utilities.

The same kind of strategy has also recently been employed by DuMouchel and Harris (1983) for combining evidence from cancer experiments. In addition, the random-coefficient framework has been used as the basis for a comparative analysis of fertility patterns across 30 developing countries by Hermalin and Mason (1980) and Mason, Wong, and Entwistle (1984). Although the fertility analyses are based on data from the World Fertility Survey (see the article by Macura and Cleland, 1985, in this volume) and not from randomized experiments, the exemplary attention to integration of qualitative evidence, social and demographic theory, and empirical data could serve as a model for future comparative analyses of randomized social experiments. For further insight on combining evidence from multiple sources see Mosteller and Tukey (1982, 1984).

6.2. Evolutionary Designs

The procedures we have been describing that arise in the social sciences combine results of experiments as found in the literature, and hence as carried out by various investigators. Thus they embody only a disorderly kind of cumulation. Would it not be more useful to plan for a more orderly cumulation and for more systematic methods of sequential knowledge building? One such systematic experimental program is known as evolutionary operation (EVOP)—e.g. see Box and Wilson (1951), Box and Draper (1969), and Box (1978)—and is most usually used in designing experiments (as in industrial applications) for which the outcomes are available rapidly. The EVOP philosophy is to use the results of a subexperiment both to check the appropriateness of the model specified for that subexperiment and to suggest the most efficient way to improve that model for a subsequent subexperiment. (A key feature in EVOP is that the changes made are sufficiently small that the production of on-specification material continues; the analogue in a social experiment setting is unclear.) Similar evolutionary procedures have long been used in animal breeding programs (Cochran, 1951) and also in educational research under the rubric of "formative" (rather than "summative") evaluation (Scriven, 1969). An example is the work of Tharp and Gallimore (1979), who applied a sequential approach to explore educational alternatives for underachieving Hawaiian native children. Compared to EVOP as applied to large-scale experiments, the Tharp–Gallimore project was much smaller in scale, used less formal and less complicated statistical technology, and admitted more intuitive elements into its evidential base. But the basic idea of letting data from one phase of the investigation formally shape succeeding phases is similar, and the method has led to the development of educational programs that appear to work.

It has been proposed (Madansky, 1980) that sequential evolutionary designs, which are somewhere between the broad general-purpose designs that assume little or nothing about the response function and the tightly specified optimal designs, might be adapted to the kinds of social experiments described in this paper. These adaptive designs are suggested particularly by the fact that an intervention may only have a substantial impact on a particular subpopulation, whose importance is detected during early (preliminary) analysis of the data. This information could be used to target a second allocation of cases into an experiment which is refocused to validate further the impact of the intervention on that more narrowly defined subpopulation. This might be viewed as an instance of improvement by means of selection (see e.g. Cochran, 1951), where one really wants to ascertain in which subgroup a particular intervention will have maximal response. A corollary advantage of adaptive designs is that with a fixed total budget of allowable individuals in an experiment, more of the cases can be allocated to treatments and subpopulations where there appear to be genuine effects to measure. A full assessment of the potential gains and losses of adaptive vs. one-shot *a priori* fixed plans in a specific setting has yet to be carried out. In our view this is a central statistical research problem, and information available from the income-maintenance or health-insurance experiments, for example, could provide a basis for such an assessment.

Any serious attempt to apply adaptive designs, in the spirit of EVOP, to social experiments would certainly further increase the already long time necessary to obtain results, for each of the subexperiments would have to run for some time before its outcome could be determined for use in planning the next subexperiment. Indeed, funding agencies and policymakers can be expected to be reluctant to sign a blank check for an unspecified series of experiments.

A modification of the usual EVOP procedure might meet this objection if it embodied more careful contingency planning in which a series of "if, then" scenarios would sketch out possible experimental paths and treatment modifications. A great deal of basic statistical research would then be required to determine the properties of such an approach. But if this research could be accomplished, the modified EVOP strategy approach would be useful in estimating the real responses to possible variations in policies, in the long run, while offering some interim results that could be useful in more immediate policy planning.

Each of the major experiments discussed in Section 3 had a fixed design which dictated the allocation of individual units (persons, households, etc.) to different treatment and control groups subject to cost constraints and a desire for balance and minimum bias. In some cases the allocations were also optimized for precision of comparisons within a multiple regression model. In addition to the fact that the *a priori* modeling framework on which the designs were based tended not to be grounded in a formal economic or sociological theory, the experiments ran over several years in nonstationary environments, thereby further suggesting that adaptive designs might be fruitful alternatives to consider for future social experiments.

6.3. Nonstationary Environments

Because social experiments run for several years in a nonstationary environment, the impact of this nonstationarity on experimental designs requires some focused commentary. In most of the literature on randomized designs it is presumed that the conditions under which the study takes place are constant or, if time varying, at least under the control of the experimenter. The dramatic shift in labor-market conditions in Seattle, Washington during the Seattle Income Maintenance Experiment as a result of massive layoffs at the Boeing Aircraft Company is an example of the imposition of a nonstationary shift in community conditions, far beyond the control of the experimenters. It is this kind of dynamic which requires monitoring and, to the extent possible, formal statistical modeling if we are to assess its impact on key dependent variables—such as labor supply—in an experiment. Two of the questions that arise for labor-force experiments or any social experiment which must be conducted, from the point of view of the experimenter, in random nonstationary settings are:

(i) Should there be a shift in the experimental protocol to compensate for the environmental changes, and how can such action be rationalized?

(ii) With defensible theories about the impact of a rapidly rising local unemployment rate on individual job-search behavior, can we develop good designs for experiments in which the allocation of persons to treatments is staged over time but that still lead to defensible comparisons between treatment and control groups? A similar question arises for utilization of health services, and also for propensity to shift housing in a national environment of rapidly fluctuating interest rates even when guaranteed protection is provided in an experimental protocol, etc.

The intrinsically dynamic character of these questions automatically suggests the need for consideration of longitudinal designs in the planning of social experiments. If optimal designs, analogous to the regression designs used in the income-maintenance and health insurance experiments, are to be employed, then some *a priori* specification of classes of descriptive and/or behavioral based stochastic process models is required. With this requirement we approach some nearly uncharted statistical territory. In particular, the literature on optimal designs for even restricted classes of time-homogeneous process models is very limited, and it pays virtually no attention to covariates of the kind that have played a central role in the planning of major social experiments to date. However, adaptive designs for counting processes (as used in the HIS), diffusion processes, and mixed discrete-state–continuous-state vector processes present the principal research challenge for more flexible social experimentation.

Appendix A. New Issues that Arise in Formulating and Estimating Choice-Theoretic Models

In Section 5.3 we presented a prototypical example of a duration model specification arising from a formal behavioral theory. The principal feature of this model is that it is qualitatively different in functional form from any member of a broad class of commonly used statistical models. Consequently, conclusions about the influence of observed covariates on the primary dependent variable in a social experiment can differ completely, depending upon which modeling framework one imposes in an analysis. Because few specifications derived from a formal theory have been used in analyses of social experiments to date, we elaborate in somewhat more detail on the underlying assumptions and distinguishing features of such a theory-based modeling approach.

Consider a time-homogeneous environment in which agents are assumed to be income maximizers. For an unemployed individual engaged in a job search, at constant cost c per unit of time, we assume that he/she is confronted with job offers at the jump times of a Poisson process with parameter λ and that successive wage offers are independent random variables from a common wage distribution $F(w)$, common to all agents. Once refused, a given offer is no longer available to that agent, and we assume that on-the-job search is not allowed. Finally, we assume an instantaneous constant rate of interest, r.

In this setting, the value of search, V, is determined via Bellman's optimality principle for dynamic programming (see Heckman and Singer, 1984, for details) and satisfies the relation

$$c + rV = \frac{\lambda}{r} \int_{rV}^{\infty} (w - rV)\, dF(w) \quad \text{for } V > 0. \tag{A.1}$$

The quantity rV is the reservation wage, i.e. the minimal wage offer that the agent will accept to terminate the search process. The optimal search policy is to accept the first wage offer w for which $w \geq rV$. The probability that an offer is unacceptable is thus $F(rV)$. In terms of these quantities, the duration of search, T, is exponentially distributed with hazard rate

$$h(t) \equiv \lambda(1 - F(rV)) \quad \text{for } t > 0. \tag{A.2}$$

In a more realistic nonstationary environment where offers occur at the jump times of an inhomogeneous Poisson process with intensity function $\lambda(t)$, the value of search becomes time dependent and the duration of search is described by

$$\mathbf{P}_{\tau_1}(T > t) = \exp\left(- \int_{\tau_1}^{\tau_1 + t} \lambda(z)[1 - F(rV(z))]\, dz \right) \tag{A.3}$$

for a spell of unemployment which began at calendar time τ_1 (see Heckman and Singer, 1984, for further details on such models).

Let us focus on the expression (A.2). In most empirical analyses it is plausible that c, r, λ, and F all depend on observed and unobserved explanatory variables. Introducing such variables into an econometric specification, however, raises several issues that more conventional analyses, based on descriptive models, tend to ignore:

(i) Current microeconomic theory provides no guidance on the functional forms of the functions c, r, λ, and F other than the restriction given by Equation (A.1). Both model identifiability criteria and parameter estimates tend to be very sensitive to the choice of functional form.

(ii) In order to impose restrictions produced by economic theory, it is necessary to solve the nonlinear equation (A.1). The constraint $V > 0$ is critically important, since $V \leq 0$ implies that an unemployed individual will not search. Furthermore, restrictions such as (A.1) virtually guarantee that proportional-hazard specifications—a standard family of descriptive model used to study the impact of covariates on durations of spells of unemployment (see Cox, 1972, or Kalbfleisch and Prentice, 1980)—will *not* be produced by economic theory.

(iii) In the search model without unobserved variables, $w \geq rV$ is an essential piece of identifying information. In a model with an unobservable θ, the restriction $w \geq rV$ is replaced with an implicit-equation restriction on the support of the distribution of θ. In particular, for an observation with accepted wage w and reservation wage $rV(\theta)$, the admissible support set is $\{\theta : 0 \leq rV(\theta) \leq w\}$. Thus, in a duration model produced from economic theory, not only is the conditional hazard $h(t \mid \mathbf{x}, \theta)$—where $\mathbf{x}$ is a vector of observed convariates—unlikely to be of the proportional-hazard functional form, but the support of the distribution of θ also will depend on parameters of the model. This is a qualitatively different kind of specification from the usual descriptive statistical models where the support of the distribution of θ is fixed *a priori*, and the class of distributions utilized in empirical studies is chosen on the basis of analytical and computational convenience alone. In addition, a new research area arises immediately from these considerations, since virtually nothing is known about model identification and properties of estimators for distributions of unobservables with implicitly defined support sets.

The important feature of this example for the analysis of data collected in social experiments is the clarification of the considerable microeconomic theory and statistical technology that must yet be developed in order to provide *defensible* behavioral interpretations for parameters estimated in models incorporating important observed and unobserved covariates.

Appendix B. Optimal Designs for Restrictive Classes of Descriptive Stochastic Process Models

The ideal form of data to be used in the development of statistical models for microdynamic processes is a continuous record over a substantial time interval $[0, T]$. The mechanism for collecting such information in either social experiments or social observational studies is usually the administration of an interview at a discrete series of *equidistant* time points $0 = t_0 < t_1 < \cdots < t_n \leq T$, with $t_{i+1} - t_i = \Delta$ for $i = 1, 2, \ldots, n-1$, and then the construction of a continuous record on one or more variables through retrospective questions. Depending on the particular variable, retrospective questions may lead to substantial measurement error in the responses, thereby leaving the investigator with confidence only in information concerning events (or states of occupancy) occurring within very close proximity of the interview date. In this situation, we are faced with the issue of "optimal" timing of interview dates to facilitate parameter estimation and other forms of statistical inference within restrictive, but *a priori* plausible, classes of stochastic process models. This design issue was relevant for each of the social experiments discussed in Sections 3 and 4; however, the regression models imposed as the central analytic framework at the outset of planning virtually forced the data-collection designs to bypass important dynamic considerations of the sort discussed in the review paper by Titterington (1980) on optimal design for dynamic systems.

We illustrate the kinds of design considerations that are important in the planning stage of social experiments when a restrictive class of stochastic process models is set forth and defended as a reasonable class of models within which to embed comparisons of interest. To focus attention on basic principles we suppose that the pure birth processes are viewed as a reasonable class of models to describe a counting process. This reduces our estimation problem to consideration of a single parameter, λ, governing the evolution of a continuous-time Markov process with intensity matrix $\mathbf{Q}$ having entries

$$q_{ij} = \begin{cases} i\lambda & \text{if } j = i + 1 \\ -i\lambda & \text{if } j = i \\ 0 & \text{otherwise} \end{cases}, \qquad i = 0, 1, 2, \ldots. \tag{B.1}$$

If observations on sample paths $X(t)$, $0 \le t \le T$, are restricted by cost considerations to consist of $\{X(t_k)\}$, $0 = t_0 < t_1 < \cdots < t_n \le T$, with no retrospective information, then we ask for a specification of $t_1, t_2, \ldots, t_n$ such that the Fisher information about λ in the sample is maximized, and then refer to such a design as optimal for given n and T. Becker and Kersting (1983) show that:

(i) If $\lambda T \le 1$, then event history data, optimal point sampling, and equidistant sampling yield nearly equivalent information; however, not much can be learned about λ in this situation.
(ii) If $\lambda T > 1$, then the payoff for optimal point sampling is substantial relative to equidistant sampling, as reflected in Table 3.
(iii) The observation times for optimal point sampling are given, approximately, by

$$t_i = \frac{3}{\lambda} \ln\left(1 + \frac{i}{n}\left[\exp\left(\frac{\lambda T}{3}\right) - 1\right]\right), \qquad 0 \le i \le n. \tag{B.2}$$

Since, in the best of situations, we will not know λ precisely, either we can specify a distribution for λ and then use a Bayesian criterion, or we can attempt to constrain λ *a priori* to lie in an interval $[\Lambda_1, \Lambda_2]$. In the latter case, our objective is not to stray too far from the optimal-design results for the true

Table 3. Loss of Information if One Uses the Equidistant Design or Optimal Point Sampling Relative to Continuous Observation in $[0, T]$[a]

	$\lambda T = 5$		$\lambda T = 10$		$\lambda T = 20$	
n	IO/IC	IE/IC	IO/IC	IE/IC	IO/IC	IE/IC
2	.737	.601	.651	.171	.648	.005
5	.953	.921	.923	.724	.927	.304
10	.988	.971	.981	.921	.980	.724

[a] IC = Fisher information for continuous observation; IE = Fisher information for $t_i = iT/n$, $i = 0, 1, \ldots, n$, in equidistant sampling; IO = Fisher information for optimal point-sampling design.

but unknown value of λ. One way to proceed is to choose observation times $\{t_i\}_{i=0}^n$ using expression (B.2) for a value of $\lambda^* \in [\Lambda_1, \Lambda_2]$ such that

$$\min_{\lambda \neq \lambda^*} \frac{I_0(\lambda^*; \lambda)}{\mathrm{IO}(\lambda^*)} \tag{B.3}$$

remains above a preassigned but achievable level. Here $I_0(\lambda^*; \lambda)$ is the Fisher information in a sample using an optimal design for a parameter value λ^* when λ is the true value, and $\mathrm{IO}(\lambda^*) = I_0(\lambda^*; \lambda^*)$ is the Fisher information in the optimal point-sampling design with λ^* as the given true parameter value. Of course what we are aiming for is that value λ^* for which

$$\max_{\lambda^* \in [\Lambda_1, \Lambda_2]} \min_{\lambda \neq \lambda^*} \frac{I_0(\lambda^*; \lambda)}{\mathrm{IO}(\lambda^*)} \tag{B.4}$$

is achieved.

Although the above discussion is specialized to pure birth processes, the considerations leading to selection of a longitudinal design are quite general and in need of development for broader and more elaborate classes of models.

Acknowledgements

In the process of gathering materials and writing drafts of this manuscript we received the generous cooperation of so many individuals that it is now difficult to construct a list of those to whom we are indebted. In particular, the authors of various reports, analyses, and critiques of the experiments discussed in this paper were generous well beyond reasonable expectations, supplying us with mountains of documents. Shan Nelson-Rowe and Beth Stasny helped us by summarizing various materials. We also received valuable comments on earlier drafts of this paper from Dennis Aigner, Anthony Atkinson, Barry Bye, James Heckman, William Kruskal, Carl Morris, Joseph Newhouse, S. James Press, and John Rolph.

This research was supported in part by the National Science Foundation under Grant No. SES-8119219 to Carnegie-Mellon University, and Grant No. SES-8119138 to the Research Foundation of the State University of New York, and in part by the National Institutes of Health under Grant No. NIH-1-RO1-HD16846-01.

Bibliography

Acton, J. P. et al. (1982). *Promoting Energy Efficiency Through Improved Electricity Pricing: A Mid-Project Report*, N-1843-HF/FF/NSF. Santa Monica, CA: The Rand Corporation.

Acton, J. P., Manning, W. G., Jr., and Mitchell, B. M. (1978). *Lessons to be Learned from the Los Angeles Rate Experiment in Electricity*. Report R-2113-DWP. Santa Monica, CA: The Rand Corporation.

Acton, J. P. and Mitchell, B. M. (1979). Evaluating Time-of-Day Electricity Rates for Residential Customers. Report R-2509-DWP. Santa Monica, CA: The Rand Corporation.

Aigner, D. J. (1979). "A brief introduction to the methodology of optimal experimental design." *J. Econometrics*, **11**, 7–27.

Aigner, D. J. (1982). "The residential electricity time-of-use pricing experiments: What have we learned?" In J. A. Hausman and D. Wise (eds.), *Social Experimentation*, NBER conference volume.

Aigner, D. J. and Leamer, E. (1984). "Estimation of time-of-use pricing response in the absence of experimental data: An application of methodology of data transferability." *J. Econometrics*, **26**, 202–227.

Archibald, R. W. and Newhouse, J. P. (1980). *Social Experimentation: Some Whys and Hows*. Report R-2479-HEW. Santa Monica, CA: The Rand Corporation.

Atkinson, A. C. (1985). "An introduction to the optimum design of experiments." In this volume, Chapter 20.

Berry, S. H. (1979). *Conducting a Survey Using the Client's Staff: Evaluation of Interviewer performance in the Electricity Rate Study*. Report R-2223-DWP. Santa Monica, CA: The Rand Corporation.

Becker, N. G. and Kersting, G. (1983). "Design problems for the pure birth process." *Adv. in Appl. Probab.*, **15**, 255–273.

Bliss, C.I. (1952). *The Statistics of Bioassay*. New York: Academic.

Borjas, G. J. and Heckman, J. J. (1978). "Labor supply estimates for public policy evaluation." In *Proceedings of 31st Annual Meeting of Industrial Relations Research Association*, 320–331.

Boruch, R. F. (1975). "On common contentions about randomized field experiments." In R. F. Boruch and H. W. Riecken (eds.), *Experimental Tests of Public Policy*. Boulders, CO: Westview, 108–145.

Box, G. E. P. (1978). "Experimental design: Response surface methodology." In W. H. Kruskal and J. M. Tanur (eds.), *International Encyclopedia of Statistics*, **1**. New York: Free Press, 294–299.

Box, G. E. P. and Draper, N. R. (1969). *Evolutionary Operation: A Statistical Method for Process Improvement*. New York: Wiley.

Box, G. E. P. and Wilson, K. B. (1951). "On experimental attainment of optimal conditions." *J. Roy. Statist. Soc. Ser. B*, **13**, 1–45.

Bradbury, K. and Downs, A. (eds.) (1981). *Do Housing Allowances Work?* Washington, D.C.: Brookings Institution.

Cain, G. G. and Watts, H. W. (eds.) (1973). *Income Maintenance and Labor Supply*. Chicago: Markham.

Campbell, D. T. and Stanley, J. C. (1963). "Experimental and quasi-experimental designs for research on teaching." In N. L. Gage (ed.), *Handbook of Research on Teaching*. Chicago: Rand McNally, 171–246.

Cochran, W. G. (1937). "Problems arising in the analysis of a series of similar experiments." *J. Roy. Statist. Soc. Suppl.*, **4**, 102–118.

Cochran, W. G. (1951). "Improvement by means of selection." In *Second Berkeley Symposium on Mathematical Statistics and Probability*, **3**, Berkeley, CA: Univ. of California Press, 449–470.

Cochran, W. G. (1954). "The combination of estimates from different experiments." *Biometrics*, **7**, 101–129.

Cohen, E. R., Crowe, K. M., and Dumond, J. W. M. (1957). *Fundamental Constants of Physics*. New York:McGraw-Hill.

Conlisk, J. and Watts, H. W. (1969). "A model for optimizing experimental designs for estimating response surfaces." In *Proceedings of the Social Statistics Section, American Statistical Association*, 150–156.

Conlisk, J. and Watts, H. W. (1979). "A model for optimizing experimental designs for estimating response surfaces." *J. Econometrics*, **11**, 27–42.

Conlisk, J. and Kurz, M. (1972). *The Assignment Model of the Seattle and Denver Income Maintenance Experiment*. Research Memo 15. Menlo Park, CA: Stanford Research Institute, Center for the Study of Welfare Policy.

Cosslett, S. R. (1983). "Distribution-free maximum likelihood estimator of the binary choice model." *Econometrica*, **51**, 765–782.

Cox, D. R. (1972). "Regression models and life tables (with discussion)." *J. Roy. Statist. Soc. Ser. B*, **34**, 187–220.

Cox, G. M. (1958). "Forward." In *Experimental Designs in Industry*, V. Chew (ed.), New York: Wiley.

Cronbach, L. J. (1982). *Designing Evaluations of Educational and Social Programs*. San Francisco: Jossey-Bass.

Duan, N., Manning, W. G., Jr., Morris, C. N., and Newhouse, J. P. (1983). "A comparison of alternative models of the demand for medical care." *J. Business and Economics*, **1**, 115–126.

DuMouchel, W. H. and Harris, J. E. (1983). "Bayes methods for combining results of cancer studies in humans and other species (with discussion)." *J. Amer. Statist. Assoc.*, **78**, 293–315.

Ericsson, N. R. and Hendry, D. F. (1985). "Econometric modeling of the market for new housing in the United Kingdom." In this volume, Chapter 11.

Ferber, R. and Hirsch, W. Z. (1979). "Social experiments in economics." *J. Econometrics*, **11**, 77–115.

Ferber, R. and Hirsch, W. Z. (1982). *Social Experimentation and Economic Policy*. Cambridge: Cambridge U.P.

Finney, D. J. (1945). "The fractional replication of factorial arrangements." *Ann. Eugenics*, **12**, 291–301.

Fisher, R. A. (1935). *The Design of Experiments*. Edinburgh: Oliver and Boyd.

Fisk, J. D. and Roth, D. M. (1980). *Work Disincentives and Income Maintenance Programs: Review of the Empirical Evidence*. Congressional Research Service, Libary of Congress, Congress of the United States.

Gallant, A. R. and Koenker, R. W. (1983). *Some Welfare Econometrics of Peak-Load Pricing of Electricity: A Continuous Time Approach*. Unpublished technical report. Murray Hill, N. J.: Bell Laboratories.

Gilbert, J. P., Light, R. J., and Mosteller, F. (1975). "Assessing social innovations: An empirical base for policy." In C. A. Bennett and A. A. Lumsdaine (eds.), *Evaluation and Experiment: Some Critical Issues in Assessing Social Programs*. New York: Academic, 39–193.

Glass, G. V., McGaw, B., and Smith, M. L. (1981). *Meta-analysis in Social Research*. Beverly Hills, CA: Sage.

Haggstrom, G. W. (1976). "The pitfalls of manpower experiments." In H. W. Sinaiko and L. A. Broedling (eds.), *Perspectives on Attitude Assessment: Surveys and Their Alternatives*. Champaign, IL: Pendleton.

Haggstrom, G. W., Blaschke, T. J., Chow, W. K., and Lisowski, Wm. (1981). *The Multiple Option Recruiting Experiment*: Report R-2671-MRAL. Santa Monica, CA: The Rand Corporation.

Hannan, M., Tuma, N., and Groeneveld, L. (1976). *The Impact of Income Maintenance on the Making and Breaking of Marital Unions: Interim Report*. Research Memorandum 28. Menlo Park, CA: Stanford Research Institute, Center for the Study of Welfare Policy.

Hannan, M., Tuma, N., and Groeneveld, L. (1977). "Income and marital events: Evidence from an income maintenance experiment." *Amer. J. Sociology*, **82**, 1186–1211.

Hausman, J. A. (1981). "The design and analysis of social and economic experiments." *Proc Internat. Statist. Inst., 43rd Session*, **49** (Book 2), 1084–1108.

Heckman, J. J. (1978). "Comment on 'The labor supply response of wage earners,' by O. Ashenfelter." In J. Palmer and J. Pechman (eds.) *Welfare in Rural Areas*. Washington: Brookings Institution, 138–147.

Heckman, J. and Singer, B. (1984). "Econometric duration analysis." *J. Econometrics*, **24**, 63–132.

Hedges, L. V. and Olkin, I. (1981). "Analyses, reanalyses, and meta-analysis: Review of Glass, McGaw, and Smith." *Contemporary Education Review*, **1**, 157–165.

Hendricks, W. and Koenker, R. (1979). "Demand for electricity by time-of-day: An evaluation of experimental results." Paper presented at Rutgers Conference, *The Demand for and Pricing of Public Utility Services*, October 1979.

Hermalin, A. and Mason, W. (1980). "A strategy for the comparative analysis of WFS data with illustrative examples." In *The United Nations Program for Comparative Analysis of World Fertility Survey Data*. New York: United Nations Fund for Population Activities, 90–168.

Immer, F. R., Hayes, H. K., and Powers, L. (1934). "Statistical determination of barley varietal adaptation." *J. Amer. Soc. Agronomy*, **26**, 403–419.

Kalbfleisch, J. D. and Prentice, R. L. (1980). *The Statistical Analysis of Failure Time Data*. New York: Wiley.

Keeler, E. and Rolph, J. (1982). *The Demand for Episodes of Medical Treatment*. Report R-2829-HHS. Santa Monica, CA: The Rand Corporation.

Keeley, M. C. and Robins, P. K. (1978). *The Design of Social Experiments: A Critique of the Conlisk–Watts Assignment Model*. Research Memorandum 57. Menlo Park, CA: SRI International.

Kershaw, D. N. and Fair, J. (1976). *The New Jersey Income-Maintenance Experiment. Vol. 1, Operations, Survey, and Administration*. New York: Academic.

Lancaster, T. and Nickell, S. (1980). "The analysis of reemployment probabilities for the unemployed." *J. Roy. Statist. Soc. Ser. A*, **143**, 141–165.

Lillard, L. A. and Action, J. P. (1980). *Seasonal Electricity Demand and Pricing Analysis with a Variable Response Model*. Report R-2425-DWP. Santa Monica, CA: Rand Corporation.

Lillard, L. A. and Aigner, D. J. (1984). "Time-of-day electricity consumption response to temperature and the ownership of air conditioning appliances." *J. Business and Economic Statistics*, **2**, No. 1, 40–54.

Madansky, A. (1980). "Response surface exploration in social experiments." Paper presented at Second Annual Public Policy Conference, Boston, MA, October 24–25, 1980.

Manning, W. G., and Acton, J. P. (1980). *Residential Electricity Demand Under Time-*

of-Day Pricing: Exploratory Data Analysis from the Los Angeles Rate Study. Report R-2426-DWP/HP. Santa Monica, CA: Rand Corporation.

Manning, W. G., Mitchell, B. M., and Acton, J. P. (1979). "Design of the Los Angeles peak-load pricing experiment for electricity." *J. Econometrics*, **11**, 131–194.

Mason, W., Wong, G., and Entwistle, B. (1984). "Contextual analysis through the multilevel linear model." In S. Leinhardt (ed.), *Sociological Methodology 1983–1984*. San Francisco: Jossey-Bass, 72–103.

Meier, P. (1975). "Statistics and medical experimentation." *Biometrics*, **31**, 511–529.

Morris, C. (1979). "A finite selection model for experimental design of the health insurance study." *J. Econometrics*, **11**, 43–61.

Mosteller, F. (1978). "Errors: Nonsampling errors." In W. H. Kruskal and J. M. Tanur (eds.), *International Encyclopedia of Statistics*, **1**. New York: Free Press, 208–229.

Mosteller, F. and Tukey, J. W. (1982). "Combination of results of stated precision: I. The optimistic case." *Utilitas Mathematics*, **21**, 155–178.

Mosteller, F. and Tukey, J. W. (1984). "Combination of results of stated precision: II. A more realistic case." In P.S.R.S. Rao and J. Sedransk (eds.), *W. G. Cochran's Impact on Statistics*. New York: Wiley, 223–252.

Neuberg, L. (1985). *Social Control Experimentation: Towards a Rationalist and Historical Critique of Empiricist Methodology*. Cambridge, England: Cambridge U.P.

Newhouse, J. P. (1974). "A design for a health insurance experiment." *Inquiry*, **11**, 5–27.

Newhouse, J. P., Marquis, K. H., Morris, C. N., et al. (1979). "Measurement issues in the second generation of social experiments: The Health Insurance Study." *J. Econometrics*, **11**, 117–130.

Newhouse, J. P., Manning, W. G., Morris, C. N., et al. (1981). "Some interim results from a controlled trial of cost-sharing in health insurance." *New England J. Medicine* **305**, 1501–1507.

Pottinger, G. (1983). "Explanation, rationality, and microeconomic theory." *Behavioral Sci.*, **28**, 109–125.

Riecken, H. W., Boruch, R. F., Campbell, D. T., Caplan, N., Glennan, T. K., Pratt, J. W., Rees, A., and Williams, W. (1974). *Social Experiments: A Method for Planning and Evaluating Social Programs*. New York: Seminar Press.

Rivlin, A. M. (1974). "How can experiments be more useful?" *Amer. Econ. Rev.*, **64**, 346–354.

Rosenthal, R. and Rubin, D. B. (1978). "Interpersonal expectancy effects: The first 345 studies." *Behavioral and Brain Sciences*, **3**, 377–415.

Rossi, P. H., Berk, R. A., and Lenihan, K. J. (1980). *Money, Work, and Crime*. New York: Academic.

Rossi, P. H. and Lyall, K. C. (1976). *Reforming Public Welfare: A Critique of the Negative Income Experiment*. New York: Russell Sage Foundation.

Scriven, M. (1969). "Evaluating educational programs." *Urban Rev.*, **3**, 20–22.

Simon, H. A. (1957). *Models of Man*. New York: Wiley.

Student (1931). "The Lanarkshire milk experiment." *Biometrika*, **23**, 398–406.

Tharp, R. G. and Gallimore, R. (1979). "The ecology of program research and evaluation: A model of evaluation succession." In L. Sechrest et al. (eds.), *Evaluation Studies Review Annual*, **4**. Beverly Hills, CA: Sage.

Titterington, D. M. (1980). "Aspects of optimal design in dynamic systems." *Technometrics*, **22**, 289–299.

University Group Diabetes Program (1970). "A study of the effects of hypoglycemic agents in vascular complications in patients with adult-onset diabetes." *Diabetes (Suppl. 2)*, **19**, 747–830.

U.S. Department of Energy (1978). *Analytical Master Plan for the Analysis of Data from the Electric Utility Rate Demonstration Projects.* Prepared by Research Triangle Institute, Research Triangle Park, NC.

CHAPTER 13

The Development of Sample Surveys of Finite Populations

Morris H. Hansen, Tore Dalenius, and Benjamin J. Tepping

Abstract

This paper discusses selected topics in the theory and methods of sample surveys which have as their objectives to provide inferences about some characteristics of a finite population (as distinguished from inferences about a causal system which may have produced certain characteristics of the finite population). The paper gives an account of some early developments, considers criteria to guide the choice of sample design, and considers various aspects of probability-sampling designs, including nonsampling error and total survey design. Finally, the paper discusses the role of models in the theory and practice of sample surveys for making inferences about a finite population, and lists some areas for research and development.

1. Introduction

In this review paper, we discuss selected topics in the theory and methods of sample surveys which have as their objectives to provide inferences about some characteristics of a finite population. This is to be distinguished from inferences about a causal system which may have produced certain characteristics of the finite population.

Sections 2–4 give an account of some early developments. Section 5 considers criteria to guide the choice of sample design, and Section 6 focuses on various aspects of probability-sampling designs. Section 7 broadens the perspective by considering nonsampling errors. The discussion naturally leads to the notion of total survey design. In Section 8 we discuss the role of models in the theory and practice of sample surveys for making inferences about a finite population, and Section 9 mentions some areas for research and development.

Key words and phrases: foundations, history of sampling, models in sampling, nonsampling error, probability sampling.

2. The Infancy of Survey Sampling

2.1. Early Developments

The history of survey sampling has been reviewed in a number of papers, including Jensen (1926), Stephan (1948), Sukhatme (1966), and Kruskal and Mosteller (1980). We find in the earlier history, until the late 1880s, a multitude of examples of *applications*, especially to human populations. While considerable progress had been made in probability theory and statistical theory by that time, little attention was paid to the development or application of *theory* to survey sampling. An early contribution was due to Laplace. He estimated the population of France from reported births for all areas and counts of inhabitants in a purposive sample of parishes, using the theory of ratio estimation. He also provided a measure of the sampling error under simplifying assumptions (Cochran, 1978).

In the late 1800s and the early 1900s, some changes began to take place. The International Statistical Institute played an instrumental role in promoting the development of sample surveys. A growing but still small number of statisticians began to develop theory and methods for what was typically referred to as "the representative method"; for a review of different meanings given to that term, see Kruskal and Mosteller (1980). We mention briefly some contributions of A. N. Kiaer, A. L. Bowley, and A. A. Tschuprow.

As a Director of the Norwegian Central Bureau of Statistics, Kiaer had ample opportunities to develop and apply the representative method, and he vigorously advocated its use at some early ISI sessions. In Kiaer (1897), such techniques as stratified sampling and multistage sampling were discussed.

Bowley was a British academician. He played a decisive role in persuading the ISI to endorse Kiaer's ideas in a resolution passed in 1903, and elaborated (Bowley, 1913) on the merits of the representative method in the context of a large-sample survey.

Tschuprow, a professor at the Technical Institute of Petrograd, played an active role in the ISI discussions of the representative method. He developed some early theory of sampling from a finite population, including a solution of the problem of the optimum allocation of a sample to strata (Tschuprow, 1923).

2.2. The ISI Commission

The endeavors of such early contributors and a few others reached a climax with the establishment in 1924 of the ISI Commission, which presented its "Report on the representative method in statistics" in 1926. Attached to the report were four technical memoranda. We summarize below the considerations put forward in the report and the resolution passed by ISI.

The Commission defined the representative method to denote two methods "both of which are aiming at securing a representative sample, in order to make it possible to generalize the results obtained from a partial investigation" (Jensen, 1926, p. 360). The method of "random selection" called for using the elements of the population as sampling units and selecting a sample of them by a procedure which gave to each element in the population the same probability of inclusion in the sample. The method of "purposive selection," on the other hand, called for using groups of elements as sampling units and selecting a sample of groups in such a way that the sample yielded the same averages or proportions as found in the population, for certain characteristics known prior to selecting the sample.

In Bowley (1926), three major contributions appeared. First, the notion and role of what is nowadays called a "frame" were discussed, i.e., a complete list (or equivalent) of the units in the population that are to serve as sampling units. Second, a theory for proportionate stratified sampling was developed. Third, a theory for purposive selection (as defined by the Commission) was developed; the backbone of the theory was the sampling of groups of elements and the exploitation of correlations between the controls and the variables being investigated.

The Commission presented a resolution, which was passed at the 1926 ISI session. It suggested that—irrespective of which of the two methods was used for an investigation—the procedure should be applied two or more times to provide a check on its performance. It also suggested that the report presenting the results should present a detailed account of the sample-selection scheme.

3. The Beginning of a New Era

3.1. Developments at the Rothamsted Experimental Station

R. A. Fisher at the Rothamsted Experimental Station was presented with the problem of analyzing data from field trials that had been carried out. This led to the development of theory for experimental design and analysis.

Fisher's theory for the design of experiments identified six "principles," including replication, local control, and randomization. The use of randomization served to eliminate the bias of the selection procedure and, in conjunction with replication, enabled the sample to become "self-contained"; that is, it enabled estimation of the sampling error using only the data collected. Local control (i.e., stratification) served to reduce the sampling error. Yates (1946) and other colleagues at Rothamsted made applications of these principles to the collection of data in sample surveys for estimating crop yields. The applications showed the need for and led to extension of methods and theory to sampling clusters of elements and to multistage sampling. The sampling units considered were of the same size, as was quite natural in agricultural experiments.

3.2. Neyman's Contributions to Survey Sampling

At a meeting of the Royal Statistical Society in London in 1934, J. Neyman presented his now classical paper on what he called "the two different aspects of the representative method." The origin of this paper went back to work he had done in Poland; he was also influenced by the earlier work of the ISI Commission and the developments at the Rothamsted Experimental Station (see Reid, 1982).

Neyman's paper has played a paramount role in promoting theoretical research, methodological developments, and applications of what is now known as probability sampling. The paper also contains an incisive comparison of purposive selection and random sampling. Although Neyman did not reject purposive selection outright, he did critically assess the basic assumptions which must be met if it is to give satisfactory results. He emphasized that the regression of the estimation variable on the control variable must be linear (or nearly so) and concluded: "I think it is rather dangerous to assume any definitive hypotheses concerning the shape of the regression line" (Neyman, 1934, pp. 576–577).

Neyman went on (p. 585) to define what he termed "*a representative method of sampling* and a consistent means of estimation." He wrote:

> I should use these words with regard to the method of sampling and to the method of estimation, if they make possible an estimate of the accuracy of the results obtained in the sense of the new form of the problem of estimation *irrespectively of the unknown properties of the population* studied.
>
> We have seen that the method of random sampling allows a consistent estimate of the average X whatever the properties of the population. Choosing properly the elements of sampling we may deal with large samples, for which the frequency distribution of the best linear estimates is practically normal, and there are no difficulties in calculating the confidence intervals.

The "new form of the problem of estimation" to which he refers involves the concept of confidence intervals, which were defined for the first time in this paper. He also emphasized that the important consideration was that valid confidence intervals be computed for the particular sample-selection and estimation procedures used, without requiring that the estimation procedure be "best" in some sense or that the confidence intervals be the shortest possible.

In addition to advancing new principles for survey sampling in his 1934 paper, Neyman contributed new sample designs. One example is his theory for optimum allocation of sampling units to strata which he developed independently of Tschuprow, whose earlier result appears to have been overlooked at that time by statisticians generally. Another example is his theory for sampling clusters of elements, with random sampling of the clusters for the sample. He indicated the importance of designing the sample to include a large number of small clusters rather than a few large ones; in that context he also discussed the use of ratio-type estimators and presented needed theory.

In 1937, W. E. Deming arranged for Neyman to give a series of lectures in Washington, D.C. One of these lectures dealt with sampling of human populations. A mimeographed edition of the lectures appeared in 1938, of which Neyman (1952) is a revised and much enlarged version. In the course of one of the lectures, M. Friedman and S. Wilcox presented a problem concerning what has become known as "double sampling" or "two-phase sampling." At the time of the lecture Neyman did not offer a solution. He later considered it in some detail and presented the theory of double sampling in Neyman (1938). In that paper Neyman introduced cost functions, in what was one of the forerunners of the joint use of cost and variance functions to optimize sample design within a specified class of designs.

4. Impact of the New Developments

In what follows, we shall pay special attention to the impact in the United States. The growing use of sample surveys in the 1930s reflected the needs of programs such as the Works Progress Administration (WPA) and others. The fiasco of the *Literary Digest* poll and the success of the Gallup poll in forecasting the 1936 presidential election helped focus attention on the potential of survey sampling and the need for critical assessment of the methods used (see Stephan, 1948).

The 1937 Enumerative Check Census of Unemployment was a national area sample that made use of probability sampling, including ratio estimation with confidence intervals (Dedrick and Hansen, 1938). The success of this survey served to establish a widely accepted precedent. In 1938 the WPA initiated a nation-wide Sample Survey of Unemployment, using multistage sampling with large primary sampling units (Frankel and Stock, 1942).

In the late 1930s, the Bureau of the Census embarked upon a developmental and research program in survey sampling. In the 1940 Census of Population, sampling was used to extend the amount of information collected per unit cost (Stephan et al. 1940). In 1942, the Sample Survey of Unemployment was transferred to the Census Bureau, where some far-reaching changes were made in the design. The changes included the extension of the use of probability sampling to all stages of sampling, the introduction of unequal probabilities at the various stages of sample selection, and the use of ratio estimation adapted to the specifics of multistage sampling (Hansen and Hurwitz, 1943).

A significant portion of the focus began to be placed on the problems of design to facilitate control of nonsampling errors.

Also in the late 1930s, the U.S. Department of Agriculture initiated a research program under the leadership of C. F. Sarle; as part of that program, a research group, including A. J. King and R. J. Jessen, was established at the Statistical Laboratory of the Iowa State University. W. G. Cochran, who had worked with F. Yates and others at Rothamsted, joined that group in 1939 and made several important contributions while there. See, e.g., Cochran (1942).

The impact of the new developments outside of the United States was remarkably early and strong in India, thanks to the efforts of P. C. Mahalanobis at the Indian Statistical Institute and P. V. Sukhatme at the Indian Council of Agricultural Research. While adapting the new procedures for probability-sampling designs to agricultural surveys, they also made important contributions to the theory and methods for control of nonsampling errors in these surveys.

The influence of the new developments in Europe is traceable in France (Thionet, 1946), West Germany (Anderson, 1949; Kellerer, 1953), and Sweden (Dalenius, 1957), to give but a few examples.

5. Criteria for Sample Design

There are different schools of thought about the appropriate criteria for the choice of sample design. For example, the Bayesian approach begins with the assumption of prior probability distributions for the parameters of the population being studied and uses the results of the sample survey to modify the prior into a posterior distribution. This may or may not involve taking account of the way in which the sample is selected. Another school of thought assumes a model of the population to be studied (which may be based on information available from earlier or preliminary studies and on information from the observed sample). The method of selecting the sample may or may not depend on such a model. The method of making inferences, given the sample, is then chosen to minimize the variance or the mean squared error of estimates under that model. On the other hand, others take the position, as we do, that, while advance study to determine approximate characteristics of the population to be sampled is essential to guide the choice of a reasonably efficient design, inferences should be independent of assumptions about the population (or at most should involve only weak assumptions), and inferences about the precision of estimates should depend jointly on the way in which the sample was selected and on the mathematical form of the estimator. It is then necessary that the selection of the sample be such that probabilities can be associated with every possible sample and that the measures of precision be based on these probabilities.

Except as indicated otherwise, the discussion in this paper is confined to probability samples, that is, the use of a sample-selection plan and estimators such that every element of the population has a known positive probability of being selected and such that confidence intervals can be computed that, for reasonably large samples, will include the parameters being estimated with probability closely approximating the specified level. The term "sample design" refers to both the sample-selection plan and the estimation procedure.

In choosing among alternative probability-sampling designs two fundamental criteria should be observed. First, the design chosen should be operationally feasible; that is, it should be possible to carry out the plan without serious deviation from specifications within the available resources, and within needed time schedules. Second, the design should be chosen so as approximately to minimize the variances or mean squared errors of estimates, and therefore the lengths of confidence intervals for given levels of effort. In practice it is usually not possible to achieve the minimum, even for a single statistic to be estimated, but statistical theory provides guidance for approximating the minimum, that is, for choosing an approximately optimum design. This means searching for and making effective use of relevant available resources such as existing data, prior experience with respect to variance components and unit costs from related studies, lists or maps, and manpower, as well as statistical theory.

Often a sample survey is to yield many statistics, and the optimum design as a result will not be the same for every statistic. This problem can sometimes be dealt with by focusing on one or a few of the most important statistics and finding a design that does reasonably well for each of them. Fortunately, in many practical circumstances approximately optimum designs will not differ greatly. Also, because the optima are broad, practitioners can often achieve approximately optimum designs even though unit costs and variance components are only roughly approximated when the sample survey is being designed. In practice, criteria are applied not only to the control of sampling errors but also to the control of other sources of errors in surveys.

6. Features of Sample Design

As noted above, a sample design includes both a sample-selection plan and an estimation procedure. Each of these has several features. Thus the sample-selection plan must specify such features as the sampling units to be used, the stratification, the allocation of the sample to various stages of sampling or to strata, etc. The estimation procedure and the sample-selection plan are interdependent. The design may involve choices in the use of auxiliary information in various aspects of sample design. For example, the estimation procedure may make use of auxiliary information in ratio or regression estimators, and the sample-selection plan may make use of auxiliary information in defining strata and sampling units, and in other ways.

6.1. Definition of Sampling Units—Cluster Sampling and Multistage Sampling

From the infancy of survey sampling it was recognized that the sampling unit might be either the element for which measurements were desired (e.g., individual people, families, etc.) or well-defined clusters of the elements. Further, the final sample of clusters might consist of all elements in a sample of clusters (such as government units), or of subsamples from an initial sample of clusters, possibly with one or more intermediate stages of sampling. We mention some basic principles concerning the choice of sampling unit.

If one has ready access to a list of the elements in the population, it might at first appear that the element should serve as the sampling unit. Indeed, this is the desirable approach if there are no substantial costs involved in collecting the information from such a sample and in controlling the quality of the data collected. But if the data are to be collected by personal interview or personal observation, it may be exceedingly and often prohibitively costly to use such a sample, especially if the units are spread over a large geographic area. Thus, we consider drawing samples consisting of clusters of neighboring units, for which lists are available or can readily be created. Such clusters may be sampled with all units in each cluster included in the final sample, or subsampling may be done within the selected first-stage units. However, such clustering almost always results in increases in variance for a given number of elements in the sample, since observations on neighboring units are usually positively correlated. The variance, $\sigma_{\bar{x}}^2$, of an estimated mean $\bar{x}$ with a clustered sample (with or without subsampling) is approximately

$$\sigma_{\bar{x}}^2 \doteq \frac{\sigma^2}{n}\{1 + \rho(\bar{n} - 1)\},$$

where ρ is the intraclass correlation among elements within clusters, $\bar{n}$ is the average number of elements in the sample per cluster, n is the total number of elements in the sample, and σ^2 is the variance among elements. It follows that even if ρ is small, but positive, the increase in variance from cluster sampling can be substantial if $\bar{n}$ is large. Since costs are reduced by clustering for a given number of elements in the sample, one chooses a design taking joint account of variances and costs which, for example, approximately minimizes the variance for a given total cost.

If no list of elements is available from which to draw the sample, it is necessary to create such a list or to use some kind of cluster sampling using clusters for which a list is available or can readily be created, possibly with multistage sampling. A sample of such clusters is drawn in one or more stages of sampling, and all or a subsample of the elements in the finally selected clusters then constitute the survey sample. Theory, along with empirical studies of approximate unit costs and variances for the various approaches and stages of sampling, provides powerful aid in arriving at approximately optimum designs. Guided by these, the final choices depend on the experience and judgment of the designer.

Neyman (1934) indicated that if compact clusters were to be the sampling units, they should be so defined as to be roughly equal in size. He also used the theory of ratio estimation to help reduce the contribution to the variance from variation in cluster size. In sampling compact clusters, the efficiency is sensitive to the average cluster size, and it becomes important to choose as sampling units clusters of roughly the optimum size. The development of empirical variance and cost functions for sampling such compact units guides the choice of an approximately optimum size of units. See, e.g., Smith (1938), Jessen (1942), and Mahalanobis (1946).

For surveys involving relatively small samples in which the data must be collected over a large geographic area (such as nation-wide surveys in the United States) it has often been found advantageous to use multistage sampling and relatively large political units (for which extensive statistics are available to be used in sample design) as first-stage units. A common approach in this case is to draw a sample of such areas, to draw an intermediate sample of smaller areas in one or more stages within the selected first-stage units, and then to list the final-stage sampling units (e.g., households) and include all or a sample of the units in the final sample.

6.2. Stratified Sampling

The use of stratification is standard in nearly all surveys. In practice one can lose nothing and sometimes gain substantially by stratifying the population and allocating the sample to strata proportionately to the numbers of sampling units in the strata. This is called proportionate stratified sampling. The gains are substantial if strata can be defined that are relatively homogeneous internally with respect to the characteristics being estimated and with substantial variation between the stratum means.

For certain types of populations additional gains can be obtained by using approximately optimum allocation of the sample to strata. Let σ_h^2 be the unit variance in stratum h, and assume that the variable costs of a sample survey can be approximated by the simple cost function $\sum C_h n_h$, where n_h is the total number of units in the sample and C_h is the cost of including a unit in the sample from stratum h. Then, approximately, the optimum allocation of the sample is to choose the n_h proportionate to $N_h \sigma_h / \sqrt{C_h}$.

If one is sampling farms, business establishments, or political units to estimate totals, means, or ratios of variables whose distribution in the population is highly skewed, such as employment, production, or sales, then substantial gains through reduced variance of estimates can ordinarily be achieved by approximating optimum allocation. However, one can also lose, as compared to proportionate stratified sampling, by using disproportionate allocation of the sample to strata. For example, when several or many variables are involved, some but not all of which are highly skewed, disproportionate stratified sampling designed to yield worthwhile gains for some variables may yield substantial losses for others as compared to proportionate stratified sampling.

Simple approximate rules have often proved useful in approximating optimum allocation when the σ_h are unknown. Such approximations are supported to some extent by theory and by a number of empirical studies (see, e.g., Hansen and Hurwitz, 1949, and Westat, 1982).

Stratification practice and theory have been extended considerably to serve various design needs. Some examples are (i) sample allocation to estimate more than one parameter (Dalenius, 1957); (ii) construction of strata and choice of the number of strata (Dalenius, 1957; Cochran, 1961); and (iii) poststratification, including multistage poststratification, with simple and complex initial sample-selection plans (Hansen et al., 1953, Chapters 5, 9 and 11; U.S. Bureau of the Census, 1978).

A further extension of the basic idea of stratification is to introduce additional controls beyond those that can be achieved by selecting only one unit per stratum. Some examples are Latin-square selection (Frankel and Stock, 1942), deep stratification (Tepping et al., 1943), and controlled selection (Goodman and Kish, 1950). However, unless the sampling is replicated, these procedures do not allow consistent estimates of the variances of the resulting estimates.

6.3. Sampling with Unequal Probabilities

Tschuprow (1923) and Neyman (1934) showed that it may be advantageous to use different probabilities of selection for different strata. In Hansen and Hurwitz (1943), the idea of using unequal selection probabilities was extended to the selection of units within strata. Thus, in principle, for a population of N units, there could be N different probabilities of selection. In the theory as given by Hansen and Hurwitz, the variances and variance estimates assumed sampling with replacement. The theory was later supplemented by Hansen and Hurwitz (1949) to include approximating the optimum probabilities of selection assuming sampling with replacement.

The theory for sampling with unequal probabilities was extended by Horvitz and Thompson (1952) to sampling without replacement. A problem with the Horvitz–Thompson variance estimator is that the variance estimates are unstable and can be negative. Stimulated by the developments by Horvitz and Thompson, many theoretical and applied papers have appeared on sample selection and variance estimation for selection with unequal probabilities without replacement. See, e.g., Yates and Grundy (1953), Brewer (1963), and Durbin (1967). A comprehensive treatment is given by Brewer and Hanif (1983).

Sampling with unequal probabilities is particularly important in multistage sampling, e.g. as in the Current Population Survey carried out by the U.S. Bureau of the Census. The design calls for selection of an initial sample of large primary sampling units (such as counties or combinations of countries), followed by subsampling within the selected primary sampling units at each successive stage with unequal probabilities in such a way as to yield samples of elements with uniform overall probabilities of selection.

The appropriate use of unequal probabilities often has the effect of reducing the variance for a given number of units in the sample. It also has the effect of controlling the variability arising from the unequal sizes of sampling units without actually stratifying by size, and thus allows more stratification on other variables to reduce variance. Also, it is used to reduce variability in workloads assigned to data collectors.

6.4. Systematic Sampling

In the simple and widely used case, systematic sampling consists of taking every kth unit from an ordered population frame, starting with a random number between 1 and k. It easily follows that the sample mean is an unbiased estimate of the population mean. As pointed out by Madow and Madow (1944), the sample of n elements amounts to a sample of one cluster of the k possible clusters. Thus, unless the systematic sample is replicated, consistent estimates of variance are not available.

A common practice for estimating the variance is to assume that systematic sampling is equivalent to simple random sampling within strata. This practice has been demonstrated through simulation studies to be quite acceptable in situations where the serial correlations are quite low between neighboring units in the frame. However, Osborne (1942) empirically demonstrated that in some real problems (in this case estimating land coverage) serial correlations are high, and the assumption of simple random sampling within strata encompassing adjacent pairs in the sample had the effect of overestimating the variance of the sample mean by factors of 2 to 4 as compared with the variances obtained by replicating the systematic samples.

Madow and Madow (1944) provided insight by modeling populations and examining the characteristics of systematic sampling under alternative models. Cochran (1946) formulated a mathematical model of populations of a type widely observed, and examined systematic sampling under the assumption that the finite population was a sample from such a "superpopulation."

6.5. Two-Phase or Double Sampling

The basic idea of two-phase sampling (also called double sampling) is to draw an initial "large" sample for which certain information is obtained at low cost. Then a subsample is selected from the initial sample and additional information is obtained at a higher cost per unit. The information from the large sample may be used to stratify in drawing the subsample (Neyman, 1938), to provide ratio or regression estimators (Cochran, 1939), or in other ways. It will pay to use two-phase sampling only if the unit cost for the initial sample is sufficiently low relative to the unit cost of the information obtained in the subsample, and if the gains from using the information in the initial sample are sufficiently large.

6.6. Multiframe and Network Sampling

Sometimes it is convenient or necessary to use two (or more) overlapping frames from which to sample. For example, in drawing a sample of farms (agricultural holdings), complete lists are often not available, and a procedure sometimes used is to draw a sample of small areas and to include in the sample all or a subsample of the farms with headquarters in the sampled areas. This sample might be supplemented by a sample from an available or specially prepared list of very large farms. In such a design the farms on the list have multiple probabilities of selection. The probabilities can be ascertained by matching the area-sample farms against the complete list of large farms. An alternative is to exclude from the area sample any farms on the list. Efficiency can sometimes be increased by approximately optimum weighting of the averages for subsets of the population that are sampled from different lists (Hartley, 1974); however, such a procedure may only be feasible when only one or a small number of estimates are to be made.

The use of multiple probabilities arises in many different forms and circumstances. For example, elements of the population may be associated with more than one sampling unit. An illustration is a sample of families obtained by selecting a sample of school children. Sirken (1970) has referred to such sampling procedures as "network sampling," and has provided some simple rules for determining the probabilities of selection. See also Westat (1982).

An alternative approach to sampling the desired units with multiple probabilities of selection is to sample by a "unique rule," so that, for the school-children illustration above, a family is retained in the sample only if the eldest school child of the family is in the sample. A unique-selection rule may provide either a more or less efficient sample than a multiple-selection rule in terms of information (least variance) per unit of cost.

6.7. Sampling on Successive Occasions

The U.S. Bureau of the Census conducts monthly sample surveys of the population to make monthly estimates of the labor force and of characteristics such as employment, unemployment, and so forth, in total and by age, sex, and other characteristics. The purpose is to estimate both the values at a point or interval of time, and the magnitudes of changes in those values over time. These surveys are illustrative of many that produce estimates for changing populations. Such surveys need to be distinguished from longitudinal studies in which individual units in a population are followed over time and changes in their characteristics are measured. We discuss here the first of these problems.

Sampling on successive occasions involves a number of special problems in theory and practical survey design. The problem of multiple statistics arises not only in making an estimate for different characteristics, but in estimating different relationships for the same characteristics. For example, for a particular characteristic, interest may be in estimates of averages or aggregates each month, in changes from month to month, in changes from the corresponding month a year ago or at other intervals, and in aggregates across time periods, such as annual aggregates.

Other problems include the extent to which the samples should be independent, identical, or overlapping in successive periods, how the information from the prior series of months can be used to improve the estimates for the current month, and related questions.

Jessen (1942) proposed a solution to the problems of optimum allocation of the sample and estimation for each of two occasions, and this sparked later developments at the U.S. Bureau of the Census on practical methods that have had extensive application there (Hansen et al., 1953, pp. 490–503). Patterson (1950) found a general solution to the problem of finding a simultaneous optimum for monthly totals, month-to-month changes, annual aggregates, etc. However, his solution involves revision of all prior estimates at each new time period, which reduces its practical usefulness. Yates (1981) and Cochran (1977) provide additional discussion of the problem.

The U.S. Bureau of the Census developed and applied a procedure using what it called "composite estimators" that did not require revision of prior estimates. Also, with some loss of precision, the procedure avoids the confusion that arises when, for example, estimates of month-to-month changes or aggregates across months are not the differences or sums of the monthly estimates. It involved the use of rotating and partially overlapping samples, the amount of rotation or overlap depending on the application and on relevant optimum design considerations. In one application, a monthly survey of retail establishments, 50 percent of the sample is replaced each month. The composite estimator is the weighted mean

$$x''_t = w\, x''_{t-1} \frac{x'_{(I)t}}{x'_{(I)t-1}} + (1 - w) x'_t,$$

where x''_t and x''_{t-1} are the composite estimates of sales for months t and $t - 1$, x'_t is the estimate based on the full sample for month t, and $x'_{(I)t}$ and $x'_{(I)t-1}$ are similar to x'_t except that they are based only on the matched sampling units in the sample in months t and $t - 1$. Thus, the composite estimator does not require revision of prior estimates.

The composite estimator is a weighted mean of two estimates of x_t, with respective weights of w and $1 - w$. The weights can be chosen to be optimum for the estimate of the monthly level, or of the month-to-month change, or of the change from same month a year ago, or of the annual level, etc. In practice, they are chosen to be a compromise among these that gives a reasonably good result for each, but with special emphasis jointly on month-to-month change and monthly level. In this survey the procedure provides substantial variance reductions in estimates of both monthly level and change because the correlations of sales for individual establishments over various time intervals are high. Theory and empirical results are given in Hansen et al. (1955), Woodruff (1959), and U.S. Bureau of the Census (1978).

The use of procedures in which the sample is rotated, with matched and unmatched portions, often results in rotation-group biases, that is, differences in the observed average values of responses from sampling units in the matched portion from the average values for sampling units in the unmatched portion (see Bailar, 1974). These may result from biases in one or the other group, or both. They tend to affect the composite estimator more or less consistently over time. In practice, the advantages of using rotating samples with a composite estimator are often substantial over procedures using a fixed sample, which also are subject to such biases but do not provide the means for measuring them.

6.8. Estimation Procedures

With probability sampling, estimates of aggregate population values can be made by weighting the sample observations by the reciprocals of their probabilities of selection. Such estimators are unbiased and are widely used. Often, improved estimates can be made if auxiliary information is available about the sampling units and/or elements. Much of the development of methods and theory of survey sampling has focused on the options for making efficient use of such information. A variety of estimation procedures are available.

Consider the case where the total Y is to be estimated and the total X for an auxiliary variable is known prior to the survey. In the survey both variables are observed and are positively correlated. Let us denote the unbiased estimators by y' and x'. Alternative estimators of Y may be derived from the general expression

$$y^* = y' + \theta(X - x')$$

by a proper choice of θ. For $\theta = 0$, y^* becomes the unbiased estimator. For $\theta = y'/x'$, y^* becomes $(y'/x')X = rX$, the ratio estimator of a total. We note also that the ratio estimator in the form $r = y'/x'$ as an estimator of $R = Y/X$ is almost universally used and useful in its own right. The ratio estimator is not unbiased (i.e., its expected value is not necessarily equal to R), but any bias becomes trivial for sufficiently large samples, and thus it is a consistent estimator (see Hansen et al., 1983).

For $\theta = d$, where d is a constant chosen independently of the sample, y^* becomes $y' + d(X - x')$, the difference estimator. A particularly simple application is to set $d = 1$ when x' and y' are highly correlated and have about the same variance. We note that if the relation between the two variables is (approximately) linear, with the regression coefficient known to be B, putting $\theta = B$ yields the estimator with minimum variance in this class of estimators. In any case, the difference estimator is unbiased.

For $\theta = b$, where b is the estimate of the regression coefficient B, y^* becomes $Y^* = y' + b(X - x')$, the regression estimator (Cochran, 1942). This is also a consistent estimator, and in large enough samples has minimum variance in this class of estimators.

The regression estimator sometimes has a strikingly lower variance than the ratio estimator, especially if the variables x' and y' are highly correlated and if the relationship between them cannot be represented approximately by a line through the origin. The regression estimator can be especially useful in surveys with large samples in which only a few items are to be estimated or where it has an important advantage in variance reduction. It is less frequently used, except perhaps for selected important items, in sample surveys where large numbers of variables are estimated, in part because it does not ordinarily yield estimates with additivity, so that, for example, estimates for male and female may not add to a corresponding estimate for the total population. If y and x identify the values for individual sampling units, if the finite population is such that the regression of y on x is well approximated by a line through the origin, and if the conditional variance of y within narrow ranges of x is approximately proportionate to x, then the variance of the ratio estimator is equal to or less than that of the regression estimator. The choice between the ratio and regression estimators is often made after the survey data are available.

Theory for the regression and ratio estimators is presented e.g. by Hansen et al. (1953, 1983), Sukhatme and Sukhatme (1970), Cochran (1977), Yates (1981), and others.

6.9. Variance Estimation

For linear estimators such as sample means or estimated totals that can be expressed in the form $x' = \sum a_i x_i$, where i indexes the units in the sample and the a_i are constant independent of the sample, unbiased estimation of the variance of x' is fairly simple. The problem of estimating the variance is more difficult for nonlinear estimators, such as estimators that involve ratios of random variables.

6.9.1. *Linearization for Estimating Variances of Nonlinear Estimators.* Often, even with fairly simple statistics such as the ratio estimator $r = y'/x'$, no unbiased variance estimator exists except in special circumstances. For reasonably large samples, such that the coefficient of variation of the denominator (x') is not too large —say, less than 10 percent (see, e.g., Cochran, 1977, p. 163)—asymptotic linear approximations to the variance of r will yield acceptable approximations. Procedures for obtaining linear approximations to variance estimators for complex designs have been developed and are extensively used (see, e.g., Tepping, 1968; Woodruff, 1971; Fuller, 1975; and Woodruff and Causey, 1976).

6.9.2. *Replication.* Many widely used methods of estimating variances may be grouped under the rubric of replicated sampling. Such variance estimators are of interest especially because, for nonlinear estimators based on complex sample-selection plans, the development of linear approximations is often difficult. Also, replicated variance estimators may have somewhat smaller biases than linear approximations.

We distinguish true replication methods from pseudoreplication. In true replication, either independent samples of the population are drawn or the sample consists of nonoverlapping subsamples from each of which a consistent estimate of the desired population characteristic can be constructed. The interpenetrating samples introduced by P. C. Mahalanobis in the Indian National Sample Survey and the replicates recommended by Deming (1960) constitute examples. In other cases of true replication the sample is divided into random subgroups, each of which constitutes a proper sample of the population of the same design as the full sample, but of smaller sample size. In these and similar cases, a consistent estimator of the variance can be shown to be a simple function of the variance between the subsamples.

Pseudoreplication (a term introduced by McCarthy, 1966) involves overlapping subsamples. McCarthy's scheme defines a set of overlapping "half-samples" and provides an estimator of the variance of a statistic as a simple function of the differences of the sample statistics computed for each half-sample. In the paper referred to, McCarthy shows how to select a balanced subset of the half-samples which, at least for linear statistics, contains all of the information provided by the complete set of half-samples. The balanced half-sample estimators of variance considerably simplify the computations and may also reduce biases (although these will be trivial for each of the variance estimators we have discussed with reasonably large samples) and have therefore proved very useful. Frankel (1971), Mellor (1973), Kish and Frankel (1974), and Krewski and Rao (1981) have given interesting evaluations of the technique. Another pseudoreplication variance-estimation procedure [called the "jackknife" variance estimator by Tukey (1958) because of its simplicity and general utility] forms replicates by dropping one unit (or some units) at a time from the sample. Thus, it is based on larger replicates than balanced half-samples, with some consequent advantages. Both procedures are readily adaptable to multistage stratified samples. Frankel (1971) and Mellor (1973) include evaluation of the jackknife, and Efron (1982) provides an extensive discussion of the jackknife and its generalizations, together with a list of appropriate references.

6.9.3. *Generalized Variances.* Whether simple or more complex variance estimators are involved, the number of statistics derived from a survey for which variance estimates are desired may be many and time schedules short. For such surveys direct variance computation for large numbers of statistics may cause intolerable delay. Also, the direct variance estimates may in certain circumstances be subject to large sampling error. Frequently in such circumstances a generalized variance-estimation approach is used. Statistics may first be classified into a few broad classes for which the design effects [the ratio of the variance to that for a simple random sample of elements of the same size, as discussed in Kish (1965)] are regarded as substantially different on the basis of empirical evaluation. For each such class, a relatively small subset of estimates may be selected. For these estimates, the variances are estimated by one or more of the devices described above and used to estimate the parameters of a generalized function which, for example, may relate the variance of the estimates to the level of the estimate. The results are examined, and if necessary an improved classification is attempted. After an acceptable curve is obtained, it is used to approximate and present variances for that class of statistics. Usually variances are estimated directly for the more important individual statistics.

6.9.4. *Advance Variation Estimation.* Efficient survey design calls for access to prior information on the contributions to variance of various design factors. Sometimes such variance components are speculated from approximate subject-matter knowledge and knowledge of various types of distributions or of simple and rough approximations to them (see, e.g., Deming, 1960, Chapter 14, and also Section 6.2 above). Commonly, for a nonrecurrent survey, estimates of these contributions are "borrowed" from previous surveys of a somewhat similar population. For a repetitive survey the situation is far better; estimates of variance in one round are useful in improving the design for subsequent rounds.

7. Total Error Measurement and Control—the Problem of Nonsampling Errors

The development of methods and theory of survey sampling reviewed above focused primarily on measuring and controlling sampling error. Studies of the impact of nonsampling error, including error due to nonresponse, early indicated that these may be important sources of uncertainty, and may in some instances contribute much more than sampling error to the total error.

7.1. Survey Models

In the last four decades efforts have been made to develop methods and theory for allocating resources to the control of various sources of error, thus providing a guide to what is referred to as "total survey design." A survey model provides a decomposition of the mean squared error of an estimator to reflect the impact of specific sources of error, just as sampling theory makes it possible to decompose the sampling variance into components such as within- and between-primary-sampling-unit components.

According to the Census Bureau model (Hansen et al., 1961, 1965), the total error of an estimate $\bar{x}$ of the parameter $\bar{X}$, as given by the mean square error (MSE), is decomposed into a sampling variance component, a nonsampling-error variance component (also referred to as a response variance component), an interaction component, and a bias component. The nonsampling-error variance component reflects errors due to the original collection of data, editing, coding, processing, and other sources, and may be decomposed also to reflect specific sources. If the sample is designed appropriately, each of the variance components of the MSE may be estimated from the sample (see Bailar and Dalenius, 1969; Fellegi, 1964). Approximate estimates of the bias component are sometimes possible by supplementary studies, analyses, and comparisons with data from other sources. Such estimates, along with cost and operating considerations, provide guidance in total survey design. The problem then is to find cost-effective procedures for dealing with the various sources or error. Some of these include better questionnaire design, training, and improved control of data collection and of data-processing operations.

Of course treatment of nonsampling errors involves not only careful control of data-collection procedures, but also of editing and data-processing procedures, in order to identify inconsistent or exceedingly unlikely entries and take action on these. Today, much of this editing and imputation is done by computers rather than by clerks, with provision for manual intervention as needed. For some principles of computer application see Pritzker et al. (1966).

7.2. Nonresponse

Nonresponse is a potentially serious source of nonsampling error. Nonresponse may be unit nonresponse (no data are collected for some units of the sample, or such incomplete data that the return is not regarded as usable), or item nonresponse (in some usable returns one or more data are missing or otherwise not usable).

The primary procedure for control of nonresponse is a sufficient follow-up to obtain a high rate of response. This is facilitated by proper sample design, as well as proper training, administration, and quality control. Where follow-up is costly as compared to initial data collection (e.g., with personal follow-up after initial mail or telephone data collection), an approximately optimum sampling procedure can be used for subsampling nonrespondents (see Hansen and Hurwitz, 1946).

In most surveys, some nonresponse ordinarily will remain, and imputation for both item and unit nonresponse is sometimes applied. It is important to recognize that imputation for nonresponse does not solve the problem as it would be solved if acceptable original responses were obtained. However, the use of direct imputation procedures is often far better than simply ignoring the problem, which, in practice, ordinarily is equivalent to implicit imputation of the average result from all respondents. An examination and evaluation of existing procedures, recommended alternative procedures, and proposals for additional research are given by a recent comprehensive study and report of the Panel on Incomplete Data of the Committee on National Statistic (1983).

For other references on the problem of nonsampling error see Neter and Waksberg (1964), Mosteller (1978), and Fienberg and Tanur (1983).

8. The Foundations of Survey Sampling

8.1. The Debate About the Foundations

In recent years the approach to survey design characterized by the use of probability sampling has been the subject of a debate which at times has been vivid and heated. Godambe (1955) formulated a general mathematical theory for survey sampling from finite populations. This was followed by debate on a number of issues, some of which are discussed by Royall (1970), Smith (1976), and others, and considered at some length and summarized in Cassel et al. (1977). It is beyond the scope of this paper to discuss all of the principal issues. Instead we briefly introduce two of them.

8.1.1. *Best Linear Unbiased Estimates.* Godambe (1955) showed that, in the context of inference to a finite population, no uniformly best linear unbiased estimator exists. This result has been interpreted by many to imply that the approach which originated with Neyman's 1934 paper must be wrong. This criticism is unwarranted, as it overlooks a crucial point: in Godambe (1955) "linear" is defined differently than as used by Neyman (1934) and by many others, and as defined in Section 6.9 above. The issue of the role of "best" estimation is still an active one.

8.1.2. *Model-Dependent Survey Sampling.* Some have advocated a model-dependent (as distinguished from model-based) approach to sample design. More specifically, from their point of view, the finite population about which inferences are to be made is viewed as a sample from an infinite superpopulation. The formal representation of the superpopulation is referred to as a model, which is given a decisive role in the choice of estimators and, for some, in the choice of a sample-selection plan. Thus, for some, accepting a model-dependent approach may lead to the rejection of the use of random sampling in favor of some version of purposive selection. In any case, the view held by the proponents of a model-dependent approach to survey sampling is clearly not compatible with the principles of probability sampling as we have defined it.

8.2. An Evaluation of the Debate

In this section we briefly comment on the two points presented in Section 8.1. We refer to Hansen et al. (1983) for a fuller discussion.

8.2.1. *Choice of Estimator.* We emphasize that survey statisticians applying probability-sampling methods typically do not confine themselves to unbiased estimators of the parameters but rather use estimators that are consistent under probability sampling. They use confidence intervals that are appropriate to the estimators, and choose sampling methods and estimators on the basis of cost and other considerations in an effort to achieve approximately maximum information per unit of cost. No claims are made or can be made that the "best" in a general sense is achieved. Hence Godambe's interesting result does not create any practical difficulty. The survey sampling practitioner will rather try to choose the better among the available alternatives and will present confidence intervals that are appropriate to the estimator chosen.

8.2.2. *Use of Models.* We consider separately the problem of making inferences to a finite population and of making inferences about a causal system.

In making inferences to a finite population the model-dependent approach is ordinarily either equivalent or superior to the probability-sampling approach *if* the assumed model, in fact, holds true. But if the model is not fully realistic, the model-dependent approach may result in misleading inferences. From the viewpoint of the practicing survey statistician, it is important to realize that a deviation between the model and the population too small to detect on the basis of a few hundred observations may, in fact, cause serious difficulty. [For a simple illustration see Hansen et al. (1983).] Moreover, the model-dependent approach does not fare well in a comparison with the probability-sampling approach in a survey where many characteristics are to be estimated. When it is necessary to produce estimates of a large number of population characteristics on tight time schedules, it is simply not analytically and computationally feasible to create models for each of them.

In making inferences about the causal system that generated the finite population at hand—or, alternatively, to make predictions about future realizations of that causal system—there is, however, no alternative to the use of a model-dependent approach. Even so, the application of probability-sampling methods in obtaining descriptive statistics can provide valuable information.

On the other hand, we emphasize that superpopulation models are often used to great advantage in the context of probability sampling for a finite population. Such a model can provide an effective guide to sample design within the framework of probability sampling. In order to distinguish this use from the model-dependent approach, we refer to it as model-*based* probability sampling. A model of the population may, for example, suggest how to assign approximately optimum allocation and how to choose an estimator among the available consistent estimators. The resulting variances will be smaller if the model used is indeed a reasonably accurate description of the population than if it is not. In any case, the estimators are consistent, as are the estimators of their variances. On the contrary, if a sample design is used that is optimum under an assumed model, without conforming to the requirements of probability sampling, the outcome may be seriously biased estimates of the characteristics of the finite population, with the sizes of the bias unknown, and perhaps with confidence intervals that make the results appear more accurate than they really are. This is illustrated by Hansen et al. (1983).

9. Areas for Future Research and Development

The profound advancement of the methods and theory of survey sampling in the last 50 years has not made further research and development unnecessary. There is considerable scope for improving existing methods and developing new methods and theory. In this section we give some examples of suggested areas for future research and development.

9.1. Improving Measurement Methods

If surveys are to be useful, they must provide statistics which help policy makers, administrators, researchers, and other users address their problems. This has received considerable attention, and solutions are not easy. Nevertheless, we believe that still more attention should be given to the determination of what we try to measure, i.e., the specification of the relevant statistics. A problem often exists in choosing between attempting to measure a concept that would be highly relevant if measured well, but that can only be measured poorly (if at all), and choosing an alternate proxy measure that is less relevant, but that can be defined and measured reasonably well. We believe that both theory and empirical studies can provide guidance in making such choices. Related areas are also rich in possibilities for improvement; we give two illustrations.

9.1.1. *Editing and Imputation.* Access to a computer makes it technically simple to automate the editing and imputation operations, with summaries and selected cases printed out for personal review and intervention as desired. This does not make it unnecessary to develop theory on which to base these operations. We give two examples of areas for research and development. One is the need for theory and summarized empirical experience for defining or dealing with outliers. The traditional theory of outliers appears to be inapplicable: it is commonly concerned with the presence of one or a few outliers in a set of n observations, all of which measure the *same* quantity. But in the survey situation, we have one observation for each of n elements, each observation measuring a *different* quantity, i.e., a value for a different element. A second possible improvement would be to develop a philosophy, theory, and empirical studies for the use of redundancy to provide data for internal checks; this would, of course, reduce the number of characteristics which can be included in a survey.

9.1.2. *Nonresponse.* The remarks above about the need for developing theory for editing and imputation are directly applicable to nonresponse. If it is to be possible to make useful imputations for nonresponse (that is, imputations which can reasonably be expected to reduce the bias which may be introduced by nonresponse), it is necessary to develop and apply a realistic model of the mechanism which generated the nonresponse. Simple models are widely used, implicitly or explicitly. In recent years, various models have been suggested, some of which are complex to apply. Additional evaluation is needed. This subject is discussed by the Panel on Incomplete Data of the Committee on National Statistics (1983). Much remains, however, to be done in this area; an especially difficult and challenging problem concerns the estimation of the parameters of any such model on the basis of incomplete data.

9.2. Improving Sample Design Methods and Theory

Many estimators in current use are nonlinear. This is true, of course, of estimators of ratios, but also of other estimators. As a consequence, exact theory is not available about their properties (such as variances and biases). We have had to be satisfied with approximate results based on empirical investigations. To a considerable extent, these results are inconclusive. To remedy this situation, the results available should be compiled and assessed, and theory extended, paying special attention to the circumstances in which the approximation yields satisfactory results and those in which it does not perform well.

There is a similar situation with respect to the distributional properties of estimators. Several empirical investigations show that for "sufficiently large samples" the commonly used estimators of means, totals, and other functions of the observations are approximately normally distributed. But it is not clear what constitutes a "sufficiently large sample" to justify the use of the normal distribution, for example when calculating confidence intervals. Again, it may be useful to compile and assess the results available. It seems conceivable that it may prove possible to classify populations and estimators by a few broad types, thus providing guidance to what constitutes (in a specific application) a sufficiently large sample.

9.3. Computer-Assisted Telephone Interviewing (CATI)

Telephone interviewing is—for some, but clearly not all, types of surveys—a rather effective means of data collection. With computer-assisted telephone interviewing (CATI), the data collection may be combined with the editing and keying. Thus, CATI not only guides the interview but provides the benefits of computer editing at the time of interviewing, so that corrective action can be taken by interaction with the respondent, speeding up the operations. This approach is promising and is now being applied, but still has its problems and needs further evaluation and development. CATI has promising potential also in circumstances where the respondent cannot be reached by telephone, or where personal interviewing or observation is desirable. As technical advances in design of portable computers occur and prices are reduced, which have been happening at remarkable rates, we anticipate that computers that are sufficiently portable and low-cost for such applications will be available.

9.4. Effects of Privacy

The concern with privacy—enhanced by computer developments—has led to legislation and regulations that may be unbalanced and that sometimes seriously reduce the ability to do effective statistical studies, with consequent losses to society. Additional research is desirable on the felt needs for privacy and on techniques for providing protection without unduly sacrificing needed information. Currently, in an effort to ensure privacy, legislation and regulations sometimes result in ensuring privacy in situations that are of no concern or of only minor concern to the individual, but at substantial cost in the ability, for example, to take effective surveys and to develop and produce statistical information for research. Such research, as it succeeds over time, has substantial implications for the health and welfare of individuals and of the community. We believe this area needs review, and anticipate that a better balance will be achieved. The benefits may be higher-quality survey results, and reduced costs, without improper invasions of privacy.

9.5. Exploiting Satellite Photography

Aerial photographs have had a major role in sample surveys over the years, especially in agricultural and land-use surveys, and in defining sampling units of desired sizes and characteristics in rural areas. A promising beginning has been made in using satellite photography for surveys of biological populations. Keyfitz (1980), among others, has suggested the use of satellite photography for surveys of *human* populations. One possible application is for developing and updating a directory of structures by certain size dimensions to serve as a frame. This area is currently being explored for some applications; we believe that it may have extensive applications, in both developing and developed parts of the world. Further study is needed.

9.6. Improving the Tools of Total Survey Design

The objective of survey design—of which the sample design is but a part—is to strike a reasoned balance between the accuracy aimed at and the cost of the survey, subject to time schedules or other restraints. The realization of such an objective depends upon certain prerequisites. Thus, we must have a theory (often called a survey model), which decomposes the total error into components that can be estimated by appropriate surveys or experiments. Moreover, we must have estimates of those components, and means of controlling their values and doing so in a cost-effective way. Finally, we must have a methodology for determining the total survey design to be used in a given application.

Significant progress has indeed been made along these lines in the last 30 years, but more work is needed, especially with respect to the methodology just mentioned. Survey design typically advances by developing alternative procedures for each one of a sequence of steps and then making a choice among them. A promising approach is to adapt "systems analysis" to this task. While this has been done to a degree, much more remains to be accomplished.

Acknowledgement

The authors thank the editors and the referees for their helpful suggestions.

Bibliography

Anderson, O. (1949). *Über die repräsentative Methode und deren Anwendung auf die Aufbereitung der Ergebnisse der bulgarischen landwirtsch. Betriebszahlung vom 31. Dezember 1926*. München: Fachausschuss für Stichprobenverfahren der Deutschen Statistischen Gesellschaft.

Bailar, B. A. (1974). "The effects of rotation group bias on estimates from panel surveys," *J. Amer. Statist. Assoc.*, **70**, 23–30.

Bailar, B. A. and Dalenius, T. (1969). "Estimating the response variance components of the U.S. Bureau of the Census' survey model." *Sankhyā Ser. B*, **31**, 341–360.

Bowley, A. L. (1913). "Working-class households in reading." *J. Roy. Statist. Soc.*, **76**, 672–701.

Bowley, A. L. (1926). "Measurement of the precision attained in sampling." *Proc. Internat. Statist. Inst.*, **XII**, lère livre, 6–62.

Brewer, K. (1963). "A model of systematic sampling with unequal probabilities," *Austral. J. Statist.*, **5**, 5–13.

Brewer, K. and Hanif, M. (1983). *Sampling with Unequal Probabilities*. Lecture Notes in Statistic, vol. 15, New York; Springer-Verlag.

Cassel, C.-M., Särndal, C. E., and Wretman, J. H. (1977). *Foundations of Inference in Survey Sampling*, New York: Wiley.

Cochran, W. G. (1939). "The use of the analysis of variance in enumeration by sampling." *J. Amer. Statist. Assoc.*, **24**, 492–510.

Cochran, W. G. (1942). "Sampling theory when the sampling-units are of unequal sizes." *J. Amer. Statist. Assoc.*, **37**, 199–212.

Cochran, W. G. (1946). "Relative accuracy of systematic and stratified random samples for a certain class of populations." *Ann. Math. Statist.*, **17**, No. 2, 164–177.

Cochran, W. G. (1961). "Comparison of methods for determining stratum boundaries." *Proc. Internat. Statist. Inst.*, **38**, No. 2, 245–358.

Cochran, W. G. (1977). *Sampling Techniques*, 3rd edn., New York: Wiley.

Cochran, W. G. (1978). "Laplace's ratio estimator." In H. A. David, (ed.), *Contributions to Survey Sampling and Applied Statistics*. New York: Academic.

Dalenius, T. (1957). *Sampling in Sweden. Contributions to the Methods and Theories of Sample Survey Practice*. Stockholm: Almqvist & Wiksell.

Dedrick, C. L. and Hansen, M. H. (1938). *Census of Partial Employment, Unemployment and Occupations: 1937, Vol. IV, The Enumerative Check Census*, Washington: U.S. Government Printing Office.

Deming, W. E. (1960). *Sample Design in Business Research*. New York: Wiley.

Durbin, J. (1967). "Design of multi-stage surveys for the estimation of sampling errors." *Appl. Statist.*, **16**, 152–164.

Efron, B. (1982). *The Jackknife, The Bootstrap and Other Resampling Plans*. Philadelphia: SIAM.

Fellegi, I. (1964). "Response variance and its estimation." *J. Amer. Statist. Assoc.*, **59**, 1016–1041.

Fienberg, S. E. and Tanur, J. M. (1983). "Large-scale social surveys: Perspectives, problems and prospects." *Behavioral Sci.*, **28**, 135–153.

Fisher, R. A. (1925). *Statistical Methods for Research Workers*. Edinburgh: Oliver and Boyd.

Frankel, M. R. (1971). *Inference from Survey Samples*. Ann Arbor, MI: Inst. for Social Research.

Frankel, L. R. and Stock, J. S. (1942). "On the sample survey of unemployment." *J. Amer. Statist. Assoc.*, **37**, 77–80.

Fuller, W. A. (1975). "Regression analysis for sample survey." *Sankhyā, Series C*, **37**, 117–132.

Godambe, V. P. (1955). "A unified theory of sampling from finite populations." *J. Roy. Statist. Soc. Ser. B*, **17**, 269–278.

Goodman, R. and Kish, L. (1950). "Controlled selection—a technique in probability sampling." *J. Amer. Statist. Assoc.*, **45**, 350–372.

Hansen, M. H. and Hurwitz, W. N. (1943). "On the theory of sampling from finite populations." *Ann. Math. Statist.*, **14**, No. 4, 333–362.

Hansen, M. H. and Hurwitz, W. N. (1946). "The problem of non-response in sample surveys." *J. Amer. Statist. Assoc.*, **41**, 517–529.

Hansen, M. H. and Hurwitz, W. N. (1949). "On the determination of optimum probabilities in sampling." *Ann. Math. Statist.*, **20**, No. 3, 426–432.

Hansen, M. H. Hurwitz, W. N., and Bershad, M. A. (1961). "Measurement errors in censuses and surveys." *Proc. Internat. Statist. Inst.*, **38**, No. 2, 359–374.

Hansen, M. H., Hurwitz, W. N., and Madow, W. G. (1953). *Sample Survey Methods and Theory* (**I** and **II**). New York: Wiley.

Hansen, M. H., Hurwitz, W. N., Nisselson, H., and Steinberg, J. (1955). "The redesign of the census current population survey," *J. Amer. Statist. Assoc.*, **50**, 701–719.

Hansen, M. H., Hurwitz, W. N., and Pritzker, L. (1965). "The estimation and interpretation of gross differences and the simple response variance." In *Contributions to Statistics Presented to Professor P. C. Mahalanobis, on the Occasion of his 70th Birthday*. Oxford: Pergamon, 111–136.

Hansen, M. H., Madow, W. G., and Tepping, B. J. (1983). "An evaluation of model-dependent and probability-sampling inferences in sample surveys." *J. Amer. Statist. Assoc.*, **78**, 776–793. Comments and rejoinder, 794–807.

Hartley, H. O. (1974). "Multiple frame methodology and selected applications." *Sankhyā Ser. C*, **36**, 99–118.

Horvitz, D. G. and Thompson, D. J. (1952). "A generalization of sampling without replacement from a finite universe," *J. Amer. Statist. Assoc.*, **47**, 663–685.

Jensen, A. (1926). "Report on the Representative Method in Statistics." *Proc. Internat. Statist. Inst.*, **XXII**, lère livre, 359–439.

Jessen, R. J. (1942). *Statistical Investigation of a Sample Survey for Obtaining Farm Facts*. Ames, Iowa: Agricultural Experiment Station, Iowa State College of Agriculture and Mechanic Arts, 304.

Kellerer, H. (1953). *Theorie und Technik des Stichprobenverfahrens*. München: Einzelschriften der Deutschen Statistischen Gesellschaft.

Keyfitz, N. (1980). "Satellite photography can improve population statistics." *Findings* (A Quarterly Summary of Findings from Social Policy Research). Cambridge, MA: Abt Associates Inc.), **2**, No. 7, 2–3.

Kiaer, A. N. (1976). "The representative method of statistical surveys" (original in Norwegian; translation for new edition). *Kristiania Videnskabsselskabets Skrifter, Historisk-filosofiske Klasse*, 1897, No. 4 (Oslo: Statistisk Sentralbyra), 37–56.

Kish, L. (1965). *Survey Sampling*. New York: Wiley.

Kish, L. and Frankel, M. R. (1974). "Inference from complex samples." *J. Roy. Statist. Soc. Ser. B*, **36**, 1–37.

Krewski, D. and Rao, J. N. K. (1981). "Inference from stratified samples: Properties of the linearization, jackknife and balanced repeated duplication methods." *Ann. Statist.*, **9**, 1010–1019.

Kruskal, W. H. and Mosteller, F. (1979). "Representative sampling, III: The current statistical literature." *Internat. Statist. Rev.*, **47**, 245–265.

Kruskal, W. H. and Mosteller, F. (1980), "Representative sampling, IV: The history of the concept in statistics, 1895–1939," *Internat. Statist. Rev.*, **48**, 169–195.

Madow, W. G. and Madow, L. H. (1944). "On the Theory of Systematic Sampling, I." *Ann. Math. Statist.*, **15**, 1–24.

Mahalanobis, P. C. (1946). "On large-scale sample surveys." *Philos. Trans. Roy. Soc. London Ser. B*, **231**, 329–451.

McCarthy, P. J. (1966). *Replication: An Approach to the Analysis of Data from Complex Surveys*. Vital and Health Statistics, Ser. 2, No. 14. Washington: U.S. Government Printing Office.

Mellor, R. W. (1973). *Subsample Replication Variance Estimators*. Unpublished doctoral dissertation. Harvard Univ.

Mosteller, F. (1978). "Nonsampling errors." In W. M. H. Kruskal and J.M. Tanur (eds.), *The International Encyclopedia of Statistics*. New York: Free Press, 208–229.

Neter, J. and Waksberg, J. (1964). "A study of response errors in expenditure data from household interviews." *J. Amer. Statist. Assoc.*, **59**, 18–55.

Neyman, J. (1934). "On the two different aspects of the representative method: The method of stratified sampling and the method of purposive selection." *J. Roy. Statist. Soc.*, **97**, 558–606. Discussion, 607–625.

Neyman, J. (1938). Contribution to the theory of sampling human populations." *J. Amer. Statist. Assoc.*, **33**, 101–116.

Neyman, J. (1952). *Lectures and Conferences on Mathematical Statistics and Probability*. Washington: Graduate School, U. S. Department of Agriculture.

Osborne, J. G. (1942). "Sampling errors of systematic and random surveys of cover-type areas." *J. Amer. Statist. Assoc.*, **37**, 256–264.

Panel on Incomplete Data, Committee on National Statistics (1983). *Incomplete Data in Sample Surveys. Vol. 1, Report and Case Studies* (Wm. G. Madow, H. Nisselson, and I. Olkin, eds.); *Vol. 2, Theory and Bibliographies* (Wm. Madow, I. Olkin, and D. Rubin, eds.); *Vol. 3, Proceedings of the Symposium* (Wm. Madow and I. Olkin, eds.). New York: Academic.

Patterson, H. D. (1950). "Sampling on successive occasions with partial replacement of units." *J. Roy. Statist. Soc. Ser. B*, **12**, 241–255.

Pritzker, L., Ogus, J., and Hansen, M. H. (1966). "Computer editing methods—some applications and results." *Proc. Internat. Statist. Inst.*, **39**, No. 1, 442–466.

Reid, C. (1982). *Neyman from Life*. New York: Springer.

Royall, R. M. (1970). "On finite population sampling theory under certain linear regression modes." *Biometrika*, **57**, 377–387.

Sirken, M. G. (1970). "Household surveys with multiplicity." *J. Amer. Statist. Assoc.*, **65**, 257–266.

Smith, H. F. (1938). "An empirical law describing heterogeneity in the yields of agricultural crops." *J. Agricultural Sci.*, **28**, 1–23.

Smith, T. M. F. (1976). "The foundations of survey sampling: A review." *J. Roy. Statist. Soc. Ser. A*, **139**, 183–204.

Stephan, F. F. (1948). "History of the uses of modern sampling procedures." *J. Amer. Statist. Assoc.*, **43**, 12.

Stephan, F. F., Deming, W. E., and Hansen M. H. (1940). "The sampling procedure of the 1940 Population Census." *J. Amer. Statist. Assoc.*, **35**, 615–630.

Sukhatme, P. V. (1966). "Major developments in sampling theory and practice." In F. N. David (ed.), *Research Papers in Statistics: Festschrift for J. Neyman*. New York: Wiley, 367–409.

Sukhatme, P. V. and Sukhatme, B. V. (1970). *Sampling Theory of Surveys with Applications*, 2nd edn. Ames, IA: Iowa State U.P.

Tepping B. J. (1968). "The estimation of variance in complex surveys." *Proc. Social Statist. Sect., Amer. Statist. Assoc.*, 11–18.

Tepping, B. J., Hurwitz, W. N., and Deming, W. E. (1943). "On the efficiency of deep stratification in block sampling." *J. Amer. Statist. Assoc.*, **38**, 93–100.

Thionet, P. (1946). *Méthodes Statistiques Modernes des Administrations Federales aux Etats-Unis*. Paris: Hermann.

Tschuprow, A. A. (1923). "On the mathematical expectation of the moments of frequency distributions in the case of correlated observation." *Metron*, **2**, 461–493, 646–680.

Tukey, J. W. (1958). "Bias and confidence in not quite large samples" (abstract). *Ann. Math. Statist.*, **29**, 614.

United States Bureau of the Census (1978). *The Current Population Survey: Design and Methodology*. Washington: U.S. Government Printing Office.

Westat, Inc. (1982). *Nonresidential/Commercial Building Energy Consumption Feasibility Study*. Unpublished report. Available from National Technical Information Service, Washingtion, D.C.

Woodruff, R. S. (1959). "The use of rotating samples in the Census Bureau's monthly surveys." *Proc. Social Statist. Sect. Amer. Statist. Assoc.*, 130–138.

Woodruff, R. S. (1971). "A simple method for approximating the variance of a complicated estimate." *J. Amer. Statist. Assoc.*, **66**, 411–414.

Woodruff, R. S. and Causey, B. D. (1976). "Computerized method for approximating the variance of a complicated estimate." *J. Amer. Statist. Assoc.*, **71**, 315–321.

Yates, F. (1946). "A review of recent statistical developments in sampling and sampling surveys." *J. Roy. Statist. Soc.*, **109**, 12–43.

Yates, F. (1949). *Sampling Methods for Censuses and Surveys*. London: Charles Griffin.

Yates, F. (1981). *Sampling Methods for Censuses and Surveys*, 4th edn., London: Charles Griffin.

Yates, F. and Grundy, P. M. (1953). "Selection without replacement from within strata with probability proportional to size." *J. Roy. Statist. Soc. Ser. B*, **15**, 253–261.

CHAPTER 14

On Some New Probabilistic Developments of Significance to Statistics: Martingales, Long Range Dependence, Fractals, and Random Fields

C. C. Heyde

Abstract

Methodology in statistics has traditionally relied on independence based theory and its utilization through strategic transformation. However, new developments in probability and stochastic processes foreshadow increasing departure from this tradition. In this paper a brief discussion is given of the topics of martingales, processes with various asymptotic independence properties, fractals, and random fields. The relevance of these processes for modeling purposes is sketched.

1. Introduction

A centenary offers a particular opportunity to take views both in retrospect and prospect, and we shall begin with some remarks of a historical nature before turning our attention to new developments and their potential. We wish to emphasize the role of the ISI in fostering dialogue between different branches of statistics.

The forerunner to the ISI was a series of eight international statistical congresses. The first of these was held in Brussels in 1853, and the last in Paris in 1878. Their chief object was to promote the organization of official statistics and to facilitate comparability of reports from the various statistical offices, and their demise came as a result of dissent over efforts to bind governments by the resolutions of their meetings. A detailed account of the congress period is given in Westergaard (1932, Chapter 14); for further perspective on this period and the formation of the ISI see Nixon (1960, Part I).

The congresses were oriented towards official statistics, but some had a significant scientific component. Nevertheless, around the end of the congress period it was widely felt that an organization should be created which was

Key words and phrases: fractals, history of ISI, long range dependence, martingales, maximum likelihood estimators, random fields, self-similar processes, time series.

more effective than the congresses in uniting the various branches of statistics. This was very much in mind when the ISI was founded in 1885.

Furthermore, since its creation the ISI has, from time to time, given particular attention to what is often described as the gulf between theory and practice in statistics. There was considerable consideration of these issues in the period after the second World War, and the most recent attention given to the subject resulted in a comprehensive report produced by a Committee on the Integration of Statistics which was delivered to the 42nd Session of the ISI in Manila in 1979 and appeared in Volume 48 of the *Bulletin of the ISI*.

An important consideration in the integration of statistics, and one which was not specifically addressed in the abovementioned report, is the differences in the natural evolution of techniques within the various parts of the subject. In particular, there is a strong tendency for probability, at least as it is often taught, to evolve into a subfield of pure mathematics conducted in a specialized language. It is, however, restrained in this by an ever diversifying contact with applications and consequent pressure from users. Attention to the usefulness of the subject is consequently of fundamental importance.

It is our objective in this paper to describe briefly some recent developments in probability and stochastic processes which seem likely to have considerable practical importance as well as theoretical significance. Our point of departure from the classical framework is the thesis that dependence is endemic and that the statisticians' traditional reliance on theory based on independence is an unnecessary and undesirable constraint.

We shall begin our account with the topic of martingales and, more generally, processes with strong asymptotic independence properties. This leads on to processes with long range dependence, to fractals, and lastly to random fields. Some emphasis is given to the relevance of these processes for modeling purposes.

The mathematical prerequisites of these topics are somewhat beyond the normal requirements for statistical practitioners, and this highlights another difficulty in the integration of statistics. The topics to be introduced here are important and offer rich opportunities for statisticians, for example, through the development of appropriate inferential methodology. However, they do represent a significant technical challenge in addition to the scientific one.

2. Limit Theory Beyond Independence

The key prototypes to much of the large sample theory in statistics have been the strong law of large numbers (SLLN) and the central limit theorem (CLT) for sums of independent random variables. The SLLN provides results of strong consistency type, and the CLT provides confidence intervals. Modern developments, however, have shown that generalizations of these results hold under a very wide variety of asymptotic independence type conditions and many of these extensions are of considerable practical significance. This remark applies particularly to the generalizations to martingales, for reasons which we shall now outline.

First we must recall the definition of a martingale. Let $(\Omega, \mathscr{F}, P)$ be a probability space: Ω is a set, $\mathscr{F}$ a σ-field of subsets of Ω, and P a probability measure defined on $\mathscr{F}$. Let I be an ordered set of integers, and $\{\mathscr{F}_n, n \in I\}$ be an increasing sequence of σ-fields of $\mathscr{F}$-sets. Then, the sequence $\{Z_n, n \in I\}$ of random variables on Ω is called a *martingale* if it satisfies the following three conditions:

(i) Z_n is measurable with respect to $\mathscr{F}_n$,
(ii) $E|Z_n| < \infty$,
(iii) $E(Z_n \mid \mathscr{F}_m) = Z_m$ a.s. for all $m < n$, $m, n \in I$.

In many cases of interest $\mathscr{F}_n$ can be interpreted as the history of the process $\{Z_n\}$ up to and including time n.

Now let $\{S_n, \mathscr{F}_n, n \geq 1\}$ be a zero mean martingale, and write $S_n = \sum_{i=1}^n X_i$, $n \geq 1$. Note that a finite variance martingale has uncorrelated increments for, taking $i > j$,

$$E(X_i X_j) = E(X_j E(X_i | \mathscr{F}_{i-1})) = E(X_j \{E(S_i | \mathscr{F}_{i-1}) - S_{i-1}\}) = 0.$$

It turns out that the key sufficiency conditions for SLLN and CLT results have natural extensions to the martingale context. Many detailed results are given in Hall and Heyde (1980); Chapter 1 in particular giving a useful perspective, while later chapters give many detailed results. For our present purposes, however, some basic results will suffice. References can be found in Hall and Heyde (1980, pp. 9, 10).

In the case of the SLLN the most basic result for sums of independent random variables is arguably the Kolmogorov criterion that if the X_i are independent with zero means and finite variances, then $\sum b_n^{-2} E X_n^2 < \infty$ for some sequence $\{b_n\}$ of positive constants with $b_n \uparrow \infty$ ensures that $\lim_{n \to \infty} b_n^{-1} S_n = 0$ a.s. However, this result continues to hold in the case where $\{S_n\}$ is a martingale.

In the case of the CLT there is the following result due to Brown in 1971. Again let $\{S_n, \mathscr{F}_n\}$ denote a zero mean martingale whose increments have finite variance. Write

$$V_n^2 = \sum_1^n E(X_i^2 | \mathscr{F}_{i-1}) \quad \text{and} \quad s_n^2 = E V_n^2 = E S_n^2.$$

If

$$s_n^{-2} V_n^2 \to 1$$

in probability and

$$s_n^{-2} \sum_{i=1}^n E(X_i^2 I(|X_i| \geq \varepsilon s_n)) \to 0$$

as $n \to \infty$ for all $\varepsilon > 0$ [$I(\cdot)$ denotes the indicator function], then $s_n^{-1} S_n$ converges in distribution to the unit normal law. This result reduces to the sufficiency part of the standard Lindeberg–Feller result in the case of independent random variables, as the condition on $s_n^{-2} V_n^2$ is then trivially satisfied.

Martingale limit results are potentially useful in any large sample context in which conditional expectations, given the past, have a simple form. This is a consequence of the simplicity and tractability of the martingale defining property. If $\{Z_n, n \geq 0\}$ is any sequence of random variables with finite means, then

$$\left\{\sum_{i=1}^{n} [Z_i - E(Z_i|Z_{i-1}, \ldots, Z_1)]\right\}$$

is a martingale relative to the sequence of σ-fields generated by Z_i, $i \leq n$.

Applications of martingale limit results based on this construction abound, but martingales also occur naturally in a wide variety of important contexts, and many examples are given in texts such as Karlin and Taylor (1975, Chapter 6). A particularly significant one is that the derivative of a log likelihood is a martingale under mild regularity conditions, and consequently martingale limit theory has an important role to play in an investigation of the properties of maximum likelihood estimators for general processes.

To see how this works we consider a sample $X_1, X_2, \ldots, X_n$ from some stochastic process whose parameter depends on a single parameter θ lying in an open interval. Let $L_n(\theta)$ be the likelihood function associated with $X_1, X_2, \ldots, X_n$, and suppose that $L_n(\theta)$ is differentiable with respect to θ. Suppose in addition that if $P_n(X_1, \ldots, X_n) = L_n(\theta)$ is the joint probability (density) function of $X_1, \ldots, X_n$, then $\sum_{x_n} P_n(x_1, \ldots, x_n)$ ($\int P_n(x_1, \ldots, x_n) dx_n$) can be differentiated with respect to θ under the summation (integration) sign. Then, $\{d \log L_n(\theta)/d\theta, n \geq 1\}$ is a martingale with respect to the (past history) σ-fields generated by $X_1, \ldots, X_k$, $k \geq 1$. Partial Taylor series expansion of $d \log L_n(\theta)/d\theta$ provides a route to the study of the properties of the maximum likelihood estimator, and the framework is a natural extension of the familiar one for the classical situation of random sampling [independent and identically distributed (i.i.d.) X_i], in which case the increments of $d \log L_n(\theta)/d\theta$ are i.i.d. with zero mean. A detailed discussion of this approach is given in Basawa and Prakasa Rao (1980, Chapter 7) and Hall and Heyde (1980, Chapter 6). Substantial progress has been made in extending the classical theory of inference, but many significant questions remain largely unresolved.

Another important example occurs in the study of the stationary linear process

$$\begin{gathered} x(n) - \mu = \sum_{j=0}^{\infty} \alpha(j)\varepsilon(n-j), \qquad \sum_{j=0}^{\infty} \alpha^2(j) < \infty, \qquad \alpha(0) = 0; \\ E\varepsilon(n) = 0, \qquad E\{\varepsilon(m)\varepsilon(n)\} = 0, \quad m \neq n. \end{gathered} \tag{2.1}$$

Every stationary purely nondeterministic process with finite variance may be represented in this form with the $\varepsilon(n)$ as the (linear) prediction errors.

Now, the classical theory of inference for the process (2.1) requires that the $\varepsilon(n)$ be i.i.d. with zero mean and variance σ^2. However, subject to some reasonable additional conditions on variances, the classical theory still holds if the independence assumption is replaced by the condition

$$E(\varepsilon(n)|\mathscr{F}_{n-1}) = 0 \quad \text{a.s.} \qquad \text{for all } n, \tag{2.2}$$

$\mathscr{F}_n$ being the σ-field generated by $\varepsilon(m)$, $m \le n$. This requirement, namely that the (linear) prediction errors $\varepsilon(n)$ are martingale differences, has a simple and natural interpretation, for then $\mathscr{F}_n$ is also the σ-field generated by $x(m)$, $m \le n$, and

$$\varepsilon(n) = x(n) - E(x(n)|\mathscr{F}_{n-1}).$$

Now $E(x(n)|\mathscr{F}_{n-1})$ is the best predictor of $x(n)$. The martingale condition (2.2) is thus equivalent to the condition that the best linear predictor is the best predictor (both in the least squares sense). This is a very natural requirement for use of a linear model.

It is shown in Hannan and Heyde (1972) that the classical inferential theory for the process (2.1), based on the sample autocorrelations, continues to hold under the condition (2.2) together with $E(\varepsilon^2(n)|\mathscr{F}_{n-1}) = \sigma^2$ a.s.; see also Hall and Heyde (1980, Chapter 6). Naturally limit theorems for martingales play a key role in this robustness study.

The above discussion refers to traditional linear time series methodology, which, until recently, had represented an overwhelming proportion of the work in the area. This theory has proved extremely useful, but nonlinear relationships are endemic in most scientific disciplines and need to be treated. Comparatively little success has yet been achieved in this area partly because of the lack of tractable prototype models, but it is undergoing rapid development.

One useful class of nonlinear models which has been receiving considerable attention is based on the bilinear form

$$\sum_{j=0}^{p} \alpha_j x(t-j) + \sum_{j=0}^{q} \beta_j \varepsilon(t-j) + \sum_{k=0}^{Q} \sum_{l=1}^{P} \gamma_{kl}\varepsilon(t-k)x(t-l) = 0, \tag{2.3}$$

where p, q, P, Q are finite, and it is assumed that

$$\alpha_o = \beta_o = 1, \qquad \sum_{j=0}^{p} \alpha_j z^j \neq 0, \quad |z| \le 1,$$

and that the $\varepsilon(t)$ are i.i.d. with zero mean and variance σ^2 (e.g. Granger and Andersen, 1978). The basic questions are (i) whether there exists a stationary solution to (2.3) in terms of the $\varepsilon(s)$, $s \le t$, and (ii) if such a stationary solution exists, whether the $\varepsilon(t)$ are the nonlinear prediction errors [i.e. whether $\varepsilon(t) = x(t) - E(x(t)|\mathscr{F}_{t-1})$, where $\mathscr{F}_t$ is the σ-field generated by $x(s)$, $s \le t$]. Rather little progress has as yet been made on these questions; for very recent contributions see Hannan (1982), Quinn (1982), and references therein. Problem (ii) is, of course, a martingale question, and inference for the process (2.3) is conveniently approached using martingale limit theory. This general class of problems seems bound to attract substantial attention in the near future.

There are many situations of asymptotic independence (short range dependence) which are not covered directly by martingale methodology but for which it nevertheless provides what is arguably the most powerful general tool available for establishing results of SLLN and CLT type. Strong arguments for this point of view are given in Hall and Heyde (1980, Chapter 1), and, as noted previously, SLLN and CLT type results have a crucial role, since they respectively provide the basis for strong consistency and confidence intervals for parameter estimators.

For example, many limit results are based on mixing conditions of asymptotic independence, the most common of which are those called uniform mixing and strong mixing. Let $\{X_n, n \geq 1\}$ be a sequence of random variables, and write $\mathscr{B}_r^s$ for the σ-field generated by $\{X_r, \ldots, X_s\}$, $1 \leq r \leq s \leq \infty$. The sequence $\{X_n\}$ is called *uniform* [or *strong*] *mixing* if there exists a positive integer N and a function f defined on the integers $n \geq N$ such that $f \downarrow 0$ and for all $n \geq N$, $m \geq 1$, $A \in \mathscr{B}_1^m$, and $B \in \mathscr{B}_{m+n}^\infty$,

$$|P(A \cap B) - P(A)P(B)| \leq f(n)P(A)$$

[or

$$|P(A \cap B) - P(A)P(B)| \leq f(n)$$

respectively].

Standard results for sums of independent random variables continue to hold with minor modification in the mixing sequence context provided that $f \downarrow 0$ with sufficient rapidity.

Mixing conditions are often difficult to check in practice, especially when the rate at which $f \downarrow 0$ is crucial and needs to be estimated. However, as indicated earlier, most results based on mixing conditions can be obtained from others based on martingales or so-called mixingales which are of wider applicability.

The concept of mixingales was introduced by McLeish (1975) as a hybrid of martingales and mixing sequences. Let $\{X_n, n \geq 1\}$ be a sequence of random variables with zero mean and finite variance on a probability space $(\Omega, \mathscr{F}, P)$, and let $\{\mathscr{F}_n, -\infty < n < \infty\}$ be an increasing sequence of sub-σ-fields of $\mathscr{F}$. Then $\{X_n, \mathscr{F}_n\}$ is called a *mixingale* (*difference*) *sequence* if, for sequences of nonnegative constants $\{c_n\}$ and $\{\psi_m\}$ with $\psi_m \to 0$ as $m \to \infty$, we have

$$E(E(S_m | \mathscr{F}_{n-m}))^2 \leq \psi_m^2 c_n^2$$

and

$$E(X_n - E(X_n | \mathscr{F}_{n+m}))^2 \leq \psi_{m+1}^2 c_n^2$$

for all $n \geq 1$ and $m \geq 0$. A discussion of mixingales is given in Hall and Heyde (1980, Chapter 2); for central limit results see McLeish (1975, 1977).

Rapid developments of substantial significance have recently been made in the provision of central limit results of broad applicability. A battery of new and very general results on CLTs for semimartingales has been obtained by Liptser and Shiryaev, and these are described in the review article Shiryaev (1981). The class of semimartingales, whose definition is rather opaque and consequently we shall omit, is very wide; it includes arbitrary random sequences and many processes in continuous time.

The above account has been concerned with the case of discrete time, but an extensive theory has also been developed in continuous time. Here martingales play a special role in the calculus of stochastic integrals (e.g. Liptser and Shiryaev, 1977). One recent development which is important from a statistical point of view is the use of martingale theory in a fundamental way to develop a comprehensive theory of point processes based on the concept of intensity (Brémaud, 1981). In particular, this new focus, which sharply contrasts with the previous standard approach via discrete random measures, facilitates the consideration of many inferential questions via likelihood ratio methods.

3. Long Range Dependence

At present there appears to be little prospect of characterizing the kind of dependence structure for, say, a stationary sequence $\{X_j, j \geq 1\}$ of random variables which allows convergence to normality of $\sum_{j=1}^{n} X_j$ when appropriately normalized. In broad terms, however, asymptotic normality requires at most weak dependence, and other possibilities emerge as the dependence strengthens.

The study of these possibilities is difficult, since we are no longer on familiar ground which allows the mimicking of independence based results. Consequently, the state of knowledge is still in the exploratory phase and fundamental new insights are awaited. A useful account of the topic is contained in Taqqu (1979).

There has also been considerable recent interest in testing for, explaining, and estimating the presence of long range dependence in empirical records or in sample functions. Long range dependence turns out to be widespread in nature and is characteristic of many hydrological, geophysical, and economic records.

A useful approach to the recognition and estimation of long range dependence was pioneered by H. E. Hurst in a series of papers on hydrology commencing in 1951. Hurst was an English engineer working in Cairo who advanced the idea of the Aswan High Dam on the river Nile, and he developed a method of evaluating the optimal size of a dam for a time horizon of d years. His work, and other more recent contributions on long range dependence, is described in Mandelbrot and Taqqu (1979) and used as a basis for the introduction of a very general technique, called R/S analysis, for use on empirical records.

For a sequence $X(t)$, $t = 1, 2, \ldots$, of observations in discrete time, we follow Mandelbrot and Taqqu (1979) and write for $t \geq 0$

$$\bar{X}(0) = 0, \qquad \bar{X}(t) = \sum_{i=1}^{[t]} X(i), \qquad \bar{X}_2(t) = \sum_{i=1}^{[t]} X^2(i),$$

$[t]$ being the integer part of t, and for lag $d > 0$ define

$$R(d) = \max_{0 \leq u \leq d} \{\bar{X}(u) - ud^{-1}\bar{X}(d)\} - \min_{0 \leq u \leq d} \{\bar{X}(u) - ud^{-1}\bar{X}(d)\},$$

$$S^2(d) = d^{-1}\bar{X}_2(d) - (d^{-1}\bar{X}(d))^2,$$

and

$$Q(d) = R(d)/S(d).$$

R is the adjusted range of $X(t)$ in the time interval $[0, d]$, and rescaling using the standard deviation S produces the R/S statistic Q. This statistic appears to be sensitive to long range correlation but robust against extreme changes in the marginal distribution of the $X(i)$.

To interpret R/S in a hydrological context we may suppose that $X(t)$ is the total discharge of a river into a reservoir during year t. Then $\bar{X}(u)$ is the total discharge in the first u years. If a time horizon of d years is set, we may suppose that a fixed quantity of water $d^{-1}\bar{X}(d)$ could be withdrawn each year from the reservoir. Then, at year u, $1 \leq u \leq d$, the amount of water in the reservoir exceeds the initial amount by $\Delta(u) = \bar{X}(u) - ud^{-1}\bar{X}(d)$, and if the reservoir is never to lack water during these d years, its initial content must be at least $-\min_{0 \leq u \leq d} \Delta(u)$. The quantity $R(d)$ then gives a useful measure for the volume of a reservoir designed to avoid overflow and to allow regular withdrawals. Rescaling using division by $S(d)$ makes Q independent of changes of scale (as well as location).

Now it turns out that, under a wide variety of distributional assumptions, there exists a real number J such that $R(d)/d^J S(d)$ converges in distribution to a nondegenerate limit law. The parameter J, which necessarily lies in $[0, 1]$, is called the *R/S exponent*.

For example, if the X's are i.i.d. with $EX^2 < \infty$, or more generally belong to the domain of normal attraction of a stable law of index α, $0 < \alpha \leq 2$, then the R/S exponent is $J = \frac{1}{2}$. Indeed, the exponent $J = \frac{1}{2}$ is ubiquitous for classical models, since it also holds when X is a (finite) ARMA (autoregressive moving average)process. Broadly speaking, $J = \frac{1}{2}$ characterizes the short range dependence situation.

Now in the case of data from the river Nile, Hurst found empirically that the appropriate J was approximately 0.7 and similarly high values are not unusual.

A convenient prototype class of processes for the modeling of situations with $J \neq \frac{1}{2}$ is provided by the self-similar processes. A continuous time process $Z(t)$, $-\infty < t < \infty$, is *self-similar* with scaling exponent H if for all $a > 0$, the finite-dimensional distributions of $Z(at)$ are the same as those of $a^H Z(t)$.

Standard Brownian motion is self-similar with $H = \frac{1}{2}$ and the corresponding R/S exponent is $J = \frac{1}{2}$. However, the Gaussian process with mean zero and stationary increments that is self-similar with parameter $H \in [0, 1]$ (called *fractional Brownian motion*) has R/S exponent $J = H$. Fractional Brownian motion has correlation function

$$r(k) \sim H(2H - 1)k^{2H-2} \quad \text{as } k \to \infty$$

and exhibits stronger long range dependence than ordinary Brownian motion for $\frac{1}{2} < H \leq 1$.

Of course in most cases of interest the cumulative process $\bar{X}(t) = \sum_{i=1}^{[t]} X(i)$ is not in itself self-similar, but it may belong to the domain of attraction of a self-similar process. This means that the process $\bar{X}(td)$, when properly normalized, converges weakly to a self-similar process as $d \to \infty$. In this case the R/S exponent J is the same as for the increments of the attracting process. A study of the domains of attraction of the self-similar processes has been provided by Taqqu (1979).

The notion of self-similarity appears to be of fundamental significance and leads into our next topic.

4. Fractals

Many spatial patterns are irregular to such an extreme degree that classical geometry provides little real help in describing their form. For example, the coastline of an island does not have a well-defined length. A piecewise linear approximation (of equal step length) to the coastline will increase without limit as the step length decreases.

Nevertheless, a coastline or similar pattern may be able to be regarded as statistically self-similar, meaning that each portion can be considered as having the same distribution as a reduced scale image of the whole. This property lies behind the concept of a fractal by which a wide variety of apparently highly irregular phenomena can be described. Not all fractals, however, need be statistically self-similar.

The concept of fractals was pioneered by B. B. Mandelbrot [see the essay monograph by Mandelbrot (1977) and references therein]. Examples given in the monograph include boundaries of coastlines and clouds, shapes of rivers and snowflakes, rough turbulent wakes, vascular networks in anatomy, clustering of stellar objects, areas of moon craters, and surface topographies. Many striking pictorial illustrations are also given.

As a prelude to defining fractals we first note that the loose notion of dimension invites several distinct approaches. For example, Brownian motion in the plane is *topologically* of dimension 1. However, being effectively plane filling it has *Hausdorff* dimension 2. A discrepancy between these two values qualifies a process as being a fractal.

The concept of dimension was vague and intuitive until the late nineteenth century. Before that time, "n-dimensional" meant that a minimum of n parameters were needed for description of the points. However, Cantor's discovery of a one-to-one correspondence between points of a line and points of a plane showed that "dimension" had nothing to do with the number of points involved. Furthermore, Peano's discovery of a continuous mapping of the interval $[0, 1]$ onto the square $[0, 1] \times [0, 1]$ showed that "dimension" could not be defined in terms of the least number of continuous parameters needed to describe a space. What was finally resorted to, and formalized by Brouwer, was an inductive definition that involved looking at the subsets of a space which disconnect it. Intuitively it goes as follows: lines, which can be divided by cuts which are not continua, will be continua of (topological) dimension one; surfaces, which can be divided by continuous cuts of dimension one, will be continua of dimension two; space, which can be divided by continuous cuts of two dimensions, will be a continuum of dimension three; and so on.

The concept of Hausdorff dimension is much more complicated. N-dimensional Lebesgue measure is not always an appropriate tool for measuring the size of erratic sets, and a class of fractional dimensional measures can be constructed which is much more flexible.

Let A be a subset of $\mathbb{R}^N$ (N-dimensional Euclidean space). The α-dimensional outer measure $S_\alpha(A)$ is defined for positive α as

$$S_\alpha(A) = \inf_{\varepsilon \downarrow 0} \inf \sum_i (\text{diameter } B_i)^\alpha$$

where, for each $\varepsilon > 0$, the second infimum is taken over all collections of open spheres $\{B_i\}$ in $\mathbb{R}^N$ whose union covers A and for which the diameter of each B_i is not greater than α. For each A in $\mathbb{R}^N$ there is a unique number α^* called the Hausdorff dimension of A defined by

$$\alpha^* = \sup\{\alpha : S_\alpha(A) = \infty\} = \inf\{\alpha : S_\alpha(A) = 0\}.$$

A detailed discussion of Hausdorff dimension is given, for example, in Adler (1981a, Chapter 8).

Formally then, fractal sets are defined as those for which the Hausdorff dimension exceeds the topological dimension. For example, the Cantor ternary set has topological dimension zero and Hausdorff dimension $(\log 2)/(\log 3)$.

Coastlines can be interpreted as approximate fractal curves whose Hausdorff dimension D exceeds unity. For example, L. F. Richardson found, from measurements, $D = 1.02$, 1.13, and 1.25 for the coasts of South Africa, Australia, and Western Britain respectively (Mandelbrot, 1967; 1977, p. 32). Other examples given by Mandelbrot (1977) include turbulence and surface topography (both producing values of D in the range $2 < D < 3$).

Descriptions using fractals are mostly qualitative at this stage, and there is considerable potential for further developments including, for example, estimation and testing procedures.

5. Random Fields

Random fields are stochastic processes of the kind $X(\mathbf{t}), \mathbf{t} \in \mathbb{R}^N$ (N-dimensional Euclidean space)—that is, processes with a multidimensional argument. These are needed, for example, for the treatment of surfaces; the rough surface of a metal can be modeled by a two dimensional random field, while a water surface requires three dimensions, the third being time.

The study of random fields has developed along two quite different pathways, corresponding to the discrete and continuous cases. The difficulties of ordering of the parameter space make all this very different from ordinary multivariate processes. The theory of the discrete case is more highly developed and has been strongly influenced by corresponding developments in statistical mechanics. It includes, in particular, the case of binary-valued Markov random fields which may be used as models for spatial point processes with interactions between the points.

A random field is said to have the Markov property if the conditional distribution of the process within a designated region, given all values taken outside the region, always depends only on those values in the boundary region. This property has been widely used in modeling, and many interesting applications have been studied in the domain of spatial point processes. A recent review of this area is given in Isham (1981).

One unusual application of direct statistical significance exploits a connection between Markov random fields on a finite set of states and log–linear interaction models for contingency tables (Darroch, Lauritzen, and Speed, 1980). The set of sites on which the random field is defined corresponds to the set of factors with respect to which objects in the experiment are to be classified, while the values of the site variables represent the factor levels. Each possible realization of the Markov random field represents a particular cell in a contingency table. The Markov random field formulation provides a very convenient framework for the examination of interaction components making up log $\mu(\mathbf{x})$, the logarithm of the probability of a random element falling in cell $\mathbf{x}$ of the table.

In contrast with the case of discrete random fields, the theory for random fields with continuous sample paths is in a much earlier stage of development but is also undergoing lively growth. A well-documented treatment of the current state of the subject, and the only one as yet in book form, has been provided by Adler (1981a). This gives emphasis to sample function behavior, level crossings, and local maxima of processes and not to applications.

The most important special class of random fields with continuous sample paths is the class of Gaussian fields—that is, the class of fields possessing finite dimensional distributions all of which are multivariate Gaussian (normal). It is not surprising that this class rather dominates the subject, because the convenient analytic form of the multivariate normal density makes it possible to obtain explicit results for Gaussian fields, and simple functions thereof, which are at present unattainable for more general processes.

There is basically a dichotomy of types of random fields with continuous sample paths. There are those for which the sample functions are smooth and possess nice continuity and differentiability properties, and there are others that tend to be extremely erratic. The study of level crossings for nice random fields involves traditional ideas of integral geometry and differential topology. However, the Hausdorff dimension, as previously discussed in Section 4, is a necessary tool for the study of level crossings for the erratic random fields. For an erratic two dimensional random fields $X(\mathbf{t})$, level sets such as

$$L = \{\mathbf{t} \in (0, 1] \times (0, 1] : X(\mathbf{t}) = 0\}$$

are often fractals. This is the case, for example, for the so-called Brownian sheet, the zero mean Gaussian field on

$$\{\mathbf{t} = (t_1, t_2) \in \mathbb{R}^2 : t_1 \geq 0,\ t_2 \geq 0\}$$

with covariance

$$E\{W(s)W(t)\} = \min(s_1, t_1)\min(s_2, t_2).$$

The choice of model in the random field context can be far from straightforward and poses interesting statistical problems. For example, it is clear that the grain and straw yield block diagrams given in Figures 1 and 2 of McBratney and Webster (1981) can be treated as random fields, but the model building question is quite unclear. Often data are in the form of families of contour lines. This is particularly the case for random field data coming from geographical disciplines. From the contour information it is necessary to model the structure of the field throughout its region of definition. A discussion of theoretical aspects of this problem in the Gaussian case is given by Wilson and Adler (1982), using an approach developed by Lindgren (1972), but the associated practical questions have hardly been touched. Furthermore, the inferential procedures that could most readily be developed would suffer by relying on asymptotics and approximations. There is undoubtedly scope for the adoption of some of the methods that have been developed for Markov random fields in the spatial point process context, but even there the approaches to the fitting of models have been rather exploratory (e.g. Isham, 1981, Section 4.3).

One context in which Gaussian random fields have been widely used is that of modeling surface roughness, for which they were introduced by Greenwood and Williamson (1966). All surfaces used in engineering practice are rough at the microscopic level, and a theoretical treatment of phenomena such as wear and adhesion requires the modeling of such surfaces.

The Gaussian random field model has been of considerable explanatory value, but recent work suggests that it is only really appropriate for freshly ground and unworn surfaces. For example, abrasion can produce marked skewness in histograms of surface heights, and this led Adler and Firman (1981) to investigate an alternative χ^2-model.

Let $X(s, t)$ represent the height of the surface over some notional base plane over which s and t range. Adler and Firman assumed that

$$X(s, t) = M - Y(s, t),$$

where M is a constant and $Y(s, t)$ is a χ_n^2-field [i.e., $Y(s, t)$ can be written in the form

$$Y(s, t) = \sum_{k=1}^{n} X_k^2(s, t)$$

where the $X_k(s, t)$ are independent Gaussian random fields with zero mean and identical covariance functions]. This model produces appropriately skewed surface height distributions and suggests that the contact area A between two surfaces should be proportional to L^α, L being the load imposed to force them together and α some constant less than unity (Adler, 1981b). The Gaussian model had suggested $\alpha = 1$, but Woo and Thomas (1980), in a recent review of experimental evidence, find a value of α around 0.76.

The χ^2 random fields, being simply related to Gaussian ones, offer reasonable analytical tractability and seem destined to provide a very useful class of models. For an engineering application to the assessment of survival of structures subject to a random load (such as the wind) see Lindgren (1980). However, there is a great need for a wider range of usable distributional forms than are covered by Gaussian fields and simple transformations thereof.

6. Concluding Remarks

The emphasis in this paper has been on recent developments and attendant opportunities. We have sought to show that probability is currently providing a wealth of relevant and stimulating theory for statisticians to take up and further develop. This, we believe, is auspicious for the future of our discipline.

Bibliography

Adler, R. J. (1981a). *The Geometry of Random Fields*. Chichester: Wiley.

Adler, R. J. (1981b). "Random field models in surface science." Invited paper presented at 43rd Session of ISI, Buenos Aires. *Proc. ISI*, **49**, 669–680.

Adler, R. J. and Firman, D. (1981). "A non-Gaussian model for random surfaces." *Philos. Trans. Roy. Soc. London Ser. A*, **303**, 433–462.

Basawa, I. V. and Prakasa Rao, B. L. S. (1980). *Statistical Inference for Stochastic Processes*. London: Academic.

Brémaud, P. (1981). *Point Processes and Queues. Martingale Dynamics*. New York: Springer.

Darroch, J. N., Lauritzen, S. L., and Speed, T. P. (1980). "Markov fields and log–linear interaction models for contingency tables." *Ann. Statist.*, **8**, 522–539.

Granger, C. W. and Andersen, A. (1978). *An Introduction to Bilinear Time Series Models*. Göttingen: Vandenhoek and Ruprecht.

Greenwood, J. A. and Williamson, J. B. P. (1966). "Contact of nominally flat surfaces." *Proc. Roy. Soc. London Ser. A*, **295**, 300–319.

Hall, P. and Heyde, C. C. (1980). *Martingale Limit Theory and Its Application*. New York: Academic.

Hannan, E. J. (1982). "A note on bilinear time series models." *Stochastic Processes Appl.*, **12**, 221–224.

Hannan, E. J. and Heyde, C. C. (1972). "On limit theorems for quadratic functions of discrete time series." *Ann. Math. Statist.*, **43**, 2058–2066.

Isham, V. (1981). "An introduction to spatial point processes and Markov random fields." *Internat. Statist. Rev.*, **49**, 21–43.

Karlin, S. and Taylor, H. M. (1975). *A First Course in Stochastic Processes*. New York: Academic.

Lindgren, G. (1972). "Local maxima of Gaussian fields." *Ark. Math.*, **10**, 195–218.

Lindgren, G. (1980). "Extreme values and crossings for the χ^2-process and other functions of multidimensional Gaussian processes, with reliability applications." *Adv. Appl. Probab.*, **12**, 746–774.

Liptser, R. S. and Shiryaev, A. N. (1977). *Statistics of Random Processes I. General Theory*. New York: Springer.

McBratney, A. B. and Webster, R. (1981). "Detection of ridge and furrow pattern by spectral analysis of crop yield." *Internat. Statist. Rev.*, **49**, 45–52.

McLeish, D. L. (1975). "Invariance principles for dependent variables." *Z. Wahrsch. Verw. Gebiete*, **32**, 165–178.

McLeish, D. L. (1977). "On the invariance principle for nonstationary mixingales." *Ann. Probab.*, **5**, 616–621.

Mandelbrot, B. B. (1967). "How long is the coast of Britain? Statistical self-similarity and fractional dimension." *Science*, **156**, 636–638.

Mandelbrot, B. B. (1977). *Fractals. Form, Chance and Dimension*. San Francisco: Freeman.

Mandelbrot, B. B. and Taqqu, M. S. (1979). "Robust R/S analysis of long run serial correlation." Invited paper presented at 42nd Session of the ISI, Manila. *Proc. ISI*, **48**, Book 2, 69–99.

Nixon, J. W. (1960). *A History of the International Statistical Institute: 1885–1960*. The Hague: International Statistical Institute.

Quinn, B. G. (1982). "Stationarity and invertibility of simple bilinear models." *Stochastic Processes Appl.*, **12**, 225–230.

Shiryaev, A. N. (1981). "Martingales: Recent developments, results and applications." *Internat. Statist. Rev.*, **49**, 199–233.

Taqqu, M. S. (1979). "Self-similar processes and related ultraviolet and infrared catastrophies." In *International Colloquium on Random Fields*, (Esztergom, Hungary, 1979). Budapest: Janós Bolyai Math. Soc.; Amsterdam: North Holland (to appear).

Westergaard, H. (1932). *Contributions to the History of Statistics*. London: King. (Reprinted by Kelley, New York, 1969.)

Wilson, R. J. and Adler, R. J. (1982). "The structure of Gaussian fields near a level crossing." *Adv. Appl. Probab.*, **14**, 543–565.

Woo, K. L. and Thomas, T. R. (1980). "Contact of rough surfaces; A review of experimental work." *Wear*, **58**, 331–340.

CHAPTER 15

Concepts of Robustness

Christopher Field

Abstract

A robust statistical procedure can be thought of as one which performs well over a range of situations and is able to stand up to a certain amount of abuse without breaking down. The development of the principal ideas of robustness is traced from about 1800 to the present, illustrating that scientists and statisticians have been concerned with the sensitivity of statistical procedures over this whole time and that some of the proposed solutions are closely related to robust estimates in use today. A brief overview of current research in robustness is given.

The word robust is often used to describe someone who is strong, vigorous, healthy. In a certain sense, this image carries over to statistics, where a robust procedure can be thought of as one which performs well over a range of situations and is able to stand up to a certain amount of abuse without breaking down. Since at least the early part of the nineteenth century statisticians and scientists have been concerned with the sensitivity of statistical procedures, particularly the mean. In an early paper, the author makes reference to an alternative for the sample mean (Anonymous, 1821, p. 189):

> For example, there are certain provinces of France where to determine the mean yield of a property of land, there is a custom to observe the yield during twenty consecutive years to remove the strongest and the weakest yield and then to take one eighteenth of the sum of the others.

In modern terminology, this procedure of removing the 5% smallest and 5% largest observations and taking the mean of the remaining observations is referred to as a 5% trimmed mean.

At about the time that the least squares criterion was being developed by Gauss, Laplace (1799) proposed the criterion of least absolute error. For the simple case of unstructured univariate observations, least squares gives the sample mean as the best summary measure, while least absolute error gives the median. The approach using least absolute errors was not developed very vigorously over the next fifty years, possibly because of its computational difficulties in comparison with least squares in such models as regression.

Key words and phrases: M-estimates, median, optimal robust procedures, outliers, robustness, trimmed mean.

From the point of view of robustness, the lack of sensitivity of the median to erroneous observations makes it an attractive alternative to the mean. The trimmed means represent a family of estimates ranging from the mean (0% trimmed mean) to the median (50% trimmed mean).

During the second half of the nineteenth century, there was considerable discussion of the rejection of outliers and the differential weighting of observations in a mean. To quote Stigler (1973),

> By the last half of the nineteenth century, weighted least squares had become a standard topic in the literature of the theory of errors, and it was a frequent practice (at least in astronomical investigations) to weight observations differently depending upon the scientist's (often subjective) estimate of the "probable error" [1] of the observation.

This practice of downweighting suspect observations is one of the main techniques used to ensure the robustness of estimates. With regard to formal outlier rejection, there were early proposals by Peirce (1852) and Chauvenet (1863). There seems to have been an implicit assumption that after the outlier test had been performed, the remainder of the data could be treated independently of what information was lost. The narrowness of this view and the faulty view that the remaining observations can be treated as a sample from a normal population were challenged at the time and led to a lively debate on the rejection of outliers. Many current robust procedures do not categorize observations as good or bad, but rather allow for a differential weighting of observations, usually with more distant observations being downweighted.

Newcomb (1886) was the first to deal seriously with the fact that in careful scientific measurements large errors occur more frequently than the normal law would indicate. Newcomb is generally regarded as the greatest American astronomer of the nineteenth century and was accustomed to using weighted means for estimation where the weights were assigned on the assessment of the relative accuracy of the observations. He proposed fitting a mixture of normals to the observations and using the Bayes estimate with respect to a uniform prior. The interested reader can examine a set of careful weighings made by the National Bureau of Standards and reproduced in Freedman, Pisani, and Purves (1978, p. 91), and note that a mixture of normals seems to fit the data very well.

Early examples of robust estimates in use today appear in the works of Smith (1888), Newcomb (1912), and Daniell (1920). All these papers were outside the mainstream of statistical thinking and seem to have had little impact on the statistical community at the time. Detailed accounts of their contributions are found in Stigler (1973, 1980).

During the early part of the twentieth century, there seems to have been little realization that somewhat slight deviations from the normal model may make classical procedures rather inefficient. For instance, in a dispute with

[1] The probable error of a symmetric distribution is half the interquartile range; for a normal distribution p.e. $= 0.6745\sigma$.

Eddington over the use of the absolute average deviation rather than the variance, Fisher noted that the variance is more efficient for normal observations but failed to note that for a very small deviation from the normal, the least absolute deviation becomes more efficient (see Huber, 1972, for example). Student (1927), E. S. Pearson (1931), and Jeffreys (1932) all were concerned with the effect of nonnormality on the performance of tests and estimates, and the emergence of distribution-free procedures in the late forties was, in part, in response to this concern.

In the 1950s, Tukey and the Statistical Research Group at Princeton began to publicize the shortcomings of the classical estimates and to develop viable alternatives to these estimates. This area of research, usually called exploratory or resistant data analysis, has been very active over the past twenty years, producing much innovative work (see Tukey, 1977, and Hoaglin, Mosteller, and Tukey, 1983, for example). In the early sixties, Huber developed a more formal approach leading to the introduction of M-estimates and a study of their optimality properties (see Huber, 1972, and Hampel, 1973, for summary articles).

To set the stage for the following paper by Jurečková, we highlight some of the motivation and main ideas inherent in the approach originated by Huber. Many statistical problems are approached by setting up a parametric model with specified distributional form such as in linear models with normally distributed errors. The classical approach to inference in such a setting is to estimate the parameters in some optimal way at the model. There is a potential problem with this approach. In most cases our model is an approximation and not strictly true. To claim that our estimate has the property of "near optimality" for the actual model which is close to the idealized model, we must rely on the principle of continuity that guarantees that a small change in the model results only in a small change in the optimum. Unfortunately many important classical optimal procedures do not have this property of continuity. A robust estimate can be thought of as one which has the property that it is continuous in the neighborhood of the idealized model. The reader familiar with mathematical modeling will notice some strong parallels with certain aspects of perturbation theory (see Saaty and Bram, 1964) where, in moving from an idealized model to the often more realistic nonlinear model, we add a small nonlinear term, examine the effect on the solution and derive new more stable solutions if necessary. In the statistical problem the idealized model in many situations involves a normal error term, which plays the same role in statistics as the straight line in the mathematics of perturbation theory. Moving to distributions in the neighborhood of the normal can be thought of as introducing small nonlinear terms.

Deviations from the model may be the result of gross errors arising from faulty measurements, recording or transcription errors, incorrect allocation of an observation, or other unexplained reasons. Rounding and grouping also can lead to deviations from the model. Hampel (1973) reports that contamination, either hidden or in the form of outliers, occurs in practice from less than

1% up to 10% of the time, and suggests that "5–10% wrong values in a data set seem to be the rule rather than the exception." Practical statisticians have long practiced outlier rejection, a robust, though not very efficient, procedure. For large and/or complex data sets or data sets which are analysed in an automated fashion on computers, it is no longer practical or possible to rely on astute statisticians to pick up all outliers. Discrepant observations remain a problem even in the relatively model-free approach of survey sampling, and although the formal theory of robustness is developed in the context of parametric models, the concerns and some of the solutions are applicable in a broader context.

To work in a specific setting, consider first the problem of estimating a location parameter θ with observations $x_1, \ldots, x_n$ from a distribution $F((x - \theta)/\sigma)$, where the form of F is not known exactly. To illustrate some of the robust estimates of location, it is worthwhile to consider a specific data set. For this purpose we use the well-known data by Cushny and Peebles (1905) on the prolongation of sleep by two soporific drugs. The ordered sample (of differences) is 0.0; 0.8; 1.0; 1.2; 1.3; 1.3; 1.4; 1.8; 2.4; 4.6. Note that our discussion here follows closely parts of Hampel (1973), and we focus on estimating the difference between the two drugs, ignoring any information about how the observations were generated. In addition to the classical estimates—mean and median—we have already discussed trimmed means, which are special cases of a class of estimates based on linear combinations of order statistics, referred to as L-estimates. A second popular class of estimates are Huber's M-estimates, which can be viewed as generalizations of maximum likelihood estimates. For location problems the M-estimate T_n solves an equation $\sum_{i=1}^{n} \psi(x_i - T_n) = 0$. The choice $\psi(x) = x$ for $|x| \leq k$, $\psi(x) = k$, $\psi(x) = k\,\mathrm{sign}(x)$ otherwise defines the prototype Huber estimate. To make estimate scale invariant, the defining equation is modified to

$$\sum_{i=1}^{n} \psi\left(\frac{x_i - T_n}{s}\right) = 0,$$

where s is an estimate of scale, often chosen to be the median absolute deviation (MAD). Finally we have the class of R-estimates, which are derived from rank tests. The Hodges–Lehmann estimate, which is derived from the Wilcoxon–Mann–Whitney test, is the best known R-estimate for estimation and can be computed as the median of the means of all pairs of observations.

For the Cushny and Peebles data, we obtain the following central values:

Mean	1.58
10% trimmed mean	1.40
20% trimmed mean	1.33
Median (50% trimmed)	1.30
Huber $\psi(k = 1.5, s = \mathrm{MAD})$	1.34
Hodges–Lehmann	1.32

It is interesting to note that the last five estimates (all of which are robust) cluster in the range 1.30 to 1.40, while the nonrobust mean at 1.58 is clearly separated from the others. Because of their close resemblance to maximum likelihood estimates, M-estimates have been the easiest class of robust estimates to extend to more complicated situations.

For general parametric models, there are several approaches for finding good or optimal robust procedures. The neighborhood model, advocated by Huber (1981, Chapter 4), requires efficiency over a neighborhood of the idealized model, efficiency being measured with respect to the maximum of the asymptotic variance over the neighborhood. Hampel et al. (1985) consider infinitesimal changes in the estimate (or test statistic) by computing "derivatives" in various directions from the idealized model. The derivative in direction x yields the influence function which is a measure of the effect on an estimate of contamination of the idealized model by a point mass at x. Optimal robust statistical procedures can be found by maximizing the efficiency at the model subject to bounds on the influence function. In the case of estimating location, both approaches yield essentially the same solution. As an alternative method, it has been proposed to introduce a shape parameter into the parametric model and carry out estimation over the enlarged family (see Box, 1980, for example).

In more complicated models where the optimization problems may be difficult to solve explicitly, it's possible to formulate a general principle on adapting classical procedures to make them robust. Quoting from Huber's discussion in Kleiner, Martin, and Thompson (1979), the procedure is as follows:

> The principal idea is that one can robustify essentially any classical (nonrobust) procedure by applying it, instead of to the original measurements Y, to a "cleaned" process CY, where "cleaning" refers to a sophisticated form of outlier modification and possibly rejection.
>
> This CY may be defined in various ways. Both for conceptual and particularly for computational reasons it is advantageously described as the fixed-point of an iteration, which starts from some approximation to CY (usually Y itself), then determines a fitted process FY, a residual process $R = Y - FY$, a "metrically Winsorized" residual process $WR = cs\psi(R/(cs))$, with scale s estimated from WR itself and, finally, an improved version of the "cleaned" process, $CY = FY + WR$.

Robustness as we have discussed it so far can be thought of as protection for deviations from the underlying model distribution. For this case, there is a well-developed theory for both estimation and testing that yield procedures which give up a certain percentage in efficiency when the model is exactly correct but maintain high efficiency or power in a neighborhood where the model is only approximately true. The loss of efficiency at the model is small (between 5 and 10%) for the location problem. A largely unsolved question concerns the robustness of procedures against deviations from the hypothesis of independence. There has been relatively little work in this area, although recently Graf (1983) has developed some robust estimation procedures to handle data exhibiting long range correlations.

As mentioned earlier for the case of estimating location, there are three types of estimates commonly used: M-, L-, and R-estimates. It has been established for some time that these estimates are closely related asymptotically. In the paper following, Professor Jurečková demonstrates the extent to which the estimators are equivalent for finite n. Knowledge of these relationships enables us to see how properties of one of the estimates translate into properties of corresponding estimates from the other classes.

Bibliography

Anonymous (1821). "Dissertation sur la recherche du milieu le plus probable," *Ann. Math. Pures et Appl.*, **12**, 181–204.

Box, G. E. P. (1980). "Sampling and Bayes inference in scientific modelling and robustness," *J. Roy. Statist. Soc. Ser. A*, **143**, 383–430.

Chauvenet, W. (1863). *A Manual of Spherical and Practical Astronomy*, Vol. 2, Philadelphia: Lippincott.

Cushny, A. R. and Peebles, A. R. (1905). "The action of optical isomers. II. Hyoscines." *J. Physiol.*, **32**, 501–510.

Daniell, P. J. (1920). "Observations weighted according to order." *Amer. J. Math.*, **42**, 222–236.

Freeman, D., Pisani, R., and Purves, R. (1978). *Statistics*. New York: Norton.

Graf, H. P. (1983). *Long Range Correlations and Estimation of the Self-Similarity Parameter*. Published Ph.D. Thesis. Zurich: E.T.H.

Hampel, F. R. (1973). "Robust estimation: a Condensed partial survey." *Z. Wahrsch. Verw. Gebiete*, **27**, 87–104.

Hampel, F. R., Ronchetti, E. Rousseeuw, P. J., an Stahel, W. (1985). *Robust Statistics: The Approach Based on the Influence Function*. New York: Wiley (to be published).

Hoaglin, D. C., Mosteller, F., and Tukey, J. W. (1983). *Understanding Robust and Exploratory Data Analysis*. New York: Wiley.

Huber, P. J. (1972). "Robust statistics: A review." *Ann. Math. Statist.*, **43**, 1041–1067.

Huber, P. J. (1981). *Robust Statistics*. New York: Wiley.

Jeffreys, H. (1932). "An alternative to the rejection of outliers." *Proc. Roy. Soc. Ser. A*, **137**, 78–87.

Kleiner, B., Martin, R. D., and Thompson, D. J. (1979). "Robust estimation of power spectra." *J. Roy. Statist. Soc. Ser. B*, **41**, 313–351.

Laplace, P. S. de (1979). *Traité de Mécanique Céleste*, Paris: Gauthier-Villars.

Newcomb, S. (1886). "A generalized theory of the combination of observations so as to obtain the best result." *Amer. I. Math.*, **8**, 343–366.

Newcomb, S. (1912). "Researches on the motion of the moon, II." In *Astronomical Papers*, **9**. U.S. Nautical Almanac Office, 1–249.

Pearson, E. S. (1931). "The analysis of variance in cases of non-normal variation." *Biometrika*, **23**, 114–133.

Peirce, B. (1852). "Criterion for the rejection of doubtful observations." *Astronom. J.*, **2**, 161–163.

Saaty, T. L. and Bram, J. (1964). *Nonlinear Mathematics*. New York: McGraw-Hill.

Smith, R. H. (1888). "True average of observations?" *Nature*, **37**, 464. Reproduced in *Biometrika* (1980), **67**, 219–220.

Stigler, S. M. (1973). "Simon Newcomb, Percy Daniell, and the history of robust estimation 1885–1920." *J. Amer. Statis. Assoc.*, **68**, 872–879.

Stigler, S. M. (1980). "Studies in the history of Probability and Statistics XXXVIII, R. H. Smith; a Victorian interested in robustness." *Biometrika*, **67**, 217–221.

"Student" (1927). "Errors of routine analysis." *Biometrika*, **19**, 151–164

Tukey, J. W. (1977). *Exploratory Data Analysis*. Reading, MA: Addison-Wesley.

CHAPTER 16

Robust Estimators of Location and Their Second-Order Asymptotic Relations

Jana Jurečková

Abstract

Let $X_1, \ldots, X_n$ be independent random variables, identically distributed according to the distribution function $F(x - \theta)$, where θ is the parameter to be estimated. F is generally unspecified; we only assume that F has a symmetric density f. Three broad classes of robust estimators of θ, these of M-estimators, L-estimators, and R-estimators, are first briefly described. Denoting these estimators M_n, L_n, and R_n, respectively, we give sufficient conditions under which these estimators are asymptotically equivalent in probability, i.e., $\sqrt{n}(M_n - L_n) \xrightarrow{P} 0$, etc., as $n \to \infty$. These relations are supplemented by the rates of convergence in most cases.

1. Introduction

We shall follow the general ideas of robust estimation sketched in the preceding paper by C. Field and illustrate them on an important special problem: that of estimating the location of an unknown symmetric distribution. $X_1, \ldots, X_n$ will be independent observations from a population with the distribution function (*d.f.*) $F(x - \theta)$, where θ is the parameter to be estimated. F is an unknown member of a family $\mathscr{F}$ of symmetric *d.f.*'s.

Among various types of robust estimators, three broad classes play the most important role: M-estimators (estimators of maximum-likelihood type), introduced by Huber (1964); R-estimators (estimators based on signed-rank tests), introduced by Hodges and Lehmann (1963); and L-estimators (linear combinations of order statistics). These estimators, though defined in different ways, follow similar ideas, and for large sample sizes the three classes approach each other. Any of these three types of estimator is defined through some "weight function" which determines its efficiency and robustness properties. We shall show that by selecting these weight functions in a proper way, we may get asymptotically equivalent estimators.

This fact was first observed by Jaeckel (1971), who established a close

Key words and phrases: α-trimmed mean, asymptotic equivalence of estimators, Hodges–Lehmann estimator, Huber's estimator, L-estimator, M-estimator, R-estimator, second-order asymptotic relations of estimators.

connection between M- and L-estimators of location. The asymptotic relations between M- and L-estimators were later studied by the author (Jurečková, 1977) in the linear regression model and by Riedl (1979) in the location model. The relations between M- and L-estimators were studied by the author (Jurečková, 1982, 1983a), by Rivest (1982), and by van Eeden (1983).

Once we know that two estimators are asymptotically equivalent as $n \to \infty$, we may be interested in the rate at which their difference tends to zero. The asymptotic relations supplemented by the rate of convergence we shall call the *second-order asymptotic relations*. Such relations of M-estimators and R-estimators were studied by Hušková and Jurečková (1981, 1985). Second-order asymptotic relations of M- and L-estimators partly follow from Jaeckel (1971) and later on were studied by Jurečková (1982, 1983a) and by Jurečková and Sen (1984) (including the linear regression model).

Why do we study the relations of various kinds of estimators, and why is it important to know that two estimators are asymptotically equivalent? With the knowledge of such relations, we may select the estimator which best fits the specific situation. The asymptotic equivalence may also carry some specific features from one type of estimator to another. Each of the three classes has some advantages as well as disadvantages: M-estimators have attractive minimax properties, but they are generally not scale-equivariant and have to be supplemented by an additional estimator of scale which may spoil the minimax properties. L-estimators are computationally appealing, although they were only recently successfully extended to the linear model. R-estimators retain the advantages and disadvantages of the rank test on which they are based.

We shall first describe (Section 2) the basic types of robust estimators. Then we shall characterize the asymptotic relations of these estimators in some simple cases.

2. *M*-, *R*-, and *L*-Estimators

Let $X_1, \ldots, X_n$ be independent random variables, identically distributed according to the *d.f.* $F(x - \theta)$, where θ is the parameter to be estimated. F is generally unspecified; we shall first only assume that F has a symmetric density f, i.e. $f(x) = f(-x)$, $x \in R^1$. Some other regularity conditions will be imposed on F in various contexts.

We get an M-estimator of θ if we minimize, with respect to t,

$$\sum_{i=1}^{n} \rho\left(\frac{X_i - t}{s_n}\right) := \min, \tag{2.1}$$

where ρ is some (usually convex) function and s_n is an estimator of scale, based on $X_1, \ldots, X_n$ and satisfying

$$s_n(X_1 + a, \ldots, X_n + a) = s_n(X_1, \ldots, X_n), \qquad a \in R^1,$$
$$s_n(bX_1, \ldots, bX_n) = bs_n(X_1, \ldots, X_n), \qquad b > 0. \tag{2.2}$$

If ρ is convex, we may express (2.1) also as an equation

$$\sum_{i=1}^{n} \psi\left(\frac{X_i - t}{s_n}\right) = 0. \tag{2.3}$$

The M-estimator M_n is often defined directly as an appropriate solution of (2.3). If ψ is a nondecreasing function, M_n can be expressed as

$$M_n = \tfrac{1}{2}(M_n^+ + M_n^-), \tag{2.4}$$

where

$$M_n^- = \sup\left\{t : \sum_{i=1}^{n} \psi\left(\frac{X_i - t}{s_n}\right) > 0\right\},$$
$$M_n^+ = \inf\left\{t : \sum_{i=1}^{n} \psi\left(\frac{X_i - t}{s_n}\right) < 0\right\}. \tag{2.5}$$

In a more general case, M_n is defined as that solution of (2.3) which is nearest to some consistent initial estimators T_n of θ (we always assume that ψ takes on positive as well as negative values).

The class of M-estimators was originated by Huber (1964) [see also Huber (1972, 1973, 1977), among others]. The properties of M-estimators are thoroughly studied in Huber's monograph (1981). If the density f of F is absolutely continuous and we put $\psi = -f'/f$, we get the maximum-likelihood estimator of θ. We get the sample mean if we put $\psi(x) \equiv x$. If we do not know the exact form of F, we try to follow the ideas mentioned in the paper of C. Field and select ψ accordingly. ψ is closely connected with the influence function of M_n; hence only bounded functions can lead to robust estimators. The most well-known ψ-function which leads to a minimax solution over a neighborhood of the normal distribution in the finite-sample as well as in the asymptotic case (under $s_n \equiv 1$) is of the form

$$\psi(x) = \begin{cases} x & \text{if } |x| \leq c, \\ c \operatorname{sign} x & \text{if } |x| > c, \end{cases} \tag{2.6}$$

where $c > 0$ is a fixed constant.

The R-estimators were suggested by Hodges and Lehmann (1963) for estimation of location in one- and two-sample models. They were extended by Adichie (1967), Jurečková (1971), Koul (1971), and Jaeckel (1972) to the linear regression model. The R-estimator of θ in the one-sample location model is defined as

$$R_n = \tfrac{1}{2}(R_n^- + R_n^+), \tag{2.7}$$

where

$$R_n^- = \sup\left\{t : \sum_{i=1}^{n} \operatorname{sign}(X_i - t)\,\varphi^+\left(\frac{R_{in}^+(t)}{n+1}\right) > 0\right\},$$
$$R_n^+ = \inf\left\{t : \sum_{i=1}^{n} \operatorname{sign}(X_i - t)\,\varphi^+\left(\frac{R_{in}^+(t)}{n+1}\right) < 0\right\}, \tag{2.8}$$

where $R_{in}^+(t)$ is the rank of $|X_i - t|$ among $|X_1 - t|, \ldots, |X_n - t|$, $i = 1, \ldots, n$, and $\varphi^+(u)$, $0 < u < 1$, is a nondecreasing score-function. If $\varphi^+(u) = u$, $0 < u < 1$ (Wilcoxon one-sample test), then

$$R_n = \operatorname*{med}_{1 \le i \le j \le n} \frac{X_i + X_j}{2}, \tag{2.9}$$

the Hodges–Lehmann estimator. We get the sample median if we put $\varphi^+(u) \equiv 1$. For a general φ^+, R_n must be calculated iteratively.

We now turn to L-estimators. Let $X_{n:1} \le \cdots \le X_{n:n}$ be the order statistics corresponding to $X_1, \ldots, X_n$. The L-estimator L_n of θ is defined as

$$L_n = \sum_{i=1}^{n} c_{ni} X_{n:i}, \tag{2.10}$$

where $c_{n1}, \ldots, c_{nn}$ are given constants, usually such that $c_{ni} = c_{n,n-i+1} \ge 0$, $i = 1, \ldots, n$, and $\sum_{i=1}^{n} c_{ni} = 1$. Simple examples are the sample mean $\bar{X}_n$ and the sample median $\tilde{X}_n$. The class of L-estimators is computationally more appealing than that of M-estimators and R-estimators, yet competes well from the standpoint of robustness and efficiency. If we wish to get a robust L-estimator, insensitive to extreme observations, we put $c_{ni} = 0$ for $i \le k_n$ and $i \ge n - k_n + 1$, with a proper k_n. Typical examples of such estimators are the α-trimmed mean,

$$L_n = \frac{1}{n - 2[n\alpha]} \sum_{i=[n\alpha]+1}^{n-[n\alpha]} X_{n:i}, \tag{2.11}$$

where $0 < \alpha < \frac{1}{2}$ and $[x]$ denotes the largest integer k satisfying $k \le x$, and the α-Winsorized mean,

$$L_n = \frac{1}{n}\left\{[n\alpha] X_{n-[n\alpha]} + \sum_{i=[n\alpha]+1}^{n-[n\alpha]} X_{n:i} + [n\alpha] X_{n:n-[n\alpha]+1}\right\}. \tag{2.12}$$

A broad class of L-estimators (2.10) which often appear in applications is of the form

$$L_n = \frac{1}{n}\sum_{i=1}^{n} J\left(\frac{i}{n+1}\right) X_{n:i} + \sum_{j=1}^{k} a_j X_{n:[np_j]}, \tag{2.13}$$

where $J(u)$, $0 < u < 1$, is a weight function, satisfying $J(u) = J(1-u) \ge 0$, $0 < u < 1$, and $p_1, \ldots, p_k$, $a_1, \ldots, a_k$ are given constants satisfying $0 < p_1 < \cdots < p_k < 1$, $p_j = 1 - p_{k-j+1}$, $a_j = a_{k-j+1} > 0$, $j = 1, \ldots, k$. Then L_n is of the form (2.13) with c_{ni} equal to $n^{-1} J(i/(n+1))$ plus an additional contribution a_j, if $i = [np_j]$ for some $j (1 \le j \le k)$. The function J is usually assumed to be smooth. Equation (2.13) is thus a combination of two types of L-

estimators. In many cases, the estimator under consideration is just of one type.

Rivest (1982) combined M- and L-estimators into one form, which he called *L-M-estimators*. Such an estimator is defined as a solution T_n of the equation

$$\sum_{i=1}^{n} J\left(\frac{i}{n+1}\right)\psi\left(\frac{X_{n:i}-t}{s_n}\right)=0 \tag{2.14}$$

with respect to t; for $J(t) \equiv 1$ the L-M-estimator reduces to an M-estimator, while it is equal to an L-estimator for $\psi(x) \equiv x$, $x \in R^1$.

3. Asymptotic Relations of M- and L-Estimators

Let $X_1, X_2, \ldots$ be independent random variables, identically distributed according to *d.f.* $F(x-\theta)$ such that $F(x)+F(-x)=1, x \in R^1$. Let L_n be an L-estimator of θ of the form (2.10) or, more specifically, (2.13). Under general conditions, we are able to find an M-estimator M_n such that

$$n^{1/2}(L_n - M_n) \xrightarrow{p} 0 \quad \text{as } n \to \infty. \tag{3.1}$$

The closer two estimators are to each other, the faster will be the convergence in (3.1). The rate of this convergence depends on the functions ψ and J, on the continuity and discontinuity of ψ, etc. Under the mildest conditions we get the rate $L_n - M_n = O_p(n^{-3/4})$; this notation means that $n^{3/4}(M_n - L_n)$ is asymptotically bounded in probability as $n \to \infty$. In other words, there exist positive C and positive integer n_o for every $\varepsilon > 0$ such that $P(n^{3/4}|M_n - L_n| > C) < \varepsilon$ for all $n > n_o$. Under stronger conditions we are able to establish the rate $M_n - L_n = O_p(n^{-1})$. The lower rate appears if the second component of L_n in (2.13) does not vanish; the ψ-function of the M-estimator counterpart has some jump points.

The basic method for deriving the rates of convergence is in a convenient representation of the estimator as a sum of independent random variables with a remainder term of a convenient order. Let us describe some representations of M-estimators. In the following, we consider an M-estimator as a solution of equation (2.3) with $s_n \equiv 1$.

(i) Let M_n be an M-estimator generated by a smooth function ψ: we shall assume that ψ is nondecreasing, is skew-symmetric $[\psi(-x) = -\psi(x)$, $x \in R^1]$, and has two bounded derivatives at all but a finite number of points of R^1. Moreover, we shall assume that

$$\int \psi^{12}(x)dF(x) < \infty, \qquad \gamma = \int f(x)d\psi(x) > 0, \tag{3.2}$$

$$\int |\psi''(x+t)|dF(x) < C \quad \text{for } |t| \le \delta, \quad \text{for some } \delta > 0 \text{ and } C > 0. \tag{3.3}$$

Then we have the following representation for M_n:

$$n^{1/2}(M_n - \theta) = n^{-1/2}\gamma^{-1} \sum_{i=1}^{n} \psi(X_i - \theta) + O_p(n^{-1/2}). \tag{3.4}$$

The proof of (3.4), as well as of the following representations, can be found in Jurečková (1980).

(ii) The functions of the type (2.6) are constant outside a bounded interval. If ψ is nondecreasing, is skew-symmetric, and has two bounded derivatives in the interval $(-c, c)$, and if $\psi(x) = \psi(c)$ sign x for $|x| \geq c (c > 0)$, then the representation (3.4) can be proved also for M_n generated by ψ.

(iii) Let ψ be a step function satisfying

$$\psi(x) = \alpha_j \quad \text{if } s_j < x < s_{j+1}, j = 0, \ldots, k, \tag{3.5}$$

where $-\infty = s_o < s_1 < \cdots < s_k < s_{k+1} = \infty, -\infty < \alpha_o \leq \cdots \leq \alpha_k < \infty$, (at least two α's different), and

$$\psi(s_j) = \frac{\alpha_j + \alpha_{j-1}}{2}, \qquad j = 1, \ldots, k. \tag{3.6}$$

The simplest M-estimator generated by a ψ of this kind is the sample median $\tilde{X}_n$ with $\psi(x) = -1$ if $x < 0$ and $\psi(x) = 1$ if $x > 0$. Moreover, the logarithmic derivative of the density of a mixture of several exponential distributions is of the type (3.5); hence the corresponding maximum-likelihood estimator is an M-estimator generated by a step function. We shall see that the L-estimators asymptotically equivalent to such M-estimators are linear combinations of several single sample quantiles [and thus coincide with the second component of (2.13)]; examples of such L-estimators are the *trimean* $L_n = \frac{1}{3}\{X_{n:[n/4]+1} + \tilde{X}_n + X_{n:n+[n/4]}\}$, and the *Gastwirth* (1966) *estimate*

$$L_n = 0.3X_{n:[n/3]+1} + 0.4\tilde{X}_n + 0.3X_{n:n-[n/3]}, \tag{3.7}$$

among others.

If M_n is an M-estimator generated by ψ of (3.5) and (3.6), then, provided F has two bounded derivatives f and f' in a neighborhood of $s_1, \ldots, s_k$ and provided

$$\gamma = \sum_{j=1}^{k} (\alpha_j - \alpha_{j-1}) f(s_j) > 0, \tag{3.8}$$

we have the following representation for M_n:

$$n^{1/2}(M_n - \theta) = n^{-1/2}\gamma^{-1} \sum_{i=1}^{n} \psi(X_i - \theta) + O_p(n^{-1/4}) \tag{3.9}$$

as $n \to \infty$.

(iv) If ψ is a sum of two components, $\psi = \psi_1 + \psi_2$, where ψ_1 satisfies the assumptions under (i) or under (ii), and ψ_2 is the step function of (3.5) and

(3.6), then the relevant M-estimator M_n can be again represented as in (3.9) with γ of the form

$$\gamma = \int f(x)\,d\psi_1(x) + \sum_{j=1}^{k} (\alpha_j - \alpha_{j-1}) f(s_j). \tag{3.10}$$

We now incorporate the representations (i)–(iv) in the relations of M_n to appropriate L-estimators of θ.

We start with the simplest case when M_n is the M-estimator generated by the step function (3.5) and (3.6) and assume that, for the sake of symmetry, $\alpha_j = -\alpha_{k-j+1}$, $s_j = -s_{k-j+1}$, $j = 1, \ldots, k$. Let F satisfy the assumptions under (iii). Then we have the following theorem.

Theorem 1. *Under the above assumptions,*

$$M_n - L_n = O_p(n^{-3/4}) \quad \text{as } n \to \infty, \tag{3.11}$$

where L_n is the L-estimator $L_n = \sum_{j=1}^{k} a_j X_{n:[np_j]}$ with

$$p_j = F(s_j), \quad a_j = \gamma^{-1}(\alpha_j - \alpha_{j-1}) f(s_j), \qquad j = 1, \ldots, k, \tag{3.12}$$

and γ is given by (3.8).

Proof. Let F_n be the empirical *d.f.* corresponding to $X_1, \ldots, X_n$, i.e.

$$F_n(x) = \frac{1}{n} \sum_{i=1}^{n} I[X_i \le x] \tag{3.13}$$

($I[A]$ is the indicator of the set A), and define the empirical quantile function $Q_n(t)$, $0 \le t \le 1$, as

$$Q_n(t) = \begin{cases} X_{n:i} & \text{if } \dfrac{i-1}{n} < t \le \dfrac{i}{n},\ i = 1, \ldots, n, \\ 0 & \text{if } t = 0. \end{cases} \tag{3.14}$$

We may put $\theta = 0$ without loss of generality. From the symmetry of F, of the s_j's, and of the α_j's, and using the representation (3.9), we get

$$n^{1/2}(M_n - L_n) = n^{1/2}\gamma^{-1} \sum_{j=1}^{k} (\alpha_j - \alpha_{j-1}) [F_n(s_j) - F(s_j) + f(s_j)(Q_n(F(s_j)) - s_j)] + O_p(n^{-1/4}), \tag{3.15}$$

which is $O_p(n^{-1/4})$ by Kiefer (1967). □

Remark 1. *Theorem* 1 (*similarly to other analogous results*) *does not enable us to calculate the value of the asymptotically equivalent L-estimator, once we have calculated the M-estimator, unless we know $F(s_j)$ and $f(s_j)$, $j = 1, \ldots, k$. Such results rather indicate the close relations of the two types of estimator;*

namely, (3.11) *shows that linear combinations of single sample quantiles are close, for large sample sizes, to the M-estimators generated by the step functions.*

Remark 2. *The assumption that at least the first derivative of F is positive and finite in a neighborhood of* $s_1, \ldots, s_k$ *is crucial for* (3.11). *The singular points at which this is not the case change the rate of consistency of the corresponding M- and L-estimators. The behavior of estimators in such nonregular cases is studied by Akahira* (1975a, b), *Akahira and Takeuchi* (1981), *Ibragimov and Hasminskii* (1979), *and Jurečková* (1983b).

Let us now consider the relation of two important estimators: Huber's M-estimator generated by the ψ-function (2.6), and the α-trimmed mean (2.11). One might intuitively expect that the Winsorized mean rather than the trimmed mean would resemble Huber's estimator (cf. Huber, 1964). Bickel (1965) was apparently the first to recognize the close connection between Huber's estimator and the trimmed mean. Jaeckel (1971) found the L-estimator counterpart to the given M-estimator such that the difference between the two is $O_p(n^{-1})$ as $n \to \infty$, provided ψ generating the M-estimator (and hence J generating the L-estimator) is sufficiently smooth. However, the function (2.6) is not smooth enough to satisfy the proof of Jaeckel's Theorem 2. The author proved this special relation using a different method (see Jurečková, 1982). The result is formulated in the following theorem.

Theorem 2. *Let* $X_1, X_2, \ldots$ *be independent observations identically distributed according to the d.f.* $F(x - \theta)$ *such that* $F(x) + F(-x) = 1$, $x \in R^1$, *and that F satisfies the following conditions:*

(a) *F has an absolutely continuous density and finite and positive Fisher information, i.e.*

$$0 < I(f) < \int \left(\frac{f'(x)}{f(x)} \right)^2 dF(x) < \infty.$$

(b) $f(x) > a > 0$ *for all x satisfying*

$$\alpha - \varepsilon \leq F(x) \leq 1 - \alpha + \varepsilon, \qquad 0 < \alpha < \tfrac{1}{2}, \quad \varepsilon > 0.$$

(c) $f'(x)$ *exists in the interval* $(F^{-1}(\alpha - \varepsilon), F^{-1}(1 - \alpha + \varepsilon))$.

Then

$$L_n - M_n = O_p(n^{-1}) \quad \text{as } n \to \infty \tag{3.16}$$

where L_n *is the* α*-trimmed mean and* M_n *is the M-estimator generated by* ψ *of* (2.6) *with* $c = F^{-1}(1 - \alpha)$.

Combining Theorem 2 with (3.11), we easily get the M-estimator counterpart of the α-Winsorized mean (2.12). Under the assumptions of Theorem 2,

$$L_n - M_n = O_p(n^{-3/4}) \quad \text{as } n \to \infty, \tag{3.17}$$

where L_n is the α-Winsorized mean and M_n is the M-estimator generated by the function

$$\psi(x) = \begin{cases} F^{-1}(\alpha) - \dfrac{\alpha}{f(F^{-1}(\alpha))} & \text{if } x < F^{-1}(\alpha), \\ x & \text{if } F^{-1}(\alpha) \le x \le F^{-1}(1-\alpha), \\ F^{-1}(1-\alpha) + \dfrac{\alpha}{f(F^{-1}(\alpha))} & \text{if } x > F^{-1}(1-\alpha). \end{cases} \tag{3.18}$$

Assume that the L-estimator L_n coincides with the first component in (2.13), i.e., L_n is generated by the function $J(t)$, $0 < t < 1$; assume that $J(t)$ is continuous in (0, 1) and has a bounded derivative up to a finite number of points of (0, 1), and that $J(t) = 0$ for $0 < t < \alpha$ and $1 - \alpha < t < 1$. The M-estimator counterpart to such an L-estimator was found by Jaeckel (1971); more precisely, if $F(x) + F(-x) = 1$, $x \in R^1$ and $\int [F(x)(1 - F(x))]^{1/2}\, dx < \infty$, then

$$L_n - M_n = O_p(n^{-1}) \quad \text{as } n \to \infty, \tag{3.19}$$

where M_n is the M-estimator generated by the function

$$\psi(x) = \int_0^1 J(t)\,(I[F(x) \le t] - t)\, dF^{-1}(t), \qquad x \in R^1. \tag{3.20}$$

Combining the representation (3.4) with the result of Helmers (1981), we can show that ψ of the form (3.20) provides an M-estimator counterpart also for an L-estimator generated by J, possibly positive for all $t \in (0, 1)$ (under more restrictive conditions on F).

4. Asymptotic Relations of *M*- and *R*-Estimators

Let $X_1, X_2, \ldots$ again be independent observations identically distributed according to the *d.f.* $F(x - \theta)$ such that $F(x) + F(-x) = 1$, $x \in R^1$. Let M_n be the M-estimator and R_n the R-estimator of θ. The author proved (Jurečková, 1977) that in the linear regression model, under quite general conditions,

$$n^{1/2}(M_n - R_n) \xrightarrow{p} 0 \quad \text{as } n \to \infty \tag{4.1}$$

if and only if

$$\psi(x) = a\varphi(F(x)) + b \quad \text{for all } x \in R^1, \quad \text{where } a > 0,\ t \in R^1, \tag{4.2}$$

where ψ is the function generating the M-estimator and φ is the score function of the R-estimator. The result was then extended by Riedl (1979) to the location model. We are again interested, for the location model, in how fast convergence can be in (4.1).

According to (2.7) and (2.8), the R-estimator R_n is defined through the statistic

$$S_n(t) = n^{-1/2} \sum_{i=1}^{n} \operatorname{sign}(X_i - t)\, \varphi^+\left(\frac{R_{ni}^+(t)}{n+1}\right), \qquad t \in R^1. \tag{4.3}$$

Assume that φ^+ is a nonconstant, nondecreasing, and square-integrable function on (0, 1) such that $\varphi^+(0) \geq 0$. Put

$$\varphi(u) = \begin{cases} \varphi^+(2u-1) & \text{if } \frac{1}{2} \leq u < 1, \\ -\varphi(1-u) & \text{if } 0 < u < \frac{1}{2}. \end{cases} \tag{4.4}$$

Let us first consider the case that φ^+ is a step function, namely

$$\varphi^+(u) = \alpha_j \quad \text{if } u_j < u \leq u_{j+1}, \qquad j = 0, 1, \ldots, k, \tag{4.5}$$

where $0 = u_0 < u_1 < \cdots < u_{k+1} = 1$ and $0 \leq \alpha_0 \leq \cdots \leq \alpha_k$, at least two of the α's being different. Then $\varphi(u)$ of (4.4) is again a step function,

$$\varphi(u) = \begin{cases} \alpha_j & \text{if } \dfrac{u_j+1}{2} < u \leq \dfrac{u_{j+1}+1}{2}, \\[2ex] -\alpha_j & \text{if } \dfrac{1-u_{j+1}}{2} \leq u < \dfrac{1-u_j}{2}, \end{cases} \tag{4.6}$$

$j = 0, 1, \ldots, k$. Denote

$$\frac{u_j+1}{2} = p_j, \qquad j = 0, 1, \ldots, k. \tag{4.7}$$

For the sake of symmetry we may redefine $\varphi(u)$ as $(\alpha_j + \alpha_{j+1})/2$ [*or* $-(\alpha_j + \alpha_{j-1})/2$] at the points of discontinuity. Then we have the following representation of the R-estimator R_n generated by φ^+ with the aid of a sum of independent random variables:

Theorem 3. *Assume that the d.f. $F(x)$ is symmetric, is continuous, and has two bounded derivatives in neighborhoods of $F^{-1}(p_0), \ldots, F^{-1}(p_k)$, and that*

$$\gamma = \sum_{j=1}^{k} (\alpha_j - \alpha_{j-1}) f(F^{-1}(p_j)) > 0. \tag{4.8}$$

Let R_n be an R-estimator generated by φ^+ of (4.5). Then

$$n^{1/2}(R_n - \theta) = n^{-1/2}\gamma^{-1} \sum_{i=1}^{n} \varphi(F(X_i - \theta)) + O_p(n^{-1/4}). \tag{4.9}$$

Theorem 3, as well as Theorems 4 and 5, were proven by Hušková and Jurečková (1985).

Combining Theorem 3 with the representation (3.9) of the M-estimator generated by a step function and with Theorem 1, we get the following corollary about the asymptotic relations of M-R-L-estimators.

Corollary 1. *Under the assumptions of Theorem 3,*

$$R_n - M_n = O_p(n^{-3/4}) \quad \text{as } n \to \infty, \tag{4.10}$$

where M_n is an M-estimator generated by the function ψ satisfying

$$\psi(x) = \begin{cases} \alpha_j & \text{if } F^{-1}(p_j) < x \leq F^{-1}(p_{j+1}), j = 0, 1, \ldots, k, \\ -\psi(-x) & \text{if } x < 0. \end{cases} \tag{4.11}$$

Moreover, if L_n is the L-estimator of the form

$$L_n = \sum_{j=1}^{k} a_j X_{n:[np_j]} \tag{4.12}$$

with $a_j = (1/\gamma)(\alpha_j - \alpha_{j-1}) f(s_j), j = 1, \ldots, k$, then also

$$L_n - R_n = O_p(n^{-3/4}). \tag{4.13}$$

If the score function φ^+ is sufficiently smooth, we can find an M-estimator counterpart such that $R_n - M_n$ is of higher order than in (4.10). Let us illustrate this situation with two examples.

We shall first assume that φ^+ satisfies the following set of conditions:

(1) φ^+ has the second derivative $\varphi^{+(2)}$ on (0, 1) such that

$$|\varphi^{+(2)}(u)| \leq K(1-u)^{-q} \quad \text{for some } K > 0, 0 < q < 2, \tag{4.14}$$

and for all $u \in (0, 1)$.

Then we could establish an asymptotic representation of R_n up to a higher-order term; but we should still put some more smoothness conditions on F, namely

(2) F has an absolutely continuous density f such that

$$\left| \frac{f'(F^{-1}(u))}{f(F^{-1}(u))} \right| \leq K(u(1-u))^{-\frac{1}{2}+\varepsilon}, \qquad 0 < u < 1, \tag{4.15}$$

for some $K > 0$, $\varepsilon > 0$; moreover, there exists a $\delta > 0$ for every $\eta > 0$ such that

$$\int (F(x)(1-F(x)))^{1-\eta}\, dF(x \pm \delta) < \infty. \tag{4.16}$$

Under these conditions, we get the following representation of R_n:

Theorem 4. *If F is symmetric and satisfies conditions* (4.15) *and* (4.16), *then, provided the score function φ^+ of the R-estimator R_n satisfies* (4.14),

$$n^{1/2}(R_n - \theta) = n^{-1/2}\gamma^{-1}\sum_{i=1}^{n}\varphi(F(X_i - \theta)) + O_p(n^{-1/2}) \tag{4.17}$$

as $n \to \infty$, where

$$\gamma = \int_0^1 f(F^{-1}(t))\,d\varphi(t) \tag{4.18}$$

and φ is given by (4.4).

Combining Theorem 4 with the representation (3.4) of the M-estimator generated by the smooth ψ-function, we get the following corollary about the asymptotic relation between R_n and M_n:

Corollary 2. *Under the assumptions of Theorem 4,*

$$R_n - M_n = O_p(n^{-1}) \quad \text{as } n \to \infty, \tag{4.19}$$

where M_n is an M-estimator generated by ψ satisfying $\psi(x) = K\varphi(F(x)), x \in R^1$, for some $K > 0$.

The condition (4.14) does not yet cover the case $\varphi^+(u) = \Phi^{-1}((u+1)/2)$, $0 < u < 1$, where Φ is the standard normal *d.f.* The corresponding R-estimator is an inversion of the van der Waerden signed-rank test of symmetry; such an R-estimator is asymptotically equivalent to the sample mean (and hence asymptotically efficient) provided F is normal. The following theorem covers this important case.

Theorem 5. *Let F be a symmetric d.f. satisfying the conditions* (4.15) *and* (4.16), *and let R_n be an R-estimator generated by a function φ^+ that has a second derivative $\varphi^{+(2)}$ on* (0, 1) *satisfying*

$$|\varphi^{+(2)}(u)| \leq K(1-u)^{-2},\ 0 < u < 1, \text{ for some } K > 0. \tag{4.20}$$

Then

$$n^{1/2}(R_n - \theta) = n^{-1/2}\gamma^{-1}\sum_{i=1}^{n}\varphi(F(X_i - \theta) + O_p(n^{-1/2}\log n) \tag{4.21}$$

as $n \to \infty$, where

$$\gamma = \int_0^1 f(F^{-1}(t))\,d\varphi(t) \tag{4.22}$$

with φ given by (4.4).

Again, combining Theorem 5 with (3.4), we get the following corollary.

Corollary 3. *Under the assumptions of Theorem 5,*

$$R_n - M_n = O_p(n^{-1}\log n) \quad \text{as } n \to \infty, \tag{4.23}$$

where M_n is an M-estimator generated by the function ψ satisfying $\psi(x) = K\varphi(F(x))$, $x \in R^1$, for some $K > 0$.

Let us illustrate the results on some special cases.

(i) The Hodges–Lehmann estimator (2.9) is an R-estimator generated by $\varphi^+(u) = u$, $0 < u < 1$; hence $\varphi(u) = 2u - 1$, $0 < u < 1$. Then

$$R_n = M_n + O_p(n^{-1}) \quad \text{as } n \to \infty, \tag{4.24}$$

where M_n is an M-estimator generated by the function

$$\psi(x) = 2F(x) - 1, \qquad x \in R^1, \tag{4.25}$$

and which is thus equal to the solution of the equation

$$\frac{1}{n}\sum_{i=1}^{n} F(X_i - t) = \frac{1}{2} \tag{4.26}$$

with respect to t. Moreover,

$$R_n = L_n + O_p(n^{-1}) \quad \text{as } n \to \infty, \tag{4.27}$$

where L_n is an L-estimator generated by the function $J(u) = cf(F^{-1}(u))$, $0 < u < 1$, for some $c > 0$.

(ii) If R_n is an R-estimator based on the van der Waerden test with $\varphi^+(u) = \Phi^{-1}((u+1)/2)$, $0 < u < 1$, then

$$R_n = M_n + O_p(n^{-1}\log n), \tag{4.28}$$

where M_n is the M-estimator generated by $\psi(x) = \Phi^{-1}(F(x))$, $x \in R^1$, which is a solution of the equation

$$\sum_{i=1}^{n} \Phi^{-1}(F(X_i - t)) = 0 \tag{4.29}$$

with respect to t. Specifically, M_n coincides with the sample mean for a normal F.

(iii) If M_n is the Huber M-estimator generated by the ψ of (2.6), then $M_n - L_n = O_p(n^{-1})$, where L_n is a proper trimmed mean. Moreover, both these estimators are asymptotically equivalent to R_n generated by the function

$$\varphi^+(u) = \begin{cases} F^{-1}\left(\dfrac{u+1}{2}\right) & \text{if } 0 \le u \le 2F(c) - 1, \\ \text{c} & \text{if } u > 2F(c) - 1, \end{cases} \tag{4.30}$$

but the order of $M_n - R_n$ (and that of $L_n - R_n$) is still an open question.

5. Concluding Remarks

(i) We have seen that, generally, given an estimator of one type, there are estimators of the other two types such that all three estimators are asymptotically equivalent in probability as $n \to \infty$. In many cases we were able to provide an order for this equivalence. The correspondence between the pertaining weight functions J, ψ, and φ depends on the distribution function F which is generally unspecified; thus, we are not able to calculate the numerical values of the other estimators, once we know the numerical value of one estimator. The results rather indicate the types of estimators which belong to each other, for instance, the trimmed mean and the Huber estimator; the Hodges–Lehmann estimator, the M-estimator with $\psi(x) = 2F(x) - 1$, and the L-estimator with $J(t) = cf(F^{-1}(t))$, $c > 0$; etc.

(ii) The lower order $O_p(n^{-3/4})$ in the discontinuous case seems to suggest that we should prefer estimators with smooth weight functions. On the other hand, the M-estimators (and R-estimators) generated by step functions ψ (or φ^+) and the L-estimators consisting of linear combinations of single sample quantiles are more universal; they are consistent and asymptotically normal under weaker conditions on F (e.g., it is sufficient to assume that F is smooth in a neighborhood of the jump points of φ).

Bibliography

Adichie, J. N. (1967). "Estimate of regression parameters based on rank tests." *Ann. Math. Statist.*, **38**, 894–904.

Akahira, M. (1975a). "Asymptotic theory for estimation of location in non-regular cases, I: Orders of convergence of consistent estimators." *Rep. Stat. Appl. Res. JUSE*, **22**, No. 1.

Akahira, M. (1975b). "Asymptotic theory for estimation of location in non-regular cases, II: Bounds of asymptotic distribution of consistent estimators." *Rep. Stat. Appl. Res. JUSE*, **22**, No. 3.

Akahira, M. and Takeuchi, K. (1981). *Asymptotic Efficiency of Statistical Estimators: Concepts and Higher Order Asymptotic Efficiency*. Lecture Notes in Statistics, **7**. New York: Springer-Verlag.

Andrews, D. F., Bickel, P. J., Hampel, F. R., Huber, P. J., Rogers, W. H., and Tukey, J. W. (1972). *Robust Estimation of Location: Survey and Advances*. Princeton: Princeton U. P.

Bickel, P. J. (1965). "On some robust estimates of location." *Ann. Math. Statist.*, **36**, 847–858.

Bickel, P. J. (1973). "On some analogies to linear combinations of order statistics in the linear model." *Ann. Statist.*, **1**, 597–616.

Bickel, P. J. (1976). "Another look at robustness: A review of reviews and some new developments." *Scand. J. Statist.*, **3**, 145–168.

Carroll, R. J. (1978). "On almost sure expansions for M-estimators." *Ann. Statist.*, **6**, 314–318.

Chernoff, H., Gastwirth, J. L., and Johns, M. V. (1967). "Asymptotic distribution of linear combinations of order statistics." *Ann. Math. Statist.*, **38**, 52–72.

Csörgő, M. and Révész, P. (1981). *Strong Approximations in Probability and Statistics.* Budapest: Akadémiai Kiadó.

David, H. A. (1970). *Order Statistics.* New York: Wiley.

Gastwirth, J. (1966). "On robust procedures." *J. Amer. Statist. Assoc.*, **61**, 929–948.

Helmers, R. (1981). "A Berry–Esséen theorem for linear combinations of order statistics." *Ann. Probab.*, **9**, 342–347.

Hodges, J. L. and Lehmann, E. L. (1963). "Estimates of location based on rank tests." *Ann. Math. Statist.*, **34**, 598–564.

Huber, P. J. (1964). "Robust estimation of a location parameter." *Ann. Math. Statist.*, **35**, 73–101.

Huber, P. J. (1972). "Robust statistics: A review." *Ann. Math. Statist.*, **43**, 1041–1067.

Huber, P. J. (1973). "Robust regression: Asymptotics, conjectures and Monte Carlo." *Ann. Statist.*, **1**, 799–821.

Huber, P. J. (1977). *Robust Statistical Procedures.* Philadelphia: SIAM.

Huber, P. J. (1981). *Robust Statistics.* New York: Wiley.

Hušková, M. (1982). "On bounded length sequential confidence interval for parameter in regression model based on ranks." *Coll. Math. Soc. János Bolyai*, **32**, 435–463.

Hušková, M. and Jurečková, J. (1981). "Second order asymptotic relations of *M*-estimators and *R*-estimators in two-sample location model." *J. Statist. Planning and Inference*, **5**, 309–328.

Hušková, M. Jurečková, J. (1985). "Asymptotic representation of *R*-estimators of location." In *Proceedings of the 4th Pannonian Symposium.* Amsterdam: North-Holland (to appear).

Inagaki, N. (1974). "The asymptotic representation of the Hodges–Lehmann estimator based on Wilcoxon two-sample statistics." *Ann. Inst. Statist. Math.*, **26**, 457–466.

Ibragimov, I. A. and Hasminskii, R. Z. (1981). *Asymptotic Theory of Estimation.* New York: Springer-Verlag.

Jaeckel, L. A. (1971). "Robust estimates of location: Symmetry and asymmetric contamination." *Ann. Math. Statist.*, **42**, 1020–1034.

Jaeckel, L. A. (1972). "Estimating regression coefficients by minimizing the dispersion of the residuals." *Ann. Math. Statist.*, **43**, 1449–1458.

Jung, J. (1955). "On linear estimates defined by a continuous weight function." *Ark. Math.*, **3**, 199–209.

Jurečková, J. (1971). "Nonparametric estimate of regression coefficients." *Ann. Math. Statist.*, **42**, 1328–1338.

Jurečková, J. (1977). "Asymptotic relations of *M*-estimates and *R*-estimates in linear regression model." *Ann. Statist.*, **5**, 464–472.

Jurečková, J. (1980). "Asymptotic representation of *M*-estimators of location." *Math. Operationsforsch. Statist. Ser. Statist.*, **11**, 61–73.

Jurečková, J. (1982). "Robust estimators of location and regression parameters and their second order asymptotic relations." In *Proceedings of the 9th Prague Conference on Information Theory, Statistical Decision Functions and Random Processes.* Dordrecht: Reidel, 19–32.

Jurečková, J. (1983a). "Winsorized least-squares estimator and its *M*-estimator counterpart." In P. K. Sen (ed.). *Contributions to Statistics: Essays in Honour of Norman L. Johnson.* Amsterdam: North-Holland, 237–245.

Jurečková, J. (1983b). "Asymptotic behavior of *M*-estimators in non-regular cases." *Statistics & Decisions*, **1**, 323–340.

Jurečková, J. and Sen, P. K. (1981a). "Invariance principles for some stochastic processes relating to M-estimators and their role in sequential statistical inference." *Sankhya*, **A43**, 190–210.

Jurečková, J. and Sen, P. K. (1981b). "Sequential procedures based on M-estimators with discontinuous score-functions." *Journ. Statist. Planning and Inferences*, **5**, 253–266.

Jurečková, J. and Sen, P. K. (1984). "On adaptive scale equivariant M-estimators in linear models." *Statistics & Decisions*, Supplement Issue No 1, 31–46.

Kiefer, J. (1967). "On Bahadur's representation of sample quantiles." *Ann. Math. Statist.*, **38**, 1323–1342.

Koenker, R. and Bassett, G. (1978). "Regression quantiles." *Econometrica*, **46**, 33–50.

Koul, H. L. (1971). "Asymptotic behavior of a class of confidence regions based on ranks in regression." *Ann. Math. Statist.*, **42**, 466–476.

Lloyd, E. H. (1952). "Least squares estimation of location and scale parameters using order statistics." *Biometrika*, **34**, 41–67.

Portnoy, S. L. (1977). "Robust estimation in dependent situations." *Ann. Statist.*, **5**, 22–43.

Riedl, M. (1979). "M-estimators of regression and location." Unpublished Thesis. Prague: Charles Univ. (In Czech.)

Rivest, L. P. (1982). "Some asymptotic distributions in the location-scale model." *Ann. Inst. Statist. Math.*, **A34**, 225–239.

Ruppert, D. and Carroll, R. J. (1980). "Trimmed least-squares estimation in the linear model." *J. Amer. Statist. Assoc.*, **75**, 828–838.

Sarhan, A. E. and Greenberg, E. Q. (eds.) (1962). *Contributions to Order Statistics*. New York: Wiley.

Serfling, R. J. (1980). *Approximation Theorems of Mathematical Statistics*. New York: Wiley.

Shorack, G. R. (1969). "Asymptotic normality of linear combinations of functions of order statistics." *Ann. Math. Statist.*, **40**, 2041–2050.

Shorack, G. R. (1972). "Functions of order statistics." *Ann. Math. Statist.*, **43**, 412–427.

Stigler, S. M. (1969). "Linear functions of order statistics." *Ann. Math. Statist.*, **40**, 770–788.

Stigler, S. M. (1973). "The asymptotic distribution of the trimmed mean." *Ann. Statist.*, **1**, 472–477.

Stigler, S. M. (1974). "Linear functions of order statistics with smooth weight function." *Ann. Statist.*, **2**, 676–693.

van Eeden, C. (1983). "On the relation between L-estimators and M-estimators and asymptotic efficiency relative to the Cramér–Rao lower bound." *Ann. Statist.*, **11**, 674–690.

CHAPTER 17

Mathematical Statistics in the Humanities, and Some Related Problems in Astronomy[1]

David G. Kendall

Abstract

The author discusses a number of problems with which he has been concerned arising in archeology and astronomy. The first is a problem of archeological seriation and its generalizations to higher dimensions, which require the construction of maps from scrappy information. The second is the problem of testing a one-dimensional set of observations for the presence of an underlying unit of length. The third is the problem of testing two-dimensional sets of observations for the presence of too many triple or higher-order alignments among the points. These last two problems arise in the first instance in neolithic archeology but have since found interesting applications in several other fields. In particular, the problem of testing for alignments is currently of great importance in quasar astronomy. Alignment testing is a problem concerned with shape, and this leads on to an outline of a general approach to the statistics of shape which has been developed recently. Finally the author discusses a different astronomical problem in directional statistics where some of the participating so-called "directions" are actually elements of projective spaces. Here the substantive question is one of testing the significance of the evidence for a hitherto unsuspected asymmetry in the universe (Kendall and Young, 1984).

Despite fashionable remarks about literacy and numeracy, we all know, I hope, that mathematicians are concerned with structure rather than with numbers. Each evolving discipline ultimately develops structures, which soon become complicated, and then it is not uncommon for difficult mathematical problems to arise which have not occurred elsewhere.

For the last twenty years I have spent a large part of my time acting as an informal consultant to workers in the humanities. One's primary duty is to give satisfaction to one's client, and to find answers to the questions he asks, rather than to those which one wishes he had asked, though it does often happen that a joint reformulation of the problem is a necessary first step. At the end of such a task one is left with a double satisfaction: the pleasure gained from working with remote technicalities in an unfamiliar discipline, and the

Key words and phrases: alignment, indirection, map, quantum, quasar, seriation, shape.

[1] This is an extended version of the author's Presidential Address to the Sections for Mathematics and Physics at the meeting of the British Association for the Advancement of Science, in Liverpool, 1982.

excitement of being the first to lay one's hands on a novel patch of mathematical structure, later to develop into a permanent feature of the mathematical scene. As I will show, these humanely generated mathematical artefacts can turn out to be relevant in other contexts.

The first item in my casebook (Kendall, 1971a) was a request for an evaluation of Flinders Petrie's *sequence-dating* technique in archeology. He was concerned, about the year 1899, with the need to provide predynastic Egypt with at least a "serial" chronology, to be derived from information about the relative frequency of occurrence of different varieties of pottery in each of a (very large) collection of graves. Effectively he presented the evidence in what we should now call matrix form, so that we have a matrix **A** in which the (i, j)th element tells us about the occurrence of variety j in grave i.

Initially the rows and columns of the matrix are each arranged in an irrelevant order, and we are required to rearrange the *rows* (that is, the *graves*) so that as far as possible the new arrangement of entries in every column makes chronological sense. Thus in an extreme case the matrix may only have elements 1 (denoting presence) and 0 (denoting absence), and we may wish so far as possible to pack the 1's together in every column into a single bunch. When the number of rows is in the hundreds, this presents a task demanding something quicker than manual sorting.

It was therefore satisfying to be able to prove a mathematical result which indicates that, at least in favorable circumstances, the relevant information can be coded into a square matrix **B**, with elements

$$b_{ij} = \sum_k w_k \min(a_{ik}, a_{jk}),$$

where the w's are positive weights associated with the columns k (the varieties of pottery). We can think of b_{ij} as the strength of *similarity* between grave i and grave j, so that the larger b_{ij} happens to be, the closer together we should try to place these two graves in our sequence.

There are at least two ways of carrying out this program. The first makes use of the technique of *nonmetric multidimensional scaling*, originally developed by psychologists, which here constructs a (say, two-dimensional) map for the graves with distances d_{ij} roughly in an inverse order relationship with the b_{ij}'s.

The other (see e.g. Hill, 1974; Bølviken et al., 1982) uses what is called *correspondence analysis*, which we can view either as a novel sort of eigenvalue technique, or (and this I prefer) as the final result of a random process in which the graves pull themselves into a pattern under the competing influence of elastic strings of strengths b_{ij} tying them together, and an overall mutual repulsion. For a famous archeological dataset I show in Figure 1 the result of such an analysis by the first of these two methods.

Notice that we can easily identify from the computed output the two *ends* of the series of graves. While the intermediate sequencing can be read off from the diagram, it could also be determined in a variety of ways by comparing each grave with the two ends. This simple remark will cease to apply if most of

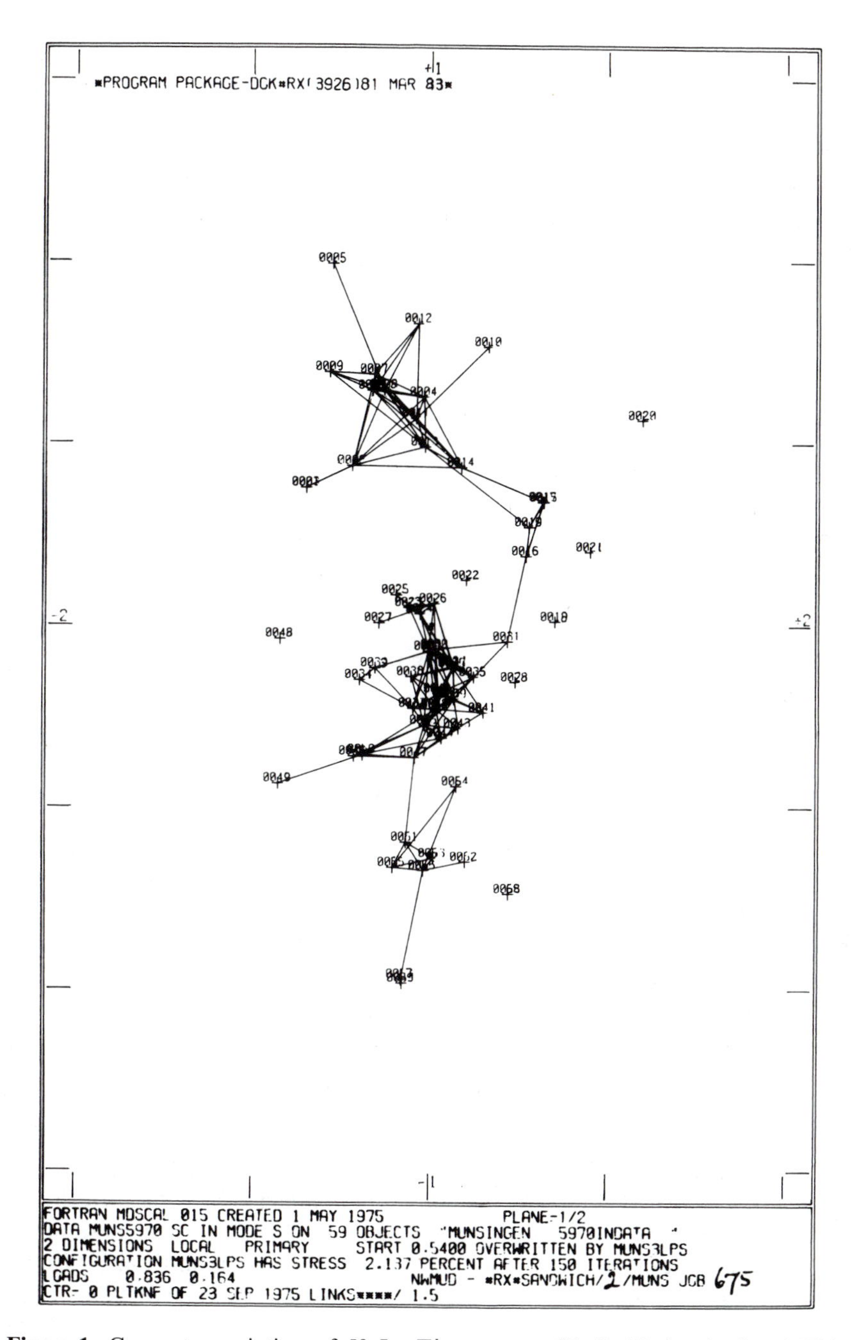

Figure 1. Computer seriation of 59 La Tène graves. (F. R. Hodson's data.) This automatic procedure generated in $1\frac{1}{2}$ minutes a sorting of the incidence matrix **A** almost as good as that arrived at by traditional archeological procedures.

the b_{ij}'s are zero, but one can then proceed similarly by using higher-order matrices analogous to **B** (Kendall, 1971b). When the number of graves is very large indeed, special techniques (Wilkinson, 1974) are required, and there are still a number of open problems.

Petrie (who used quite different methods) called this the problem of sequence *dating*, but today we prefer to speak of *seriation*, both because we will not know *a priori* which end of the sequence is "early" and which end is "late," and also because the seriation found may not be chronological at all; it could be sociological or geographical, or indeed all such effects may be involved at once and then a one-dimensional interpretation becomes inadequate. It is best to think of such techniques as providing one with an objective map of the data, through which they are encouraged to speak to us directly. The message received may be far from clear, but it provides an objective starting point which will normally form the basis for a more traditional solution.

Similar generalized "mapping" problems arise in other contexts, and I will just mention one or two of these with which I have been concerned. For example, it has been shown that we can make a map of a rural district by using the numbers of intervillage marriages recorded in parish registers (Kendall, 1971c). This is a surprising fact, because prior to modern times most marriages in rural areas were between parties from the same village. When suitable data is available I hope to attack the same kind of problem using similarities between local stocks of family names. This could be relevant to an archival approach to the identification of lost settlements. Another problem involving medieval data is that of mapping the open fields of a manor from abuttals listed in terriers and surveys; a preliminary study using a cartulary and survey of about 1402 A.D. has led to promising results (Kendall, 1975). Further examples could be drawn from molecular genetics and from developmental biology.

Let us now go back to archeology and consider the neolithic stone circles of Scotland. Professor Alexander Thom (1967) has suggested that these were planned using a common unit of length, so that a circle of a desired size could be constructed by first agreeing upon the size of its radius expressed in numerical terms. When this hypothesis was first proposed about thirty years ago, it was not very well received, and was answered by questions which begged other questions, such as "how could numeracy precede literacy?" But the growing volume of data and the development of techniques adapted to deal with it by Broadbent and others was accompanied by a change in attitude, and there are now many who would agree with me that the evidence (Kendall, 1977) from such preliminary studies as have been made is sufficiently strong to justify repeating the analysis using a more extensive survey designed from the outset not to prejudge the issue, and with a proper control over all the numerous and distinct contributory sources of variation and error.

An important consideration is the size of the stones themselves. These are so large that we can scarcely hope confidently to detect the difference between informal planning and more accurate planning (e.g. between planning by "pacing," and planning by the use of a measuring rod). But for me the primary question is not *how* the measurements were made, but *whether* the constructors agreed on the diameter of the circle before it was laid out, by quoting an integer number of units, informal or not.

How does one test this so-called "quantum hypothesis"? There is room here only to give a brief description of my own method. If X_j is the diameter of the jth of the N circles, then

$$\phi(t) = \sqrt{\frac{2}{N}} \sum_{1}^{N} \cos(2\pi X_j t)$$

will be likely to have a large positive peak (when plotted against t) at $t = 1/q$, if a quantum unit q has been employed. So what we must do is to plot $\phi(t)$ against t, and then locate and assess for significance the height of the *highest* peak in an *agreed* search range $[t_0, t_1]$ corresponding to extreme credible quantal values

$$q_0 = \frac{1}{t_1} \quad \text{and} \quad q_1 = \frac{1}{t_0}.$$

With $q_0 = 2$ feet and $q_1 = 10$ feet, the "cosine quantogram" for Thom's data (Figure 2) shows a marked peak at $q = 5.44$ feet. Testing this for significance is a delicate matter; we therefore use a data-based simulation test derived from random corruptions of the data. This indicates significance at a P-level of about $\frac{1}{120}$.

After this work was done I was asked by an art historian to help him to evaluate the working unit employed by a medieval Italian architect. That investigation is still in progress, but it is clear that the same methods can be tried.

There are other mathematical problems concerning neolithic stone monuments, and the one that I find most fascinating is purely geometrical: it has been claimed that some sets of single standing stones display collinearities to an extent not attributable to chance. Figure 3 shows the locations of a famous

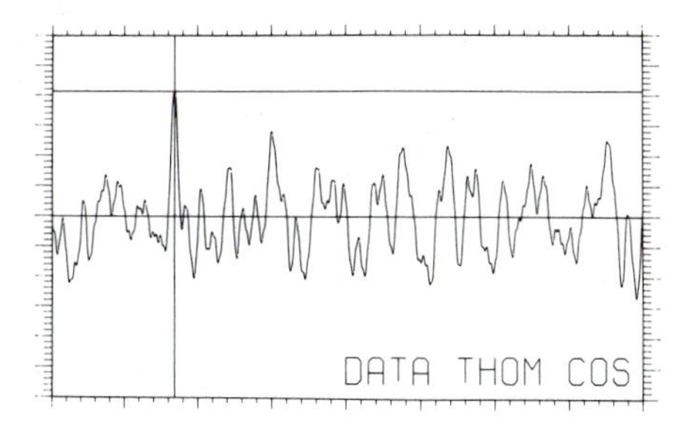

Figure 2. The "cosine quantogram" for A. Thom's data. The dominant peak at $q = 5.44$ ft ($t = 0.184$) is shown by the vertical marker. The horizontal axis ranges from $t = \frac{1}{10}$ to $t = \frac{1}{2}$.

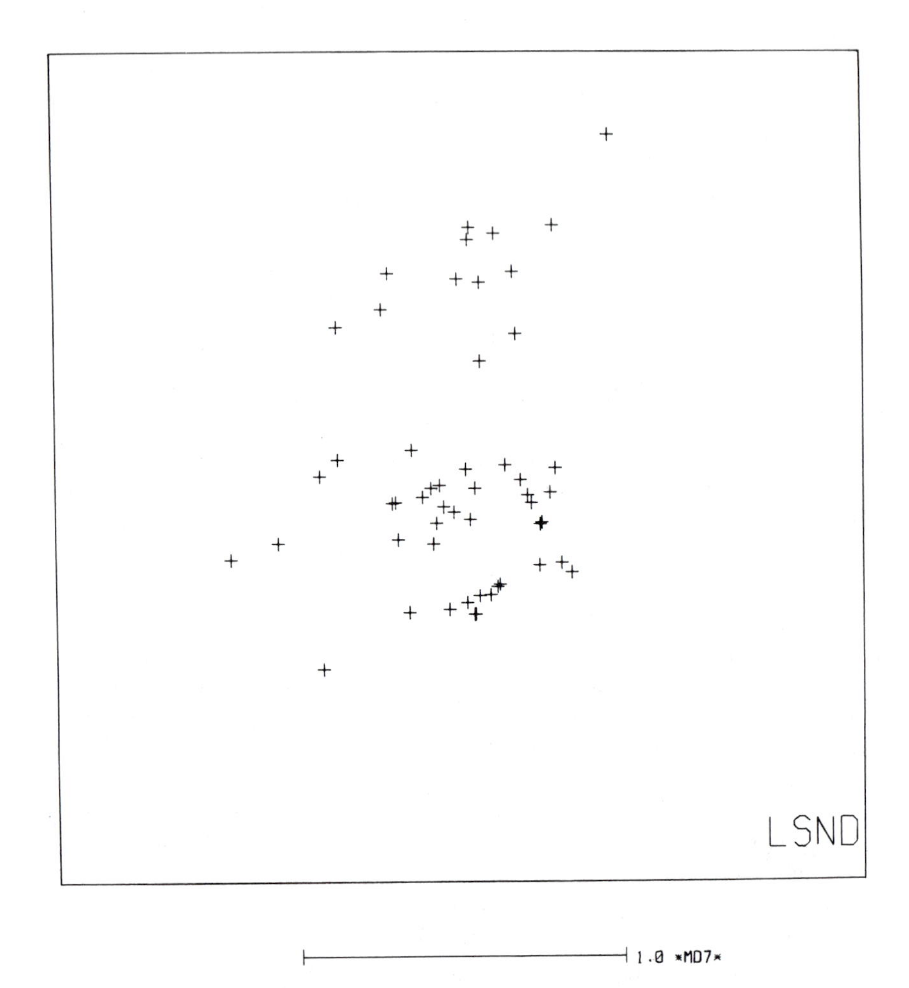

Figure 3. The positions of the 52 Land's End stones. [Reproduced by permission from Kendall and Kendall (1980).]

set of 52 standing stones near Land's End. It has been claimed that these do display "too many," and so some "planned," collinearities.

How many nearly collinear triplets can you see? While you are thinking about this, let us clear up the meaning of the phrase "nearly collinear." A number of different but closely related definitions can be used, but here at first I will follow Broadbent (1980), and say that A, B, and C are ε-collinear when the largest angle of the triangle ABC exceeds $\pi - \varepsilon$. Here ε is to be in radians, but in numerical work one more commonly uses the equivalent measure in minutes of arc. A common test value for ε is 40 minutes; let us adopt this for the moment.

We then find that there are $N = 108$ such collinear triplets. Are there too many?

It proves almost impossible to answer this question without calculation, even when we first spell out exactly what it means. As so often in matters of probability remote from everyday affairs, intuition is a most unreliable guide.

We reformulate the question as follows. Suppose that the 52 stones were shaken out of a giant's pepper pot, so as to lie randomly, independently, and uniformly in a square; what then is the number $\bar{N}$ of 40-minute-collinear triplets that you would expect to get by chance?

The answer, which may surprise you, is $\bar{N}$ = eighty-six.

This calculation was first made by Broadbent, who pointed out that a sampling fluctuation of about $1.65\sqrt{\bar{N}}$ was to be expected, so that the difference from $N = 108$ is technically "significant." However, the actual pattern of points is slightly elongated, and so one needs to make use of Broadbent's theorem that if the points lie in an $s \times 1$ rectangle, then we must increase the theoretical $\bar{N}$ by the factor

$$f = \frac{1}{2}\left(s + \frac{1}{s}\right),$$

where here $s = 1.661$. This is enough to wreck the significance! The calculation is not much affected if we replace the "rectangle" distribution by a Gaussian distribution.

At this point you may well ask, what happens if we now vary the test level from 40 minutes? Indeed we can do this, but the problem becomes harder because we now have to pay the cost of searching for the most striking result when ε is allowed to vary in a range of admissible values

$$\varepsilon_1 \leq \varepsilon \leq \varepsilon_2.$$

W. S. Kendall and I (1980) have worked out a technique for doing this, and once again the end result is not significant.

Experience with this investigation revealed the surprising poverty of our technical knowledge concerning statistical problems involving *shape*, and so it has been quite natural to turn to that wider aspect of the matter, and to attempt to fill the gap. Here we give only a sketch; more detail will be found in Kendall (1984).

What do we mean by "shape"? In this context I mean: whatever remains when everything depending on location, rotation, and size is filtered away. One therefore seeks to set up a mapping from {all possible triangles} to {all possible triangular shapes}, and to preserve as much as possible of the natural structure. For example, we should try to preserve topological and indeed metrical ("procrustean") structure, and to build a *shape space* Σ of *shape-points* σ retaining and respecting as far as possible all the symmetries of the original problem.

This is a highly technical matter, but the answer at least is simple. *The natural shape space for labeled triangles is the whole surface of a two-dimen-*

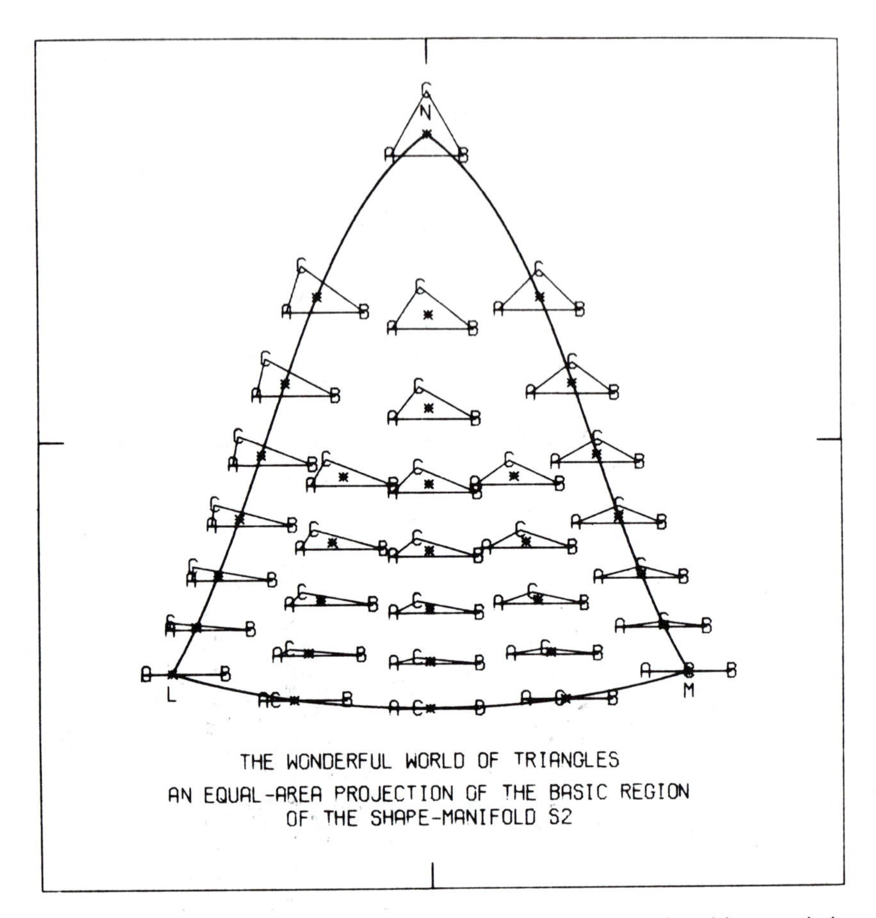

Figure 4. Thirty-two typical triangle shapes "at home." [Reproduced by permission from Kendall (1981).]

sional sphere of radius $\frac{1}{2}$. Even this simple fact is quite new, and in some ways a surprising result.

Now let us feed in some probabilistic ideas. Suppose that the triangle has random vertices A, B, and C, say independent and with isotropic Gaussian distributions. This method of generating random triangles must imply a statistical distribution for the triangle shape σ, which turns out again to be very simple: the triangle-shape points σ lie uniformly on the spherical surface. Of course we could consider other sorts of random triangle, not necessarily with independent vertices, thus getting other shape distributions.

Geographers have taught us how to map a sphere onto a plane without distortion of area. So let us use this to make an area-true symmetry-collapsed map of triangle shapes. In Figure 4 you will find such a map, with several triangle shapes "at home." The nearly collinear triangles all live along the arc labeled LM at the bottom of the "bell," and now you will see how alignment analysis has been rather neatly reduced to geometry.

So much for the shapes of triangles. The theory generalizes to sets of k points in m dimensions, but here I will only add a few rather technical words about k points in two dimensions. The shape space now becomes a version of the complex projective space CP^{k-2}, with its standard riemannian metric normed so that [as on $S^2(\frac{1}{2}) = \mathrm{CP}^1$] maximally remote points are separated by a geodesic distance $\frac{1}{2}\pi$. This space is homogeneous in relation to a large transitive group of symmetries, and there is a unique invariant measure which can be normed to become a probability measure. If the k points are statistically independent and have isotropic Gaussian distributions, then the shape point σ is distributed according to this invariant probability measure, while if they have (identical) *anisotropic* Gaussian distributions, then the density of the shape measure depends only on a single statistic $L(\sigma)$ $(0 \le L \le 1)$, which has a beautiful geometrical interpretation.

For $k = 3$, the truly collinear shapes correspond to points on a particular great circle on $S^2(\frac{1}{2})$. This I call the *collinearity locus*. For $k \ge 3$ the collinearity locus on CP^{k-2} is a copy of real projective space, RP^{k-2}, with its natural embedding; it is a compact subset of the shape space, and

$$\tfrac{1}{2}\sin^{-1} L(\sigma)$$

is the shortest geodesic distance from the shape point σ to the collinearity locus RP^{k-2}. Thus $L(\sigma)$ in the most natural possible way measures the lack of alignment in the k points determining σ.

We can infer the statistical distribution of $L(\sigma)$ in the isotropic case from work done on terrestrial magnetism by Mauchly (1940), while in the anisotropic case it is provided by a result of the present theory, which tidies up an earlier version (again in the magnetism context) due to Girshik (1941). So we have all the essentials for a natural geometric statistical test for *multiple* collinearities.

We can now use this technique on a topical problem. Arp and Hazard (1980) have reported a curious set of six quasars presenting striking collinearities. One would not worry about this much if the quasars formed part of a spatial cluster, but it appears that they do not; their red shifts differ considerably. So there is trouble for cosmologists if the alignments have to be taken seriously.

Figure 5 shows the 20 shape-points σ corresponding to the 20 triangles that one can make from the six quasars, these being positioned as in Figure 6. Notice how all the points lie near the collinearity locus for triplets, LM. How do we test this situation? Astronomers have made some rough tests, but we can now make a more appropriate and searching one.

To avoid selection errors let us *disregard* the triple collinearity ABC which first caused attention to be directed to the group of six. Thus we must restrict our test to the fourfold collinearity XYZB, and then we can use the L-test for $k = 4$. The significance level for this is

$$P = 0.000176 \times \binom{6}{4} = 0.00264.$$

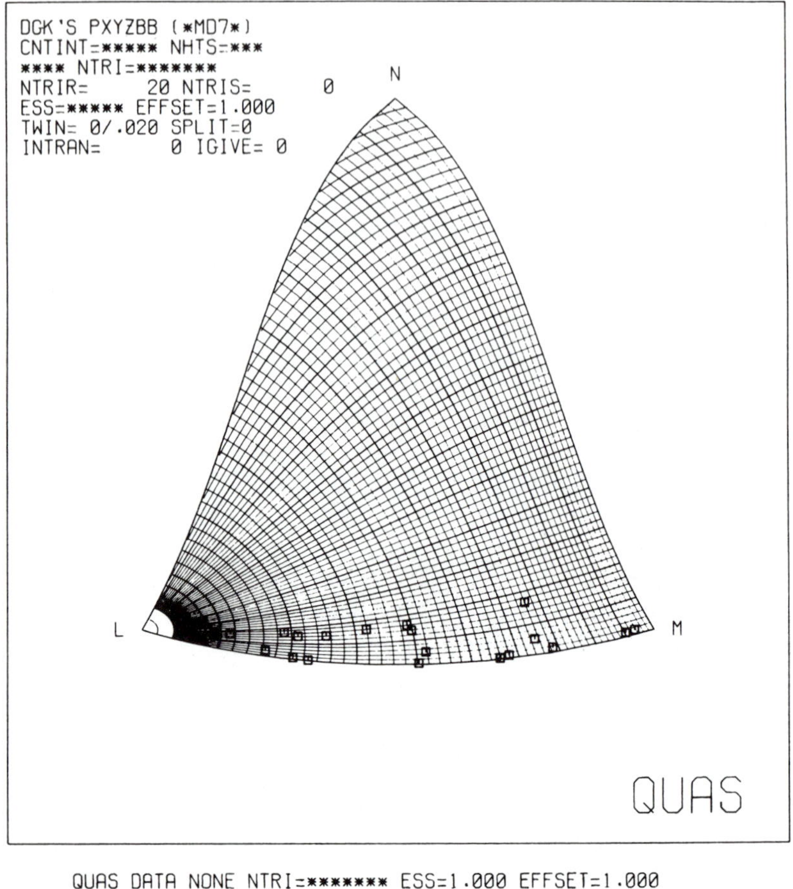

Figure 5. The 20 "quasar" triangle shapes "at home." The pencil of curves emanating from the point L are the loci of constant maximum angle for a triangle shape. [Reproduced by permission from Kendall (1984).]

This P-value is smaller than one given by Zuiderwijk (1982). However, there is need for caution. Quasars are not easy to identify, and one should beware of the consequence of stopping the search for them as soon as a striking configuration has been found. If there were more quasars in the region scanned, then the significance level P would be further eroded by the necessity of multiplying it by a larger combinatorial factor. For example, a set of 24 quasars will supply us with

$$\binom{24}{4} = 10{,}626$$

quartets, about one of which would therefore be expected to display the observed behavior.

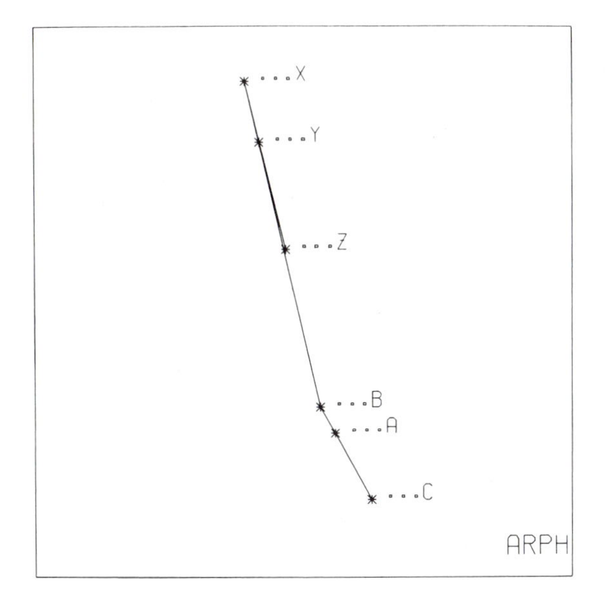

Figure 6. The positions of the six quasars in the tangent plane to the celestial sphere. [Reproduced by permission from Kendall (1984).]

Further exploration of this problem must be left to the astronomers. I mention it here in order to make my main point, which is that good applications of mathematics in one discipline can generate further applications in quite remotely different fields, and also (if one does not stifle one's natural curiosity) can lead to new developments in mathematics itself.

The statistics of shape may be seen as a natural companion to an already established topic: the statistics of directions. Here one of the earliest contributions was made by von Mises (1918) in an article concerned with the near-integral character of the atomic weights, and written a year or two before the discovery of atomic numbers and isotopes. It was in that paper that the von Mises distribution,

$$\frac{\exp(\alpha\cos\theta)\,d\theta}{2\pi I_o(\alpha)} \qquad (\theta \in S^1),$$

first appeared, and ultimately this has had a tremendous influence on all subsequent work (indeed, it lies at the root of the quantum-hunting investigation reported on above). But statisticians did not devote serious attention to directional questions until 1953, when Fisher introduced what is in fact the generalization of the von Mises distribution to directions in three dimensions, this therefore being a probability law on S^2; and Fisher's work had an origin in the humanities: it was concerned with the measurement of the fossilized geomagnetic field in archeological pottery fragments, and has led in its turn to an important technique for archeological dating.

Very recently the need has become apparent for yet another development, which we may call the statistics of indirections. Here by *indirection* we mean the ambiguous directional information given to us by an *undirected* line. If we think of a direction as a point on the circle S^1, then we must think of an indirection as a point on the projective space RP^1 obtained from S^1 by the identification of antipodes. It will be recalled that we can transform RP^1 to S^1 by throwing away one semicircle and joining up the two ends of the other, and this simple idea lies at the heart of all indirectional calculations. To round off the present essay I give a brief sketch of a very recent piece of work by G. A. Young and myself which perhaps will generate further indirectional studies in the humanities.

Here again the data come from astronomy, and the objects being investigated are double radio sources. Most of these appear to consist of a compact core accompanied by two more or less antipodal jets which define a geometrical major axis for the double source, that axis being undirected and so determining an indirection in the tangent plane to the celestial sphere. There is however also a second indirection to be observed at the same locus; this is defined by the magnetic polarization in the signals as received at the earth. In general the two indirections are not parallel, and we can measure the discrepancy between them by the acute angle Δ which they embrace, this being counted as positive if (say) the rotation from the geometric axis to the polarization axis is anticlockwise about the line of sight, and as negative otherwise. Notice that Δ must lie between -90 and 90 if we measure in degrees, and that $-90 = +90$ in this context because there is only one way in which two indirections can be perpendicular to one another. Such Δ-values have been measured in about 100 cases; they vary considerably, and have a marginal distribution which is roughly von Mises in form, save that here we must use the indirectional version

$$\frac{\exp(\alpha \cos 2\Delta)\, d\Delta}{\pi I_o(\alpha)} \qquad (-\tfrac{1}{2}\pi < \Delta \leq \tfrac{1}{2}\pi)$$

with a value for α of about 0.7.

Now Birch (1982) noticed the extraordinary fact that in some very rough sense there is a congregation of positive Δ-values in one celestial hemisphere and a congregation of negative values in the other. This was quite unforeseen by astronomers and caused some concern, if not disbelief. Birch had pointed out that *one* explanation would be a rotation of the Universe as a whole. In a critical review Phinney and Webster (1983) queried the statistical significance of the Birch effect, and at the same time proposed that, if it should turn out to be significant, a less disturbing explanation might lie in imperfect correction for the modifications in the polarization geometry introduced by the magnetic fields of plasma sheets relatively close to the observer. These last two authors therefore invited Mr. Young and me to try to assess the significance of the Birch effect.

Because of the remark about the equivalence of $\Delta = +90$ and $\Delta = -90$, it is clear that we will do better to take the primary variable to be $\sin 2\Delta$, a transformation very natural in any case because of the indirectional character of the problem. So let $\mathbf{p}$ denote a line-of-sight unit vector pointing to any one source, and let $f(\mathbf{p})\,d\mathbf{p}$ denote the source distribution. After much experiment we decided to adopt a model represented by the probability law

$$f(\mathbf{p})\,d\mathbf{p}\frac{\exp(\alpha\cos 2\Delta + (\boldsymbol{\lambda}\cdot\mathbf{p})\sin 2\Delta)\,d\Delta}{\pi I_o(\sqrt{\alpha^2 + (\lambda_1 p_1 + \lambda_2 p_2 + \lambda_3 p_3)^2})}.$$

Here the first (directional) factor gives the distribution of the source over the celestial sphere, while the second (indirectional) factor gives the conditional distribution of Δ in $\mathbf{p}$-dependent form. If we write β for the modulus of the vector $\boldsymbol{\lambda}$, then when $\beta = 0$ we have independence, and the Δ-distribution will be the indirectional version of the von Mises distribution with parameter α. But if β is greater than zero, then there will be a topographic regression of the indirection Δ on the direction $\mathbf{p}$, controlled by the vector $\boldsymbol{\lambda}$, whose magnitude β will express the strength of the Birch effect and whose direction $\boldsymbol{\lambda}/\beta$ will determine the positive pole of the topographic dependence.

The reader will see that the way is now clear to set up a likelihood-ratio test for $\beta = 0$, against $\beta > 0$, and it is very important that the f-distribution cancels out at this point. (It would have been difficult to specify a reasonable form for f, because it is principally controlled by the distribution of observatories over the surface of the earth.) Some care has to be taken, before maximizing the numerator likelihood with its four parameters $(\alpha, \lambda_1, \lambda_2, \lambda_3)$, to check (as is the case) that the maximum is unique and occurs at a point of differentiability.

As a first step one can think of α and β as small and perform a first-order calculation. This leads to the test statistic

$$T_1 = \tfrac{1}{4}\mathbf{v}'\mathbf{B}\mathbf{v},$$

where

$$v_i = 2\sum_{r=1}^{N} p_i^{(r)} \sin 2\Delta^{(r)},$$

and where $\mathbf{B}$ is the reciprocal of the matrix $\mathbf{A}$ with components

$$a_{ij} = \sum_{r=1}^{N} p_i^{(r)} p_j^{(r)};$$

this reciprocal will exist provided that the sources do not lie on a common great circle on the celestial sphere. As usual, T_1 will have asymptotically a $\frac{1}{2}\chi^2$ distribution, in this case with 3 degrees of freedom, but we are scarcely at the asymptote in the present problem. Accordingly we prefer a data-based simulation test, and the corresponding P-value was found by making up 10,000 random reassignments of the observed set of Δ's to the observed set of $\mathbf{p}$-

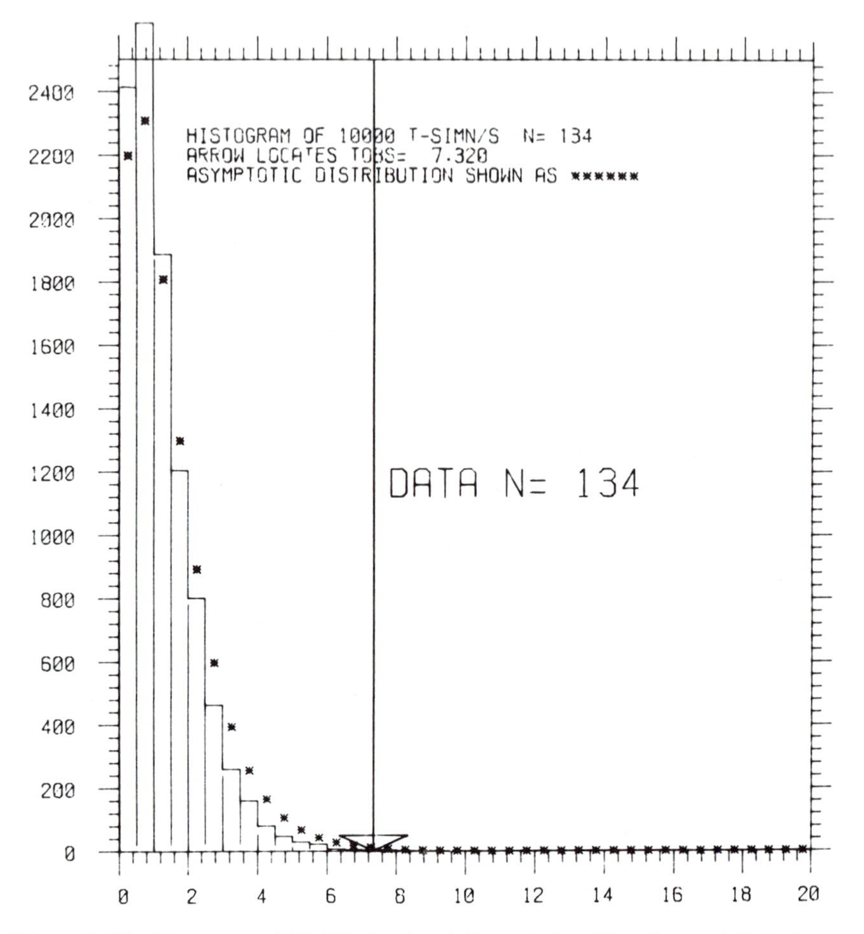

Figure 7. The histogram of 10,000 simulated T_1-statistics. The observed T_1 is shown by an arrow, and points on a matched $\frac{1}{2}\chi_3^2$-distribution are shown by asterisks. [Reproduced by permission from Kendall and Young (1984).]

vectors. Figure 7 shows the histogram for the associated 10,000 simulated T_1-values, and also the observed T_1-value, which stands very far out in the tail ($T_1 = 7.32$, $P = 0.0008$). The same calculation gives preliminary estimates for α and λ, and these can now be used to start off an exact iterative calculation using the precise formulae for the likelihoods. This gives the final test-statistic as $T = 8.02$, with much the same P-value as before (but less convenient to calculate, because now the repetitions of the exact maximization are very time-consuming). The exact estimates of the parameters were

$$\alpha = 0.727, \qquad \beta = 0.933, \qquad \text{pole} = \text{RA } 13^{\text{h}}30^{\text{m}}, \text{ Decl. } 37\overset{\circ}{.}4 \text{ S.}$$

It seems that the Birch effect does indeed exist, and calls for explanation, even if this is something less dramatic than universal rotation.

Acknowledgements

I should like to thank my colleague Alastair Young for allowing me to include the last few paragraphs which sum up our recent joint research. It seemed particularly worthwhile doing this because our substantive results (Kendall and Young, 1984) appeared in an astronomical rather than a statistical periodical.

Bibliography

Arp, H. and Hazard, C. (1980). "Peculiar configurations of quasars in two adjacent areas of the sky." *Astrophys. J.*, **240**, 726–736.

Birch, P. (1982). "Is the universe rotating?" *Nature*, **298**, 451–454.

Bølviken, E. et al. (1982). "Correspondence analysis: An alternative to principal components." *World Archaeology*, **14**, 41–54.

Broadbent, S. R. (1980). "Simulating the ley hunter." *J. Royal Statist. Soc. Ser. A*, **143**, 109–140.

Fisher, R. A. (1953). "Dispersion on a sphere." *Proc. Roy. Soc. London Ser. A*, **217**, 295–305.

Girshik, M. A. (1941). "The distribution of the ellipticity statistic L_e when the hypothesis is false." *Terrestrial Magn. and Atmos. Elec.*, **46**, 455–457.

Hill, M. O. (1974). "Correspondence analysis: A neglected multivariate method." *Appl. Statist.*, **23**, 340–354.

Kendall, D. G. (1971a). "Seriation from abundance matrices." In F. R. Hodson, D. G. Kendall, and P. Tautu (eds.), *Mathematics in the Archaeological and Historical Sciences*. Edinburgh U.P., 215–252.

Kendall, D. G. (1971b). "Abundance matrices and seriation in archaeology." *Z. Warsch.*, **17**, 104–112.

Kendall, D. G. (1971c). "Maps from marriages." In F. R. Hodson, D. G. Kendall, and P. Tautu (eds.), *Mathematics in the Archaeological and Historical Sciences*. Edinburgh U.P., 303–318.

Kendall, D. G. (1975). "The recovery of structure from fragmentary information." *Philos. Trans. Roy. Soc. London Ser. A*, **279**, 547–582.

Kendall, D. G. (1977). "Hunting quanta." In *Proceedings of the Symposium to Honour J. Neyman*. Warsaw PWN,: 111–159.

Kendall, D. G. (1981). "The statistics of shape." In V. Barnett (ed.), *Interpreting Multivariate Data*. Chichester: Wiley, 75–80.

Kendall, D. G. (1984). "Shape-manifolds, procrustean metrics, and complex projective spaces." *Bull. London Math. Soc.*, **16**, 81–121.

Kendall, D. G. and Kendall, W. S. (1980). "Alignments in two-dimensional random sets of points." *Adv. Appl. Probab.*, **12**, 380–424.

Kendall, D. G. and Young, G. A. (1984). "Indirectional statistics, and the significance of an asymmetry discovered by Birch." *Monthly Notices Roy. Astron. Soc.*, **207**, 637–647.

Mauchly, J. W. (1940). "A significance test for ellipticity in the harmonic dial." *Terrestrial Magn. and Atmos. Elec.*, **45**, 145–148.

Phinney, E. S. and Webster, R. L. (1983). "Is there evidence for universal rotation?" (together with a reply by Birch). *Nature*, **301**, 735–736.

Thom, A. (1967). *Megalithic Sites in Britain.* Oxford: Clarendon.

von Mises, R. (1918). "Über die 'Ganzzähligkeit' der Atomgewichte und verwandte Fragen." *Phys. Z.*, **19**, 490–500.

Wilkinson, E. M. (1974). "Techniques of data analysis: Seriation theory." *Archaeophysika*, **5**, 1–142.

Zuiderwijk, E. J. (1982). "Alignments of randomly distributed objects." *Nature*, **295**, 577–578.

CHAPTER 18

Reflections on the World Fertility Survey

Miloš Macura and John Cleland

Abstract

A total of 62 countries comprising about 40 per cent of the world's population participated in the World Fertility Survey (WFS). The WFS is thus by far the most ambitious project ever undertaken by ISI and probably the largest social survey ever attempted. In terms of sampling and data collection, the methodology of the project was exemplary. The data-processing record is less satisfactory, because of an initial failure to foresee the complexity of the issues and the magnitude of the practical problems. In its later years, the project made an unexpectedly major contribution to analytic methodology. So far, the main contribution to knowledge has been to confirm the downward trend in fertility that characterized much of Asia and Latin America in the 1970s, and to highlight the contrast with Africa where both fertility and the desire for large numbers of children remain high. So far, important new insights into the causes of fertility change have not emerged, but attempts to synthesize the vast range of WFS findings have only just begun, and it is thus premature to make any final assessment of this aspect of the project.

1. Historical Perspective

It is a mere coincidence that the International Statistical Insitute is approaching its centenary jubilee at the time of completing the World Fertility Survey—its largest and most successful scientific undertaking over the last century. It is, however, a happy coincidence which provides an opportunity for quiet reflection on the WFS and its connection with the Institute: connection not only in terms of organization, but also in human and scientific terms.

Indeed, the WFS has absorbed a good deal of the Institute's energy and talent. To carry it out the ISI had to engage in a complex partnership with many national and international agencies. The WFS was not a short-term project but a large operation lasting for more than ten years. If we look at it in a long-term perspective, we realize that its existence extended over 10 per cent of the Institute's life span, that it covered about 60 per cent of the nations represented in the Institute's membership, and that its substance was a part of

Key words and phrases: demographic transition, household roster, machine editing, maternity history, questionnaire design, sample clusters, sampling errors, sampling frame.

the wider study on population traditionally pursued since the 1853 International Statistical Conference in Brussels.

The need for international statistical collaboration, felt in Europe by the mid nineteenth century, and the circumstances under which the International Statistical Institute was established in 1885, were analysed by Friedrich Zahn (1934). He also extensively discussed the questions of international statistics connected with the emergence of the League of Nations in 1919.

Statistical activities of the League, which gradually developed between the two world wars, did not affect the work of the Institute. But a major change in the international statistical setting took place in 1947, when new functions concerning official statistics were established at the United Nations. This matter was discussed on several occasions in technical journals and was ably summarized by Nixon (1960). We do not need, therefore, to deal with those aspects of the Institute's history, However, it should be mentioned that, once the responsibility for official statistics had been assumed by the United Nations, the ISI had to adjust its role and activity to the evolving international statistical system. Both Armand Julin and Stuart A. Rice, the outgoing and the incoming president of the ISI at that time, agreed that it would be appropriate for the Institute to maintain its character as an international academy while accepting an unofficial leadership in the development and improvement of the world's statistics.

For the sake of clarity, we should be reminded that for a number of decades the ISI was the only institution to guide and promote the development of national and international statistics—administrative, applied, and scientific. Continuing the effort initiated by the International Statistical Conference, which met on nine occasions from 1853 to 1876 and which covered many major fields of statistical organization, theory, and applied statistics, the ISI was established as an international academy which also undertook the difficult task of developing and promoting official statistical work. According to provisional rules which were absorbed in the Statutes in 1887, the aim of the Institute was "the development of the progress of administrative and scientific statistics." This objective was to be achieved by introducing uniformity of methods, compilation of statistical publications with a view to international comparability, international publication of data, and the promotion of "general appreciation of statistical science ... to stimulate the interest of governments and individuals in the study of social phenomena." The Institute had no administrative power delegated to it and therefore proceeded "by inviting the attention of governments to the various problems capable of solution by statistical observation" (Zahn, 1934).

Quality and comparability of data was a major concern. Thus, in his inaugural address to the first International Conference (1853, Brussels), Quetelet stated:

> Chaque science a débuté par des méprises, souvent même par de deplorables abus. Ce qui peut nous étonner, ce n'est pas que la statistique ait erré; mais que, si prés de sa naissance, elle ait déjá compris sa mission et senti le besoin de

> régulariser sa marche. Ce congrès, si je ne me trompe, commencera pour elle une èra nouvelle; la statistique entre dans la même phase que plusieurs autres sciences, ses soeurs ainées, qui ont apprecié, comme elle, les besoins d'adopter une langue commune et d'introduire de l'unité et de l'ensemble dans leurs recherches. (ISI, 1853.)

We should be also reminded that one of the highest priorities was the development of demographic statistics and studies. Quetelet, to whom we owe the idea of international statistical organization, was among those who have insisted that population be studied as a multidisciplinary phenomenon. Population was continuously on the agenda of the International Statistical Conferences and continued to be on the agenda of the ISI. In the first volume of the *Bulletin de l'Institut International de Statistique*, in which the foundation of the Institute was discussed by von Neumann Spallart, papers were published on the population of ancient Rome by Beloch; on vital statistics in Europe and the USA by Rawson; on the sex and age distribution of population by Perozzo; and on Italian emigration to other European countries, on the education of populations, together with a comprehensive study of the worlds territory and population, by Lavasseur, supplemented with statistical tables for each European country (ISI, 1886).

These early concerns set the mold for later decades, for demographic statistics and studies were always in the focus of ISI interest. At the 1872 Conference in St. Petersburg, a first resolution was adopted on the substance of topics to be included in population censuses. This was supplemented with a tabulation programm adopted at the 1887 session of the Institute in Rome. Plans were worked out for a world population census to be organized in 1900. These were discussed by the Institute and the International Conference on Hygiene and Demography but did not materialize at that time for reasons we can well understand today (Zahn, 1934). Progress was made in vital and migration statistics as well, so that von Mayr could propose publication of an international annual report on vital and migration statistics. This was accepted by the Institute at its Bern session in 1895. It is interesting that von Mayr also proposed the creation of an International Demographic Statistics Office to be in charge of compiling and publishing the annual reports (ISI, 1896).

Development of methods, promotion of population and vital statistics, and compilation and publication of data became eventually a major function of the ISI. Among publications in population statistics, mention should be made of *Annuaire International de Statistique*, partie "Demographie", tomes I–V with an Annexe (La Haye, 1917 through 1921) and *Aperçue de la démographie des divers pays du monde 1929–1936* (La Haye, 1939). In addition, many substantive studies were produced for and discussed at the biennial sessions of the ISI, covering a range of demographic topics. The Institute had thus established demography an important component of its work. This tradition persisted throughout the years and events, and continued to be one of the preoccupations of the Institute.

2. The Origin of the Survey

Consistent with its new role in the postwar world statistical system, the ISI has shifted its activity towards theory and methodology, application in new fields, and education and training areas in which exploration of new frontiers was essential. Yet it did not lose its interest in the practical issues of census taking, sampling and surveys, vital statistics, and industrial, agricultural, and other statistics, which were now compiled, processed, and published by statistical agencies of the United Nations system. There has developed an apparent contradiction between the ISI membership's interest in practical statistical issues and the ISI terms of reference and program. In this connection the need has been felt to reexamine the aims of the Institute embodied in its Statutes of 1947. Of special concern were subparagraphs relative to the ISI task of "promoting the use in all countries of the most appropriate statistical methods" and "furthering international comparability of statistical data."

The experience of two decades appeared to be sufficient to make it possible for a Reappraisal Committee to suggest that "only two courses lie open before the Institute: to resign itself to a continuation of activities on the present scale regarding itself as an international academy composed of an elite membership of a few hundred people; or to reshape its work in a fundamental way by attempting to meet the challenge of the present and future needs of the profession." The Committee suggested that over and above activities already well established, the ISI should engage—through its research staff—in developing methods and techniques for research on economic, demographic, and social phenomena, studying certain issues of international interest, setting up basic courses of education and training, etc. (report, 1968). At the 1969 session in London the report was discussed and adopted. In addition to organizational aspects, the following areas of possible study were mentioned: population explosion and birth control, rain stimulation, food supply, theory of communication, etc. (ISI, 1969). Thus the ISI was ready, psychologically if not yet organizationally and financially, to assume a new responsibility for the advancement of statistics and research.

At the same time, the need was increasingly felt for reliable fertility data, both national and international. Following the 1966 United Nations resolution on population and development, there was little doubt that population change was of the utmost concern to the international community. Fertility was considered important in many countries in which it was high, contributing to an excessive population growth, but also to countries in which it was too low, below the replacement level. Moreover, fertility appeared to be a key variable in national population policies and family-planning programs. Hence a strong interest was aroused in reliable fertility data and their adequate interpretation, capable of improving existing knowledge and policies.

The idea of an international inquiry on fertility was proposed by a small group of leading American demographers. Exploration of the proposal continued following an exchange of views on possible involvement in an inquiry of the ISI with representatives of the United States Agency for International Development (USAID) and the United Nations Fund for Population Activities (UNFPA). Consultation took place during and after the 1971 Session in Washington, D. C. and at the ISI Headquarters in the Hague. Following a detailed study of needs for a world fertility survey, the work done so far, and an examination of the relevant conditions involving the Institute and the interested agencies, it was decided to initiate the project. While the start was modest and cautious, as in many other research undertakings, it soon became clear that a major international inquiry had begun. It happened that the Chairman of the ISI Reappraisal Committee, Maurice Kendall, was appointed as Project Director of the Survey.

As a project of ISI, undertaken in collaboration with the United Nations and the International Union for the Scientific Study of Population, and with financial support of UNFPA, USAID, and other agencies, the World Fertility Survey became operative in 1973. It was structured so as to rely on a central staff in London, flexible regional support, and, as a key component, national survey organizations. The initial phase did not last long, owing to the work done in 1972. The ISI members received a first report on the WFS at the 1973 session in Vienna. In his presidential address, Petter Jakob Bjerve referred to the complexity of the world population situation, to the persistent high fertility in the Third World, and to the understanding of which the Institute would contribute in a world wide inquiry in fertility. By starting the project, the ISI was implementing also the desire of its 1969 London Session to involve the organization in productive statistical and research work (Bjerve, 1973).

3. Aims and Organization

The explicit aims of the project were threefold: to assist interested countries to describe and interpret their population's fertility by conducting scientifically designed surveys, to enhance national capability to undertake demographic surveys, and to produce internationally comparable data on human fertility.

As participation in the program was, of course, voluntary, the level of response was a critical determinant of its potential success. Accordingly, WFS devoted considerable energy during 1973 and 1974 to establishing its international credibility, publicizing its plans, and encouraging participation. The results far exceeded most initial expectations. As shown in Table 1, a total of 62 countries, comprising about 40 per cent of the world's population, took part. Of these, 42 were eligible, as developing countries, for substantial financial assistance from UNFPA, USAID, or other donors and/or for technical support from WFS central staff. The remaining 20 developed countries conducted their surveys in a more autonomous and less coordinated fashion. The geographical distribution of participants is reasonably dispersed, and only the absence of a few of the most populous nations—China, India, Brazil, and the Soviet Union—detracts from the global coverage of the survey.

Table 1. Some Characteristics of WFS Surveys

	Estimated population size in mid 1981 (millions)	Year of field-work	Year of publica-tion[a]	Achieved sample sizes		Sample universe (Individual):	
				House-holds	Individual women	Age Limits	Marital[b] status
			Africa				
Nigeria	79.7	1981–82	1984	8,600	9,700	15–49	All
Morocco	21.8	1980	1984	17,100	5,800	15–50	All
Sudan (North)	19.6	1978	1982	12,000	3,100	<50	EM
Kenya	16.5	1977–78	1980	8,900	8,100	15–50	All
Ghana	12.0	1979–80	1983	6,000	6,100	15–49	All
Cameroon	8.7	1978	1983	37,900	8,200	15–54	All
Ivory Coast	8.5	1980–81	1984	3,800	5,200	15–50	All
Tunisia	6.6	1978	1982	5,700	4,100	15–49	EM
Senegal	5.8	1978	1981	18,000	4,000	15–49	All
Benin	3.8	1981–82	1984	20,000	4,000	15–49	All
Mauritania	1.7	1981	1984	14,800	3,500	15–50	EM
Lesotho	1.4	1977	1981	18,200	3,600	15–49	EM
			Asia and Pacific				
Indonesia	148.8	1976	1979	10,200	9,200	15–50	EM
Japan	117.8	1974	1976	5,000	2,900	<50	EM
Bangladesh	92.8	1975–76	1979	5,900	6,500	<50	EM
Pakistan	88.9	1975	1976	4,900	5,000	<51	EM
Philippines	48.9	1978	1979	12,700	9,300	15–49	EM
Thailand	48.6	1975	1977	4,300	3,800	<50	EM
Iran	39.8	1977	—	5,700	4,900	15–50	EM
Korea, Rep.	38.9	1974	1978	20,900	5,400	<51	EM
Sri Lanka	15.3	1975	1978	8,100	6,800	<50	EM
Nepal	14.4	1976	1977	5,700	5,900	15–49	EM
Malaysia	14.3	1974	1977	7,800	6,300	<50	EM
Fiji	0.6	1974	1976	4,900	4,900	15–49	EM

			Caribbean				
Haiti	6.0	1977	1981	3,000	3,400	15–49	EM
Dominican Rep.	5.6	1975	1976	10,900	3,100	15–49	All
Jamaica	2.2	1975–76	1979	4,600	3,100	15–49	All[c]
Trinidad and Tobago	1.2	1977	1981	4,600	4,400	15–49	All[c]
Guyana	0.9	1975	1979	4,400	4,600	15–49	All[c]
			Europe				
Italy	57.2	1979	1982	NA	5,500	18–44	EM
United Kingdom	55.9	1976	1979	14,000	6,600	16–49	All
France	53.9	1977–78	1979	14,100	3,000	20–44	All
Spain	37.8	1977	1978	NA	6,300	15–49	EM
Poland	36.0	1977	1980	NA	9,800	<45	CM
Yugoslavia	22.5	1976	1980	NA	8,100	15–49	CM
Romania	22.4	1978	1981	NA	10,100	15–49	CM once
Czechoslovakia	15.4	1977	1979	NA	3,000	18–44	CM once
Netherlands	14.2	1975	1978	NA	4,500	[d]	CM
Hungary	10.7	1977	1979	10,000	4,000	<40	CM
Portugal	10.0	1979–80	1983	10,900	5,100	15–49	EM
Belgium	9.8	1975–76	1978	NA	4,900	16–44	All
Bulgaria	8.9	1976	—	13,400	6,900	15–44	EM
Sweden	8.3	1981	1982	NA	5,000	20–44	All
Switzerland	6.3	1980	1981	NA	600	[e]	CM
Denmark	5.1	1975	1979	NA	5,200	18–49	All
Finland	4.8	1977	1980	NA	5,400	18–44	CM once
Norway	4.1	1977–78	1981	NA	4,100	18–44	All
			Middle East				
Turkey	46.2	1978	1980	5,100	4,400	<50	EM
Egypt	43.5	1980	1983	10,100	8,800	<50	EM
Syria	9.3	1978	1982	14,700	4,500	<50	EM
Yemen, Arab Rep.	5.4	1979	1984	13,300	2,600	<51	EM
Israel[f]	3.9	1973–74	1976	NA	10,700	<60[g]	EM
Jordan	3.3	1976	1980	14,500	3,600	15–49	EM

Table 1. (continued)

	Estimated population size in mid 1981 (millions)	Year of field-work	Year of publication[a]	Achieved sample sizes		Sample universe (Individual):	
				Households	Individual women	Age Limits	Marital[b] status
North and South America							
United States	229.8	1976	1977	33,000	8,600	15–44	EM
Mexico	68.2	1976–77	1979	13,100	7,300	20–49	All
Colombia	27.8	1976	1978	9,800	5,400	15–49	All
Peru	18.1	1977–78	1979	7,400	5,600	15–49	EM
Venezuela	15.5	1977	1981	8,600	4,400	15–44	All
Ecuador	8.2	1979–80	1984	5,800	6,800	15–49	All
Paraguay	3.3	1979	1981	4,000	4,600	15–49	All
Costa Rica	2.3	1976	1978	4,200	3,900	20–49	All
Panama	1.9	1975–76	1978	4,800	3,700	20–49	All

[a] For countries in Africa, Asia and the Pacific, the Caribbean, the Middle East, and South America publication refers to a substantial report, containing 200 to 300 tables of figures and accompanying text. For European countries, publications took more varied and often fragmented forms. The year of publication indicates the date of release of at least an appreciable set of results.

[b] All = All women regardless of marital status; EM = ever married women; CM = currently married women; CM once = currently married women still in their first marriage.

[c] Excluding full time school girls.

[d] All women married between 1963 and 1973 irrespective of age.

[e] All women married between 1970 and 1979 irrespective of age.

[f] Survey details relate to survey of Jewish women only.

[g] The upper limit was 60 for women of Asian or African origin and 75 for women of European or American origin.

The ISI made special efforts to encourage participation of these large countries, but met only limited success. India agreed to conduct a number of state-level surveys and maintained a loose technical liaison with WFS headquarters. Regrettably, the survey instruments chosen in India differ in certain key respects from the standard WFS instruments, with attendant loss of comparability. China formally requested participation in the WFS program too late for inclusion; however, plans have been made to conduct several WFS-type surveys in 1984–1985 with ISI technical assistance and thus the most glaring defect in global coverage will eventually be remedied.

The initiative for embarking upon a survey and the ultimate responsibility for its successful implementation lay with government departments, usually statistical offices of individual countries. The contribution of WFS, however, was not restricted to the provision of technical assistance upon request, but took the form of a partnership, a shared commitment to the achievement of a common goal. Involvement by WFS staff in developing-country surveys started at the very beginning with planning and budgeting, and close contact was maintained throughout all stages, including reporting and dissemination of main results and subsequent in-depth analyses of particular topics. Typically, this amounted to a dozen or so visits by WFS staff, each of several weeks' duration, over the three-year life span of each survey. It is this close contact over a long period, coupled with the demonstration effect of implementing a high quality survey, that represents the main transfer of expertise. We suggest that this form of indirect training has made a large impact on subsequent survey procedures in many developing countries, and in this way the WFS has contributed to enhancement of local capability. More obvious forms of institution building, such as the purchase of computer hardware, long-term overseas training, and the design of master sample frames, lay outside the scope of the WFS and were left to other agencies.

Though the United Nations appointed regional liaison officers, and resident advisers were recruited in a few countries, the WFS remained a centralized organization, with a highly mobile staff who returned to London headquarters after each assignment. As the volume of work increased, staff numbers grew to a peak of nearly 50 professionals in 1980, declining to half this number as the program drew to a close in mid 1984. Overall supervision of the WFS was exercised by a Programme Steering Committee, composed of representatives of the funding agencies, other international organizations, and eminent individuals. Further guidance in the formative years, 1973 to 1977, was provided by a Technical Advisory Committee.

The organization and operational strategy of the WFS, with its centralized staff and its sense of joint responsibility for the surveys themselves, may be viewed as a natural consequence of the aims of the project. Perhaps only such a posture could have ensured the successful completion of all surveys, the maintenance of sufficiently high standards for the production of reliable national results, and the necessary degree of standardization for cross-national comparability; but the strategy has attracted criticism. Centralization, it has been argued, may be detrimental to the creation of regional expertise; standardization implies an insensitivity to local situations; the scientific standards have been too high and expensive to act as a realistic model for future surveys in poorer countries; the concept of partnership has not encouraged the growth of national self-sufficiency in survey taking; and national priorities have been distorted by the lure of readily available funds.

The cost of the WFS program has provoked more adverse comment than any other aspect. The total sum absorbed by WFS is approximately US$40-million, a large slice of the total social research funds available for this topic in the decade 1974–84; but it would be a mistake to divide this amount by the number of funded surveys and conclude that each cost a million dollars. WFS activities spread well beyond survey implementation, as a glance at its publication list will reveal. The operational in-country costs of WFS surveys were probably not excessive by comparison with other surveys. On the other hand, the amount and costs of the technical assistance provided by WFS were perhaps too high; these costs often equaled the total in-country costs. The complexity and perfectionism of WFS methodology are largely to blame. As we shall see, much has been gained by this approach, but the price has been high.

4. Methodology: Development, Application, and Appraisal

The priority of the WFS program has always been the expeditious production of substantive findings, a mandate that precluded methodological development and experimentation on an appreciable scale. The initial phase of the enquiry therefore took the form of a synthesis and refinement of the best features of previous work rather than the forging of radically new procedures, though, as will be discussed below, more innovation was possible in data processing and analysis than in data collection.

4.1. Data-Collection Instruments

The volume of recent experience on which the WFS could draw was considerable. A growing awareness of the consequences of rapid population growth in the Third World had provided the incentive and funds for a multitude of fertility and family-planning surveys, including several attempts to produce systematic recommendations for their content and conduct, most notably *Variables and Questionnaires for Comparative Fertility Surveys* (UN, 1970) and *A Manual for Surveys of Fertility and Family Planning: Knowledge, Attitudes, and Practice* (Population Council, 1970). The former was the point of departure for the development of WFS questionnaires; early in 1973, Ryder and Westoff, who had been long associated with fertility surveys in the USA, were commissioned to make the first drafts, and after several revisions in the light of comments by over 200 researchers, the essential features were agreed on by the end of the year.

The contents of the inquiry reflect pragmatic rather than theoretical considerations. Although some individuals wanted the substance of the questionnaire to be determined by a coherent theoretical framework, others successfully argued that no acceptable theory existed and that attention should be restricted to variables that had been tried and tested in earlier surveys. In consequence, the core questionnaire is characterized by a cautious consensus rather than by novelty, and by an emphasis on thorough demographic description rather than on the testing of causal hypotheses.

The basic or core data-collection instruments comprise a household schedule and an individual questionnaire for women in the reproductive age range. The household schedule consists essentially of a roster of household members, together with such details as age, sex, and marital status of each member. One of its main purposes is to identify women eligible for the individual interview but, in countries lacking recent vital data, provision was made for expansion of the contents to include summary measures of fertility and mortality. In such instances it was envisaged that an enlarged household sample would be used to allow computation of acceptably precise vital rates for population subgroups. In the event, 15 of the 42 developing-country surveys, mostly in Africa, employed an enlarged household sample. Sample sizes may be found in Table 1.

The WFS designed two model questionnaires: one for conditions prevailing in low-fertility, developed countries (which we shall not discuss further) and another for conditions in high-fertility, developing countries. The main topics covered in the individual questionnaire for developing countries were: complete maternity and marriage histories; knowledge and use of contraceptive methods; socioeconomic characteristics such as education, employment, residence, and religion of both husband and wife; and individual preferences concerning family size. Only the last topic, fertility preferences, provoked serious controversy; this concerned the extent to which child bearing was the product of conscious rational decision making and hence on the legitimacy of questions on individual preferences. The solution to this disagreement lay in flexibility. Attitudinal questions were kept to a minimum in the core questionnaire, but participating countries were urged to use an expanded set of questions (termed the Fertility Regulation Module) with a stronger emphasis on reproductive motivation. A total of 24 countries used this module. However, even with these additions, the questionnaire developed by the WFS diverges most markedly from earlier models in the relative absence of attitudinal items. The other crucially important departure from most earlier work was the inclusion of questions on breastfeeding, which were to prove an invaluable asset.

The schedule and questionnaires were pretested in 1974 on a national scale in Fiji and on a smaller scale in India, Zaire, and Colombia. The following year they were published, first in English and subsequently in the other official languages of the WFS—Spanish, French, and Arabic. Though subject to a myriad of adaptations to local circumstances, their contents have proved remarkably resilient. Only one significant set of changes was made, and this was in response to the view of the Programme Steering Committee that items on contraceptive availability and induced abortion should be added. Their suggestions were piloted and assessed in three countries (Rodriguez, 1977) and recommendations published (WFS, 1977). The manner and speed with which the WFS was able to react testifies to the growing confidence of the project staff.

The household schedule and core questionnaire represents only the basic minimum of common content across countries, to ensure comparability of key results. Participating countries were encouraged to expand the contents in accordance with their own priorities and circumstances. To assist them, WFS started in 1974 to develop and pretest a number of questionnaire modules: sets of questions on particular topics that could be added to, or integrated with, the core.

A total of eleven modules were considered. Three of these never advanced beyond preliminary drafting, and a fourth was abandoned after a pretest had revealed serious problems of respondent incomprehension and resentment. This left a residue of seven modules, covering the following areas of interest: fertility regulation (mentioned above); economic determinants of fertility; the influence on fertility of community factors such as the presence of schools and electricity; mortality; induced abortion; family planning; and noncontraceptive mechanisms of fertility regulation such as postpartum sexual abstinence and temporary spousal separations. These modules, published in 1977, have proved popular; in nearly every survey, at least one has been used entirely or in part. Most Asian and Latin American countries have adopted the fertility-regulation and family-planning modules, while African countries have preferred to focus on noncontraceptive mechanisms of fertility control and on mortality.

It is arguable whether this policy of a common mandatory core accompanied by a constellation of voluntary modules struck the right balance between the potentially conflicting aims of international comparability and national priorities and circumstances. However, it is clear that a degree of standardization was necessary to accomplish the task. If each survey had been unique in its central characteristics, the program could not have been completed without an even greater input of technical assistance.

In the design of both modules and the core questionnaire, WFS strove to produce instruments that could be administered verbatim to respondents in their own language. This policy necessitated both the introduction of elaborate skip instructions which complicated the structure and increased the length of the documents, and the need to translate into local languages. A maximum of ten local-language versions were used, with an average of about four per survey in Africa. In the latter respect, the WFS was truly pioneering;

many countries, particularly those in sub-Saharan Africa, had relied hitherto on colonial linguae francae. Unfortunately, it was beyond the scope of the WFS to assess experimentally the advantages, in terms of data quality and comparability, of this approach over alternative less expensive and less structured instruments, but undoubtedly there were substantial benefits. However, in our view, the WFS may have taken the verbatim principle rather too far. Analyses of tape-recorded interviews (e.g. Thompson et al., 1982) indicate that much of the discussion between respondent and interviewer deviates from the printed content of questionnaires. This suggests that good training and supervision of interviewers is more crucial than elaborate questionnaires.

4.2. Sample Design and Data Collection

While intercountry comparability of results demands a certain standardization of substantive content, it does not necessarily imply standardization of sample designs. The only WFS principles here were that samples should be drawn on a strict probability basis and that they be nationally representative. The first principle has never been violated, but there have been a few exceptions to the second, in response to overriding local considerations.

Despite the absence of any technical need for standardization, the common substantive content, field strategy, problems, and constraints in developing countries encouraged the development of a general approach which is described in the *Manual of Sample Design* (WFS, 1974). In particular, the prior decision that interviewers should work in teams and the difficulties of localized travel in many developing countries argued in favor of geographically small clusters with a dense sample of units within each. These considerations led to a type of design which, at its simplest, involved only the following stages: a selection of the smallest available areal units (usually census enumeration districts) implicitly or explicitly stratified by region and urban–rural character; a separate listing operation of dwellings or households within each area or cluster; a selection of listed dwellings or households, and finally the interviewing of all selected households and all resident eligible women. To maximize sampling efficiency and to obtain a reasonably uniform sample "take" per cluster, the first stage of selection was typically with probability proportional to size, and the second stage inversely proportional, so as to yield a self-weighting sample of between 20 and 50 women per cluster. For an average WFS sample size of 5000 women this implies a total of between 200 and 300 clusters. In Asia, the Middle East, and most European countries the eligibility criteria for the individual interview were based on age (usually 15 to 49 years) and marital status (currently or formerly married), while in Latin America, sub-Saharan Africa, and the Caribbean, where reproduction is less confined to marriage, the criterion of marital status was usually dropped (see Table 1).

Naturally, local circumstances have necessitated considerable departures from the simplified model described above. In most developed countries, sampling frames of dwellings, households, or individuals were used. In developing countries also, the WFS, wherever possible, avoided the expense of a special listing operation by making use of frames from earlier surveys. In 15 of the 42 developing-country surveys, a frame of dwellings or clusters was available from a recently conducted demographic or other household survey. In surveys using an enlarged household sample, the household survey usually took the place of the listing operation. Conversely, in other countries, listing had to be preceded or accompanied by extensive mapping, either because the areal units of the frame were large and needed splitting to reduce listing costs, or because boundaries were not defined with sufficient clarity.

The experience of the WFS in sample design has already been distilled (e.g. Verma, 1980; Verma et al., 1980; Scott and Harpham, 1984). Perhaps the most important lesson learnt is that it is feasible to design and implement probability samples on a national scale in all developing countries, given the will to do so and the necessary resources. The successful fulfillment of this primary aim has allowed the computation of sampling errors for all surveys with a program, CLUSTERS, specially developed by WFS staff (Verma and Pearce, 1980), an achievement rarely attempted even in developed countries. Indeed, the WFS program has created the largest and richest array of data ever assembled on sampling errors. In contrast, relatively little attention has been paid to nonsampling error, though useful studies of response reliability have been conducted (e. g. O'Muircheartaigh and Marckwardt, 1980).

The implications for sample design of the use of the team approach to interviewing have already been mentioned. The major reasons for this arrangement, which was followed in all but a couple of developing-country surveys, were that it facilitated strict quality control of work by team supervisors and team editors, ensured the physical safety of staff and maintained morale, and finally made optimal use of vehicles for transporting field workers in rural areas. In a major departure from customary survey practice, WFS insisted that the individual interviews be conducted by females, on the assumption that women in most societies would be reluctant to discuss the intimate subject matter of the inquiry with men. As few statistical departments in developing countries maintain a regular female field force, special recruitment and a three-week training course were necessary. WFS recommendations in these and other aspects of fieldwork and training are contained in four issues of the *Core Documentation*, published in 1974 and 1975.

It may be correctly inferred from this brief description that the main consideration of WFS field strategy with its exceptionally long training period and high ratio of supervisors to interviewers was to ensure the collection of data of the highest possible quality. The detailed data evaluations indicate that this goal was achieved; WFS survey data have nearly always proved to possess greater validity than those from previous surveys and censuses and in several instances have revealed unexpected defects in vital registration systems. Perhaps more surprisingly, fieldwork generally encountered few major problems. To be sure, the usual difficulties of floods, vehicle breakdowns, delayed payments, and so on hampered progress, but in no case did fieldwork have to be terminated before completion, and usually the work finished more or less on schedule.

Equally important are the high response rates obtained, typically in the range of 85 to 95 per cent in developing countries. In a detailed analysis, Marckwardt (1984) adduces evidence that WFS's insistence on obtaining high response rates through callbacks was justified. Households and individual women who required more than one visit for successful interview differ appreciably in their characteristics from those successfully interviewed on the first visit. At the same time, worrying coverage errors are apparent in a number of surveys. This reflects both mapping and listing defects, and the tendency to misdeclaration of female age on the household roster so as to render individual women ineligible for the detailed interview and thus reduce interviewers' workloads.

4.3. Data Processing

While it may be claimed with considerable justification that the data collection phases of WFS surveys are models of excellence, the record of the WFS in data processing is less satisfactory. The machine editing of data files proved particularly problematic and over five times as much technical assistance from WFS staff was required for data processing as for fieldwork. Most of the major delays in survey execution occurred at this stage. An average of 14 months was needed for machine editing and correction, against an initial average planned allocation of only 2.5 months. These delays, in conjunction with similar delays at the tabulation, report writing, and printing stages, resulted in a doubling of the planned 15-month average span between the end of the fieldwork and publication of the main results.

Apart from an obvious failure of the WFS to anticipate the complexity of data editing, there are a number of reasons for this inability to meet self-imposed targets [see Otto and Rattenbury (1984) for details]. Among these are the nature of the work itself, intellectually unrewarding yet demanding painstaking precision; the frequent diversion of senior national survey staff to new projects; ill-maintained hardware and scarcity of local programmers; the complexity of the questionnaire; and the perfectionist policy of WFS that all identifiable and correctable errors be removed, a policy clearly more justifiable in view of the repeated, worldwide use of WFS data files than it would be in more normal circumstances. A unique study has recently shown that WFS's emphasis on editing was indeed excessive (Pullum et al., 1984). Comparisons of substantive results, including the use of multiple-regression analysis of individual responses, derived from unedited and edited versions of the same country data sets revealed very few differences. The conclusion is clear: the cost of perfectionist editing far exceeded the gains.

Lack of suitable software was another problem. An expert working group meeting, convened in 1974, noted the absence of a survey edit package which met WFS requirements of good documentation, ease of use, and portability across a range of machines. Plans were made for a collaborative effort to remedy this deficiency by developing a new package, but these came to naught, and in the following year, the WFS decided to adopt the edit package CONCOR, despite the fact that it could only be used on the IBM 360/370 series. For the minority of statistical offices without access to this series, ad hoc editing programs were written.

The dearth of suitable software was less marked for tabulation than for editing, and the WFS had little hesitation in choosing the US Bureau of Census package COCENTS as its main instrument for producing the tables needed for the main survey report. Its somewhat cumbersome nature was greatly improved in 1977 when the WFS programming staff developed COCGEN, a program which generates COCENTS commands for user-supplied table specifications. Since that time, the production of camera-ready tabulations has been a highly automated procedure. COCGEN and CLUSTERS are only two of a number of programs developed by WFS staff which are already proving useful outside the WFS program itself. Among these may be mentioned NUPTIALS (fitting a model nuptiality schedule to survey data), DEIR (editing and inputing dates in event histories), FERTRATE (computation of fertility rates), and WFSSPSS (conversion of variable descriptive information on data dictionaries into SPSS data descriptive commands). Such programs are an important part of the WFS legacy to the scientific community and are available in the last of the WFS Basic Documentation Series (WFS, 1984).

4.4 Reporting and Analysis

In order to expedite the production and dissemination of findings, the WFS initially envisaged a two-stage system: a first report, following the demographic tradition of detailed numerical description, would consist of a large number of standarized two - or three-way cross-classifications of data derived from the core questionnaire, with a short accompanying text, outlining the methodology of the survey and the key findings. A consultant was engaged in 1974 to develop precise specifications and by 1975 a working version was agreed on, though the final document was not published until 1977.

While the first report was considered to be essentially descriptive, the second report was to be analytical in nature, drawing on refined statistical and demographic techniques. It soon became apparent, however, that analysis was not amenable to the approach adopted for the first report. The lack of a single theoretical framework, the range of subject matter, and variations in quality of data and in national priorities precluded the evolution of a standard set of detailed guidelines. Instead, the Technical Advisory Committee decided in 1975 that a short document should be written, outlining the aims and principles of analysis [it was later published as *Strategies for Analysis of WFS Data* (WFS, 1978)], and that this should be supplemented by a series of technical bulletins to assist national analysts in the application of particular techniques. By the end of 1983, a total of ten such bulletins had been published by the WFS, ranging in scope from such well-known techniques as path analysis and life-table analysis to more narrowly focused concerns, such as the computation of fertility indices from WFS-type surveys. The wide use of these documents for teaching purposes testifies to their success.

Partly because of the absence of a clear WFS policy towards analysis, but also because of lack of local facilities and skills, the initial response of participating countries to analytical opportunities was disappointing. By 1976 it was apparent that the WFS would need to play a more positive role if the full potential of the data was to be exploited. This realization found expression in a sequence of policy initiatives during the years 1976–1978. In 1976, survey directors were sent a list of priority topics for analysis. This was followed in late 1977 with a recommendation by the Programme Steering Committee that the WFS commission a series of 12 illustrative analyses, from WFS staff and outside experts, working in close collaboration with national staff. These were eventually published in the WFS Scientific Report Series between 1979 and 1982. Together they make an impressive body of work, with several important new contributions to the analysis of cross-sectional fertility survey data and a strong emphasis on the integration of statistical and demographic techniques, always an aim of the WFS.

In 1978, the analytical strategy was further broadened. It was decided to encourage countries to hold national meetings with the main objective of disseminating to policy makers the findings of the first report. A subsidiary purpose was to gather together research staff from a number of institutions and thereby formulate an integrated program of national analysis. The first such meetings, held in 1978, were so successful that they subsequently became a routine feature.

In the same year, the first analysis workshop was held at the International Institute for Population Studies in Bombay for Asian countries that had participated in WFS. This occasion clearly demonstrated the advantages of a group approach over individual, isolated analyses, and the WFS has increasingly relied upon this strategy to achieve results. Most notable has been a series of seven evaluation workshops starting in 1979, held at WFS headquarters, using a newly acquired Hewlett-Packard 3000 computer. These enabled analysts from developing countries to spend three months assessing the quality of the data collected in their respective surveys, working under the close guidance of WFS staff. During the course of these workshops, a battery of external and internal checks on consistency and plausibility of data was evolved. The enhanced understanding of the strengths and limitations of demographic survey data represents one of the major legacies of the WFS Project.

All these initiatives taken to encourage full analysis of data sets have borne fruit. By the end of 1983, over 500 national analysis projects had been completed or were in progress, the majority being executed by developing-country researchers. The common criticism of survey researchers that they never fully analyse the information that they have collected cannot be leveled at the WFS. Furthermore this activity represents a transfer of analytic skills on a vast scale.

While the prime focus has been on national analysis, the unique advantage of the WFS operation over previous survey research on fertility behavior lies in the production of comparable data for a large number of countries, at totally different stages of demographic modernization and with sharply contrasting social and cultural systems. It is the cross-cultural nature of the project that offers the main potential gains in broad understanding of reproductive behavior. The United Nations Population Division in conjunction with Specialized Agencies and Regional Commissions plays a central role in comparative analysis, and important work is also taking place at universities throughout the world. However, the part played by the WFS itself has been crucial. In particular, an increasingly important archive section in London has carried the responsibility of creating, checking, maintaining, and distributing well-documented, standardized data files. By the end of 1983, over 1000 data files had been distributed to more than 300 institutions. The WFS also started in 1979 a new publication series of Cross National Summaries, designed to disseminate key results and act as a starting point for more elaborate comparative analyses.

The quickening tempo of research activity culminated in the World Fertility Survey Conference, held in London in July 1980 and attended by 600 people from 90 countries. The main purpose was to give a public account of the findings of the WFS to date, but there was also a strong emphasis on methodological issues and ample opportunity for discussion of the future of the Project. The three-volume *Record of Proceedings* undoubtedly will constitute a major intellectual resource for many years (ISI, 1981). However, at that time, the results for only half of the 42 participating countries were available and much important analytical work remained to be done. The final major meeting convened by the WFS in London in April 1984 provided a better opportunity to assess the impact of the WFS, both in methodological and in substantive terms. The eventual publication of the presented and background papers for this 1984 Symposium will be indispensable by comparison with the proceedings of the 1980 Conference.

5. Review of Findings

Undoubtedly the most significant substantive contribution of the WFS surveys in the developing world has been to provide trustworthy and detailed descriptions of fertility and its major direct determinants, namely sexual exposure, birth control, and breastfeeding. In countries with good vital registration systems or previous national surveys, such as the Republic of Korea, Malaysia, Philippines, Sri Lanka, and Costa Rica, the main descriptive utility has been to update, confirm, and augment previous evidence. In many other countries, however, prior information at the national level was either outdated, unreliable, or nonexistent. In these circumstances the provision of demographic facts was invaluable; in such diverse countries as Nepal, Syria, and Senegal the survey enabled confident estimation of the level of fertility to be made for the first time, while in others such as the Dominican Republic, Peru, Indonesia, and Morocco, recent and sometimes unexpected fertility declines were identified and measured.

The overall quality of the data collected by the WFS came as a surprise to many demographers. A decade ago, it was commonly assumed that trustworthy information on births and deaths in developing countries could only be gathered by dual-record systems and/or longitudinal survey designs. In terms of future data collection, the lasting contribution of WFS has been to demonstrate that single-round surveys, if conducted to high standards, can provide reasonable demographic estimates. In this regard, the decision to include a complete retrospective maternity history in the core questionnaire has been fully vindicated.

Fortunately the data-collection phase of the WFS (1974–1982) coincided with the emergence of new downward trends in natality of profound significance for the future of mankind. The WFS was in a unique position to document the rise in marriage ages, the growth of modern contraception, and the concomitant decline in fertility that has characterized much of Asia and Latin America during the last 15 years. A very crude measure of fertility change is given in Table 2 by the difference between the total fertility rate in the period immediately prior to the survey and the cumulative number of births to women aged 40 to 44. The exceptions to these prevailing forces are clearly identified. In Nepal and Bangladesh, the surveys indicated that, despite government efforts, contraception had remained at an insignificant level of use and fertility had changed little. In Pakistan, a slight fertility decline can be attributed to rising marriage age rather than to family planning, which was still confined to a tiny minority.

The demographic situation in sub-Saharan Africa, revealed by the surveys, was markedly different from that of other regions. In such countries as Kenya, Benin, the Ivory Coast, and Senegal, extremely high and unchanging levels of fertility were apparent; in Lesotho the level, though lower, showed no signs of actual or incipient decline, while in Cameroon fertility was probably increasing in response to improving health conditions. Only in Ghana was there any suggestion of a slight decline. Moreover, the reproductive aspirations of African respondents indicated no latent desire for smaller families (see Table 2).

In the Arabic-speaking world, the picture was mixed. Well-established declines were evident in Egypt and Tunisia, and a recent decline appeared to

Table 2. Demographic Indicators for WFS Participant Countries[a]

	Total fertility rate (1)[b]	Mean no. children ever born to women aged 40–44 (2)[c,i]	Female mean age at marriage (3)[d]	Percentage currently using contraception (4)[e]	Mean length of breast-feeding (5)[f,i]	Mean desired expected family size (6)[g,i]	Infant mortality rate (7)[h]
Yemen, Arab Rep.	8.5	6.5	17	1	10.6	5.4	162
Kenya	8.3	7.6	19	9	15.7	7.3	87
Jordan	7.6	8.4	22	25	11.1	6.3	66
Syria	7.5	7.4	22	30	11.6	6.1	65
Ivory Coast	7.4	6.7	18	2	17.5	8.4	113
Senegal	7.2	6.8	18	5	18.5	8.9	112
Benin	7.1	6.1	18	20	19.2	7.5	108
Ghana	6.5	6.1	19	10	17.9	6.1	73
Cameroon	6.4	5.2	18	2	17.5	8.0	105
Mauritania	6.3	5.9	19	1	15.6	8.8	90
Pakistan	6.3	6.9	20	5	19.0	4.2	139
Mexico	6.2	6.6	22	30	9.0	4.5	72
Nepal	6.2	5.6	17	2	25.2	4.0	142
Bangladesh	6.1[j]	7.1	16	8	28.9	4.1	135
Sudan (North)	6.0[j]	6.2	21	8	15.9	6.4	79
Morocco	5.9	7.1	21	19	14.2	5.0	91
Tunisia	5.9	6.5	24	32	14.0	4.2	80
Lesotho	5.8	5.3	20	5	19.5	6.0	126
Dominican Rep.	5.7	6.4	21	32	8.6	4.7	89
Nigeria	5.7	5.2	19	6	u	u	82
Peru	5.6	6.3	23	31	13.1	3.8	97
Haiti	5.5	5.6	22	25	15.5	3.6	123
Ecuador	5.4	6.4	22	34	12.3	4.1	76
Egypt	5.3	6.3	21	24	16.3	4.1	132
Philippines	5.2	6.4	25	36	13.0	4.4	58
Guyana	5.0	6.3	20	32	7.2	4.6	58

Jamaica	5.0	5.4	18	39	8.1	4.1	43
Paraguay	5.0	5.8	22	47	11.4	5.3	61
Colombia	4.7	6.1	22	43	9.2	4.1	70
Indonesia	4.7	5.2	19	26	23.6	4.3	95
Malaysia	4.7	6.0	23	33	5.8	4.4	36
Thailand	4.6	5.9	23	33	18.9	3.7	65
Venezuela	4.5	6.1	22	60	7.4	4.2	53
Korea, Rep.	4.3	5.1	23	35	16.3	3.2	42
Turkey	4.3	5.9	21	50	u	3.0	133
Fiji	4.2	6.1	22	41	9.9	4.2	47
Panama	3.8	5.6	21	54	7.4	4.3	33
Sri Lanka	3.8	5.3	25	32	21.0	3.4	60
Israel	3.7	3.6[k]	23	70[k,l]	NA	3.5[k]	23
Costa Rica	3.3	6.1	23	64	5.0	4.7	53
Trinidad and Tobago	3.3	5.2	20	60	8.0	3.8	41
Spain	2.7	3.1	24	51	NA	2.8	11
Romania	2.5	2.4	20	58	NA	2.2	30
Czechoslovakia	2.4	2.3	21	95	NA	2.4	20
Bulgaria	2.3	2.1	21	76	NA	1.9	23
Yugoslavia	2.3	2.8	21	55	NA	NA	36
Hungary	2.2	NA	21	74	NA	2.1	26
Poland	2.2	2.8	22	75	NA	2.5	24
Portugal	2.2	2.8	23	66	3.1	2.4	26
Japan	2.1	2.2	24	67	NA	NA	11
Denmark	1.9	2.6	25	63	NA	2.4	11
France	1.9	2.8	23	71	NA	2.5	11
Norway	1.8	2.8	22	71	NA	2.5	11
United States	1.8	3.3	21	70	NA	2.6	15
Belgium	1.7	2.6	21	85	NA	2.3	14
United Kingdom	1.7	2.6	22	77	NA	2.3	14
Finland	1.7	2.7	23	80	NA	2.5	12
Italy	1.7	2.4	22	78	NA	2.4	15
Netherlands	1.7	NA	22	75	NA	2.4	11
Sweden	1.7	2.1	24	72	NA	2.5	7
Switzerland	1.5	NA	23	70	NA	2.2	9

Table 2. (continued)

[a] ordered by level of current fertility.

[b] Column 1: The total fertility represents the completed fertility of a hypothetical woman who experienced throughout her reproductive lifetime the fertility rates prevailing at a specified period. For European countries, the USA, Japan, and Israel, the rate is derived from registration data for the calendar year of survey fieldwork. For other countries, the rate is based on WFS data, averaged for the five years preceding the survey (Alam and Casterline, 1984).

[c] Column 2: For European countries and the USA the data refer to currently married woman in undissolved first marriages. For other countries they refer to all women, regardless of marital status. In all cases, the source is the WFS Survey.

[d] Column 3: The singulate mean age at first marriage is computed from the proportions of women who are single, by current age. For European countries, the USA, Japan, and Israel, the sources are the most recent census or survey prior to 1980, as compiled by the Population Reference Bureau (1981). For other countries, the source is the WFS (Smith, 1980).

[e] Column 4: The figures refer to the percentages of all currently married women who reported that they were using any method of contraception at the time of the survey. The source is the WFS (Sathar and Chidambaram, 1984; Berent, 1982).

[f] Column 5: The means are computed from WFS data on proportions of all live births in the five years preceding the survey who were still breastfeeding at survey date (Ferry and Smith, 1983). Questions on breastfeeding were not included in developed-country surveys.

[g] Column 6: For most European countries and the USA the mean refers to the sum of previous live births and additional children expected, for women aged less than 45 years whose first marriage is still intact (Berent, 1983). For other countries, the mean refers to the total desired family size of all currently married women, typically in response to the question "If you could choose exactly the number of children to have in your whole life, how many children would that be?" (Lightbourne, Singh, and Green, 1982).

[h] Column 7: For European countries, the USA, Japan, and Israel, the infant mortality rate is taken from the United Nations Statistical Yearbook and relates to the calendar year of Survey field work. For other countries, the rate is derived from the WFS by averaging over the five years preceding each survey (Rutstein, 1983).

[i] u = unavailable; NA = not asked or not applicable.

[j] There is evidence that these rates are underreported.

[k] Refers to the Jewish population only.

[l] This figure is for ever use rather than current use.

have occurred in Morocco. In Jordan, Syria, North Sudan, and North Yemen, however, fertility still remained at very high levels. In Jordan and Syria, there were clear signs of change, with widespread practice of contraception and rising female age at marriage. It seemed, though, that these fertility-depressing forces had been offset by declines in the length of breastfeeding. Though the WFS data on breastfeeding, being cross-sectional, cannot provide direct evidence of such trends, there were in these two countries particularly sharp differences in breastfeeding habits between urban and rural couples and between educated and uneducated women which were indicative of historical decline in the length of breastfeeding.

Breastfeeding data have proved of immense interest and importance not only for an understanding of the demographic situation in the Middle East but in all developing regions. Indeed, in this regard, the WFS has helped to remedy one of the glaring weaknesses in our understanding of human fertility. Though the link between lactation and the postponement of ovulation following childbirth had been known for some time, hitherto there had little empirical evidence concerning national or subnational patterns of breastfeeding and no acceptable methodology for quantifying the fertility-reducing impact of lactation. Happily the advent of the data coincided with the development of an appropriate model (Bongaarts, 1978), and the combination has led to a major advance in our understanding of fertility trends and differentials.

It can now be demonstrated, much more clearly than before, that modernization often sets in motion counteracting influences on fertility. By raising the age at marriage and the propensity to adopt birth control practices it acts to depress fertility, while by eroding customs of prolonged lactation (and other traditional fertility-reducing behavior such as postpartum sexual abstinence and polygyny) it is conducive to an increase. The net outcome naturally depends on the precise strengths of the various relationships, which themselves change over time.

Lesthaeghe (1982) has argued that this new perspective helps to solve the enigma, observed particularly in Latin America, that high fertility persisted for a considerable length of time despite the spread of contraception, before falling suddenly. The underlying mechanism, he suggests, is that declines in lactation completely offset the initial rise in birth control; however, as soon as the trends towards earlier weaning leveled off, further increments in birth-control practice were translated into an abrupt fertility decline.

In the same way, previously puzzling irregularities in differential fertility can now be better understood. For instance, the finding that, in several Asian and African countries, couples with a few years of schooling have slightly higher fertility than those with no formal education can now be at least partially explained in terms of the fact that their slightly higher level of contraception is insufficient to compensate the pronatalist impact of earlier weaning. Such insights do much to discredit interpretations of levels of marital fertility solely in terms of volitional factors, motivated by parental demand for children.

Notwithstanding the fact that modernization invariably brings about a decline in the length of breastfeeding and is associated in Africa with a shortening of the postnatal taboo on sexual intercourse, alarm that major surges in fertility may thereby result appears to be unwarranted. While large increases in marital fertility due to the erosion of traditional restraints may have been common in the first half of this century, Singh et al. (1984) found little evidence of such trends in the contemporary Third World. Rather, declines in breastfeeding were typically accompanied by increases in contraception and rising age at marriage, with the net result that fertility remains steady, or falls, or rises only slightly.

Because so much more was known about contraception than about breastfeeding prior to the WFS, the survey findings on this topic have been less exciting. Nevertheless the great wealth of information has offered genuinely new insights. It has now been established beyond doubt that a rural, uneducated background does not necessarily preclude the adoption of modern contraception by couples. Also contrary to popular belief, a substantial minority of women in developing countries appear to use contraception for spacing rather than limitation. Even more important from the point of view of practical policy is new evidence, based on analyses of WFS data (e.g. Chidambaram and Mastropoalo, 1980; Jones, 1984) that proximity to sources of family-planning advice and supplies does facilitate contraceptive use in rural areas, especially of methods such as the pill and condom which require continuous supply. The effect of proximity on contraceptive practice is weak, however, and the results lend little support for family-planning policies based solely or largely on the principle of supply saturation.

Substantial advances have also been registered in our knowledge of the third major direct determinant of fertility, sexual exposure (as measured by the existence of formal and informal unions). A unique array of data on divorce, separation, widowhood, and remarriage has been assembled. One unexpected finding is that marital dissolution in most developing countries is followed rapidly by remarriage and thus the effect on fertility of appreciable intercountry variations in the incidence of dissolution is minimal.

The relationship between age at first marriage and subsequent fertility has long been a subject of study, because of its potential relevance to population policy. Analyses of WFS data have shown clearly that postponement of marriage until age 20 has no effect on subsequent fertility and that marriage ages between 20 and 25 are consistent with family sizes of four to six children (McDonald et al., 1981). Strategies of population control based solely on postponement of marriage are thus unlikely to succeed.

Satisfactory explanations of human reproduction have to take into account the motivations and conscious intentions of individuals. This is also an area of crucial concern for those concerned with policies designed to influence current fertility levels or which need to anticipate future trends. Consequently, WFS data on fertility preferences have received close analytical attention. From the many interesting findings, perhaps two deserve to be mentioned for their special importance. First, survey data on the desired size of families indicate preferences in most developing countries for three to four children, a level of reproduction well above that required to bring about an eventual end to population growth. Second, there is reasonably convincing evidence that unwanted pregnancies are still common in the contemporary Third World. Quite apart from its effect on the welfare of individual families, the prevention of all unwanted births would have an appreciable demographic impact. In many of the countries studied, the birth rate would fall by between 6 and 15 births per 1000 population, implying a substantial reduction in the rate of population increase (Lightbourne et al., 1982). These findings suggest that there is still an appreciable unmet need for birth control. This accords with the finding that large proportions of women report that they want no more children yet are practicing no method of contraception. There are, of course, a host of possible reasons for this apparent inconsistency between professed wishes and behavior, on which the WFS surveys have shed some light. However, the interesting point to have emerged is that the level of unfulfilled need does not decline as the national level of contraceptive practice increases. The explanation for this paradox appears to be that declines in desired family size accompany the growth of contraception and therefore demand keeps slightly ahead of implementation (Westoff and Pebley, 1981).

Thus far we have indicated some of the more striking advances in our knowledge of fertility patterns and their direct determinants, that can be attributed to WFS. The contribution to an understanding of the indirect underlying social, economic, and institutional causes is by comparison meager. The lack of an adequate theoretical framework and the concomitant atheoretical nature of WFS inquiries, plus inherent limitations of social surveys, are among the main reasons for this relative failure. Though analyses of particular countries have revealed much about the diffusion of change across social strata and the precise ways in which modes of family formation have shifted, the evidence has not enabled us to account for the persistence of high fertility in certain societies or the timing of change in others. Studies of the demographic transition in Europe have shown this to be a topic of intractable complexity, and thus is it not surprising that a full understanding of parallel processes in other regions still eludes us.

Taken as a whole, however, the evidence from the WFS is more consistent with ideational than with structural or microeconomic theories of change. Female employment has not emerged as a strong determinant of reproductive behavior (UN, 1981), nor is there any evidence that the switch from family enterprises to new forms of production is crucial to fertility transition (Rod-riguez and new forms of production is crucial to fertility transition (Rodriguez and Cleland, 1981). Community measures of agricultural modernization and other forms of development do not appear to be strongly associated with fertility change (Casterline, 1983). Furthermore, the sheer pace of fertility change over the last decade in countries such as Thailand and Costa Rica runs counter to the more narrowly deterministic of microeconomic theories. Attempts to synthesize the massive and detailed evidence of the WFS and link it to other data sources has only just begun, but it is likely that major theoretical upheavals will ensue.

Finally we may note one unanticipated contribution to knowledge. The surveys have yielded a rich array of high-quality data on infant and child mortality. Our knowledge of mortality levels, trends, and correlates has been vastly improved. The most significant of all the findings concerning childhood mortality has been the strong relationship between birth spacing and risks of death. Though this link has been known for at least 50 years and was one of the original cornerstones of the family-planning movement, its empirical investigation in the Third World had hitherto been strangely neglected. In two multinational studies (Rutstein, 1983; Hobcroft et al., 1983), WFS staff have demonstrated that the length of the interval preceding the birth of a child exerts a major influence on that child's survivorship, not only in infancy but up until the age of five years. The relative risks of death are several times greater for a child born within two years of an elder sibling than for a child born after a longer interval. For survivorship in childhood, long preceding intervals of four or more years are more advantageous than moderate intervals of two or three years.

These comparative results have been buttressed by a number of detailed national studies in which the link between spacing and mortality has generally persisted after controls for such possibly confounding factors as birth order, maternal age, socioeconomic status, and breastfeeding practices. For Pakistan, Cleland and Sathar (1984) showed that early death of the preceding child did not alter the effect of birth-interval length; this suggests that depletion of maternal nutrition leading to low-birth-weight babies may be the main causal mechanism, rather than competition for and parental care between two living children of similar age. However, such depletion does not appear to be cumulative or irreversible. When the length of the immediately preceding interval was controlled, no effect of earlier spacing patterns was observed.

These important new findings, with their obvious policy implications, came late in the WFS program. No doubt others are in store. Many of the African and Arab surveys have only just been completed and detailed analysis is yet to commence. Though the WFS program itself ended to mid 1984, its impact on demographic understanding will grow for several more years. Thus this chapter should not be regarded as an epitaph but as an interim report on achievements.

Bibliography

Alam, I. and Casterline, J. B. (1984). *Socio-economic Differentials in Recent Fertility.* WFS Comparative Studies, No. 33.

Berent, J. (1983). *Family Planning in Europe and USA in the 1970's.* WFS Comparative Studies, No. 20.

Berent, J. (1983). *Family Size Preferences in Europe and USA: Ultimate Expected Number of Children*. WFS Comparative Studies No. 26.

Bjerve, P. J. (1973). "Presidential address to the 39th Session of the ISI." *Proc. Internat. Statist. Inst.*, **45**, Book 1, 51–63.

Bongaarts, J. (1978). "A framework for analysing the proximate determinants of fertility." *Population and Development Review*, **4(1)**, 105–132.

Casterline, J. B. (1983). "Community effects on fertility." Paper presented at WFS Seminar on Collection and Analysis of Data on Community and Institutional Factors, London, June 1983.

Chidambaram, V. C. and Mastropoalo, L. (1982). "Role of WFS data in the analysis of family planning programs." In A. I. Hermaline and B. Entwistle (eds.), *The Role of Surveys in the Analysis of Family Planning Programs*. Liege: Ordina Editions, 279–312.

Cleland, J. G. and Sathar, Z. (1984). "The effect of birth spacing on childhood mortality in Pakistan." *Population Studies* (in press).

Ferry. B. and Smith, D. P. (1983). *Breastfeeding Differentials*. WFS Comparative Studies, No. 23.

Grebenik, E. (1981). *The World Fertility Survey and its 1980 Conference*. London: World Fertility Survey.

Hobcraft, J., McDonald, J., and Rutstein, S. O. (1983). "Child-spacing effects on infant and child mortality." *Population Index*, **49**, 585–618.

International Statistical Institute (1853). *Compte rendue du traveaux du Congrès Général de Statistique Reuni and Bruxelles les 19, 20, 21 et 22 septembre 1853*. Bruxelles, 23.

International Statistical Institute. (1886, 1896, 1969). *Bulletin de l'Institut International de Statistiques*.

International Statistical Institute (1981). *World Fertility Survey Conference July 1980. Record of Proceedings*. Voorburg, Netherlands: International Statistical Institute.

Jones, E. F. (1984). *The Availability of Contraceptive Services*. WFS Scientific Reports, No. 56.

Lesthaeghe, R. (1982). "Lactation and lactation related variables, contraception and fertility: An overview of data problems and world trends." Paper prepared for World Health Organization and the U. S. National Academy of Sciences Seminar on Breastfeeding and Fertility Regulation, Geneva, February 1982.

Lightbourne, R. E., Singh, S., and Green, C. P. (1982). "The World Fertility Survey." *Population Bull.*, **37**, No. 1. Population Reference Bureau.

Marckwardt, A. M. (1984). *Response Rates, Callbacks and Coverage: The WFS Experience*. WFS Scientific Reports, No. 55.

McDonald, P. E., Ruzicka, L. T., and Caldwell, J. C. (1981). "Interrelations between nuptiality and fertility: The evidence from the World Fertility Survey." In *WFS Conference July 1980. Record of Proceedings*, **2**. Voorburg, Netherlands: International Statistical Institute, 77–146.

Nixon, J. W. (1960). *A History of the International Statistical Institute 1885–1960*. The Hague.

O'Muircheartaigh, C. A. and Marckwardt, A. M. (1980). "An assessment of the reliability of WFS data. In *WFS Conference, July 1980: Record of Proceedings*, **3**. Voorburg, Netherlands: International Statistical Institute, 313–388.

Otto, J. and Rattenbury, J. (1984). "WFS data processing strategy." Paper presented at WFS Symposium, London, April 1984.

Population Council (1970). *A Manual for Surveys of Fertility and Family Planning: Knowledge, Attitudes, and Practice*. New York.

Population Reference Bureau (1981). *Fertility and the Status of Women*. Wallchart.

Pullum, T. W., Ozsever, N., and Harpham, T. (1984). *An Assessment of the Machine Editing Policies of the World Fertility Survey*. WFS Scientific Reports, No. 54.

Report of the Reappraisal Committee (1968). Voorburg, Netherlands: International Statistical Institute Mimeo.

Rodriguez, G. I. (1977). *Assessing the Availability of Fertility Regulation Methods: Report of a Methodological Study*. WFS Scientific Reports, No. 1.

Rodriguez, G. and Cleland, J. (1981). Socio-economic determinants of marital fertility in twenty countries: A multivariate analysis." In *WFS Conference July 1980: Record of Proceedings*, *2*. Voorburg, Netherlands: International Statistical Institute, 325–414.

Rutstein, S. O. (1983). *Infant and Child Mortality: Levels, Trends and Demographic Differentials*. WFS Comparative Studies, No. 24.

Sathar, Zeba and Chidambaram, V. C. (1984). *Differentials in Contraceptive Use*. WFS Comparative Studies, No. 36.

Scott, C. and Harpham, T. (1984). "Major issues of survey and sample design." Paper presented at WFS Symposium, London, April 1984.

Singh, S., Casterline, J. B., and Cleland, J. G. (1984). "the proximate determinants of fertility: Sub-national variations." *Population Studies.*, **39**, 113–135.

Smith, D. P. (1980). *Age at First Marriage*. WFS Comparative Studies, No. 7.

Thompson, L., Ali, N., and Casterline, J. B. (1982). *Collecting Demographic Data in Bangladesh: Evidence from Tape-Recorded Interviews*. WFS Scientific Reports, No. 41.

United Nations (1970). *Variables and Questionnaire for Comparative Fertility Surveys*. U. N. Document ST/SOA/SER.A/45. New York.

United Nations (1981). *Variations in the Incidence of Knowledge and Use of Contraception: A Comparative Analysis of World Fertility Survey Results for Twenty Developing Countries*. U.N. Document ST/ESA/SER.R/40. New York.

Verma, V. (1980). "Sampling for National Fertility Surveys." In *WFS Conference, July 1980. Record of Proceedings*, **3**. Voorburg, Netherlands: International Statistical Institute, 389–454.

Verma, V. and Pearce, M. C. (1980). *Users Manual for Clusters*. London: World Fertility Survey.

Verma, V., Scott, C., and O'Muircheartaigh, C. (1980). "Sample designs and sampling errors for the World Fertility Survey." *J. Roy. Statist. Soc. Ser. A*, **143**, No. 4, 431–473.

Westoff, C. F. and Pebley, A. R. (1981). "Alternative measures of unmet need for family planning in developing countries." *Internat. Family Planning Perspectives*, **7**, No. 4, 126–136.

World Fertility Survey (1974). *Manual of Sample Design*. WFS Basic Documentation, No. 3.

World Fertility Survey (1975). *World Fertility Survey, The First Three Years: January 1972 to January 1975*. Voorburg, Netherlands: International Statistical Institute.

World Fertility Survey (1977). *Modifications to the WFS Core Questionnaire and Related Documents*. WFS Basic Documentation, No. 10.

World Fertility Survey (1984). *WFS Software Users Manual*. WFS Basic Documentation, No. 12.

World Fertility Survey (1978). *Strategies for Analysis of WFS Data*. WFS Basic Documentation, No. 9.

Zahn, Friedrich (1934). *50 annees de l'Institut International de Statistique*. La Haie, 180 pp.

CHAPTER 19

Economic and Social Statistics for Comparative Assessments

E. Malinvaud

Abstract

Confronted with a request for comparative assessments about complex economic and social phenomena, official statisticians must reach a degree of accuracy that cannot be taken for granted. The initial choice of concepts and classifications plays a fundamental role. In principle statistical operations should be harmonized so as to serve well for comparisons, but this is not always feasible, in particular for international comparisons. Reconciliation of available data often meets with serious difficulties. Diffusion of useful and valid results may require some data analysis treatment or imply delicate decisions. These points are discussed and brought out in a number of examples.

1. Introduction

Our societies are complex and find it difficult to know their own evolution. Economic and social statistics provide the main source of knowledge in this respect. By comparing the values obtained for the same characteristic in two more or less distant periods, in two distinct countries, or in two separate social groups, one aims at improving one's understanding of the facts.

Statisticians have the function of determining what data ought to be available, how they should be collected and processed, and to which valid inferences they can lead. Experience shows that trying to achieve rigor and efficiency in fulfilling this function raises many problems, some of which are of a rather fundamental nature. Indeed, statistics must be available for use by different people and are requested to be objective. But the questions to which users want to find answers vary, as well as their degree of prior knowledge on the subject matter and their level of competence in dealing with data. Hence, potential sources of misunderstanding exist.

This article aims at describing how economic and social statisticians respond to this challenge and how they contribute to the improvement and objectivization of knowledge in present societies. It stresses the roles of the initial and final phases of the statistical operations, leaving aside what can be considered as the technology of the statistical production process. By focusing

Key words and phrases: classifications, comparisons, diffusion, harmonization, official statistics, reconciliation.

attention on comparative assessments, it limits itself to one particular type of inference, but certainly the most frequent one and one of the most delicate to perform.

The main conclusion to be argued here follows from the realization that economic and social phenomena are typically found to be more complex and less clearcut than one had first thought. As a consequence, valid comparative assessments impose a higher degree of accuracy than the layman may believe to be necessary. In particular, the initial choice of concepts always deserves a very careful examination, so that they are adequate to the questions being asked. Moreover, when the values taken by some variable in two distinct entities have to be compared, identical definitions and procedures for data collection ought in principle to be applied in the various entities concerned. But this is not always feasible, and the reconciliation of available data coming from various origins, while often unavoidable, is the source of many difficulties that are usually underestimated. In these circumstances, delicate decisions have to be made concerning the diffusion of statistical information to the general public, since most of this diffusion is bound to be unsophisticated.

Section 2 will survey the main issues in broad terms. The choice of concepts will be examined in Section 3. Section 4 will consider the way in which data sources may be harmonized or at least reconciled. The last section will briefly deal with the reporting of statistics and with their actual use for comparisons.

2. The Issues

Comparative assessments are answers to questions such as the following ones: By how much is the cost of living higher today than it was a year ago? What are the poorest regions in the European Economic Community? Is real output per person higher in Western Germany than in France? Are incomes more equally distributed in Italy than in Spain? Have young workers a higher risk of unemployment than old ones? Is the decline in fertility more rapid for educated Indian women than for uneducated ones? Has social mobility increasd during the past twenty years?

2.1. Understanding the Question

The best answer to give to such a question usually depends on a good understanding of the motivation behind it, since the question itself often is only an approximate message about what is required. We shall stress here the rather general viewpoint of "improvement of knowledge"; the questions will be supposed to be asked by a number of people who want to know better the world in which they live. But this viewpoint clearly simplifies the issue and should not be allowed to hide certain important aspects of the problem.

The search for a better knowledge is always directed, since it is motivated by some concern. People looking for statistical facts may either have a direct interest in them or want to introduce them as explanatory elements in their study of some phenomenon; but in any case they have some purpose in mind. The claim of serving the improvement of knowledge would then not be an excuse for an unselective and loose reference to available data.

When confronted face to face with a particular user asking a specific question, the statistician has first to clarify the purpose that is intended to be served and then to present whatever evidence is pertinent for this purpose. It often happens that he cannot obtain the figures that would be most appropriate; he then has to present imperfect substitutes and to explain whatever meaning they may have with respect to the specific query.

In general the statistician does not have such a direct relationship with the users, who moreover pursue somewhat different purposes. He must then know what are the prevailing motivations. Before presenting any new result, he must explain what kind of purpose they may concern; he may then provide several responses, appropriate to alternative conceptions of the question.

Or course, cases also occur in which no misunderstanding can arise about what is wanted. This applies in particular when some specific statistic is being used for the implementation of contracts between parties (when a payment is indexed) or for the application of some institutional rule (such as allocation of some public fund between local communities on the basis of their populations at the last census). But the latter cases again pose a problem that is relevant in the present context, namely whether the statistics being thus commonly used for comparative assessments provide a good reference for the calculations which they serve and for their ultimate aim.

Looking at the questions listed at the beginning of this section, one might easily comment about the context in which each one of them usually occurs; but this would take us too far astray. The reader may easily imagine a diversity of relevant situations in order to stimulate his or her thinking.

2.2 Looking for Accurate Answers

For the general discussion of the subject, it is important to note that three aspects are present in all the above questions and must be faced seriously: (i) Notions are used which obviously require careful definitions: cost of living, poverty, real output, degree of equality of an income distribution, risk of unemployment, and so on. (ii) Valid answers cannot be expected to be found without a careful and extensive phase of data collection. (iii) The data will have to be selected, aggregated, and compared in a way which must be appropriate to the notions being used and the questions being asked.

Difficulties and misuse of statistics may originate in each one of these three stages. It often happens that concepts are loosely defined, badly understood, or inadequately used. One often hears assessments relying on data that are far too imprecise to support them, and one may even read publications in which, for lack of a better alternative, two different data sources, each one of them providing only quite imperfect proxies, are directly used for the two entities to be compared without careful examination and correction of the biases that may occur. One also often witnesses innocent or intended distortions in the selection of the aggregate statistics being quoted. Such mistakes or distortions may concern the reliance on one particular index, the choice of the reference periods or of the reference populations, the decision to quote the unadjusted or adjusted figures even when adjustments are commonly or easily made in order to eliminate particular effects, and so on.

Confronted with these difficulties and misuses of statistics, economic and social statisticians not only have considerably improved the collection of data, but have also followed and continue to follow two principles: to *promote standards of rigor* in the quantitative appraisal of the phenomena and to *promote harmonization* of the concepts and methods used for this appraisal. Some general comments are in order at this stage about each one of these two principles, beginning with the second one.

Statistics produced by one organization at different times or for different domains must be harmonized; statistics produced by various agencies must similarly be harmonized: indeed, they must be appropriate for validating the comparative assessments which they are expected to serve, and for feeding the models within which phenomena will be studied. Full satisfaction of this requirement would mean that the same definition ought to be used everywhere and at all times for the same concept; it would also mean that the definitions of different concepts ought to be consistent with one another. Of course, such a situation will never be reached, and it would indeed not be satisfactory either, because it would imply a lack of adaptability to the particular needs felt at various times and places. Statistical systems must be periodically revised in order to fit the needs of the time; they must also allow for some flexibility so that the specific needs of each domain are not neglected, even when they conflict with the principle of full harmonization. Hence, some reconciliation between data obtained from various sources will always have to occur and may call for great statistical expertise; later we shall have to consider how this reconciliation can best be made.

The requirement for harmonization, however, imposes a high degree of coordination between various statistical systems. This explains why economic and social statisticians must closely collaborate across the barriers of institutions and of international borders, and this not only for the rapid diffusion of methodological advances in statistical techniques. Indeed, international collaboration developed early among statisticians. For a long time the International Statistical Institute devoted most of its efforts to the harmonization of statistics and was the main institution dealing with it. Nowadays, international governmental organizations are rightly concerned with this statistical harmonization, each one in its own field of competence. They have taken over most of what the International Statistical Institute was formerly doing in this respect. But with the explosion of statistical programs and with the rapid progress of statistical methodology, the need for free international scientific and technical exchanges has increased tremendously, so that the Institute has in fact greatly expanded its activities.

The principle of promoting standards of rigor in the quantitative appraisal of phenomena is more difficult to explain. It was recently dealt with at length by B. A. Bailar (1983) in her discussion on how to appraise the quality of statistical data. It implies in the first place continuous and stringent efforts to make economic and social statistics appropriate to the descriptive or prescriptive uses that they will serve. This is why statisticians must work in very close contact with the sciences concerning the subject matter of their field of investigation: they must be economists or sociologists when they build up conceptual framework within which the quantitative measurement of particular phenomena will take place; they must be associated with at least some research programs so as to evaluate correctly the needs for new sets of detailed or aggregated data.

2.3. Collecting the Data

But most efforts for the promotion of rigor in socioeconomic statistics still concern the technical operations involved in the production of data. One can rightly speak in this respect of a *methodology of official statistics*, which is largely distinct from the methodology of mathematical statistics and has its own requirements. Dealing with it in this article would lead us to consider the whole subject of economic and social statistics and would require too much space. In order to concentrate on what is most specific to comparative assessment, we shall leave aside all questions dealing with technical efficiency in the production of data. The following sections accordingly discuss what comes before or after the collection of data.

We can, however, get a taste of what is involved in the methodology of official statistics if we briefly survey the various stages through which production of data has to proceed.

Assuming that the question to be investigated and the definition of the variables to be observed have already been decided, statisticians must often choose between various ways of collecting the relevant data. A serious discussion weighing the advantages and deficiencies of the alternative approaches has to occur, based on experience and reflection. As an example, the statistics on wages and salaries may be based on surveys addressed to employers or addressed to employees, or they may still come from the processing of information collected for administrative purposes, usually taxation. More generally, one often has to choose between deep and precise investigations with incomplete coverage of the population and less ambitious inquiries permitting full coverage. The best solution usually consists in using two or several approaches simultaneously, while carefully defining their respective roles, and in framing each of them accordingly.

The exact design of the census, survey, or extract from administrative records has then to be decided. The range of feasible patterns is particularly wide when a sample survey is planned. This flexibility, as well as problems concerning inference from the results of sample surveys, explains why the methodology for designing these surveys has developed into a special discipline, drawing heavily on mathematical statistics. The requirements of comparative assessments have then to be explicitly considered. When the problem consists in following a phenomenon through time, one has in particular to think of whether and how the units observed in successive surveys are to be replaced.

A number of questions, again to be decided at the planning stage, have an apparently more practical nature but actually are subtle enough to require expert examination. How should a questionnaire be phrased? What kind of test should be performed in order to check that misunderstandings or systematic distortions of answers are kept to a minimum? How shall one proceed in order to avoid nonresponses as much as possible? How shall one correct for missing answers, particularly when they do not concern the same units in successive investigations intended for comparative purposes?

Finally, the control, editing, processing, tabulation, storage, and retrieval of the data involve many operations that have to be carefully planned and carried out if statistics are to be accurate. The emergence and subsequent progress of computer technology have led and still continually lead to important revisions of earlier practices. Some of the problems to be tackled in these operations are considered within the field of statistical computing, broadly understood.

Leaving the technology of statistical production aside, we shall first concentrate our attention here on the initial stage of any comparative assessment, namely the selection and exact definition of the variables that will be measured.

3. The Choice of Concepts

Laymen have difficulty realizing how much the quality of the representation given by statistics depends on good definitions and good classifications. They have difficulty understanding why statisticians spend so much time and effort before agreeing on a system of basic concepts, on the definition of an aggregate, or on a classification. But from time to time the public becomes aware of one or another of the problems involved.

In order to deal with these problems, it will be convenient here to consider examples and special topics rather than to attempt a general treatment in abstract terms. This will hopefully make the questions more lively and exhibit their relevance. Such an approach implies that our coverage of the subject can hardly be complete; but this cost seems to be quite worth accepting. Examples will of course be taken from the issues that most often concern public opinion:

measurement of the price level, of real output, of unemployment, and of income inequalities or disparities.

3.1. Basic Concept

Around 1980, U.S. citizens learnt that the rate of increase of their official consumer price index was quite sensitive to the vicissitudes of monetary policy because of the role played by current mortgage interest rates in the evaluation of the price of owner-occupied housing [see A. S. Blinder (1980) for instance]. Some of them also learnt that in most other countries interest rates did not enter into the computation of the consumer price index because interest payments were not considered as a consumption expenditure but as a transfer related to the management of their wealth by households. The issue was fully reexamined at the Bureau of Labor Statistics and since January 1983 the U.S. consumer price index for all urban consumers has been compiled officially with a rental equivalent measure of owner-occupied housing, so that it is no longer directly affected by changes of interest rates [see R. F. Gillingham (1983) and J. L. Norwood (1983) for more details on this issue]. At this occasion, the attention of the people was brought to one of the many questions that have to be faced and solved when a system of economic statistics is being built, long before the actual details of economic evolution can be known.

What was involved in this particular case was the exact coverage to be assigned to consumption when defining the consumer price index. Consumption is indeed one of the main conceptual categories appearing in the description of economic activity, along with investment, income, production, and so on. Economic statisticians progressively realized that they had to be more precise in their definition of these notions than ordinary language and even economic theory were. Nowadays, the main conceptual system used for economic statistics is embodied in the framework of national accounting, which has been examined and discussed at length before being made the object of international recommendations, such as those constituting the United Nations System of National Accounts (see United Nations, 1970).

Similar difficulties and risks of ambiguity also occur in social statistics. Public discussions in many countries of the world often require a knowledge of the exact number of unemployed people. It appears that, at any given time and in any given country, the evaluation of this number has varied substantially, depending not only on the source of information selected but also, for a given source, on minutiae of the definition chosen for unemployment. One then easily understands the risks faced in comparisons through time and space.

The definition of unemployment currently used by statisticians does not directly follow from a literal interpretation of the word, but rather from the concern with some social and policy issues that are relevant in modern economies. People are considered as being "unemployed" if they are not working for a gain and would like to be. But this principle alone does not precisely determine the number of unemployed; its application is subject to a large margin of uncertainty as long as no rule is given on how to decide exactly (for instance) when someone is to be considered as wishing to work; the present rule, namely that he or she must be "actively looking for a job to be taken immediately," has itself to be made more precise with respect to what is meant by "immediately" (usually within a week) and by "actively looking" (usually having made during the preceding month a move such as visiting a potential employer, writing a letter of application, or inquiring at a labor exchange office). Such a rule, whose practical implementation from available information cannot be perfect in all cases and is even sometimes quite loose, has progressively emerged as a result of a long process in which many people and meetings were involved. [See Chapter 2 in *National Commission on Employment and Unemployment Statistics* (1979) for a brief history of the emergence of employment and unemployment statistical concepts; see also "Thirteenth International Conference of Labor Statisticians" (1983) for the present international recommendations.]

Assessments concerning the inequality of incomes provide so many cases of possible misunderstandings that dealing with all of them would require a full article. As an example, Figures 1 and 2 both concern incomes of manual workers ("ouvriers") and high level employees ("cadres supérieurs") in France in 1975. One notes that the median income of high level employees is 3.1 times larger than that of manual workers according to Figure 1, but only 2.3 times larger according to Figure 2; moreover, in Figure 1 the distribution of the logarithm of income is more skew for high level employees than for manual workers, whereas the opposite holds in Figure 2.

The reason for these differences is that Figure 1 concerns the annual labor income earned by individual workers, whereas Figure 2 concerns the annual disposable income per adult in the households in which individual workers live. [See C. Baudelot and O. Choquet (1981) for exact definitions.] Which one of these two concepts is the proper one of course depends on the question being asked. But it is fair to say that most of the published statistics refer to the concept used for Figure 1, that most of the questions asked by the public at large implicitly refer to the concept used for Figure 2, and that most of the time the latter questions are being answered by unselective reference to published statistics.

Actually, the difficulties are greater than is suggested by this case, because some questions concern not annual incomes but rather something like a life income or the income earned during a "normal" year. The difficulties culminate when one refers to statistics on low pay as providing a measure of the extent of poverty; not only does the interruption of earnings through unemployment or sickness not neccessarily lead to poverty, but also, in a substantial proportion of cases, low pay, even for full time work, corresponds to a transitory situation due to apprenticeship, to the initial phase of a new employment or to acceptance of an occasional job, as shown for instance by C. Baudelot (1981).

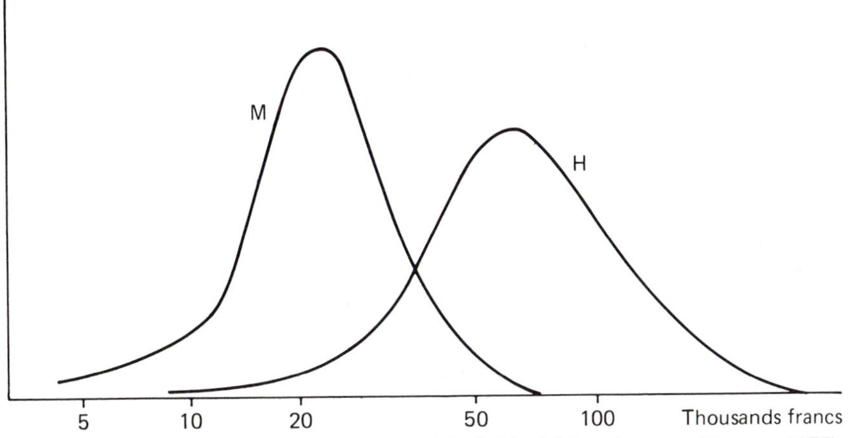

Figure 1. Distribution according to annual individual labor income (France, 1975). M = manual workers; H = high level employees.

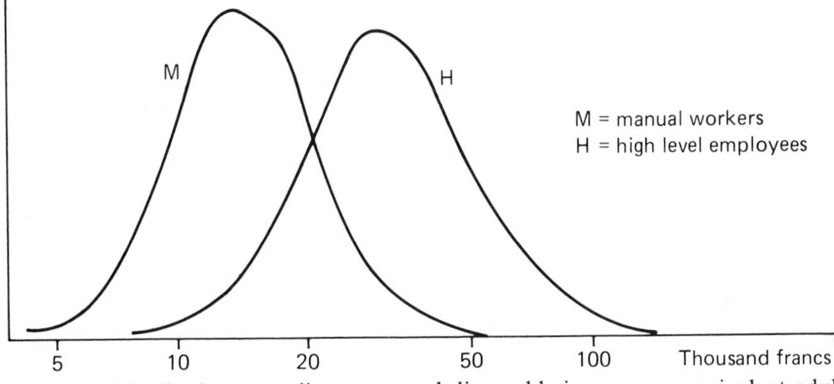

Figure 2. Distribution according to annual disposable income per equivalent adult (France, 1975). M = manual workers; H = high level employees.

Whereas in the two preceding cases—the definition of consumption and the definition of unemployment—a long maturation took place, resulting in international recommendations that public statistical offices aim at following, the same thing cannot be said about the measurement of income inequality. Experienced statisticians working in this field of course have clear concepts in their mind and are precise in reporting the results of statistical inquiries bearing on this topic. Economists and sociologists also have carefully studied the various concepts that must be distinguished and are respectively relevant for various aspects of the subject [see A. B. Atkinson (1975) in this connection]. But the process of collective education leading to the wide acceptance and understanding of a conceptual framework is less advanced than for national accounting aggregates. This case is typical of a situation to be found also in many other fields of socioeconomic statistics.

3.2. Classifications

Figures 1 and 2 are also suggestive of the role played by classifications. Indeed, what is involved is the measurement of income disparities within the population of wage and salary earners; the two groups "manual workers" and "high level employees" are taken to be typical of situations at both ends of the social scale (this will be true whether or not other groups are added so as to cover the whole population concerned). Very often the comparison refers not to the full statistical distributions but only to one characteristic of their location such as the median income within each group.

Now it is clear that the ratio between the two median incomes depends on the definitions being used for the groups: how extensive are the categories of "manual workers" and "high level employees"? Though for manual workers no confusion is likely to arise due to the fact that they contain skilled as well as unskilled workers, the case of foremen is less obvious, and the separation between skilled workers and technicians is not always clearcut in modern industry. On the other hand, how broad the category of "high level employees" should be cannot be decided except by a convention that has to be commonly accepted or at least commonly known.

Reflection upon this case brings out two important ideas. In the first place, classifications are part of the language, as are conceptual definitions; good transmission of knowledge cannot be effective unless a stable system of classifications is commonly used. In the second place, statistical classifications often result from an attempt at making precise distinctions that correctly translate notions with vague contents; but the question may always be raised whether the translation is a good one.

Indeed, the two categories of "manual workers" and "high level employees" belong to a classification of the whole population into socioeconomic groups; this classification is intended to provide the basis for any breakdown into broad social classes. One does not see how to find a better general and workable principle for assigning an individual to a social class than to refer either to his present or past employment or to the present and past employment of members of his household, parents, relatives, etc.; but one must recognize that this principle may be applied more or less well and that it seems to neglect some aspects that have sometimes been associated with the notion of social classes,such as for instance common conditions of living or acceptance of a common ideological core.

One cannot overemphasize the need for a classification of the population into socioeconomic groups. Many questions concerning modern societies indeed refer to the notions that it makes precise, for instance: how does behavior differ from one class to another? What are the main dimensions of social inequality? Has social mobility increased? It is interesting to note that the emergence of such classifications in official statistics is nevertheless quite recent, perhaps because ad hoc treatment of questions such as the preceding ones could be found as long as only a few research workers were dealing with them; but this is no longer the case. [See A. Desrosières, A. Goy, and L. Thévenot (1983) for the French experience and reflections on classifications into socioeconomic groups.] Of course, other classifications for statistical purposes have existed for a long time and have been the object of periodic revisions or reorganizations: classifications of commodities, of industries, of urban or rural communities, and so on.

An important concern is whether the classifications used in the processing and presentation of statistics properly exhibit the heterogeneity that is present in the raw data. The dimensions and locations of this heterogeneity are not known *a priori* in some cases; they may then remain hidden if the data are stored and published according to a grouping that has not identified these dimensions and locations. In other cases the sources of the heterogeneity are known, but the classifications being used may not be fully adequate and blur a good deal of the disparities; for instance, supposedly rural areas may contain newly built settlements of an extending suburban district.

3.3 Aggregates and Indicators

If some difficulties occur in the choice of definitions to be given for the basic concepts, other difficulties occur when one must choose definitions for the aggregates or indicators to be computed from the data, in order to summarize the information in some important respect.

The usefulness of economic indices concerning output, prices, international trade, and so on has been recognized for a long time, so that there is a long tradition, in official statistics, of the regular computation and publication of many such indices. Undoubtedly they serve the function for which they have been introduced, namely to give a measure of the overall evolution of some economic magnitude, which concerns many operations at the micro level. Statisticians are clear on the precise scope of each of these indices and try to define it so that it agrees with the most common uses. One must note, however, that no "ideal" mathematical formula has been found for the computation of indices from the micro data, so that the formulae applied in practice remain somewhat conventional. One must then experiment from time to time with alternative formulae, so as to compare the results with those normally produced. Fortunately, the sensitivity of the numerical results to the exact choice of the formula is typically found to be small. There are nevertheless some exceptions, of which a typical one will be presented at the end of this subsection.

The regular publication of economic indices by official statisticians has been familiar for more that fifty years, but the setting up of national accounts and the presentation of the main aggregates derived from them spread more recently. The principle of measuring and aggregating economic assets and operations by exclusive reference to their monetary value is not only convenient but also appropriate in most cases for most uses. Exceptions, however, exist when either there is no transaction that would directly give the evaluation (production for own use, equipment that has been in service for some time, etc.) or the price used in the transaction is not a reliable yardstick (underpriced public utilities, labor service provided by a young worker at the beginning of his career in a permanent job, etc.). Moreover, the determination of volume aggregates "at constant prices" raises exactly the same problems that concern the choice of production and price indices.

The choice of good indicators for some important social phenomenon may similarly be less easy than it appears at first sight. For instance, an indicator intended to measure the trend of human fertility or mortality in a country must be more subtle than a simple ratio of the number of births or deaths to the size of the population. Changes in the age distribution of women in the country or of the whole population induce variations in the number of births or deaths, quite independently of any basic change in fertility behavior or in exposure to mortality. Thus demographers have adopted the practice of computing indicators that correct for such changes in age composition.

As another example, one may think it natural to use the unemployment rate (number of unemployed people over number of person in the labor force for the same demographic group) in order to compare the risks of unemployment facing various groups. One then notes for instance that in March 1970 the rate was 2.8 per cent for French men aged 18 to 24, whereas it was only 0.8 per cent for French men aged 40 to 49, a ratio of 3.5 to one (the corresponding rates in March 1980 were respectively 10.8 per cent and 2.4 per cent, a ratio of 4.5 to one). But this apparently simple answer must be qualified; in the first place, unemployment rates are much more sensitive for young men than for older ones to the precise conventions used in measuring unemployment, so that minor changes in these conventions would significantly change the ratio 3.5 to one. In the second place, rates of unemployment do not provide a complete picture of the risks of unemployment faced by various social groups. Unemployment durations are not the same for all groups; it is noteworthy that in France they are much shorter for young men than for older ones. The average duration of unemployment spells around 1970 has been evaluated by R. Salais (1974) at less than 3 months for unemployed men aged 18 to 24 as against 8 months for those aged 40 to 49. For most young people unemployment when it occurred, before taking a job or between two jobs, was a quite temporary situation; it was much more serious for most older unemployed people. The phenomenon still holds in France in the early 1980s, although not quite to the same extent. But a different situation was found for the U.S. by H. B. Clark and L. H. Summers (1979).

When one is considering essentially multidimensional phenomena, the definition to be chosen for a single numerical indicator may not be obvious.

For example, suppose one wants to measure the degree of inequality of an income distribution. Referring to Figure 1 or 2, one may want to know whether the degree of income inequality is greater for high level employees than for manual workers (if the example does not seem stimulating enough, one may imagine that two curves similar to the ones of Figure 1 have been observed for the incomes of the households of two distinct countries; one may then want to know which country has the smaller degree of income inequality). Since it is now no longer a question of evaluating the disparity between the two groups but rather of comparing inequalities within each group, one will have to eliminate the difference in location of the two statistical distributions, then compare their dispersions. Without trying to specify here any particular method for this comparison, we may be sure, looking at Figure 1 or 2, that there is a greater degree of income inequality among high level employees than among manual workers. But one often faces situations in which the case is much less clear and in which the conclusion depends on the aggregate characteristic that is chosen as a measure of the degree of inequality. There is an important literature on the proper definition of this characteristic, but no attempt will be made here to discuss it; the reader may refer to A. B. Atkinson (1975).

Special attention must be devoted to international comparisons of real product and purchasing power, a subject to which was devoted one of the two largest direct international statistical undertakings that were ever made. [See I. B. Kravis, A. Heston, and R. Summers (1982) for a complete presentation.]

The first idea that comes to mind when one attempts to compare levels of output in various countries is to refer to the results of the respective national accounts, which are of course evaluated in national currencies, and to use official exchange rates in order to convert them to a common unit. One will for instance refer to the per capita gross domestic product of India, which was evaluated at 1,220 rupees in 1975; one will convert it into 146 dollars, using the then prevailing exchange rate of 8.4 rupees for a dollar; comparison with a per capita gross domestic product of 7,176 dollars in the U.S. will lead to the conclusion that in 1975 output per person was 49 times higher in the U.S. than in India.

One may be disturbed by the fact that the above result is sensitive to erratic fluctuations of exchange rates that should have little to do with output per person. But the problem is still more serious than that, since it appears that reference to currency exchange rates systematically and strongly overstates the real output discrepancies. A direct comparison of the volume of production between the two countries evaluates output per person in the U.S. as 15.3 times higher than it is in India. The bias resulting from reference to exchange rates appears quite generally, as is clear from the small selection of results presented in Table 1.

The source of the difficulty is that the structure of relative prices systematically and strongly varies when one moves from one country to another one where output per person is higher. Currency exchange rates may be more or

Table 1. Index of Per Capita Gross Domestic Product[a]

Country	Converted at official exchange rates	Evaluated in volume
India	2.03	6.56
Colombia	7.92	22.4
Hungary	29.6	49.6
Japan	62.3	68.4
France	89.6	81.9
United States	100.0	100.0

Source: I. B. Kravis, A. Heston, and R. Summers (1982, p. 10).
[a] U.S. = 100; results for 1975.

less in line with prices of internationally traded goods, but they do not correctly convert the full range of prices confronting domestic final users of production.

However, the fact that the structure of relative prices varies from one country to another also raises problems for direct comparisons of the type reported in the last column of Table 1. When two countries are compared, one gets significantly different results depending on the common system of relative prices that one uses in computing various aggregate volumes of output: the system of the first country, that of the second, or a combination of the two. For multilateral comparisons, involving several countries simultaneously, the range of *a priori* sensible alternative methods is still wider. Depending on the method being used, the numerical result obtained will differ, but of course much less than between the two columns of Table 1. For instance, depending on whether one takes the relative prices prevailing in the U.S. or in India in 1975, one finds either 9.0 per cent or 4.1 per cent for the level of the Indian per capita gross domestic product in 1975 on the basis of 100 for U.S. The margin between the two is quite sizable, and this is in fact a fairly representative case for binary comparisons between quite different countries. The sign of the difference is systematic: using the relative prices of a country or period in a comparison with another country or period typically leads to a lower relative output for this country or period than using the relative prices of the alternative country or period.

3.4. Some Reflections

No attempt will be made to summarize here the large methodological literature that deals with this problem, including old as well as recent contributions. The fact that several distinct methods may be considered as competing without anyone of them being definitely superior is rather typical of a frequent situation in economic and social statistics: reference to a clear and proper general principle leaves some room for convention; which convention is chosen does not matter in most comparisons actually made, but only in a few doubtful cases (in such cases it should be said that heterogeneity of the situations being compared precludes an unambiguous conclusion).

It should be clear from direct reflection as well as from the preceding examples that for each phenomenon approached, extensive reference must be made to the subject matter science dealing with this phenomenon. The cost of living and real output for instance are economic concepts; it belongs to economics to clarify how they should be defined and used. The measure to be given to human fertility must be made precise by demographers, and that to be given to social mobility by sociologists. This general remark explains why, in striving to make their product more rigorous and more pertinent, economic and social statisticians have to deal much more with subject matter scientists than with mathematical statisticians.

In some cases, specific theories have developed for the purpose of clarifying the definitions to be chosen in economic and social statistics. Such specific developments seem to concern essentially the various problems raised by the aggregation involved in these statistics: index number theory, national economic accounting, the measurement of inequality, social mobility matrices, etc.

One may, however, wonder why the methodology of mathematical statistics is so little used in these questions dealing with the choice of concepts. Should not systematic reference to data play a large role when one must define a classification or an aggregate measure? One possible general criterion that could be used here would be to minimize the loss of information involved in the grouping of different units or in the substitution of one single quantity for a vector of quantities. Available data could then be used to decide which measure minimized the criterion. The difficulty with this approach and with the implementation of such an inductive procedure lies in the fact that the choice of an appropriate measure of the loss of information may raise serious difficulties and may even turn out to be practically impossible to make *in abstracto*.

Take for instance the problem of statistical classifications. One might have thought that a classification could be built inductively, starting from a very fine definition and choosing the classes at the successive levels of aggregation in such a way as to group the objects that appear to be the closest with respect to some criterion. If a classification of industrial activities were to be found, one would start from a detailed description of the activities of the existing firms; one could then adopt a rule to group the activities that are often associated within the same firms. An objective method for the construction of nested clusters could then be used. An attempt made along these lines by M. Volle et al. (1970) has not been found convincing.

Indeed, the problem with the above proposal is not mainly that it would be very hard to carry through, but that official statisticians must simultaneously take many criteria into account when building a classification, because it will be used for many different purposes during a rather long period. If an industrial classification is to be set up, one would certainly like to group activities that are associated within the same firms, but other criteria appear just as important: similarity between the outputs of the activities (i.e. substitutability between these outputs in each one of their many uses), similarity between the technologies applied (i.e., same input requirements, same conditions of work, etc.), and so on. This is why an official classification is the outcome of many compromises adopted during negotiations involving many partners.

4. Harmonizing and Reconciling Data Sources

Discussing the choice of concepts, one immediately realizes how easy it is to draw unwarranted conclusions from the comparison of two statistics. Since numerical results are sensitive to this choice and the choice cannot be expected to be made in the same way at all times and places, the risk of significant mistakes exists.

The problem is even worse because of imperfections in the collection of data. The process of observation in economic and social statistics never reaches the degree of accuracy that is claimed by the so-called "exact sciences." Most often the data collected during a survey or from an administrative record are inaccurate because of incomplete coverage of the population concerned and of inexact reporting due to misunderstanding, to oversight, or to intentional distortions on the part of the respondents. Since such imperfections vary from one operation to another, the comparison between the results of two distinct operations is likely to be subject to errors, even when the concepts used in both cases are identical. These conditions do not, of course, forbid the use of statistics for comparative assessments but make it particularly slippery. They are part of the challenge that economic and social statisticians must face, and of which they have always been aware. In particular, if they have always stressed that the methods followed in regular statistical operations should not be revised except for strong reason, it is precisely in order to avoid perturbing the comparisons through time. This behavior usually guarantees a fair degree of accuracy for the frequently made assessments concerning two periods not too far apart and the same country or region. This is why we shall focus attention now on the way in which statisticians try to deal with more risky comparative assessments, namely those concerning two or more different countries.

4.1. Harmonized Statistical Operations

The best way to proceed is to set up statistical operations that are planned to apply exactly in the same way in all countries concerned and that are implemented according to rigorous rules. The World Fertility Survey is one example of such an international statistical operation (see the article by Macura and Cleland, Chapter 18 in this volume). Two other significant examples, which will be examined here, concern employment and income statistics in the experience of the European Economic Community.

International comparisons of unemployment rates are notoriously dangerous. Each country has its own tradition for the current evaluation of the number of unemployed people, often using basic sources of information that are specific to its institutions, and often referring to a definition of unemployment that is quite different from the one contained in the internationally agreed ILO recommendation. The common practice of serious commentators and analysts when publishing side by side data on the evolution of unemployment rates in several countries is to add a footnote blankly stating that these rates are not internationally comparable. This was the practice even at OECD until recently, when the organization decided to make adjustments to some of the national figures, using a method that was elaborated at the U.S. Department of Labor.

Two main reasons explain these national differences. First, unemployment statistics are politically sensitive and revising the tradition on which they are based may appear to be impossible because of the perceived political implications. More fundamentally, except in such a large country as the United States, monthly evaluations cannot come from a purely statistical investigation. Roughly speaking, one may say that the number of people to be included in a sample survey of the labor force in order to reach a given degree of accuracy is independent of the size of the population; small countries would then have to spend for this purpose the same amount as the U.S. spends, and they cannot afford it. This being the case, monthly figures are derived almost everywhere from registered unemployment, the definition of which inevitably varies from one country to another one because the circumstances in which people are registered and classified as "unemployed" also varies, depending on how the labor market and unemployment insurance are organized.

Though one cannot entertain the idea that monthly figures should be collected in the same way in different countries, it is nevertheless conceivable to organize harmonized surveys to be made at longer intervals and providing in particular a kind of benchmark for the interpretation of the monthly figures. Such was one of the motives for the European Economic Community in launching a survey as early as 1960, and later in making regulations for a harmonized biennial labor force survey, which of course gives broader information about employment and is not restricted to unemployment [see Eurostat (1983) for the latest results]. The technical aspects of the implementation of the survey are laid down in an agreement with the national statistical institutes. On the basis of proposals from the EEC statistical office, a working party determines the content of the survey, the list of questions, and the common coding of individual replies. The national institutes are responsible for selecting the sample, preparing the questionnaires, conducting the direct interviews among households within a preassigned period, and forwarding the results to the EEC office in accordance with a standard coding scheme, using in particular common classifications. This office processes and disseminates the information, which is, however, also processed, analyzed, and disseminated in most countries by national institutes.

Actually, perfect comparability between the nine countries is not achieved, and indeed OECD adjusted figures for the unemployment rate do not exactly match with the results of the harmonized labor force survey. Perfect comparability would require a single direct survey, carried out at the same time on the basis of the same questionnaire and in accordance with a single method of recording, whereas a number of compromises must be accepted in order to accommodate to difficulties faced by national institutes. As a consequence, there are still some important differences in the holding of the survey in the nine countries. Although organized at Community level, the survey does not provide strictly comparable data but harmonized data, i.e. comparable "as far as possible" from country to country. The differences do not, however, cause sizable distortions. [This last sentence, as well as the two preceding paragraphs, directly draws on Eurostat (1977), in which the reader can find many more details, concerning in particular national specificities.]

A higher degree of harmonization was actually reached in another survey organized by the European Economic Community. To this second survey, concerning the structure of earnings and described for instance in Eurostat (1972), we now turn our attention.

Even for wage and salary earners, comparable statistics on remuneration are difficult to get. Indeed, most statistics regularly published in the various countries aim at measuring the evolution of wage rates; they refer to indicators that are relatively easy to define and observe but do not take all elements of remuneration into account, do not cover all wage and salary earners, and have no reason to coincide from one country to another. In their effort to develop a basic harmonized system for the countries of the European Economic Community, statisticians of these countries have found it useful to set up a special survey, which has been conducted in 1966, 1972, and 1978, the intention being to repeat it every sixth year.

The surveys aim both at being identical internationally and at having complete coverage. The latter, for instance, requires from a large random sample of employers to give detailed report on the remuneration of a random sample of their employees. The remuneration is that given for October 1978, including all kinds of premiums and bonuses paid monthly, and that given for the full year 1978, including moreover all kinds of supplementary payments accruing on a nonmonthly basis. Data are collected not only on the activity and size of the employer, on the individual characteristics of the employee, and on the conditions of employment and payment, but also on the period of employment and the duration of absences from work. Harmonization concerns not only the time, coverage, questionnaire, and sampling scheme of the survey, but also the classifications used for processing its results.

This example clearly shows that full harmonization is very costly. Indeed, of all statistical surveys of any size that are made in Western Europe, this one imposes by far the heaviest burden on the respondent firms. The reader will easily understand that all the information requested is not readily available and has to be gathered from various sources within the firms. This is the cost that must be borne for an exact recording of the real economic operations that are to be compared. This cost explains why the survey is not made often. (As we shall later see, the same degree of accuracy on earnings could not be expected from a survey addressed to wage and salary earners rather than to the firms employing them.)

But the example also shows the advantages gained by full harmonization: the results can be used with confidence for comparative assessments. The results moreover reveal the lack of accuracy of some comparisons that were made previously by competent statisticians on the basis of the then available data [Vlassenko (1977) makes reference to such a case]. To get the flavor of the results, one may note that the interdecile ratio of the distribution of the remuneration paid in October 1972 to wage and salary earners that were employed full time in industry during that month (apprentices excluded) was approximately 2.75 in France and Italy as against approximately 2.35 in Belgium, the Netherlands, and Western Germany. The difference, which is mainly due to the relatively low pay of manual workers in the two first countries, is significant; but it is small enough to escape accurate measurement by less careful operations than the EEC structure of earnings survey (as a comparison one may note that the interdecile ratio for French manual workers moves from 3.6 to 3.0 when one goes from the distribution given in Figure 1 to that given in Figure 2, both distributions including part time work and apprentices).

4.2. Harmonization of Concepts

Fully harmonized statistical operations cannot but be exceptional because of their cost. For the bulk of economic and social statistics harmonization is achieved less rigorously, limited as it is mainly to concepts and classifications. Indeed, if exactly the same concepts and classifications were applied everywhere and if statistics, no matter how established, were completely accurate, full comparability would be achieved. The division of labor that has spontaneously prevailed is roughly patterned according to the following principle: international harmonization concentrates on concepts and classifications, whereas national statisticians are supposed to find the best way to accuracy in measurement within their own countries.

International harmonization of concepts and classifications, which in former times was performed by the International Statistical Institute, has been the object of intensive work during the past forty years under the auspices of governmental international organizations (UN, OECD, EEC, CMEA, etc.). This work will not be described here. Notwithstanding its undeniable success, it obviously cannot be considered as complete. It is less advanced in social than in economic statistics, and for instance it has not yet fully achieved the desirable integration of the various international classifications of products and activities.

The main question that can be raised about the present state of this harmonization is, however, whether it is appropriate to the needs served by statistics, both within each country and for international comparative assessments. It has a good chance to be appropriate because it results from a long and collective process in which many have taken part. But one must recognize on the one hand that a few people have been particularly influential in this process and may have misjudged the best compromises between conflicting needs, and on the other hand that needs vary with time and that the frequency with which statistical concepts and classifications should accordingly be revised raises a difficult problem, the requirement of comparability through time being among the first to be taken into account.

Taking a very broad view, one may say that the needs of developed countries are fairly well met and that the set of concepts, the accounting framework, and the standard classifications that have been produced under the auspices of international organizations meet the main requirements requested for these countries. If comparative assessments remain sometimes difficult, notably between Western and Eastern Europe, it is more often because of a lack of published data or because of incomplete knowledge of the definition and accuracy of available data than because of insufficient efforts at harmonization.

Whether the needs of developing countries are adequately met is much less clear. Certainly the paucity of statistical data about many important features of the societies concerned is the main barrier limiting comparative assessments. But it has often been claimed that one should resist the temptation to apply directly the conceptual framework that has been worked out by statisticians of industrial countries. It has been suggested that an alternative framework could be found that would better serve the needs of developing countries.

After many years of inconclusive research in this direction, it is reasonable to say now that no such alternative solution exists. The truth is rather that aggregate economic magnitudes cannot provide for developing societies the same kind of first approximation to a proper description as they often do for developed ones. The conditions of life are less closely correlated with the earning of a monetary income; quite a few economic operations are not reflected in monetary transactions; the price system prevailing in transactions is not always well defined; the duality of the society reduces the significance of any aggregate indicator.

Let us consider for instance the results reported in Table 1 and concentrate on the last column, which is intended to approach the best possible measure of average production levels per capita. The figures given for Hungary, Japan, France, and the United States have clearly understandable meanings and provide a good basis for a first comparative assessment about production levels in these countries and hence indirectly about potential standards of living. The results for India and Colombia are of course not devoid of all interest, but one is entitled to think that they carry less information: knowing the economic difference between these two countries, which in Table 1 appears as larger than that between Hungary and the United States, gives more a stimulus toward looking at less aggregated data than a final or even provisional answer to a frequently raised question.

4.3. Reconciliation of Statistics

Notwithstanding the steady effort toward harmonization that has been made during the past decades, users do not always find harmonized data at their disposal. When trying to make comparative assessments of significant questions they often must rely on data concerning somewhat different concepts and resulting from heterogeneous statistical operations. In such cases they must try to *reconcile* the data.

The word "reconciliation" originated within the work of national statistical offices in which one often has to deal with distinct sources of information concerning the same phenomenon. The data collected from two different sources never exactly match, and indeed they hardly ever concern exactly the same variable for the same population. It is, however, always important to explain the differences, and in many cases it is found to be advisable to correct one set of data by reference to another set which has proved to be more reliable. The need for reconciliation is particularly frequent between statistics derived from administrative sources and those obtained by direct statistical operations.

The measurement of unemployment again provides a significant example. In most countries it combines reference to registered unemployment for monthly figures and to labor force sample surveys for less frequent benchmark data; the reasons for this state of affairs were given earlier in this article. Reconciliation of the two sources is important: one wants to know how best to interpret the current figures though they do not correspond to a definition of unemployment that would be appropriate for most users; one also wants to check that the trends of the monthly series do not deviate from the trends derived at longer intervals from sample surveys. These trends often differ either because of permanent discrepancies in the definitions being used or because of changes in labor and unemployment insurance regulations, which require or induce more or fewer people to register as unemployed. Valid comparative assessments between two periods must then correct for deviations that the series of registered unemployment can experience. [See in particular O. Marchand and C. Thélot (1983) for a detailed description and analysis concerning France.]

By extension, one may speak of reconciliation between available statistics bearing on the same subject and concerning different populations. To reconcile these statistics is to adjust them in such a way as to improve their comparability. The objective would of course be to insure full comparability if possible. But the adjustments that must be made to correct for differences between the concepts used and between the methods of collection are always crude, since by definition in such cases harmonized data do not exist.

In some cases reconciliation does not aim only at adjusting figures because of differences of definitions or methods of collection; it is also intended to reduce errors of observation that affect each data set taken in isolation. This applies in particular to international flows of goods, financial assets, and people. It has been noted on many occasions that the official statistics of two countries give different figures for their bilateral trade or for the flows of tourists or migrants between them. The dating of an operation and the final destination or first origin of a movement may be recorded differently in the two countries; moreover, administrative rules and arrangements for declarations are often the cause of a substantial lack of accuracy of recorded responses. [E. Veil (1981) discusses the origin of the present major lack of consistency between current balances of payments at the world level.] An attempt at reconciliation of migration flows may be found for instance in C. D. Walker and J. J. Kelly (1978).

Although it must often be made by users, and although it is even sometimes made by producers (for instance when they compile aggregate statistics or national accounts while they have no proper data for parts of the universe), reconciliation can never be as good as direct observation of harmonized concepts by identical methods. This explains why some statisticians are reluctant to embark upon reconciliations of some importance and prefer in many cases to abstain from making comparative assessments. But such abstention is fruitless, because others soon make and publish such assessments by direct comparison of unreconciled data.

Moreover, there cannot exist any general methodology to follow when reconciling statistical data. A model that permits a rough estimation of the missing or correcting items has to be worked out in each case, depending on what information is available. The only clear rule stipulates that one should report explicitly on the methods that have been used for reconciling non-harmonized data. Indeed, different users may have different views on what should be the proper way of proceeding in each particular case.

In order to get a clear understanding of the importance, role, and difficulties of reconciliation, one may refer to a study made by Malcolm Sawyer (1976) on income distribution in OECD countries. The main object was to compare the degree of inequality of household incomes in the countries concerned. Methodological questions had to be faced, which correspond to those mentioned in Section 3 of this article; but it turned out that the choice of concepts and in particular the definition given to the degree of inequality did not matter very much for the assessments that could be made. On the other hand, the basic available data were considerably different from one country to another, which was the main source of difficulty.

Indeed, the discrepancies between countries came from both tails of the income distribution, the poorest twenty per cent and the richest ten per cent, and it is well known that, for a given country, those tails are particularly sensitive to the exact definitions and methods of collection. For instance, when household sample surveys provide the basic data, the bias due to nonresponses and to absences from the enumeration list used in sampling are particularly important in the tails and are likely to vary significantly from one survey to another, the accuracy of reported incomes depends on the minutiae of the questions asked. When data coming from reports to fiscal authorities are used, corrections must be made for nontaxed incomes and for households that do not have to report. That such questions concerning the data origin can matter appears, for instance, for France, where two estimates of the same interdecile ratio given by J. Bégué (1976) evaluate it respectively at 14.6 (on the basis of a household survey) and at 20.7 (from reports to tax authorities). It also appears for the United States, where the share of the top 5 per cent of consumer units in the size distribution of total money income is evaluated respectively at 17.4 per cent by the 1972 current population survey and at 19.4 per cent by the Exact Match Project for 1972, which also used income tax returns. [See E. Budd and J. K. Salter (1981), which also contains a discussion of the comparative accuracy of various sources of income statistics. See also J. F. Ponsot (1983) for the same issue concerning France.]

Considering what data were available, Malcolm Sawyer had no alternative but to try to reconcile them as well as he could, which he did as he explained in his text and in long appendices to his article. He then had to warn the reader that, while he had aimed at improving the international comparability of the data, it was probably impossible to reach full comparability.

5. Dissemination of the Results

Assuming that data permitting comparative assessments are available, perhaps after some reconciliation, we must still consider how they are used and how the results are reported. Problems and possibilities for progress again exist at this tage.

5.1. Adjusting for Unwanted Effects

Users of statistics always have a purpose in mind, in particular when they want to proceed to comparisons. It very often turns out that this purpose requires reference not directly to the raw data, but to statistics that have to be extracted from these data. Indeed, the raw data describe a reality that is often the combined result of many causal effects, whereas users would like to know what comes from a single cause, or from only a subset of all causes. This is why further processing of data often occurs.

Seasonally adjusted series are by now quite familiar. They are intended to meet the needs of those who want to know how the current trend of a variable changes, independently of variations due to seasonal factors. In most cases seasonal adjustments do not create serious problems, notwithstanding the fact that they lack the rigorous justification one would like to find (such a justification could only come from a model that would fully explain the movements of the series). But the necessarily somewhat conventional formula used for computing adjusted series does matter in some cases—for instance for registered unemployment, because the seasonal component is sizable, the seasonal pattern often changes, and assessing trends of unemployment is politically sensitive.

A very frequent objective is to adjust for differences in the structures of two populations that are being compared. For instance, one may note that the unemployment rate is higher in one region than in another; one may then want to know whether the disparity is not only due to differences in the sociodemographic composition of the labor force and in the industrial structure of employment between the two regions. Or one may want to analyze wage rate disparities between men and women so as to measure what is really due to sex, independently of differences in qualification, seniority, industry of employment, and the like. In such cases the raw data must be fitted to a model that permits estimation of the various effects one wants to identify. Linear models, of the analysis of variance type, often prove to be adequate, as in D. Depardieu (1981). For qualitative data, which are very frequent in official statistics, it is often useful to fit log–linear models, the statistical treatment of which has been presented in Y. M. Bishop, S. E. Fienberg, and P. W. Holland (1975).

With the production of ever richer data bases and the increasing power of computers, it will probably become customary for government statistical offices to proceed to data analyses of the kind just mentioned, because they will fulfill well-identified needs. This likely development has been discussed in J. -C. Deville and E. Malinvaud (1983). Official publications should then provide an increasing amount of material that is directly appropriate to a large variety of significant comparisons.

5.2 Reporting Comparative Assessments

Official statistics are intended to be used by the whole community. This requires in principle that they be widely diffused and reported, so as to be accessible to anyone, Such a requirement would raise no problem with the scientific ethic if statistics were perfectly adequate to their uses, perfectly accurate, and perfectly well reported to final users. But one could hardly complete an article of this kind without reminding the reader that under actual conditions problems do exist. Since it is not the main subject of this paper to expand on these problems, they will only be briefly illustrated by two examples which are chosen so as to be representative of opposite situations.

In most developed countries the consumer price index is measured by official statisticians with a fairly high degree of accuracy; it is widely and regularly reported, so that it is familiar to the general public. Even in such an ideal situation, difficulties are not fully avoided. Indeed, the claim is often made in many countries that indexation of incomes to this price index compensates for inflation. But strictly speaking, the index was not defined for this exact purpose, and compensation for inflation is a very difficult concept about which economists will never end their debates. The fact that statisticians have to refer to an ideal "average consumer" and to a specific definition of "consumption" shows that they do not claim to know how much compensation anyone is entitled to; but this is often forgotten.

The ethical problem raised by this example is to decide whether, to whom, and how one should present the conceptual distinction between indexation on the consumer price index and compensation for inflation. After a close examination, serious investigators would agree that the first concept provides an approximation for the second. When the distinction is stressed to the general public, whose level of information and sophistication in economic analysis remains low, the message is misunderstood; the generally drawn conclusion is that the official price index provides a biased measure of price inflation and should therefore be mistrusted, which turns out to be more misleading than ignorance of the distinction.

As a second example one may consider the report that was issued in France on Malcolm Sawyer's study of the income distribution in OECD countries. According to the figures presented in this study, the degree of inequality of after-tax income was higher in France than in any of the eleven other countries that had been considered, including notably Italy and Spain. But, as was mentioned previously, the comparability of the data remained imperfect even after reconciliation, and it might be that the particular statistics used for France was responsible for the very poor ranking of this country.

The study was published by OECD as a technical report of its author, "in no way reflecting the position of the organization." But it happened that at the time, hot political disputes were going on in France about the extent of inequality. The French media then gave large publicity to the study, often presenting it as an official report of the international organization and omitting to reproduce the cautious methodological note published at the beginning of the text. The French government eventually presented an official protest to the OECD, arguing first that the statistical data used for France were heterogeneous with respect to those for other countries, and second that the Secretariat had not publicly denied the conclusions attributed to it by the French press. The affair went on in the media for some time.

Considering what has happened, one must agree that this type of unintended and probably unwanted reporting was not adequate. The diffusion of statistical results must take their robustness into account. Whereas the study was very interesting for specialists, its figures should neither have been directly given to the general public as final truth, nor completely dismissed as void of any meaning. Since the public was keen on comparative assessments about income inequalities between France and abroad, it should have been exposed to more balanced articles that would have conveyed more qualitative conclusions, showing in particular which uncertainties remain for a precise description of the situation.

As is suggested by these two examples, the reporting of economic and social statistics, as well as the diffusion of comparative assessments based on them, must be multifarious. Reports to qualified economists and sociologists will never be conveyed intact to a broader public. They will be either ignored or transformed along the channels of information transmission. In order to avoid distortions, statisticians must devote some of their time to try and find the best compromises for communicating statistical results to the various groups of users. This is neither an easy nor a gratifying task, but by addressing it the statistician participates in the improvement of knowledge.

Acknowledgements

Appreciation is expressed to C. Baudelot, O. Choquet, B. Grais, and J. -F. Ponsot for suggestions and other assistance provided for this article.

Bibliography

Atkinson, A. B. (1975). *The Economics of Inequality*. Oxford: Clarendon.

Bailar, B. A. (1983). "The quality of statistical data." *Proc. Internat. Statist. Inst. (Proceedings of the 44th Session, Madrid)*, **2**, Book 2, 813–835

Baudelot, C. (1981). "Bas salaires: état transitoire ou permanent." *Econ. et Statist.*, **131**, 35–49.

Baudelot, C. and Choquet, O. (1981). "Du salaire au niveau de vie." *Econ. et Statist.*, **139**, 17–28.

Bégué, J. (1976). "Remarques sur une étude de l'OCDE concernant la répartition des revenus dans divers pays." *Econ. et Statist.*, **84**, 97–104.

Bishop, Y. M., Fienberg, S. E., and Holland, P. W. (1975). *Discrete Multivariate Analysis: Theory and Practice*. Cambridge: M.I.T. Press.

Blinder, A. S. (1980). "The consumer price index and the measurement of recent inflation." *Brookings Papers Econ. Activity*, **2**, 539–573.

Budd, E. and Salter, J. K. (1981). *Supplementing Household Survey Estimates of Income Distribution with Data from Other Sources: the US Distribution of Total Money Income for 1972*. Bureau of Economic Analysis, US Department of Commerce.

Clark, K. B. and Summers, L. H. (1979). "Labor market dynamics and unemployment: A reconsideration." *Brookings Papers Econ. Activity*, **1**, 13–73.

Depardieu, D. (1981). "Où rechercher les disparités de salaire?" *Econo. et Statist.*, **130**, 45–60.

Desrosières, 'A., Goy, A., and Thévenot, L. (1983). "L'identité sociale dans le travail statistique: La nouvelle nomenclature des professions et catégories socio-professionnelles," *Econ. et Statist.*, **152**, 55–79.

Deville, J. C. and Malinvaud, E. (1983). "Data analysis in official socio-economic statistics,"*J. Roy. Statist Soc. Ser. A*, **146**, 335–361.

Eurostat (1972). *Structure of Earnings in Industry*, **1–8**. Social Statistics, Special Ser. Luxembourg.

Eurostat (1977). *Labour Force Sample Survey. Methods and Definitions*. Luxembourg.

Eurostat (1983). *Labour force Sample Survey—1981*. Luxembourg.

Gillingham, R. F. (1983). "Measuring the cost of shelter for homeowners: Theoretical and empirical considerations. *Rev. Econ. and Statist.*, **LXV**, 254–265.

Kravis, I. B. Heston, A., and Summers, R. (1982). *World Product and Income: International Comparisons of Real Gross Product*. Baltimore: Johns Hopkins U. P.

Macura, M. and Cleland, J. (1985). "Reflections on the World Fertility Survey." In this volume, Chapter 18.

Marchand, O. and Thélot, C. (1983). "Le nombre des chômeurs." *Econ. et Statist.*, **160**, 29–45.

National Commission on Employment and Unemployment Statistics (1979). Washington: U.S. Government Printing Office.

Norwood, J. L. (1983). "Problems in the measurement of consumer prices." *Proc. Internat. Statist. Inst.* (*Proceedings of the 44th Session, Madrid*), **1**, Book 1, 148–157.

Ponsot, J. F. (1983). "L'utilisation des données fiscales pour l'établissement de distributions statistiques sur les revenus en France—Méthodes et problémes," *Proc. Internat. Statist. Inst.* (*Proceedings of the 44th Session, Madrid*), **1**, Book 1, 243–262.

Salais, R. (1974). "Chômage: Fréquences d'entrée et durées moyennes selon l'enquête emploi."*Ann. INSEE*, **16–17**, 163–237.

Sawyer, M. (1976). *La Répartition des Revenus dans les Pays de l'OCDE*. Etudes Spéciales, Juillet 1976. Paris: OCDE.

Thirteenth International Conference of Labor Statisticians (1983). *Bull. Labor Statist.* (I.L.O.), **3**, 6–11.

United Nations (1970). *System of National Accounts*. Methodological Papers, Ser. F, No. 2, Rev. 3. New York.

Veil, E. (1981). *L'Ecart Statistique de la Balance Mondiale des Opérations Courantes*. Etudes Spéciales. Paris: OCDE.

Vlassenko, E. (1977). "Le point sur la dispersion des salaires dans les pays du Marché Commun." *Econ. et Statist.*, **93**, 64–72.

Volle, M. et al. (1970). "L'analyse des données et la construction des nomenclatures d'activités économiques de l'industrie." *Ann. INSEE*, **4**, 101–131.

Walker, C. D. and Kelly, J. J. (1978). *Migrations between Canada and the United Kingdom—Comparisons of Unadjusted and Adjusted Data*. CES/AC. 42/8. Geneva: United Nations.

CHAPTER 20

An Introduction to the Optimum Design of Experiments

Anthony C. Atkinson

Abstract

Optimum experimental design is introduced as arising from technological experiments, rather than from the "classical" requirements of agricultural trials. A variety of optimality criteria are explained heuristically in terms of ellipsoidal confidence regions for the parameters of a linear model. Relationships between the criteria are established, as are the requirements of response surface designs. The paper ends with sketches of applications of the theory to designs for agricultural variety trials and clinical trials, and to response surface designs. Although the paper is primarily intended to provide an introduction to the paper by Nalimov, Golikova, and Granovsky, the material is also relevant to parts of the paper by Fienberg, Singer, and Tanur.

1. Introduction

The chapter which follows (Nalimov, Golikova, and Granovsky, 1985) makes appreciable use of the theory of the optimum design of experiments. Although much of this theory had emerged by 1960, the ideas are still not familiar to all statisticians. The purpose of this note is to provide such readers with a brief introduction to the subject.

The paper is in four short sections. In the remainder of this section the context of the theory is contrasted with that of longer-established approaches to the design of experiments. In Section 2 an outline is given of some parts of the theory. Section 3 provides some comments on the contrasting work on response surface designs which is also referred to by Nalimov et al. Finally, in Section 4, mention is made of three areas in which the theory has found applications. These applications appear to indicate a synthesis between the various approaches to experimental design.

Optimal experimental design is mentioned not only by Nalimov et al. but also by Fienberg, Singer, and Tanur (1985) in their paper on social experimentation (Chapter 12 in this volume). Both sets of authors, as well as Pearce

Key words and phrases: biased coin design; design algorithms; DETMAX; exact design; graph theory; optimum block designs; optimum design; response surface designs; sequential clinical trials; variety trials.

(1985, Chapter 22 in this volume) in the context of experimental design in developing countries, mention the work of Sir Ronald Fisher. His contribution is an appropriate place to start, for Fisher's work on the design of experiments provides an outstanding example of the usefulness of the statistical approach. As Box (1979) states, "by invention of the concept of Experimental Design, Fisher promoted the statistician from a curator of dusty relics to a valued member of a scientific team." One large, important, and continuing part of research on the design of experiments can be viewed as a continuation of Fisher's work on agricultural experiments. In these "classical" designs the treatment structure is often simple, for example a few selected levels of a fertilizer or several varieties of a cereal crop. The experimental units, in this case areas of soil, may however be highly variable. Examples are described by Pearce. Under such conditions the division of experimental units into blocks is important in providing local control over error.

It may be helpful to consider the theory of the optimum design of experiments as coming, by contrast, from technological experiments, where variability of the units is virturally ignored. The difference is that between experimenting on a field and on samples of a well-mixed liquid. In technological experiments the structure of the treatments can be complicated, perhaps reflecting a mathematical model of the system under study. Usually random error enters not from the variability of the experimental units, but from random shocks to which the system is subject.

Further references to the vast literature on the design of experiments can be found in two recent review articles. Atkinson (1982a) develops some of the ideas of this paper in greater detail. In Steinberg and Hunter (1984) the emphasis is, appropriately enough for an article celebrating the quarter centenary of *Technometrics*, primarily on technological experiments.

2. Optimum Experimental Design

The kernel of the theory (Kiefer, 1959) is concerned with designs for estimation of the k parameters in the linear model $E(Y) = x^T\beta$. The approach is thus distinct from that of classical experimental design in requiring the specification of a model. For example, an agricultural experiment might be designed to provide information about the response to four levels of a fertilizer. But any model for the dependence of yield on level would typically arise during the analysis of the data, rather than being prescribed at the design stage.

It is assumed in the theory that the errors are independent with constant variance σ^2. The unknown parameters β are estimated by least squares. The variance of the estimate of β from n trials is

$$\mathrm{var}(\hat{\beta}) = \sigma^2(X^T X)^{-1},$$

where the $k \times k$ information matrix $X^T X$ is taken to be of full rank.

The values, x, of the individual explanatory variables at which experiments can be performed are subject to constraints. Often these constraints will be simple upper and lower bounds, for example the range of pressure over which an apparatus may be safely operated. Taken together these constraints define a design region $\mathscr{X}$. An experimental design consists of selecting n sets of conditions from $\mathscr{X}$, which then form the n rows of X. The experiments would normally be run in random order. If blocking is required, extra parameters are included in the model and the order of experimentation randomized within blocks.

A good design is one from which the k components of β can be precisely estimated. Because the variance of $\hat{\beta}$ is a $k \times k$ matrix, it is usually not possible to find a design for which each component is estimated with a smaller variance than for any other design. There therefore has to be a compromise between the variances and covariances of the various parameter estimates.

For least squares estimation with normal errors the contours of the sum of squares surface are ellipsoids and the confidence region for β has the form

$$(\beta - \hat{\beta})^T X^T X(\beta - \hat{\beta}) \leq \text{constant}.$$

These contours provide one motivation for several of the criteria of optimum experimental design.

The content (that is, volume) of the confidence region is proportional to $\{\det(X^T X)\}^{-1/2}$. In D-optimum designs this volume is made as small as possible, which is equivalent to maximizing the determinant $\det(X^T X)$. Two other criteria related to these contours are A-optimality and E-optimality. In A-optimum designs the trace of $(X^T X)^{-1}$ is minimized, corresponding to minimizing the sum of the variances of the parameter estimates. In the third criterion, E-optimality, the variance of the least well estimated contrast $a^T \hat{\beta}$ is maximized, subject to the constraint $a^T a = 1$. A- and E-optimum designs depend on the scale of the explanatory variables. In this they differ from D-optimum designs. However, it is usual to scale quantitative factors to lie between -1 and $+1$, so the force of this argument in favor of D-optimality is reduced.

An apparent alternative to designs for estimation of the parameters is to consider designs for the fitted response at x, $\hat{y}(x) = \hat{\beta}^T x$, which has variance

$$\operatorname{var}\{\hat{y}(x)\} = \sigma^2 x^T (X^T X)^{-1} x.$$

Designs can be found which minimize this variance at a point or over a region which need not be the design region $\mathscr{X}$. One criterion is that of G-optimality, in which the maximum of $\operatorname{var}\{\hat{y}(x)\}$ over $\mathscr{X}$ is a minimum. Another criterion is to minimize the average of the variance over $\mathscr{X}$.

These are some of the suggested criteria for optimum designs which are used by Nalimov et al. However, the strength of the theory lies not in the development of criteria but in the consequences of the formulation of experimental design as an optimization problem. For several of the criteria this optimization problem has a relatively simple structure which provides algorithms for the construction of designs.

Both for the theory and for the construction of designs it is convenient, for the moment, to move away from the n trial design represented by the matrix X. If instead we work with a measure ξ over the design region, we have the advantages of continuous rather than discrete mathematics. For an exact n trial design—that is, one for which the weights at the experimental points are integer multiples of $1/n$—the measure is denoted by ξ_n. If the integer restriction is removed, the design is denoted by ξ and is referred to as a continuous or approximate design. Such designs, examples of which are given by Nalimov et al. in their Section 2, may not be exactly realizable in practice. One way of obtaining exact designs is to round $n\xi$, yielding integer weights for the design points.

The information matrix of the approximate design is called $M(\xi)$. It follows that for an exact design

$$M(\xi_n) = n^{-1}(X^T X).$$

Also, instead of $\operatorname{var}\{\hat{y}(x)\}$ it is convenient to consider the standardized variance

$$d(x, \xi) = x^T M^{-1}(\xi) x.$$

One consequence of the approximate theory is that relationships between the various optimality criteria can be established. For example, the celebrated general equivalence theorem of Kiefer and Wolfowitz (1960) relates G- and D-optimality through the equivalence of three requirements on the optimum measure ξ^*:

(i) ξ^* maximizes $\det M(\xi)$: ξ^* is D-optimum.
(ii) ξ^* minimizes the maximum of $d(x, \xi)$ over $\mathscr{X}$: ξ^* is G-optimum.
(iii) The maximum value of $d(x, \xi^*)$ over $\mathscr{X}$ is k, the number of linearly independent parameters in the model.

One use of this equivalence theorem is in the sequential construction of D-optimum designs which can be generated by successively adding trials at the point where $d(x, \xi)$ is a maximum. More sophisticated algorithms based on this idea are referenced in §2.3 of Atkinson (1982a). A second use is to check whether a design is D-optimum by calculating the value of the variance at the design points. It should however be stressed that the equivalence between G- and D-optimum designs is for the approximate theory. For exact designs the two can be quite different. One example is quadratic regression with four observations with experimental region $\mathscr{X}$ such that $-1 \leq x \leq 1$. The exact D-optimum design consists of one trial at any two of the x-values $-1, 0$, and 1 with two observations at the third value. The exact G-optimum design has trials at four distinct points. In general, as the number of trials increases, the difference between exact G- and D-optimum designs decreases.

Equivalence theorems can also be found for the other criteria we have mentioned so far. The relationship between these criteria is expressible in terms of the eigenvalues μ_i of $M(\xi)$. In D-optimality the product of the eigenvalues, $\Pi\mu_i$, is maximized. A-optimality corresponds to minimizing $\Sigma 1/\mu_i$, whereas E-optimality finds the design for which the smallest eigenvalue is maximized. These three criteria can be combined in a simple parametric family called $\Phi(p)$ optimality (Kiefer, 1975), obtained from power transformation of the eigenvalues μ_i. In this family D-, A-, and E-optimality correspond respectively to values of 0, 1, and ∞ for p. A use of this family is to provide a parametric framework in which to study the dependence of designs on criterion. A more promising approach, mentioned by Nalimov et al., is to use subsidiary criteria to choose between designs which are almost optimum when assessed by a main criterion. Such choices are aided by the empirical observation that designs are frequently much more sensitive to the optimality criterion than are the values of the criteria themselves. These often change slowly in the neighborhood of the optimum (Kiefer, 1975, p. 278).

3. Response Surface Designs

Over the 25 years since the theory of optimum design emerged, by far the greatest effort has gone into the calculation of optimum designs for response surface models in which the response is a low order polynomial in several continuously variable factors. Often the experimental region is a cube or a sphere and the criterion is D-optimality. For first order models with interactions the resulting designs are the 2^k factorials and their fractions. In more complicated cases, the resulting approximate designs have to be converted to exact designs before they can be used.

The rounding of an approximate design may not lead to the best exact design for n trials. Algorithms for the construction of exact designs are compared by Johnson and Nachtsheim (1983). Fedorov (1972, §3.2) suggested an exchange algorithm in which one trial at a time is replaced by a better set of experimental conditions. A second family of algorithms, based on Mitchell's DETMAX (Mitchell, 1974) with improvements by Galil and Kiefer (1980), uses a series of excursions in which several trials may be added to or subtracted from designs according to a specified search pattern. These algorithms yield good designs which cannot be guaranteed to be globally optimum. Globally optimum exact designs for a variety of criteria can be found by the branch and bound algorithm of Welch (1982), who gives examples of designs for the quadratic model in 3 factors at 3 levels for 10 to 20 trials. In accordance with the idea of choice based on subsidiary criteria, the algorithm provides a list of D-optimum and nearly D-optimum designs. These designs are then compared for G-optimality by looking at the maximum value of var$\{\hat{y}(x)\}$ at the design points, and also by looking at the average value of the variance. In this way optimum design theory is used to generate a compromise design, satisfactory on several counts.

The designs found by such methods are based on the assumption that the true model is known. Nalimov et al. mention the work of Box and Draper

(1959, 1963), who start from the proposition that the fitted model is only an approximation to some more complicated reality. In selecting a design it is therefore desirable to consider the mean squared error of prediction: in addition to variance, bias must be included in the criterion. Of course, the true model will not be known; otherwise it would be fitted. But it can be approximated. For example, if a second order model is to be fitted, the bias terms can be taken as arising from a third order model. The actual bias depends on the parameters of the third order model. But Box and Draper show that rotatable designs, which are designs satisfying certain moment conditions, provide some protection against many possible forms of bias.

4. Some Applications of Optimum Design Theory

In this last section sketches are given of three nonstandard uses of optimum design theory to provide designs not easily obtained by other methods. These three examples suggest the emergence of a synthesis in which optimum design theory is seen not in antithesis to other approaches to experimental design, but as one of a number of techniques available for the solution of practical problems.

This paper began with reference to Fisher and agricultural experiments, where there is a simple treatment structure combined with experimental material which needs division into blocks. If the blocks are not large enough for each treatment to be used in each block, an incomplete block design is employed in which only some of the treatments occur in each block. Over the whole experiment each treatment occurs the same number of times, and for suitable block sizes and numbers of treatments, balanced or partially balanced designs are known in which pairs of treatments occur together in roughly the same number of blocks. The classical procedures for the construction of these and related designs use combinatorial methods. Two recent examples are Agrawal and Prasad (1982) and Cheng, Constantine, and Hedayat (1984).

For nonstandard cases designs are often not available, or if they are available, the designs need to be accessed and constructed, perhaps on a computer. Optimum design theory is useful in directing attention towards statistically important features of a design, rather than towards those of combinatorial interest.

An example is the use by John and Mitchell (1977) of the DETMAX algorithm, mentioned in Section 3, to construct optimum block designs. One finding was that the designs had a particularly simple structure when interpreted in terms of graph theory. Two developments followed from this discovery. One was the construction of optimum designs from regular graphs (Cheng and Wu, 1981). The other (Paterson, 1983) uses the counts of circuits in the graphs of designs as a surrogate for D-optimality. Candidate designs are generated in a simple manner, for example as special kinds of lattice designs. The counting of circuits is a fast way of assessing optimality, so that many designs can be compared. The results of this construction are in use as designs for statutory variety trials.

A second application of optimum design theory is to the sequential design of clinical trials. If, in a general setting, interest is in the set of contrasts $A^T\beta$ rather than in all the parameters in the model, the criterion of D-optimality becomes that of D_A-optimality, in which $\det[A^T\{M^{-1}(\xi)\}A]^{-1}$ is maximized. The analogue of the standardized variance of Section 2 can be written $d_A(x, \xi) = x^T M^{-1} A(A^T M^{-1} A)^{-1} A^T M^{-1} x$, where $M = M(\xi)$. The sequential construction of the optimum design, analogously to that for D-optimality, consists of adding the next trial where $d_A(x, \xi)$ is a maximum. Such a criterion provides an alternative to the criterion in Section 4 of Fienberg et al., which is an almost parallel development of A-optimality.

Now consider the design of a clinical trial in which patients arrive sequentially and are to be assigned to one of several treatments. Information about each patient may include a number of prognostic factors across which the trial should be balanced. The sequential construction of the D_A-optimum design of the last paragraph could be used, the matrix A allowing for the presence of the prognostic factors. In this approach the design region $\mathscr{X}$ consists of the set of treatments. But, as Efron (1971) argued in the development of biased-coin designs, some randomness in the allocation of treatments may be desirable to avoid biases and the suspicion of conscious or unconscious cheating. Atkinson (1982b) suggests a design of the biased coin type in which the probability of allocation of the treatments is proportional to $d_A(x, \xi)$. The solution provided by optimum design theory to this nonstandard problem is thus derived from the underlying statistical model.

As a last example of the application of optimum design theory we turn again to response surfaces. A disadvantage of the method of Box and Draper is that it cannot provide designs for arbitrary design regions and arbitrary numbers of trials. Welch (1983) extends the original idea of Box and Draper to consider a general form of departure from the fitted polynomial, subject to an upper bound $z_{\max}$ on the size of the departure. The mean squared error of prediction is averaged analytically over all departures to give a criterion to which the algorithms of optimum design theory apply. Approximate designs are found by algorithms similar to those for approximate D-optimum designs. For exact designs a variant of Mitchell's DETMAX is employed.

One numerical example considered by Welch is a 9 trial design for the first order model over the points of a 3^2 factorial. When $z_{\max} = 0$ the all variance design is obtained, which is as near as possible to the D-optimum 2^2 factorial. As the value of $z_{\max}$ increases, the design changes by stages to the uniform all bias design with one trial at each of the 9 experimental conditions. Since an exact design is being generated, the design does not change continuously, but only at a few values of $z_{\max}$. The resulting sequence is not however used to yield a design for a plausible value of $z_{\max}$. Rather, the efficiency of each design is calculated relative to the all-bias and all-variance designs, leading to the choice of a compromise design reasonably efficient for both extremes.

The idea of a compromise design thus arises again here as it does in the paper of Nalimov et al. A further link between their paper and these examples is the emphasis on methods of generating, storing, and retrieving designs. Suitable methods include books, computer storage, and algorithms which calculate designs tailored to specific applications.

In conclusion, in a volume of this international nature, it is appropriate to notice the geographical spread of work on optimum design. In this paper much mention has been made of the work of Kiefer, a review of which is given by Wynn (1984). Due to Kiefer's untimely death the monograph promised on p. 850 of Kiefer (1974) is unlikely to appear. But recent books on optimal design have come from Scotland (Silvey, 1980), the German Democratic Republic (Bandemer and Näther, 1980; Pilz, 1983), Czechoslovakia (Pazman, 1980), and the Soviet Union (Ermakov, 1983). The work of Professor Nalimov and his coworkers is part of an appreciable tradition, one manifestation of which is the pioneering book of Fedorov (1972).

Bibliography

Agrawal, H. L. and Prasad, J. (1982). "Some methods of construction of balanced incomplete block designs with nested rows and columns." *Biometrika*, **69**, 481–483.

Atkinson, A. C. (1982a). "Developments in the design of experiments." *Internat. Statist. Rev.*, **50**, 161–177.

Atkinson, A. C. (1982b). "Optimum biased coin designs for sequential clinical trials with prognostic factors." *Biometrika*, **69**, 61–67.

Bandemer, H. and Näther, W. (1980). *Theorie und Anwendung der optimalen Versuchsplanung II*. Berlin: Akademie Verlag.

Box, G. E. P. (1979). "Some problems of statistics and everyday life." *J. Amer. Statist. Assoc.*, **74**, 1–4.

Box, G. E. P. and Draper, N. R. (1959). "A basis for the selection of a response surface design." *J. Amer. Statist. Assoc.*, **54**, 622–654.

Box, G. E. P. and Draper, N. R. (1963). "The choice of a second order rotatable design." *Biometrika*, **50**, 335–352.

Cheng, C. -S., Constantine, G. M., and Hedayat, A. S. (1984). "A unified method for constructing PBIB designs based on triangular and L_2 schemes." *J. Roy. Statist. Soc. Ser. B*, **46**, 31–37.

Cheng, C. -S. and Wu, C. -F. (1981). "Nearly balanced incomplete block designs." *Biometrika*, **68**, 493–500.

Efron, B. (1971). "Forcing a sequential experiment to be balanced." *Biometrika*, **58**, 403–417.

Ermakov, S. M. (ed.) (1983). *Mathematical Theory of Experimental Design*. Moscow: Nauka.

Fedorov, V. V. (1972). *Theory of Optimal Experiments*. New York: Academic.

Fienberg, S. E., Singer, S., and Tanur, J. M. (1985). "Large-scale social experimentation in the U.S.A." In this volume, Chapter 13.

Galil, Z. and Kiefer, J. (1980). "Time- and space-saving computer methods, related to Mitchell's DETMAX, for finding *D*-optimum designs." *Technometrics*, **22**, 301–313.

John, J. A. and Mitchell, T. J. (1977). "Optimal incomplete block designs." *J. Roy. Statist. Soc. Ser. B*, **39**, 39–43.

Johnson, M. E. and Nachtsheim, C. J. (1983). "Some guidelines for constructing exact *D*-optimal designs on convex design spaces." *Technometrics*, **25**, 271–277.

Kiefer, J. (1959). "Optimal experimental designs" (with discussion). *J. Roy. Statist. Soc. Ser. B*, **21**, 272–319.

Kiefer, J. (1974). "General equivalence theory for optimum designs (approximate theory)." *Ann. Statist.*, **2**, 849–879.

Kiefer, J. (1975). "Optimal design: variation in structure and performance under change of criterion." *Biometrika*, **62**, 277–288.

Kiefer, J. and Wolfowitz, J. (1960). "The equivalence of two extremum problems." *Canad. J. Math.*, **12**, 363–366.

Mitchell, T. J. (1974). "An algorithm for construction of '*D*-optimal' experimental designs." *Technometrics*, **16**, 203–210.

Nalimov, V. V., Golikova, T. I., and Granovsky, Y. V. (1985). "Experimental design in Russian practice." In this volume, Chapter 21.

Paterson, L. (1983). "Circuits and efficiency in incomplete block designs." *Biometrika*, **70**, 215–225.

Pazman, A. (1980). *Zaklady Optimalizacie Experimentu*. Bratislava: Veda.

Pearce, S. C. (1985). "Agricultural experimentation in a developing country." In this volume, Chapter 22.

Pilz, J. (1983). *Bayesian Estimation and Experimental Design in Linear Regression Models*. Leipzig: Teubner.

Silvey, S. D. (1980). *Optimal Design*. London: Chapman and Hall.

Steinberg, D. M. and Hunter, W. G. (1984). "Experimental design: review and comment" (with discussion). *Technometrics*, **26**, 71–130.

Welch, W. J. (1982). "Branch-and-bound search for experimental designs based on *D*-optimality and other criteria." *Technometrics*, **24**, 41–48.

Welch, W. J. (1983). "A mean squared error criterion for the design of experiments." *Biometrika*, **70**, 205–213.

Wynn, H. P. (1984). "Jack Kiefer's contributions to experimental design." *Ann. Statist.*, **12**, 416–423.

CHAPTER 21

Experimental Design in Russian Practice

V. V. Nalimov, T. I. Golikova, and Yu. V. Granovsky

Abstract

The modern state of experimental-design theory is considered with regard to its mathematical foundations, the multitude of optimality criteria, the need to make compromise decisions when designs are constructed, the theory of factorial designs, the design of screening experiments, designs that take into account the systematic error of the proposed model, and designs for models which are nonlinear with respect to the parameters. Applications of experimental design in the USSR are reviewed. An attempt is made to show that the evolution of ideas and methods of experimental design has made the concept of a good experiment precise enough so that one can contemplate the formulation of a mathematical theory of experiment.

1. Premises for the Emergence of a Mathematical Theory of Experiment. The Content of the Theory

More than half a century has elapsed since the publication by R. A. Fisher (1918) which laid the foundation for a new attitude towards the conduct of an experiment. Then the architecture of the experiment, i.e. the way the points at which the experiment is carried out are situated in the space of independent variables, became the object of an independent study. The history of the new field, called *experimental design*, is instructive both for experimenters, to whom the discipline is related directly, and for the mathematicians who are developing it. While at the beginning experimental design seemed nothing more than a set of cookbook recipes, it has become obvious that what is taking shape is a mathematical theory of experimentation as a self-sustained section of mathematical statistics. Fisher's insight has been supported by the more recently developed mathematical theory of optimal experimental designs. It has become possible to speak of the logical foundations of experimental design (Nalimov and Golikova, 1981). The present paper will essentially be devoted to this subject.

The new theory provides us with something more than a set of recipes. It

Key words and phrases: catalogues of designs, choice of model, compromise designs, experimental designs, factorial designs, mathematical theory, nonlinear models, optimality criteria, screening experiments.

reveals how to express formally, in the language of mathematics, what a well-designed experiment, in the statistical sense, is. It allows us to formulate distinctly new problems which emerge when attempts are made to deepen and expand the concept of a good experimental architecture. Last but not least, it enables us to delineate the area in which mathematical formalism is applicable to experimental research. Like any mathematical discipline, the mathematical theory of experimental design is constructed as a deductive system. Optimality criteria of experimental design may be considered as axioms, and designs as the deduced theorems.[1]

The major results of the mathematical theory of experimental design may be formulated as follows:

1.1. The Need for a Bipartite Model

Experimental design becomes possible only after the model of the phenomenon under study is given. The experiment must be preceded by the model, for example the formalization of the problem formulation. This requirement can in no way be regarded as a sign of weakness of the theory. Applied mathematics in its essential manifestations remains a deductive science. But its deductive nature acquires here a two-step quality. On the one hand, optimality criteria are chosen *a priori*; on the other hand, the model of the phenomenon under study is also chosen *a priori*, which enables us to relate the concepts of optimal behavior to the real problem, the object of this behavior. At an early stage of the evolution of experimental-design theory, the choice of models was extremely limited, and this could produce an erroneous impression that the consultant–mathematician ignored the second level of deduction.

The requirement of the *a priori* choice of a model corresponds to the concept of science as a system of interaction between man and nature that begins with the problem formulation. The logic of questions and answers is well elaborated today. One proceeds from the fact that a mathematical model recorded by the researcher in a general form, before the parameters are numerically evaluated, is a question that the researcher poses to nature. A mathematical model of any question has two parts: an assertion part, which may be either correct or incorrect, and which makes the question possible; and the question itself, which can be either relevant or irrelevant, but never true or false. At the first stage of the research the assertion constituent is represented by the analytical form of the model: it contains prior knowledge, that is, knowledge received before the given study, about the object of the research. The requirement to estimate parameters numerically represents the question part of the model. After the parameters are numerically estimated, they pass over to the assertion part. The question part of the model now becomes the

[1] As usual, we define a *theorem* as any proved proposition or assertion of a deductive theory.

requirement to test the model against the experimental data. The final question part of the model is the requirement to interpret it.

1.2. The Mosaic Character of the Model

A difference from the structure of pure mathematics is that the axioms of the mathematical theory of experimental design do not form an internally consistent structure.[2] We may rather speak here of a certain pattern or, even better, of a *mosaic* of the initial assertions, which may even be mutually exclusive. Therefore, pragmatically, the major task of the mathematical theory of experimental design becomes the elaboration of methods and techniques enabling us to make a *compromise decision*, forming a design architecture that will to a certain extent satisfy several optimality criteria simultaneously. Formalized concepts of the optimality criteria do not thrust any unconditional decision on us; they rather serve to sharpen the researcher's intuition when choosing the architecture for experimental design.

1.3. Simultaneous Variation of Variables

An essential function of the criteria is to link the points in the space of independent variables with the obvious (according to our insight) requirement that the variance characteristics of the model follow from the geometry of the variance ellipsoid constructed in the space of the regression-coefficient variances. One of the ensuing requirements possesses a high degree of generality: the architecture of experimental designs should be such that the experimenter includes all the independent variables simultaneously in the research procedure. All the variables should be varied in each experiment. This remarkable result of the theory of experimental design, which was understood when the latter was in its early stages, contradicts the paradigmatically fixed requirement which had been reigning in science for more than two centuries. Here is how Schrödinger, an outstanding physicist of the recent past and one of the creators of quantum mechanics, formulated this requirement in the middle of our century:

> An analogy might be sought in the working of a large manufacturing plant in a factory. For developing better methods, innovations, even if as yet unproved, must be tried out. But in order to ascertain whether the innovations improve or

[2] Recall that Bourbaki in their basic article "Architecture of Mathematics," in answering the question "what mathematics is," state that the distinctive feature of mathematical science is that its knowledge is contained in mathematical structures that imply everything that later becomes explicit in theorems. It seems possible to assert that applied mathematics, even in its most thoroughly elaborated aspects, lacks mathematical structure in the sense of Bourbaki, but consists rather of a mosaic of basic propositions. This is considered in detail in the book by one of the present authors (Nalimov, 1981).

> decrease the output, it is essential that they should be introduced one at a time while all the other parts of the mechanism are kept constant. (Schrödinger, 1944, pp. 41–42).

The recommendations of the mathematical theory of experimental design are exactly the opposite.

1.4. Systematic Errors

The second, no less striking result of the evolution of the theory of experimental design is the revelation of a mechanism that produces systematic errors in the estimation of model parameters. Here one can speak both of systematic errors proper, brought about by the inadequacy in model selection, and of the distortions in the parameter estimation caused by the nonorthogonal architecture of the experimental design. Little can be done here, from a practical point of view, to improve the situation: orthogonal design is possible only for the simplest, polynomial models, and here too, in the case of a second-order polynomial, orthogonal allocation of the points sharply decreases the efficiency as evaluated by the variance criteria. But the important thing here is that the experimenter, guided by the theory, starts to comprehend the complexity of the situation. That enables him to pattern his behavior while choosing not so much the *design* as the *model*, and (what is especially significant) it allows him to be appropriately cautions in interpreting the results of the research as a whole.

1.5. Screening Experiments

The theory has provided experimenters with the previously unknown technique of *screening experiments*, in which the number of experiments is smaller than that of the variables among which the screening is carried out, and a dominant group is selected (Satterthwaite, 1959).

1.6. Summary

Summing up, we may formulate the following conclusion: the possibility for an experiment to be a *good experiment* is primarily determined by the choice of the model. Here a new problem arises immediately: what is a good model? However, being itself the result of the mathematical comprehension of what a good experiment is, this problem cannot be analyzed within the framework of mathematical formalism. In each concrete case the model properties are determined by a number of particular circumstances: on the one hand, by the level of prior knowledge, and on the other hand, by the formulation of the problem for which it is constructed. We can make here only general remarks:

If a model is a question posed by the researcher to nature, then a good model will be such that the question part is well balanced with the assertion part. If much is known *a priori*, then one can ask much, by resorting to models which are nonlinear with respect to the parameters. These allow one to penetrate into the mechanism of the phenomenon under study. If the *a priori* knowledge is little, it is equally natural to resort to simple models directed only at finding the contribution made by individual factors. However, this recommendation is too general; besides, the consultant–mathematician is always uncertain how to estimate the level of knowledge (or even the level of ignorance) of the researcher–experimenter. We unavoidably address here the deep levels of thinking preceding any formalization.

1.7. Difficulties in Application

Now a few words on the difficulties of applying the mathematical formalism in experimental research. Naturally the difficulties increase as the model grows more complicated; the more one wants to know, the more difficult the process of recognition becomes. If we resort to a manifold of simple models aimed at the evaluation of the part of individual factors, then in general all the formalized procedures enable us to obtain reliable, unambiguous results. Matters stand differently when we use models which are nonlinear with respect to parameters, as is the case e.g. in problems of chemical kinetics. The main trouble here is that the parameter estimates, despite all the sophistications of the experimental design, as a rule turn out to be correlated, their correlation coefficients being close to unity. As a result, on the one hand, all the numerical parameter estimates are very difficult to get, and on the other hand, the model becomes very sensitive to reparametrization: the slightest change in its structure sharply affects the values of all the parameters. Models obtained by different researchers, even if their initial objective (chemical) premises are only slightly different, may prove essentially different as a whole but in equally good correspondence with experiment.

The second principal difficulty concerns the procedure of discrimination.[3] In elaborating the formal aspect of this procedure, one must proceed from a rigid assumption that the rival models contain a model which could conventionally be called the true one. If that is not so, what is the sense in the discrimination procedure? The model behaving best within the narrow range of variation of the variables that is accessible in laboratory tests can behave quite differently under far-reaching extrapolation involved in the transfer of the model to industrial conditions. One is tempted to reject the basic principle of mathematical statistics and represent the research results not by the single

[3] The procedure for the design, estimation, and discrimination of models which are nonlinear with respect to parameters is described in detail in Fedorov (1972). The accompanying difficulties are described in Nalimov and Golikova (1981).

best model but by a multitude of equally good models, assuming that they will be able to sharpen and thus enrich the intuition of a chemist–researcher and make him more cautious at the stage of launching major projects. These considerations are also supported by practice.

Thus we see that the mathematical approach to the concept of a good experiment produces a theory with far-reaching nontrivial conclusions. Far from trivial are: the requirement to start the research with the formulation of the problem; the idea of the multitude of optimality criteria; the requirement of randomization and the necessity of varying all the variables in each experiment; and the necessity, in some cases, of resorting to a multitude of models when interpreting results of one experiment. (The idea of R. A. Fisher is thereby refined: there emerges the need to reduce the data by expanding the models.) To our mind, the theory of experimental design, i.e. the characterization of a good experiment, is more important than the ability to use a particular solution ensuing from the theory.

1.8. Examples

To conclude this section we give two examples of experimental design for a model which is linear with respect to the parameters.

Example 1. In the study of response surfaces, the steepest-ascent method (Box and Wilson, 1951) is often used. First a small series of experiments is carried out which represents, as a rule, a fractional replicate of a multifactor experiment; it allows one to screen out insignificant factors and estimate the response-surface gradient. Then, a few steps along the gradient are made.

Now consider the problem where this method was used to discover the optimal composition and production conditions of electrode coatings for argon arc welding of nickel, to produce a seam with the minimum content of gas pores (Novik, 1971).

Independent variables and optimization parameter. The components of the electrode coating were varied, namely, cryolite (x_1), titanium (x_2), aluminum (x_3), and sodium fluoride (x_4). Other variables were the length of the welding arc (x_5) and the time of the coating calcination (x_6). The factor x_5 had two values: "long" ($+1$) and "short" (-1). The optimization parameter (y) was the number of gas pores per 100 mm of seam length, determined by X-rays.

Experimental design. Since the experiments were labor-consuming, a $\frac{1}{8}$-scale replicate of a complete 2^6 experimental design was used. Conditions and results of the experimental series and steepest-descent method are given in Table 1.

Results. Movement along the gradient of the linear model led to a decrease in porosity. However, the best result (experiment 10, Table 1) did not yet provide the necessary quality of the seam. Therefore, another series of

Table 1. Conditions and Results of the Initial Experimental Series and Steepest-Descent Search

	Composition				Arc length	Time of calcination	Number of pores
	Cryolite	Ti	Al	NaF			
Upper level (+)	16%	6%	7.5%	8%	Long	135 min	
Lower level (−)	12%	4%	4.5%	4%	Short	105 min	
Variable	x_1	x_2	x_3	x_4	x_5	x_6	y
Experiment 1	−	−	−	−	−	−	275
2	+	−	+	−	+	−	181
3	−	+	−	+	+	−	185
4	+	+	+	+	−	−	65
5	+	−	−	+	−	+	142
6	−	−	+	+	+	+	301
7	+	+	−	−	+	+	304
8	−	+	+	−	−	+	223
9[a]	16.4	5.4	6.8	8.4	Short	100	58
10[a]	17.6	5.6	7.2	9.6	Short	90	35
11[a]	18.8	5.8	7.6	10.8	Short	80	64

[a] These design points correspond to steps in the direction opposite to the gradient (noncoded data).

experiments was carried out, close to the conditions of experiment 10. The content of titanium (x_2) was chosen to be constant (6%), and the welding was done with a short arc. By taking into account the significant decrease of the seam porosity, the intervals of variation for the rest of the factors were reduced. In the second experimental series, a $\frac{1}{2}$-scale replicate of the 2^4 design was used (8 experiments). After four experiments for the steepest descent were made, the number of pores in the two last experiments proved to be equal to zero, i.e., the seams were absolutely dense. Thus, the problem was solved in 23 experiments. In order to make poreless seams the electrode coating should contain 23–25% cryolite, 6% Ti, 9.5–10.3% Al and 10% NaF. The coatings must be calcinated for 76–80 min and welded with a short arc.

Example 2. Consider another example of the design of experiments for evaluating a second-order polynomial model. Such designs are commonly used to investigate the response surface in the vicinity of the optimum.

The optimal composition of pig iron for molds in the glass industry was investigated. It was necessary to discover a highly heatproof composition (Novik and Arsov, 1980).

Independent variables and optimization parameter. The contents of carbon (x_1), silicon (x_2), and phosphorus (x_3) were varied. The dependent variable y was the number of glass articles molded by the punch out of pig iron of the given composition until cracks appeared on the surface of the glass article.

Experimental design. A symmetrical quasi-D-optimal design for $k = 3$ was chosen (Brodsky et al., 1982). The conditions and results are given in Table 2.

Table 2. Conditions and Results of Experimental Series, Second-Order Design

	C	Si	P	No. of articles
Upper level (+)	3.7%	2.2%	0.3%	
Lower level (−)	3.2%	1.8%	0.1%	
Variable	x_1	x_2	x_3	y
Experiment 1	0	+	+	550
2	0	−	+	650
3	0	+	−	550
4	0	−	−	1050
5	+	0	+	700
6	−	0	+	500
7	+	0	−	900
8	−	0	−	750
9	+	+	0	600
10	−	+	0	500
11	+	−	0	950
12	−	−	0	800
13	0	0	0	850

After the calculations were made, the following model was obtained:

$$y = 850 + 75x_1 - 156.3x_2 - 106.3x_3 + 100x_2x_3 - 62.5x_1^2 - 75x_2^2 - 75x_3^2.$$

The model proved adequate:

$$F_{(\text{exp})} = 1.01,$$

$$F_{(.05;8;9)} = 3.2.$$

The response surface was an ellipsoid, the maximum being its center. The center coordinates were $x_{1s} = 0.60$, $x_{2s} = -2.73$, $x_{3s} = -2.53$. On the natural scale $x_1(\text{C}) = 3.6\%$, $x_2(\text{Si}) = 1.45\%$, $x_3(\text{P}) = 0\%$. From pig iron of the given composition, a punch was made by means of which 1750 glass articles were molded. Thus, the best result achieved in a series of experiments (1050 articles) was improved by 66%.

2. Choice of Optimal Designs as a Compromise Procedure

Now let us examine the development of the theory that has provided the possibility of making compromise decisions in the construction of experimental designs.

Over the 1950s through the 1970s, J. Kiefer and his school created the theory of continuous experimental design. For models which were linear with respect to parameters, he introduced criteria for optimal design which have more generality and are more in accordance with the general, though somewhat abstract, ideas of efficiency in mathematical statistics (Kiefer, 1959, 1974, 1975). These include a large group of criteria under the general heading of "$\Phi_{(p)}$-efficiency," among which, corresponding to various values of the parameter p, are such criteria as D-, A-, and E-efficiency, involving various requirements on the form and size of the ellipsoid of variance of the parameter estimates. For various types of models, analytical and numerical procedures have been elaborated for constructing optimal continuous designs (probabilistic measures on the given region of the design in R^n). Essential contributions to the theory have been made by V. V. Fedorov (1972). In particular, he elaborated the algorithm providing the numerical method for constructing continuous Φ-optimal designs where the problem of minimizing functionals of covariance matrices of the parameter estimates is reduced to a sequence of searches for the extremum in a small-dimensional space.

However, Kiefer's theory was countered by the approach of Box and his colleagues to the study of response surface designs. Their approach had by that time been made practical and distinct. Proceeding from only two optimality criteria, orthogonality and rotatability, Box proposed designs with a small number of observations divided into orthogonal blocks thus allowing one to make the model more complicated and more convenient to implement without a computer. The high practical efficiency of such designs was demonstrated by concrete examples.

In his article, Box (1978) compiled a long list of logically grounded optimality criteria which united the criteria of Kiefer, those related to minimizing errors due to incorrect choice of the model, and many desirable properties of designs ensuring their convenience in application. Practically, however, the majority of these criteria are in poor concord with one another. As a rule, continuous Φ-optimal designs, even for standard symmetrical regions of design, require, for their practical application, rounding off and consequent loss of efficiency. Furthermore, approximating designs do not usually possess symmetry or certain other convenient properties. For sufficiently complicated models (in particular, for second-order polynomial models), the majority of the criteria are mutually incompatible, i.e., in principle the designs cannot simultaneously satisfy them all.

In the latest papers by J. Kiefer and his school, the question is posed of the choice of designs which are robust with respect to the criteria of a definite class, and the efficiency of these designs with respect to satisfying another criterion is tested. However, up to now no strict mathematical formulation has been provided for the problem of constructing designs which are robust with respect to change in the optimality criteria.

A group of researchers at the Laboratory of Statistical Methods (Moscow

State University) published in 1970, in the journal *Technometrics*, a paper (Nalimov et al., 1970) where the efficiencies of designs often used in application are compared numerically with respect to a number of criteria. A few new designs were proposed as well. Characteristics of the optimal continuous designs constructed up to that time were chosen as a standard. It was shown that it is almost always possible, with a small number of observations, for standard models and design regions to find, among symmetric designs, those with a sufficiently high efficiency e.g. according to the criterion of D-optimality. It became clear that compiling catalogues of designs and their characteristics would enable one to select designs close to optimal according to many criteria and characteristics simultaneously. Soon after this paper was published, there appeared the first catalogue of designs for second-order polynomial model (Golikova et al., 1975), followed by a catalogue of the third-order designs (Merzhanova and Nikitina, 1979).

As a result, the entire approach to choosing the optimality criterion for designs changed. It became clear there was no necessity to choose one criterion and produce endless arguments to explain this choice. There emerged the possibility of choosing *compromise* designs that are sufficiently good (though not optimal) from the viewpoint of various criteria. (We act in a similar manner when making decisions in our everyday life: we choose what satisfies the entire variety of requirements, though our choice is not optimal in any unique sense). Such an approach makes the logic of constructing the theory clearer, and leads naturally to its evolution with respect to mathematically expressible criteria. However, other criteria that do not lend themselves to a generalized interpretation have to be regarded as "auxiliary" ones, though they sometimes play an important role in the problem consideration from the viewpoint of the researcher.

Table 3 gives the efficiencies of some second-order designs for the number of variables $n = 5$, and the number of model parameters $k = 21$.

The first four lines of the table represent the efficiencies of continuous optimal designs according to various statistical criteria. Recall that D-optimal designs have the least determinant of the covariance matrix of parameter estimations, A-optimal ones have the least mean variance of estimates; E-optimal ones have the least maximal eigenvalue of the covariance matrix of estimates, and Q-optimal ones have the least mean variance of the model estimate over the design region. To be used in practice, those designs need rounding off, resulting in the unavoidable loss of precision and, as a rule, symmetry.

The last line of the table corresponds to a compositional symmetric three-level design often used in applied research. It contains a number of observations slightly exceeding the number of parameters estimated ($N = 26$). The choice of this last design may be considered as a compromise, being highly efficient according to the principal statistical criteria (the efficiency is calculated per observation). This design has a number of other convenient properties as well.

Table 3. Efficiencies of Some Second-Order Designs Under Different Criteria

Design	Efficiency			
	$e^{(D)}$	$e^{(A)}$	$e^{(E)}$	$e^{(Q)}$
Continuous, *D*-optimal	1.000	0.853	0.437	0.772
Continuous, *A*-optimal	0.933	1.000	0.752	0.993
Continuous, *E*-optimal	0.718	0.820	1.000	0.853
Continuous, *Q*-optimal	0.921	0.988	0.865	1.000
Compositional symmetric three-level ($N = 26$)	0.927	0.857	0.620	0.927

In 1982 the publishing house Metallurgiya issued the book *Tables of Experimental Designs for Factorial and Polynomial Models* (Brodsky et al., 1982), which was the result of cooperation of the staffs of Moscow State University and the Institute of Automation for Nonferrous Metals. The book represents a catalogue of designs, sufficiently complete in scope and content, for the most frequently applied models. It embraces designs for factorial and polynomial models constructed in accordance with certain criteria, as well as some compiled from the literature.

The part of this book related to factorial designs can be described as constructive. The user is not provided directly with an experimental design, but is enabled to construct one satisfying various requirements on the basis of the system of tables and proceeding from precise instructions. In particular, in this chapter consideration is reduced to:

(1) models of the main effects for multilevel factors, and those with two-factor interaction for two-level factors;
(2) a maximum of 64 experiments;
(3) a maximum of 40 factors;
(4) a maximum of 8 levels.

Methods of evaluating the efficiency of the designs obtained are given according to various optimality criteria. The constructive presentation of designs enables the authors to compress in a comparatively small volume the information allowing the researchers to construct factorial designs for an enormous number of practical situations.

Such a way of presenting designs is an intermediate one between the usual cataloguing and computer construction. It has the advantages of both methods, namely, the visual presentation of the first and the immense capacity of the second.[4]

The theoretical basis for the catalogue of designs for factorial models developed in papers by R. C. Bose, D. J. Finney, S. Addelman, and B. Margolin, was explained in detail in the book by V. Z. Brodsky (1976). The classical theory of factorial design had for a long time been separated from the more modern theory of optimal experimental design developed for continous designs. In the book by Brodsky (1976), an attempt was made to regard the theory of factorial design from a single approach including the problems of optimality. It gives strict definitions of factorial designs and models which make it possible to unite several previously developed approaches within the framework of one theory. A broad range of problems concerning the construction of efficient designs for factorial models is considered. Efficiency is regarded not only from the viewpoint of the criteria traditional in the classical theory (taking into account various symmetry properties), but also from the viewpoint of modern statistical criteria.

The stage which follows the publication of tables is the compilation of a computer catalogue of factorial designs. This catalogue includes a branching system of algorithms intended for a computer. The following circumstance should be noted here. A common way of constructing a computer catalogue consists either in describing some general numerical procedure, or in creating a voluminous archive of matrices of optimal designs, or (what is practically the same) in listing analytical procedures for obtaining elements of this archive. However, it has gradually become clear that all these ways, taken separately, are unacceptable. General numerical methods are, as a rule, rather unstable when the problems are many-dimensional and lead to excessive computational time. Analytical methods, despite their great variety, do not provide solutions for a sufficiently broad class of problems.

Approaches combining numerical and analytical methods have proved to be fruitful. Such an approach is realized, for example, in the subsystem FACTORIAL DESIGN of the system EXPERIMENTER-COMPUTER (Brodsky et al., 1978).

Tables have also been published in a catalogue (Brodsky et al., 1982) of incomplete block designs: factorial designs of main effects for the two factors, one of which is the main one and the other the auxiliary one. Three types of incomplete block designs with equal block dimensions are considered: balanced incomplete block designs (BIB), partially balanced incomplete block designs with two associative classes [PBIB(2)], and cyclic designs (c). In ad-

[4] However, the following question may easily be asked: why should we publish cumbersome catalogues of designs when paper is getting scarcer while computers become more easily accessible? The answer is: it is often necessary to leaf through a catalogue and embrace the entire variety of designs and their characteristics in one glance. We are still dominated by the tradition of book-science. Even the famous *Science Citation Index* edited by E. Garfield, a cumbersome periodical whose preparation is based entirely on computer techniques, continues to be published in two versions, on paper and on tape. We statisticians also like to use books of tables.

dition, a type of incomplete block design with unequal block sizes is considered: balanced designs with unequal block sizes (BUB).

Further, the book provides the fullest possible catalogue of polynomial models of second and third order. It includes the major part of the work of Golikova et al. (1975) and Merzhanova and Nikitina (1979). First-order polynomial models are a particular case of factorial models; therefore, the corresponding designs may be found in Chapter 1 of Brodsky et al. (1982).

Second- and third-order designs are constructed on a multidimensional cube and a multidimensional ball. For the second-order model, 257 designs with up to seven independent variables are considered; for the third-order model, 141 designs with up to four independent variables are considered. All these designs are compiled from various sources and referred to in the tables of cumulative characteristics of designs.

Cumulative characteristics also contain information on various statistical properties of the variance ellipsoid of the parameter estimates obtained through the corresponding designs, on the average and maximum dispersion of the model evaluated in the given region, and on specific features of the structure of the estimates of the covariance matrix. The cumulative characteristics include various peculiarities which are usually taken into consideration when a design is selected: symmetry, compositionality, orthogonality, etc.

Tables of the cumulative characteristics represent the chief part of the catalogue of polynomial designs. They enable the user to select a compromise design taking into account principal and secondary characteristics. At the end of Chapter 3 and 4 of Brodsky et al. (1982), lists of designs are given satisfying the limitations all the principal characteristics.

Besides the tables of designs proper, the catalogue also includes tables of the corresponding covariance–correlation matrices and of the coefficients of the linear combinations of observational results that yield the parameters. The latter allow us to process observations using a small amount of calculation.

The last part of the book contains the catalogue of designs for various polynomial models for the case when the design is constructed on a multidimensional simplex. In practice such designs are used to study multicomponent mixtures. Polynomial models of two types are considered: complete (canonical) and incomplete polynomials of degree $d \leq 4$ for $n \leq 9$ independent variables. In all, 62 designs are considered, and almost all possess the properties of symmetry and optimality with respect to some criteria.

As in the previous chapters, tables of cumulative characteristics of the designs are given, as well as tables of the designs themselves. The form of presentation of covariance–correlation matrices and coefficients is slightly different from the case of second- and third-order designs. Since all the designs under consideration are symmetrical, the tabular presentation is replaced by a more general formula presentation.

All the catalogues are provided with illustrations and general recommendations concerning the choice of designs.

The cataloguing of designs makes it possible to place each new design constructed in accordance with any criterion in a series of other designs; this enables one to see its advantages with respect to the criterion chosen, reveals its other advantages and drawbacks, and determines its practical value.

The publishing house Nauka has published a monograph, *Mathematical Theory of Experimental Design* (ed. Yermakov, 1983), containing the corresponding theoretical considerations.

3. Theoretical Solution of Certain Nonstandard Problems of Experimental Design

Experimental design is so far not so thoroughly developed as to embrace in a single theoretical framework all the practically possible problem formulations. In this section, two nonstandard problem formulations are considered. Interesting new theoretical solutions have been found for them and potential practical realizations have been demonstrated, but they still could not be included in the catalogue.

3.1. The Problem of Unbiased Designs

One of the important problem formulations of experimental design relates to the choice of the optimal design in cases when a model might prove inadequate. A problem of this type was formulated for the first time by Box and Draper (1959). However, its strict formulation is due to S. M. Yermakov (1970). The problem consists in the simultaneous choice of a procedure of experimental design and a procedure of analysis such that the estimate of the selected approximating function will approximate the response surface optimally in the sense of the metric of the given linear normalization of the functional space F. We start with a subspace $R \subset F$ of functions within which it is possible to find a sufficiently good approximation for the regression function.

The choice of the subspace R is often determined by the limitations of the time and cost of the experiment. It can usually be accomplished by limiting the number of experimental points. Special cases of this problem occur when one has some prior information concerning the regression function. The classical theory of experimental design assumes this information to be sufficiently great that systematic error is absent, i.e., $F = R$. If, on the contrary, prior information is very meager, one has to solve the nonparametric problem of experimental design and analysis; in this case it is recommended to use randomized procedures. The intermediate case (to which, e.g., the Box–Draper problem formulation belongs) is the one when F is a finite-dimensional space and the dimension of F is greater than the dimension of R. Since the number of experiments is limited, it is necessary to select simpler models than would suffice to obtain the optimal approximation; hence there is systematic error. The major part of the papers in the collection of papers (*Netraditsionnye Metody*, 1980) is devoted to elaborating various aspects of this approach. The criterion is formulated for "unbiased" experimental design and analysis in various metrics for the space F. In particular, "locally unbiased" designs are given for the case when the unbiased procedures depend on the unknown parameters of the model, and methods of constructing such designs are proposed. Numerical techniques for obtaining Φ-optimal observation weights in the spectra of locally unbiased designs are also proposed. (The Φ-optimality criterion relates to some convex functionals of the parameter-estimate covariance matrix.)

Two branches of mathematics meet in such a problem formulation: numerical methods of integration (the spectra of many unbiased designs coincide with the nodes of various cubature formulas) and the statistical theory of continuous optimal design.

Since even under the assumption of the absence of systematic error, the construction of experimental designs is connected with a number of optimality criteria, Yermakov and Makhmudov (1982) suggest that in the general case one introduce vector optimality criteria and then use, for the choice of designs, certain methods of solution of multicriterion problems that have lately acquired popularity.

3.2. Problems of Screening Experiments

One of the most significant problems of experimental design is that of selecting significant factors. The problem is formulated as follows: on the basis of an experiment with a small number of trials, one has to choose a subset of independent variables out of the large number of potentially possible ones that are of interest. What we have in mind is, on the one hand, reduction of the number of independent variables and, on the other hand, the estimation of parameters of the regression model constructed on the selected space. As long ago as 1956, Satterthwaite (1959) proposed a solution to this problem by a peculiar method which was later called the *random balance method*. The main peculiarity of this method is the randomization of the procedure for selecting the factor levels in each experiment. However, American statisticians viewed that approach negatively, since the method lacked an accurate mathematical foundation and moreover was in obvious conflict with existing ideas concerning the efficiency of procedures.

In the USSR this method was treated in the monograph by Nalimov and Chernova (1965). Later there appeared publications containing the description of actual problems solved by this method. At least in some situations, it helped to solve problems which could not be solved otherwise.

In 1970, L. D. Meshalkin (1970) made an attempt to give a mathematical foundation for the random balance method in a simplified version. Having assumed the absence of random error and the ability to scan all possibilities completely, he discovered a correlation between the number of observations

(N), the number of essential effects (l), the number of suspected effects (S), and the probability of being able to construct by means of Satterthwaite's method designs which would select all the essential effects. In particular, e.g., for $l \leq 12$, $N = 23$, $S = 75$, such a design will be constructed with probability greater than 0.96.

Another formulation of a screening problem, that of M. B. Malyutov (1975), is also of interest. It deals with the construction of designs by searching for disorders. If, in a system consisting of a large number S of elements, the number of disorders reaches a certain critical level $l \ll S$, then it is necessary to find and replace disorders. The methods of selecting l disorders out of the large number S of elements may be considered as a problem in the design of screening experiments. Such a formulation is, as a matter of fact, a simplified discrete version of a screening problem. It is of interest that many results on the solution of this problem were obtained by means of the concepts of information theory. Essential discretization, precise problem formulation, and application of information theory enable us to go rather far in solving this screening problem. Random strategies for making observations have been proposed, as well as various techniques for the analysis of the experiment. Estimates have been given of the number of observations and the complexity of the process of applying these techniques.

In sequential screening, the whole experimental design is not obtained immediately, but in stages, and it may change depending on the results of observations already made. The number of experiments is not fixed beforehand. In the paper by Myatlev (1977), various optimality criteria are formulated for the procedures of sequential screening, and a review is made of theoretical results for solving the problem under various additional premises.

4. Experience in Practical Applications

We shall now try to sketch the current situation and indicate how the mathematical theory of experimental design, having achieved a certain degree of maturity, is being applied in practice.

First, a long series of publications on applications of experimental design procedures in the USSR appeared in 1963 (see Nalimov and Chernova, 1965). The set of problems and procedures used was small: the design of response-surface experiments, factorial designs of the 2^k type, steepest-ascent method, and designs in the almost stationary region for use in constructing second-degree polynomial models. All these were used for the solution of problems in chemistry and metallurgy. Mathematical models were built for rather complicated extraction processes applied to separating rare metals and obtaining optimal conditions for their separation. Such models also increase the speed publications in this field became an order of magnitude greater (200 papers) (Adler and Granovsky, 1967). Side by side with factorial design and Box–Wilson methods, use was made of random balance, the simplex procedure for searching for the optimum region, and simplex-lattice designs for research on mixtures. While the variety of applied experimental design procedures did not increase significantly, the range of their application broadened. Besides chemistry and metallurgy, they were applied to biology, the technology of building materials, enrichment of minerals, automation, electronics, and mechanical engineering. The efficiency of Box's method can be evaluated by the extent of the dependent-variable increment; for a sample of 100 problems, the increment after one series of experiments proved to be significant in half the cases.

Further progress in applied research has been related to the optimality criteria. The latter are most broadly used in choosing second-order designs for describing an almost stationary region. Previously the prevailing designs were rotatable (in 70% of problems); in the last 5–7 years compromise designs have been used more and more often (in 25%).[5] As a rule, these designs are close to D-optimal ones in the value of the information-matrix determinant and are not usually optimal with respect to other criteria. Examples of such designs are those of type B_i for three and four factors and Hartley's design for five factors (Brodsky et al., 1982). Optimality criteria are also used to choose first-order designs in response-surface experiments and designs for describing mixtures, as well as in problems of sequential design for estimating constants of models which are nonlinear with respect to parameters. Information on the optimality criteria for experimental designs has penetrated into manuals and even the popular literature (Adler et al., 1976; Shakhnazarova and Kafarov, 1978; Golikova et al., 1981; Adler et al., 1982).

Side by side with the development of the concept of a good experiment, there have emerged new uses in applied research: thermo- and electroenergetics, light industry, food processing, instrument engineering, testing and maintenance of agricultural machinery, wood technology, etc. (*Planirovanie*, 1972; Granovsky et al., 1978). In the Vth All-Union Conference on Experimental Design and Automation (1976) about 1000 persons from 400 institutions participated. The questionnaires filled in by the participants demonstrated that the most widely used designs are those for response-surface experiments (40%), those for constructing models which are nonlinear with respect to parameters (12%), factorial designs with the subsequent use of analysis-of-variance methods (10%), and screening experiments (9%). Of the problems, 50% included from 1 to 5 factors and 40% from 6 to 10 factors (Adler and Preobrazhenskaya, 1977; Markova et al., 1980).

The span of applied research may be estimated by the number of publications as well. By the beginning of the 1980s a few thousand journal articles and more than one hundred books had been published describing the solution of real problems in which experimental-design techniques were applied. The books are mainly monographs on chemistry and chemical engineering (34%), on metallurgy and metal science (13%), on mining and enrichment of minerals

[5] These data are obtained from a sample of 150 problems.

(8%), and on industrial experimentation, reliability, and quality control (8%), (Adler and Granovsky, 1977).

There are two peculiarities typical of applied problems. The first consists of the sequential application of several experimental-design procedures for the study of one object. Here is one such pattern in solving response-surface problems: rank correlation methods for the expert estimation of the factor effect; screening experiment; search for the optimum region; model construction in the almost stationary region; interpretation. The second peculiarity arises in the study of methodological problems emerging when experimental design methods are applied. These are: nonformalized solutions in the steepest-ascent method, selection of intervals for variable factors, choice of step size for moving along the gradient, interpretation of polynomial models, elimination of model inadequacy in describing the almost stationary region, etc. (Komissarova and Granovsky, 1980). Methodological problems arise not only from the peculiarities of the experimental-design methods applied, but also from the specific features of the research objects, e.g. in describing mixture diagrams with several specific points, or in the statistical estimation of constants in problems of chemical kinetics.

Now let us consider the experience of applying experimental design in two fields of chemical–technological research.

These methods have been used for more than 20 years in chemical–pharmaceutical research for optimizing the production of synthetic medicines. Only linear models for optimization have been used; the mechanism of the reactions has not been considered. According to the publications, about 10% of all research has been carried out with the application of the above mathematical methods (Veksler, 1974). The most popular are complete and fractional factorial designs of the 2^k type, second-order designs, nonsymmetrical uniform and nonuniform factorial designs, and the search for extrema by means of the simplex procedure. As early as the beginning of the 1970s experimental design began taking into account multiple optimality criteria. The number of problems involving many factors (up to twenty or even more) increased; see e.g. Veksler et al., (1974, 1975). According to M. A. Veksler (1980) the application of experimental design merely to optimize the processes under way made the output 15% greater on the average.

The second example concerns a more complicated problem: constructing a model of large-capacity catalytic processes (e.g. the production of ammonia), which provides the possibility of both designing reactors and controlling them. The first stage of the work is the selection of a catalyst. Here models which are linear with respect to parameters are used, mainly based on the random balance method. In selecting a catalyst, 15–20 components are varied. Thus, the procedure for finding new catalysts can be intensified.[6] The

[6] To emphasize the potential importance of such improvements, we note that the industrial catalyst for ammonia synthesis was eventually introduced by Haber after studying several thousand catalysts.

second stage is research on the absorption kinetics of reagents on the catalyst surface, establishing the mechanism of chemical transformations, and constructing a kinetic model of the chemical reaction (Kafarov and Pisarenko, 1980) which takes the hydrodynamics into consideration.[7] It is necessary here to deal with differential equations (whose right-hand side is nonlinear) by numerical integration, and with the parametric nonlinearity of solutions by linearization (without which the procedure of design and discrimination is impossible). Note that up to now the overwhelming majority of catalysts and mechanisms of complicated chemical reactions has been studied only in a general way. A large volume of information on physical and chemical properties of catalysts, thermodynamic characteristics of outcome reactions, and physical–chemical properties of individual reagents and their mixtures still remains insufficiently ordered theoretically. The number of intermediate reactions whose mechanisms are perhaps of interest can reach several thousands or tens of thousands. The formalization of all these data and their representation in a compact and meaningful model presents a difficult task. The modeler—commonly a researcher originally trained in chemistry—has to interact with various other professions on the basis of his own experience and intuition. This process of acquaintance and accommodation may require up to 10 years or even more. Fuzzy initial chemical concepts are overlaid by ambiguity in the technique of mathematical analysis. Also, various researchers working on the same processes may come up with quite different models. This does not seem to tell crucially on the ultimate result, since when the equipment is mounted and started, it is usually necessary to correct the project, again on the intuitive level.

Thus both the simulation and the choice of an optimal experimental design prove to be deeply immersed in heuristics. It is the ability to put up with this fact that leads to success. This is what we have actually learned to do in trying to interact with the real world via mathematics.

Bibliography

Adler, Yu. P. and Granovsky, Yu. V. (1967). "*Obzor prikladnykh rabot po planirovaniyu eksperimenta*" ("Review of applied work on experimental design"). Preprint 1, Moscow State U.P., 96 pp.

Adler, Yu. P. Markova, Ye. V., and Granovsky, Yu. V. (1976). *Planirovanie eksperimenta pri poiske optimalnykh uslovií* (*Experimental Design in Searching for Optimal Conditions*). Moscow: Nauka, 280 pp.

Adler, Yu. P. and Granovsky, Yu. V. (1977). "Methodology and practice of experimental design for a decade (a review)" (in Russian). *Industrial Laboratory*, **43**, No. 10, 1253–1259.

[7] The scale ratio for passage from the laboratory to the semiindustrial to the industrial plant is $1:10^3:10^6$.

Adler, Yu. P. and Preobrazhenskaya, G. B. (1977). "The 5th All-Union Conference on Experimental Design and Automation in Scientific Research.." (In Russian) *Industrial Laboratory*, **43**, No. 8, 1031–1032.

Adler, Yu. P. Granovsky, Yu. V. and Markova, Ye. V. (1982). *Teoriya Eksperimenta: Proshloye, Nastoyashee, Budushee* (*Theory of Experiment: Its Past, Present, and Future*). Moscow: Znaniye, 64 pp. (Novoe v zhizni, nauke, tekhnike. Ser. Matematika, Kibernetika, No. 2.)

Box, G. E. P. (1978). "Experimental design: Response surface methodology." In W. H. Kruskal and J. M. Tanur (eds.), International Encyclopedia of Statistics, **1**. New York: Free Press, 294–299.

Box, G. E. P. and Draper, N. R. (1959). "A basis for the selection of a response surface design." *J. Amer. Statist. Assoc.*, **54**, 622–654.

Box, G. E. P. and Wilson, K. B. (1951). "On the experimental attainment of optimum conditions." *J. Roy. Statist. Soc. Ser. B*, **13**, 1.

Brodsky, V. Z. (1976). *Vvedenie v Faktornoye Planirovanie Eksperimenta* (*Introduction to Factorial Experimental Design*). Moscow: Nauka, 225 pp.

Brodsky, V. Z., Brodsky, L. I., Maloletkin, G. N., et al. (1978). "On the catalogue of factorial designs for computers" (in Russian). In *Matematiko-Statisticheskie Metody Analiza i Planirovaniya Eksperimenta*. Moscow: VINITI, 6–24.

Brodsky, V. Z., Brodsky, L. I., Golikova, T. I., and Nikitina, Ye. P. (1982). *Tablitsy Planov Eksperimenta* (*dlya Faktornykh i Polinomialnykh Modeleĭ*) (*Tables of Experimental Designs for Factorial and Polynomial Models*). Moscow: Metallurgiya, 751 pp.

Fedorov, V. V. (1972). *Theory of Optimal Experiments*. New York: Academic, 292 pp.

Fisher, R. A. (1918). "The correlation between relatives on the supposition of Mendelian inheritance." *Trans. Roy. Soc. Edinburgh*, **52**, 399–433.

Golikova, T. I., Panchenko, L. A., and Fridman, M. Z. (1975). *Katalog planov vtorogo poryadka* (*Catalogue of Second-Order Designs*), Moscow U.P., **I**, 387 pp., with illustrations; **II**, 384 pp., with illustrations.

Golikova, T. I., Nikitina, Ye. P., and Teryokhin, A. T. (1981). *Matematicheskaya Statistika* (*Mathematical Statistics*), Moscow State U.P., 185 pp.

Granovsky, Yu. V., Murashova, T. I., Lubimova, T. N., and Adler, Yu. P. (1978). *Planirovanie Eksperimenta. Bibliografia Prikladnykh Rabot za 1971–1975 gg.* (*Experimental Design. Bibliography of Applied Works for 1971–1975*). Moscow: Biological and Chemical Departments, Moscow Univ. 250 pp.

Kafarov, V. V., Pisarenko, V. N., Mortikov, Ye. S., Solokhin, A. V., Kononov, N. F., Merzhanova, R. F., Papko, T. S., and Minachyov, Kh. M. (1974). "Application of random balance method to study the effect of conditions of preparation of zeolite catalysts on their strength characteristics" (in Russian). *Izv. AN SSSR Ser. Khim.*, **7**, 1627–1630.

Kafarov, V. V., and Pisarenko, V. N. (1977). "System analysis in industrial catalysis." (In Russian), Zh. Vsesoyuznogo Khim. Obshestva im. D. I. Mendeleeva, **22**, 550–555.

Kafarov, V. V., and Pisarenko, V. N. (1980). "Current state of the problem of identification of kinetic models" (in Russian). *Uspekhi Khimii*, **49**, 193–222.

Kiefer, J. (1959). "Optimum experimental designs." *J. Roy. Statist. Soc. Ser. B*, **21**, (In Russian) 272–319.

Kiefer, J. (1974). "General equivalence theory for optimum designs (approximate theory)." *Ann Statist.*, **2**, No. 5, 849–879.

Kiefer, J. (1975). "Optimal design: Variation in structure and performance under change of criterion." *Biometrika*, **62**, (In Russian) 277–288.

Komissarova, L. N., and Granovsky, Yu. V. (1980). "Experimental design in obtaining and studying inorganic substances." *Zh. Vsesoyuznogo Khim. Obshestva im. D. I. Mendeleeva*, **25**, No. 1, 62–71.

Malyutov, M. B. (1975). "On randomized design in a model of screening experiments" (in Russian). In *Planirovaniye Optimalnykh Eksperimentov*, Moscow U.P., 181–185.

Markova, Ye. V., Adler, Yu. P., and Granovsky, Yu. V. (1980). "Experimental design in chemistry" (in Russian). *Zh. Vsesoyuznogo Khim. Obshestva im. D. I. Mendeleeva*, **25**, No. 1, 4–12.

Merzhanova, R. F., and Nikitina, Ye. P. (1979). *Katalog Planov Tretyego Poryadka* (*Catalogue of Third-Order Designs*). Moscow U.P., 169 pp.

Meshalkin, L. D. (1970). "On the foundations of the random balance method." (In Russian) *Industrial Laboratory*, No. 3, 316–318.

Myatlev, V. D. (1977). "Theorems and algorithms about one pattern of a sequential search for defective elements" (in Russian). In *Teoreticheskie Problemy Planirovaniya Eksperimenta*. Moscow: Sovetskoe Radio, 70–109.

Nalimov, V. V., and Chernova, N. A. (1965). *Statisticheskie Metody Planirovaniya Ekstremalnykh Eksperimentov*. Moscow: Nauka. English translation: *Statistical Methods for Design of Extremal Experiments*, Microfilm Foreign Technology Div., Wright-Patterson AFB, OH. AD 673747, FTD.MT 23-660-67, 419B 9 Jan. 68 USA.

Nalimov, V. V., Golikova, T. I., and Mikeshina, N. G. (1970). "On the practical use of the concept of *D*-optimality." *Technometrics*, **12**, 799–812.

Nalimov, V. V., and Golikova, T. I. (1981). *Logicheskie Osnovaniya Planirovaniya Eksperimenta* (*Logical Foundations of Experimental Design*) (2nd expanded edition) Moscow: Metallurgiya, 151 pp.

Nalimov, V. V. (1981). *Faces of Science*. Philadelphia: ISI Press, 151 pp.

Netraditsionnye Metody Planirovaniya Eksperimenta (*Nontraditional Techniques of Experimental Design*) (1980). Moscow: VINITI.

Novik, F. S. (1971). *Matematicheskie Metody Planirovaniya Eksperimentov v Metallovedenii* (*Mathematical Methods of Experimental Design in Metal Science*). Moscow: Moscow Institute of Steel and Alloys, Section I. General notions of experimental design. First-order designs, 84–90.

Novik, F. S., and Arsov, Ya. B. (1980). *Optimizatsiya Protsessov Tekhnologii Metallov Metodami Planirovaniya Eksperimentov* (*Optimization of Technological Processes by Experimental-Design Methods*). Moscow: Mashinostroyenie; Sofia: Tekhnika, 232–265.

Planirovanie Eksperimenta. Ukazatel Literatury na Russkom i Ukrainskom Yazykakh za 1970–1971 gg. (1972). (*Experimental Design. Index of Literature in Russian and Ukranian for 1970–1971* [partially for earlier years] Moscow: Lenin State Library Press, 224 pp.

Rosental, I. L. (1980). "Physical regularities and numerical values of the fundamental constants." (In Russian) *Uspekhi Fizicheskikh Nauk*, **131**, No. 2, 239–256.

Satterthwaite, F. E. (1959). "Random balance experimentation." *Technometrics*, **1**, 111–137.

Schrödinger, E. (1944). *What is Life? The Physical Aspect of Living Cell*. Cambridge U.P., 91 p.

Shakhnazarova, S. A., and Kafarov, V. V. (1978). *Optimizatsiya Eksperimenta v Khimii i Khimicheskoi Tekhnologii* (*Optimization of Experiments in Chemistry and Chemical Technology*). Moscow: Vysshaya Shkola, 319 pp.

Veksler, M. A. (1974). *Primenenie Matematicheskikh Metodov Planirovaniya Eksperimenta pri Razrabotke Tekhnologicheskikh Protsessov Polucheniya i Analiza Lekarstvennykh Veshestv* (*Application of Mathematical Methods of Experimental Design in Developing Technological Processes for Obtaining and Analyzing Medicines*), Ministry of Medical Industry Vyp. 3. Moscow: 34 pp.

Veksler, M. A., Chernysh, G. P., Oger, Ye. A., Koreshkova, Ye. G., and Samokhvalov, G. J. (1975). "Application of multifactorial experimental design in selecting a catalytical system and conditions of hydriding acetylene glycole-C_{20}" (in Russian). *Khimiko-farmatsevticheskii Zh.*, No. 3, 92–96.

Veksler, M. A., Petrova, Yu. I., Geling, N. G., and Klabunovski, Ye. I. (1974). "The study of stereospecific hydriding of D-fructose on a nickel skeleton, applying the mathematical method of experimental design" (in Russian). *Izv. AN SSSR Ser. Khim.* No. 1, 53–57.

Veksler, M. A. (1980). "New philosophy of chemical–pharmaceutical research" (in Russian). *Zh. Vsesoyuznogo Obshestva im.D.I. Mendeleeva*, **25**, No. 1, 54–61.

Yermakov, S. M. (1970). "On optimal unbiased designs for regression experiments" (in Russian). In *Trudy Matematicheskogo Instituta AN SSSR*, vyp. III, 252–257.

Yermakov S. M. (ed.) (1983). *Matematicheskaya leoriya planirovaniya eksperimenta* (Mathematical Theory of Experimental Design). Moscow: Nauka, 392 pp.

Yermakov, S. M., and Makhmudov, A. A. (1982). "On multicriterion problems of regression experimental designs" (in Russian). *Dokl. AN Uz. SSR*, No. 7, 4–6.

CHAPTER 22

Agricultural Experimentation in a Developing Country

S. C. Pearce

Abstract

Agricultural experimentation in a developing country encounters a number of problems not found elsewhere. For example, skilled scientists may be few and mostly they have been trained elsewhere, which means that they are not always well equipped to deal with the special problems of their own country. There is pressure to conduct very simple experiments. Even greater difficulties arise from the specification of objectives, especially as many of the farming systems that need to be improved have a cultural or religious basis that needs to be respected.

Four topics are discussed that have special relevance to the conditions of a developing country: (1) local control of environmental variation, (2) intercropping, (3) studies of reliability, and (4) systematic designs. A further section examines the transfer of technology.

1. Introduction

The study of agricultural experimentation in developing countries has many statistical aspects. First of all, there is the question of motivation and funding. What is it intended to achieve? Who is going to pay for it, and who is expected to benefit—the people who supply the money, the people of the country in general, or the farming community in particular? If the intention is to benefit the country at large, that brings up further questions, whether, for example, the aim is to reduce dependence on imports, to earn foreign currency, or to raise nutritional levels. If, on the other hand, the farmers are to be helped, it is necessary to ask in what way—by raising their living standards, by reducing their indebtedness to moneylenders, or in some other way? All these enquiries involve a large statistical content.

The scope of this paper is much more limited, namely, to look at the experiments themselves to see what methodological problems are raised. (Indeed, the scope is even more limited because the writer's experience has been almost entirely with crops and not with animals.) In general, the problems arise from several related causes.

Key words and phrases: blocks, design of experiments, intercropping, nearest-neighbor analysis, objectives, Papadakis, reliability of yield, rows and columns, systematic designs.

1.1. Supply of Statisticians

Skilled scientists are in short supply, and that is especially true of statisticians. There is therefore pressure to find short cuts in experimental techniques. Against that, many tropical agricultural research institutes are so lavishly endowed with land as to encourage large experiments, albeit commonly of a simple kind. Further, many local scientists and almost all of those from overseas will have been trained in the conditions of a developed country, where opportunities and objectives can be very different. When dealing with estate crops grown for export, that may not matter a lot. The managers will have had scientific training, and economic considerations will be dominant and, what is more, of a familiar kind. Helping with food crops can be much more difficult, largely on account of problems of communication, e.g. lack of mutual trust as well as of language.

1.2. Specifying Objectives

An even greater difficulty with local smallholders, which arises from what has just been said, is that of getting them to specify precise objectives. For one thing, they often appear to be inarticulate and unable to explain what they do want. (Indeed, their needs may be so obvious to them that they cannot imagine that anyone needs to be told.) Also, viewing them from outside, they are clearly not motivated by the same economic considerations as farmers in the developed world. For one thing, they cannot usually tide themselves over a bad harvest by raising a loan, being perhaps in debt already. For another, it is not really sensible to cost family labor and discuss whether it could be better deployed, if the only alternative to present conditions is migration to a shanty town outside a city with factories. If they are not motivated in the same way as those more fortunate, what can be done for them? If the answer is that farming systems must be found to minimize the risk of bad harvests, that calls for experiments of a different kind from those found elsewhere.

1.3. Obstacles to Change

Allied with the difficulties of defining objectives are those of comprehending the cultural and religious background of the country concerned. Traditions of family or community life can encourage or inhibit proposed developments. An African accustomed to equate cattle with money does not take readily to a suggestion that he should cull his herds. Again, the introduction of machinery may indeed expedite ploughing when the rains start and make cultivation possible, but it may disrupt patterns of employment, either by leaving some people out of work or by calling for skills not ordinarily to be found in a village. A common problem is resistance to alterations in the farming year, e.g., times of sowing, since existing practices may well have traditional significance, probably with a religious basis. Possible improvements in agricultural practice have to be seen against these obstacles to change, which are obvious to the farming community but not necessarily to anyone else.

2. Local Control

One feature of the Fisherian experiment is its control of local variation, whether due to soil, moisture, shelter, or anything else. The favored devices are blocks, or perhaps rows and columns as in a Latin square.

At first sight it should not be difficult to divide the area of an experiment into portions such that each is fairly uniform within itself. Where land has been farmed in one piece for a long time, local differences come to be understood, though even in such conditions there can be doubts about future fertility patterns. Some causes of variation can have effects that are almost permanent, but some, like bird damage from nearby trees, can be transient. Even so, the possibility can be foreseen. In the Third World the government may form a research farm by amalgamating a lot of smallholdings and leveling the whole without regard to former roads or dwellings and without a detailed survey to show where they were. In such conditions it is not easy to form blocks. The large amount of land available can actually make matters worse by encouraging the placing of each new experiment on land that has not been used before and about which little is known. There is therefore an enhanced risk of placing blocks across fertility contours instead of along them, and so doing harm rather than good. Further, if there is need to use some complicated design, perhaps quite an unusual one, to fit the desired treatment structure to the blocks indicated, in a less developed country there may not be the statisticians and computers available.

Rows and columns are not necessarily any better. It is true that tropical agriculture sometimes lends itself to such an approach, e.g., when crops are grown on beds that run down a hill with water courses or irrigation channels between them, but if a row-and-column design is laid down on land of unknown fertility pattern, there could well be a serious failure of the model, which assumes row and column effects to be additive. If they are so in one direction, they may not be so if the orientation is altered (Pearce, 1983b).

Devices like fitting polynomials or Fourier functions to margins are no better. They save error degrees of freedom but are less flexible. One scheme that could be considered is the fitting of a paraboloidal surface to indicate underlying fertility. At least it avoids the problems of orientation that are the bane of additive models. On the other hand, it is not to be supposed that a smooth surface can fit a large area which contains good and bad patches distributed unevenly, and that is often the case when land has fairly recently been brought into cultivation. The same applies to land that is cultivated only at intervals, e.g. on a "slash and burn" basis.

Nowadays there is much interest in the method first proposed by Papadakis (1937, 1940) and since developed by Bartlett (1938, 1978) and others (Atkinson, 1969; Wilkinson et al., 1983; Martin, 1984). The writer has used it on and off for some forty years (Pearce, 1953) and likes it. There is now quite a lot of experience with it (Pearce and Moore, 1976; Pearce, 1978, 1980; Lockwood, 1980; Kempton and Howes, 1981). The idea is this: For each plot a note is made of the deviation of its yield from the mean yield of all plots receiving that treatment. Then, again for each plot, an estimate of inherent fertility is derived from some function of the deviations of its neighbors, which estimate is used as a covariate in the analysis of data. Accordingly, the yield of a plot is judged in relation to the performance of its neighbors. [Wilkinson et al. (1983) have amended the approach so as, in effect, to use running blocks.] At first sight the method is ideal for patchy ground because it requires no overall model but only a number of local ones. Unfortunately it is not as robust as might be expected. While it may be able to deal well with trends, etc., it could be dangerous with discontinuities, and that is just what experimenters fear could exist. Also, although iteration appears to improve the estimation of regression coefficients, adjusted treatment means, etc., there are suspicions that the resulting improvements are spurious. Nor are those the only problems; e.g., difficulties arise with edges and corners, where some neighbors are lacking, and with the arbitrary allocation of treatments to neighboring plots, which has led some to think of systematic designs. At the moment it would be unwise to advocate the general adoption of Papadakis' method in any of the forms that have been suggested, but the whole approach is so mentally satisfying that continuing research is both to be expected and to be desired. Before leaving the subject it may be remarked that the one-dimensional form of the analysis, in which plots are in line, is perhaps nearer solution than the two-dimensional. It has advantages of its own, e.g. the plots can be long and narrow and adjoin on the long sides, thus promoting the correlations required.

3. Intercropping

The tropical zone has one important practice that is found less frequently elsewhere, namely, intercropping, in which two species are grown in close association, e.g. in alternate rows or on some other configuration, or perhaps two seeds of different kinds are dropped into the same hole. (Actually, something similar can arise in the temperate zones, e.g. in fruit growing and in vegetable gardens, but it has not attracted much attention from statisticians.) There are two variants, namely, mixed cropping, in which two lots of seed from different species are intermingled and then sown in the ordinary way, and relay cropping, in which a second species is sown between the rows of the first before that has been harvested. Statistically both have much in common with intercropping itself, though mixed cropping has quite a literature of its own (Federer et al., 1976, Federer, 1979).

The practice often arises from the shortness of the rainy season; everything must be sown as soon as the soil is wet. There are however other reasons. One is reliability. A poor farmer must have yield of some kind or he will starve. Where two crops are grown in conditions of competition and either is an acceptable substitute for the other, the failure of one will enhance the other's yield. (Note however the assumption that the two species respond differently to the weather. They must not fail together.) Another reason is economy of land use. Where species make different demands on the soil or call for water at different times, the total productivity can be increased by growing them together. If they can substitute for one another, that is a great advantage, but if they serve different purposes and each is essential (e.g. a cereal and a pulse, or a food crop and a cash crop), the two must respond similarly to the weather. If conditions favor one species, it will compete more strongly and the other must be able to fight back. If it cannot, an essential component of yield will have been lost. It has been argued elsewhere that intercropping studies can helpfully be related to past records of weather cropping (Pearce and Edmondson, 1983). More general studies are those of Mead and Stern (1980), Pearce (1983a), and Pearce and Edmondson (1984).

Now that statisticians have turned their attention to intercropping the situation is rather confused. A paper on the subject and the discussion it provoked (Mead and Riley, 1981) show an underlying tussle between those who see it as calling for new approaches and those who think that with suitable adaptation conventional methods will suffice. (Perhaps the truth can be put either way.) In what follows it will be apparent that the present writer favors a new approach. Others may not agree.

First, the problem is essentially bivariate. There are two crops. Only if they serve the same purpose with a constant relative value can they be combined to give a univariate analysis. For example, if an investigation concerned an intercrop of two similar species like millet and sorghum and if attention were concentrated on protein content, it would be reasonable to combine yields as a single variate, but that is an unusual case. Because relative market prices change, the method cannot be used with monetary values, though some have tried.

The bivariate approach was first advanced (Pearce and Gilliver, 1978) to allow for varying objectives. The two crops from each treatment are first subjected to a transformation to generate two other variates, each with unit variance and uncorrelated. The new variates are then plotted on a diagram in which the ordinary rules of Euclidean geometry apply and to which contours can be added according to various scales of excellence. A treatment point that is well placed with regard to one set of contours may be badly placed with regard to another. At least the diagram gets the position clear. Some have objected to the approach because it assumes that the correlation between different plots of the same treatment is constant and there are cases, e.g., in trials of configurations, when that can scarcely be so. Accepting the assumption, however, the method of analysis is exact (Rao, 1952) and lends itself to clear presentation, whether of tests or of estimates (Pearce and Gilliver, 1979; Gilliver and Pearce, 1983). Recently the method has been extended to cover the case of unequal intratreatment correlations (Singh and Gilliver, 1983). It represents a most useful advance, though there are no longer any simple diagrams to assist interpretation. (Incidentally, those developments have underlined that there are really two problems. Sometimes treatments, e.g. fertilizer, are necessarily applied to both species, but other times, e.g. variety, they can be applied only to one of them.)

An interesting example of the way in which the problems are irresistably bivariate is afforded by the history of the *land equivalent ratio* (Willey and Osiru, 1972; Mead and Willey, 1980). In its origins it expressed the land saved by using an intercrop compared with obtaining the same yields from apportioning the land between sole crops. If an area of land can be expected to produce a yield of a when only the first species is grown and b when the second is used instead, and if an intercrop of the two can be expected to yield m and n of the two species respectively, then the land equivalent ratio (LER) is

$$\text{LER} = \frac{m}{a} + \frac{n}{b}.$$

That is to say, it is the area of land needed to produce the same yield from sole crops. Certainly it serves as a general indicator of land economy, but latterly the cry has been that one must know where the economy comes from, and that has meant concentrating on the two components, i.e., m/a and n/b, and that is effectively a return to the original variates, albeit scaled by the sole crop yields.

However, a more serious objection to the use of the LER is its letting the intercrop decide the problem. If a ratio of m to n in yields is required, all is well, but not if someone wants a different ratio. In that case, the land will have to be apportioned between the intercrop and the sole crop that is in short supply, which gives rise to the *effective land equivalent ratio* (ELER), discussed by Mead and Willey (1980); subsequently Riley (1984) has studied the apportionment of land between two intercrops. The same problem had been previously considered graphically using the bivariate diagram (Pearce and Gilliver, 1979). Another development is that of the *staple land equivalent ratio* (SLER) proposed by Chetty and Reddy (1984). Here the aim is to produce enough crop from one species and as much as possible of the other. In short, the LER, though valuable in some contexts, is not an absolute quantity, as was once thought, but depends upon objectives. Well used, it is a most valuable concept.

Because the components of an LER are ratios, some have had doubts about their distribution and consequently about the validity of an analysis of variance of LER. Recent work (Oyejola and Mead, 1982), however, allays many of those fears.

One difficulty in the design of an intercropping experiment concerns the number of plots to be assigned to the sole crops. Often a lot are included because the investigators fear a rejection of their conclusions if they cannot demonstrate a high LER, but that should not be necessary. If the intercrop is fixed and the question is how to fertilize it or protect it from pests, there should be no need of sole crops at all. On the other hand, given a factorial set of treatments, some belonging to one factor and some to another, not only are plots of sole crops needed, but often they should themselves carry differential treatments (Singh and Gilliver, 1983).

One of the weaknesses of research into intercropping is the difficulty of initiating new systems. Given one that already exists, experiments can indicate improved methods of fertilizing it or controlling pests or can suggest other varieties, but few appear to have succeeded in devising a completely new system *ab initio*. Pearce and Edmondson (1983) have suggest that some combinations of crops are unlikely to succeed anyway, and that for reasons already mentioned, namely, the different reaction of species to weather. They present formulae that give a rough forecast of the behavior of an intercropping system based upon historical data of the variability of the two species when grown as sole crops in the past (variances) and the extent to which they fail in the same or different years (correlation coefficients). These examples lead to a preliminary conclusion that if two species serve different purposes, any system that involves both will be unreliable unless the two react very similarly to differences between seasons. Since some traditional systems succeed nonetheless, that cannot always be so, but the warning remains.

With any intercropping system in which the species play different roles, each being essential, it is important that neither shall overcome the other; but quite a small change in conditions, e.g. more rainfall or less, can destabilize a system and lead to that undesired end. It has been suggested (Pearce and Edmondson, 1984) that no system should be recommended to the farming community until it has been tried at a range of sites chosen for their diversity. The same authors give formulae to help in the making of such forecasts. They also argue that a recommendation is much strengthened if those who make it understand how the system works. To that end they point out the potential of methods of path analysis, of which they give an illustration.

Intercropping provides perhaps the most important fresh area of research in agricultural experimentation since R. A. Fisher introduced his methods in the 1920s. (Some would add adjustment by neigboring plots in the manner of Papadakis.) It is too early to say where it will lead, but it could refresh some topics that seem to have gone stale. For example, the new attention to bivariate analysis is already leading to other applications, while competition studies acquire a deeper meaning when viewed against farming systems that depend upon them.

4. Study of Reliability

There is one especially pressing problem in a poor country, at least for the poor man. It is that of reliability. Most farmers in a developed country and large companies anywhere have the financial resources to take a risk, knowing that they can borrow money if need arises to tide them over the occasional crop failure, but a poor man starves and his family with him if he makes a mistake. People sometimes complain about the conservatism of the small farmer in an undeveloped country, but they usually give the wrong reason for his reluctance to change. He may indeed be ignorant and uneducated, but that does not mean that he is stupid. On the contrary he has the sense to avoid risks, that being a condition of his survival. Actually, in a tightly knit agricultural community an improved method may be quickly and generally adopted once initial objections have been fairly met, but they do have to be met, otherwise no one will respond.

It is this need for reliability that underlines much of the practice of intercropping. It also raises problems in the generalization of results. A farming system can be tried for three or four years at a research center and also at outstations, where conditions will approximate better to those of practical farming besides affording a spread of conditions. If all goes well, in a developed country the system will be recommended for adoption despite the possibility that in some conceivable contingency it may fail badly. That is not good enough if failure is going to mean starvation for a family or their becoming thereafter impoverished.

It is not difficult to simulate extreme conditions of weather, though the only place that the writer has seen it done is with variety trials at a farm of the University of the West Indies in Trinidad. For example, sprinkler irrigation taken to excess will simulate high rainfall, while shelter will keep the soil dry. Also fans will provide wind, gusty if need be. The difficulty comes with combining elements of weather. For example, rain does not fall from a cloudless sky and lack of rain is not usually associated with shade. Nevertheless, such studies point the way. It is not implied that such studies can simulate differences of climate or season. They can however show up varieties that will probably fail in extreme conditions. Bearing in mind how disastrous a small crop can be, no one should make a recommendation to a poor farmer without some assurance that it will succeed consistently, because minimum yields can matter more than means. That is another example of the way in which cultural differences can lead to misunderstanding. The very fact of being well educated and in receipt of a steady salary can separate a scientist from the folk he seeks to help. Being a foreigner makes things worse.

A number of developing countries have set up elaborate testing procedures for new farming systems, often with a carefully considered statistical base. In India, for example, the organization has existed for more than 30 years. A more recent scheme is that of the Phillipines (Gomez, 1983), and there are many more. Little has been written about them, which is a pity, because all developing countries need such procedures and should not be left to devise them without guidance from those with experience. The central difficulty is that a research institute can succeed with systems that cause difficulty on an ordinary farm. That raises the question of how much instruction a collaborating smallholder must be given. If he is supervised at every turn, there will still be no certainty that farmers in general will succeed with the new practice; if he is left unsupervised, a failure on his part will be attributed to the practice being itself a bad one. To avoid that dilemma, some territories, e.g., the Windward Islands, have tried staging demonstrations on smallholdings for all to see and then leaving local people to imitate if they want to, meanwhile assessing continuously the progress of the farms on which demonstrations were held. Since the owner, of all people, can be expected to know exactly what innovations had been introduced, it is his reaction that is of greatest value. Such an approach also serves to identify the sort of farmer best able to make good use of new practices—how far education helps, or leadership in local affairs or sufficient family labor. The subject is large one and deserves more attention.

5. Systematic Designs

First a distinction needs to be made. Some experiments are systematic because their designers did not know enough to have them randomized. Such instances are rare, but they do occur, though with decreasing frequency. In the systematic designs to be considered here randomization has been deliberately discarded to gain some alternative advantage. It is interesting to note that the earliest example of purposive avoidance was in the tropics (Bocquet, 1953).

The need arises whenever treatments are so diverse that some must not adjoin. An obvious example is afforded by spacing trials, which are difficult in any case, partly because plots have to be large to accommodate a range of spacings, but also because wide discard areas are needed on account of shadows, whether of sun, wind, or rain. In general shadows are more intense in tropical conditions. There, if the skies are clear, the sunlight can be strong; if they are not, there is probably heavy rain, carried sometimes in strong winds. In the temperate zones, as the name implies, the weather does not usually go to such extremes. Also, the greater simplicity of management makes such designs suitable for a developing country. Nevertheless, the chief advances in systematic spacing trials have in fact come from developed countries (Nelder, 1962; Bleasdale, 1967b; Freyman and Dolman, 1971; Rogers, 1972; Pearce, 1976).

An important point concerns the analysis of data. The whole intention is to keep similar treatments together, so the lack of randomization is serious if an analysis of variance is attempted. Also, the object is to find an optimal spacing; consequently the chief need is to compare each treatment with those most like it. A pooled error therefore underrates the precision of the important contrasts, even though they are better estimated, while overrating that of the unimportant. A better approach is to fit response curves by regressing plant data on spacing distance to find optima. Here much useful work has been done on the relationships that might be expected between, say, yield or size and spacing (Holliday, 1960; Bleasdale and Nelder, 1960; Bleasdale, 1967a; Farazdagh and Harris, 1968).

One other use of systematic designs had its origins in a problem of the tropical zone and appears to have found little application elsewhere. Where the wind blows steadily from one direction, each plot may get its pollen from one neighbor only. In a variety trial that can be a serious limitation. The problem has been studied (Freeman, 1967; Dyson and Freeman, 1968), and one solution is the use of complete Latin squares (Freeman, 1979a, 1979b, 1981). The work has been extended to cover the case where the wind cannot be relied upon (Freeman, 1969). In these designs there is the constraint that all the plots of a treatment have a complete set of neighboring treatments in certain directions. Similar attention has been given to linear experiments (Williams, 1952; Dyke and Shelley, 1976).

To return for a moment to the subject of local control and the possible use of the method of Papadakis, such designs could well be advantageous in securing a good spread of neighboring treatments for each plot. That, however, remains an open question.

In view of the importance of assessing competition for intercropping situations, some interest attaches to beehive designs (Martin, 1973; Veevers and Boffey, 1975, 1979) in which plants are surrounded by varying numbers of another species with which they are in competition. It is not apparent though that these useful designs have in fact been used in that context.

There is so much potential use for these designs in the tropics that, quite apart from their simplicity in the field, they should be much more common. The story at present is largely one of missed opportunities.

6. Transfer of Technology

One matter that lies at the heart of development schemes is the transfer of technology. It should not be assumed that equipment evolved for a developed country will be ideal for one less developed. Quite apart from questions of maintenance, there are those of suitability. Tractors exhibit both facets. They need mechanics, who may not exist locally, and they save labor, which is not much of an advantage in an area of rural unemployment. Also their introduction may need foreign currency, which could well be scarce. Much the same applies to trained staff. Those who have studied abroad may come back with expensive and exotic tastes which must be gratified if they are to stay; the total cost may well exceed any good they can do. There is much talk these days of "intermediate technology" as if the need were for steps by which poorer countries could climb to the standards taken for granted in those more fortunate. It is by no means certain that they should become latter day replicas of those countries now thought of as developed. Indeed, it is to be hoped that they will evolve into something different according to their own social and religious aspirations and so be able to make contributions that are distinctive and perhaps complementary to what already exists, thereby enriching the whole.

The extent to which statistical considerations enter into the conduct of field experiments depends upon the source of funds. Some bilateral aid, e.g., that from the U.S.A. and U.K., is given with discreet pressure that everything shall be biometrically sound. Admirable as that intention is, it can lead to designs that are very obvious and very simple, partly because the aid-receiving country may be short of statisticians, and partly because rectitude may appear to be more important than initiative in convincing distant officials that the proprieties have been observed. For the rest, statistical standards can vary a lot, as a recent excellent paper shows (Preece, 1982): some of the worst experiments derive from the bad advice sometimes given by F.A.O. to its workers in the field. (The writer's observation confirms that of Preece.) Another reason for bad experimentation is the tendency to carry out trials at a range of sites, which in itself is good, many of which are unknown to the initiator, which is bad. As a result a plan is sent off, which does not fit the site but nevertheless must be followed exactly because it is believed that local staff are not competent to judge such matters. It is here suggested that those who provide funds for research would get better value for their money, both at the time and in the future, if they would insist on good statistical standards and also ensure that sound advice was available. That could be the best way of transferring technology.

To apply those general remarks to the narrow field of statistical methods in agricultural research, it will be seen that the various topics discussed do not all tell the same story. To take first the subject of local control, traditional methods started in Europe without much examination or thought as to alternatives and were found to work. They were then transferred with even less examination or thought to other conditions. It is true that much work was done in the tropics on plot size, etc., but there seems to have been little attention given to the methods themselves or to their suitability in those conditions. The method of Papadakis was developed there—the data used by Bartlett (1938) came from the Sudan—but the approach was not taken up seriously until the developed world became interested for its own purposes. Much the same can be said of systematic designs, which also arose in the conditions of the Third World but were neglected till people in more developed countries wanted to use them.

Intercropping has always belonged primarily to the tropics, and most relevant statistical research began there. So far advances in the developed world have come from statisticians with first-hand experience of the problems, but, as with the method of Papadakis, there is a danger that the subject will become fashionable and a lot of mathematics will be generated that has really little to do with practicalities. The study of reliable farming systems likewise is primarily of importance in tropical conditions, and data have to be collected there. So far little theoretical work has been done, at least as far as experimentation is concerned.

If there is a conclusion to be drawn from all this, it is that development of statistical techniques has been almost a monopoly of the developed countries, nor is that surprising when statisticians in poorer countries are few and have little time to undertake more than the routine jobs that need doing. It is true that they are sometimes able to invite practitioners from more developed countries to visit them and such encounters can be useful to both parties, but essentially they facilitate the transfer of established technology from the country in which it was developed to countries whose needs are perhaps different. Also, the experienced practitioner on his return has probably little time to work on other people's problems, having enough of his own. In this connection there are so many theoretical topics needing examination that there might be some point in trying to stimulate the interest of academics, who have more opportunity to choose the subjects on which they conduct research. If they want problems, there is no lack of them in the Third World. Above all, however, the need is for those who can help to do so in a realistic way by discerning the problem before proffering a solution. In many ways the problems can be different.

Bibliography

Atkinson, A. C. (1969). "The use of residuals as a concomitant variable." *Biometrika*, **56**, 33–41.

Bartlett, M. S. (1938). "The approximate recovery of information from field experiments with large blocks." *J. Agric. Sci. (Cambridge)*, **28**, 418–427.

Bartlett, M. S. (1978). "Nearest neighbour models in the analysis of field experiments" (with discussion). *J. Roy. Statist. Soc. Ser. B*, **40**, 147–174.

Bleasdale, J. K. A. (1967a). "The relationship between the weight of a plant part and total weight as affected by plant density." *J. Hort. Sci.*, **42**, 51–58.

Bleasdale, J. K. A. (1967b). "Systematic designs for spacing experiments." *Expl. Agric.*, **3**, 73–86.

Bleasdale, J. K. A. and Nelder, J. A. (1960). "Plant population and crop yield." *Nature (London)*, **188**, 342.

Bocquet, M. (1953). "Note sur l'expérience de densité Marchal appliqué a l'hévéaculture." *Arch. Rubbercult.*, May 1953 Extra No., 194–199

Chetty, C. K. R. and Reddy, M. N. (1984). "Analysis of intercrop experiments in dryland agriculture." *Expl. Agric.*, **20**, 31–40.

Dyke, G. V. and Shelley, C. F. (1976). "Serial designs balanced for effects of neighbours on both sides." *J. Agric. Sci. (Cambridge)*, **87**, 303–305.

Dyson, W. G. and Freeman, G. H. (1968). "Seed orchard designs for sites with a constant prevailing wind." *Silvae Genet.*, **17**, 12–15.

Farazdagh, H. and Harris, P. M. (1968). "Plant competition and crop yield." *Nature (London)*, **217**, 289–290.

Federer, W. T. (1979). "Statistical designs and response models for mixtures of cultivars." *Agron. J.*, **71**, 701–706.

Federer, W. T., Hedayat, A., Lowe, C. C., and Raghavarao, D. (1976). "Application of statistical design theory to crop estimation with special reference to legumes and mixtures of cultivars." *Agron. J.*, **68**, 914–919.

Freeman, G. H. (1967). "The use of cyclic balanced incomplete block designs for directional seed orchards." *Biometrics*, **23**, 761–778.

Freeman, G. H. (1969). "The use of cyclic balanced incomplete block designs for non-directional seed orchards." *Biometrics*, **25**, 561–567.

Freeman, G. H. (1979a). "Some two-dimensional designs balanced for nearest neighbours." *J. Roy. Statist. Soc. Ser. B*, **41**, 88–95.

Freeman, G. H. (1979b). "Complete Latin squares and related experimental designs." *J. Roy. Statist. Soc. Ser. B*, **41**, 253–262.

Freeman, G. H. (1981). "Further results on quasi-complete Latin squares." *J. Roy. Statist. Soc. Ser. B*, **43**, 314–320.

Freyman, S. and Dolman, D. (1971). "A simple systematic design of planting density experiments with set row widths." *Canad. J. Plant. Sci.*, **51**, 340–342.

Gilliver, B. and Pearce, S. C. (1983). "A graphical assessment of data from intercropping factorial experiments." *Expl. Agric.*, **19**, 23–31.

Gomez, K. A. (1983). "Measuring potential productivity on small farms: A challenge." Invited paper presented at the 44th Session of the I.S.I., Madrid. *Proceedings*, **1**, 571–585.

Holliday, R. (1960). "Plant production and crop yield." *Nature* (*London*), **186**, 22–24.

Kempton, R. A. and Howes, C. W. (1981). "The use of neighbouring plot values in the analysis of variety trials." *Appl. Statist.*, **30**, 59–70.

Lockwood, G. (1980). "Adjustment by neighbouring plots in progeny trials with cocoa." *Expl. Agric.*, **16**, 81–89.

Martin, F. B. (1973). "Beehive designs for observing variety competition." *Biometrics*, **29**, 397–402.

Martin, R. J. (1984). "Papadakis' Method." (to appear). In *Encyclopaedia of Statistical Sciences*, New York: Wiley.

Mead, R. and Riley, J. (1981). "A review of statistical ideas relevant to intercropping" (with discussion). *J. Roy. Statist. Soc. Ser. A*, **144**, 462–509.

Mead, R. and Stern, R. D. (1980). "Designing experiments for intercropping research." *Expl. Agric.*, **16**, 329–342.

Mead, R. and Willey, R. W. (1980). "The concept of a 'land equivalent ratio' and advantages in yields from intercropping." *Expl. Agric.*, **16**, 217–228.

Nelder, J. A. (1962). "New kinds of systematic designs for spacing experiments." *Biometrics*, **18**, 283–307.

Oyejola, B. A. and Mead, R. (1982). "Statistical assessment of different ways of calculating land equivalent ratios (LER)." *Expl. Agric.*, **18**, 125–138.

Papadakis, J. (1937). "Méthode statistique pour des expériences sur champ." *Inst. Amél. Plantes, Salonique* (*Gréce*), *Bull.*, **23**.

Papadakis, J. (1940). "Comparison de différentes méthodes d'expérimentation phytotechnique." *Rev. Argentina. Agron.*, **7**, 297–362.

Pearce, S. C. (1953). *Field Experimentation with Fruit Trees and Other Perennial Plants.* Commonwealth Agric. Bur., Farnham Royal, Tech. Commun., 23, Section 48.

Pearce, S. C. (1976). Pearce (1953), 2nd edn., Section 66.

Pearce, S. C. (1978). "The control of environmental variation in some West Indian maize experiments." *Trop. Agric.* (*Trinidad*), **55**, 97–106.

Pearce, S. C. (1980). "Randomized blocks and some alternatives: A study in tropical conditions." *Trop. Agric.* (*Trinidad*), **57**, 1–10.

Pearce, S. C. (1983a). *The Agricultural Field Experiment: A Statistical Examination of Theory and Practice.* Chichester: Wiley, Section 9.8.

Pearce, S. C. (1983b). "The control of local variation in a field experiment." Invited paper presented at the 44th Session of the I.S.I., Madrid. *Proceedings*, **I**, 586–593.

Pearce, S. C. and Edmondson, R. N. (1983). "Historical data as a guide to selecting farming systems with two species." *Expl. Agric.*, **18**, 353–362.

Pearce, S. C. and Edmondson, R. N. (1984). "Experimenting with intercrops." *Biometrics*, **40**, 231–238.

Pearce, S. C. and Gilliver, B. (1978). "The statistical analysis of data from intercropping experiments." *J. Agric. Sci. (Cambridge)*, **91**, 625–632.

Pearce, S. C. and Gilliver, B. (1979). "Graphical assessment of intercropping methods." *J. Agric. Sci. (Cambridge)*, **93**, 51–58.

Pearce, S. C. and Moore, C. S. (1976). "Reduction of error in perennial crops, using adjustment by neighbouring plots." *Expl. Agric.*, **12**, 267–272.

Preece, D. A. (1982). "The design and analysis of experiments: What has gone wrong?" *Utilitas Math.*, **21A**, 201–244.

Rao, C. R. (1952) *Advanced Statistical Methods in Biometric Research.* New York: Wiley, 260.

Reddy, M. N. and Chetty, C. K. R. (1984). "Staple land equivalent ratio for assessing yield advantage from intercropping." *Expl. Agric.*, **20**, 171–177.

Riley, J. (1984) "A general form of 'land equivalent ratio'." *Expl. Agric.*, **20**, 19–29.

Rogers, I. S. (1972). "Practical considerations in the use of systematic spacing designs." *Austral. J. Expl. Agric. Anim, Husb.*, **12**, 306–309.

Singh, M. and Gilliver, B. (1983). "Statistical analysis of intercropping data using a correlated error structure." Paper contributed to 44th Session of the I.S.I., Madrid. *Proceedings*, **1**, 158–163.

Veevers, A. and Boffey, T. B. (1975). "On the existence of levelled beehive designs." *Biometrics*, **31**, 963–968.

Veevers, A. and Boffey, T. B. (1979). "Designs for balanced observation of plant competition." *J. Statist. Planning Infer.*, **3**, 325–332.

Wilkinson, G. N., Eckert, S. R., Hancock, T. W., and Mayo, O. (1983). "Nearest neighbour (NN) analysis of field experiments" (with discussion). *J. Roy. Statist. Soc. Ser. B*, **45**, 151–211.

Willey, R. W. and Osiru, D. S. O. (1972). "Studies on mixtures of maize and beans (*Phasaelus vulgaris*) with particular reference to plant population." *J. Agric. Sci. (Cambridge)*, **28**, 556–580.

Williams, R. M. (1952). "Experimental designs for serially correlated observations." *Biometrika*, **39**, 151–167.

CHAPTER 23

On the Statistical Analysis of Floods

Luis Raúl Pericchi and Ignacio Rodríguez-Iturbe

Abstract

We deal with the problem of model selection for the representation of river floods. We concentrate on methods based on the series of maximum annual floods, and we analyze some features of the parametric models most widely used in hydrology. In particular we comment on the recommendation of the U.S. Federal interagency group to use the log–Pearson III distribution as the base model for estimating flood frequencies. We mention the limitations of classical goodness of fit procedures, and we propose simple methods based on the "flood rate" in order to make an initial screening of alternative models. We analyze historical data on five rivers in different continents.

In Section 2 we outline the most common procedures. Section 3 is devoted to the study of two sets of historical data, and in Section 4 we introduce the "flood rate" criterion and we describe methods for testing and estimating it.

1. Introduction

The importance of estimating the frequency and magnitude of flood discharges in rivers is obvious in view of the losses of human lives and property which these events bring to populations all over the world. This is a major reason for the many studies by hydrologists and statisticians which try to provide sound methodologies for flood frequency analysis. All this work has not yet culminated in a universally accepted philosophy for the calculation of the probabilities of occurrence of different discharges in a river. In this complex problem much depends on the art, experience, and competence the hydrologists bring to the analysis.

In many instances the main concern of the civil engineer lies in the largest or smallest value which a design variable may take during a certain length of time. This is indeed the case in hydrology, where a flood frequency analysis is the basis for the engineering design of many reservoirs and flood control works. For these projects it is necessary to estimate the magnitudes of floods corresponding to return periods far in excess of the length of records available.

Many techniques have been used for the estimation of floods with small

Key words and phrases: Cox's hazard estimate, flood rate, Gumbel distribution, inverse Gaussian distribution, lognormal distribution, log–Pearson III distribution, Pearson III distribution, total time on test transformation.

probabilities of occurrence. In general these techniques are based on the choice of a particular probability distribution for the magnitude of floods. The available sample is then used to estimate the parameters for that distribution. Different methods may produce very different results with very different engineering, economic, and social consequences. All this prompted in the United States the appointment of a Federal interagency group whose mission was described in the report transmitted to the President by a Task Force on Federal Flood Control Policy:

> Techniques for determining and reporting the frequency of floods used by several Federal agencies are not now in consistent form. This results in misunderstanding and confusion of interpretation by state and local authorities who use the published information. Inasmuch as wider, discerning use of flood information is essential to mitigation of flood losses, the techniques for reporting flood frequencies should be resolved. (House Doc. 465, 1966, cited by Benson, 1968, p. 892.)

Easier said than done! On the basis of comparisons of the different techniques available, the Federal interagency group recommended that all government agencies adopt the log–Pearson Type III distribution as the base method for estimating flood frequencies, "with provisions for departures from the base method where justified." In fact Federal agencies in the United States are requested to use the log–Pearson Type III distribution for flood estimation in all planning activities involving water and related land resources.

The above recommendation does not, unfortunately, have much statistical or hydrological basis. The data length is in most cases too short to permit a clear-cut choice among the different probability distributions used in flood analysis. The final choice must necessarily remain a professional one.

The task of the hydrologist is to make inferences about the underlying process that generates the floods, but in addressing this problem one is faced with a number of sources of uncertainty. These may be summarized into three categories: (1) natural uncertainty, the uncertainty in the stochastic process related to the occurrence of extreme streamflows, (2) parameter uncertainty, the uncertainty associated with the estimation of the parameters of the model of the stochastic process due to limited data, and (3) model uncertainty, the lack of certainty that a particular probabilistic model of the stochastic process is true.

In the field of flood frequency analysis the last of the above uncertainties is the crucial one. The natural uncertainty cannot be reduced, and the parameter uncertainty, although very important in its own right and not independent of model uncertainty, can be faced with more established criteria. Probably, the most important aspect of parameter uncertainty in flood frequency analysis is related to the choice of the method of estimation for the parameters of a given distribution. At least on this point, the hydrologist has some guidance from well-established statistical criteria.

Most of our comments in this paper are addressed to the problem of model selection in flood frequency analysis. The *model* is fundamentally a characterization of the probability distribution of the random variable of interest. The question addressed in model selection is, "Which model among those being entertained provides the best characterization, in some statistical sense, of the process under consideration?"

This paper is not intended to be a comprehensive review of the different approaches and methodologies for flood frequency analysis. Neither does it intend to present a description of those techniques—classical and Bayesian—which are of a general character in discriminating among alternative models. Rather, we will attempt to show, in a general manner and through examples, the drastic differences in predictions the hydrologist confronts when performing flood frequency analysis with different models, even with models which seem to fit the data quite well in a traditional statistical sense. We will then proceed to suggest alternative criteria which we feel may be of great help in the problem.

2. Flood Frequency Analysis

We now outline very briefly the most common procedure—by no means the only one—for performing flood frequency analysis in hydrology. From the discharge records the largest flow in every year is selected and a series of "annual floods" is thus constructed. The hydrologist's task is then to describe the probabilistic structure of the annual flood series. This involves:

(i) The choice of a probability distribution to represent the universe of floods.
(ii) The estimation of the parameters of the probability distribution on the basis of the sample that constitutes the annual flood series.

Of course, the implicit assumption of the analysis is that the probabilistic structure one aims to find holds good beyond the end of the observation period, thus allowing the hydrologist to make calculations for discharges with very slight probability of occurrence. Thus hydrologists speak of the "hundred-year" flood or the "thousand-year" flood, the latter just being a convenient synonym for a flood whose probability in a given year is $\frac{1}{1000}$. This means that the flood has a return period of one thousand years or that on the average such a flood—or a greater one—occurs once every 1000 years.

Because of the enormous losses that extreme streamflows may produce if they exceed the design parameters in engineering works, the calculation of floods for high return periods is everyday work in hydrology. The hydrologist's work on the estimation of floods is therefore focused on the upper tail of the estimated probability distribution. It is precisely because of this that the problem of model selection for the characterization of floods is both so important and so difficult. The amount of data is seldom enough to discriminate among even two or three acceptable distributions, yet differences may be quite large in the tails of these distributions. Thus, although techniques like the χ^2 goodness of fit statistic may well be used for an initial screening of contending distributions to eliminate those which do not reproduce even the gross characteristics of the data, it may be a serious mistake to base the final decision of an extreme value problem on the value of a χ^2 statistic or any other similar quantity based on the fit of the center of the observed data.

Amongst the many models or probability distributions used for the statistical estimation of floods, only a few can claim some physical or theoretical justification. Most of them are just convenient in some sense, such as involving a small number of parameters, flexibility in adjusting to empirical data, or to some characteristic of their upper tail.

Until 1966, the most widely used distribution for flood frequency analysis was the extreme value type I distribution, also called the Gumbel distribution, followed by the lognormal (United Nations—WMO, 1967). A classical reference is Gumbel (1958). Both of the above distributions are still frequently used, along with the three parameter gamma and its logarithm, which are commonly designated in hydrology as Pearson type III and log–Pearson type III.

Of these four distributions, the Gumbel is the only one that by its nature may have a theoretical claim to suitability for extreme streamflow analysis. The Gumbel is the distribution of the largest of many independent and identically distributed random variables with a common exponential type of upper tail distribution (e.g., normal, lognormal, exponential, logistic, etc.):

$$P[Q \leq q] = \exp[-e^{-\alpha(q-u)}], \qquad -\infty \leq q \leq \infty. \tag{2.1}$$

The engineering word "many" is equivalent in this case to the more mathematical term "an infinite number of" random variables, and this gives the distribution an asymptotic character.

The mean and variance of the Gumbel distribution are

$$\begin{aligned} \mu_Q &= u + \frac{v}{\alpha} \simeq u + \frac{0.577}{\alpha}, \\ \sigma_Q^2 &= \frac{\pi^2}{6\alpha^2}, \end{aligned} \tag{2.2}$$

where v is Euler's constant. The skewness coefficient is positive and, independently of the parameters α and u, has the value

$$\gamma_1 = 1.1396. \tag{2.3}$$

The Gumbel distribution is one of the three types of extreme value distributions which in reduced form have the expressions

$$F(z) = \exp(-e^{-z}), \qquad -\infty < z < \infty \qquad \text{(Gumbel model)},$$

$$F_\alpha(z) = \begin{cases} 0 & \text{if } z < 0 \\ \exp(-z^{-\alpha}) & \text{if } z > 0, \quad \alpha > 0 \end{cases} \qquad \text{(Fréchet model)},$$

$$F_\alpha(z) = \begin{cases} \exp(-(-z)^\alpha) & \text{if } z < 0, \quad \alpha > 0 \\ 1 & \text{if } z > 0 \end{cases} \qquad \text{(Weibull model)}.$$

These models, with a convenient change of location and dispersion, can be put in the unified form given in Jenkinson (1955):

$$F(z/k) = \exp(-(1 + kz)^{-1/k}) \quad \text{for } 1 + kz > 0, \ -\infty < k < \infty, k \neq 0.$$

Under the classic asymptotic theory of extreme values, the distribution for the probabilistic character of floods would only depend on the parental distribution of the streamflow values from which the floods arise. Tiago de Oliveira (1982) has developed a statistical decision procedure to choose among the Gumbel, Fréchet, and Weibull models. The fact is that the assumptions of classic asymptotic theory are seldom fulfilled by hydrological data and thus the use of the Gumbel model—or the Fréchet or the Weibull—is subject to controversy. Nevertheless, the Gumbel distribution is the only one of the commonly used distributions in flood analysis for which one may, at least, make an analysis of the physical meaning of the assumptions involved in the choice of the model.

The parent population of random variables for which the hydrologist wants the distribution of extreme values may be the record of daily flows at a given station. But in a small basin, it would be necessary to work with instantaneous flows, given the small duration of the flood avenues, which make the daily value very different from the instantaneous peak. For the sake of illustration, let us consider the common case where the parent population is taken as the record of annual daily flows of 365 values. The population of 365 values of daily flows from which the largest value is taken may well be dependent and not large enough to be considered asymptotic. However, one could argue that, although daily flows are not independent, in an annual collection of 365 values one should be able to find a group of daily flows which can be considered independent and numerous enough so that the asymptotic condition would not become critical. This may be true with respect to the previous assumptions, but in fact there is a fairly frequent occurrence of atypically large values when fitting a Gumbel distribution—or any other—to floods.

David (1970), gives numerous references which show, under suitable conditions, that the limiting distribution of the maximum of dependent variables is the same as for independent ones. This point is also covered in detail in the book by Leadbetter, Lindgren, and Rootzén (1983). The vital assumption is, in our opinion, one that has seldom been questioned as critical in the hydrological literature, namely the assumption that the parent population is made up of identically distributed random variables. To assume that two daily flows, which may occur, let us say, on the 15th of May and on the 20th of December, are identically distributed random variables, is clearly in violation of hydrological reality. This assumption "smooths" the historical parent population by assuming that there exists not only the same type of distribution, but also equal parameters (e.g. mean and variance) for every day of the year. Under this assumption there is no provision for the fact that of two days with the same mean flow, the one with a larger variance is more likely to produce much larger floods with the same return period as the day whose flows have a smaller variance. The hydrological reality is that the combined values of mean and variance for the daily flows of a given month make some months of the year more susceptible than others to flood occurrence.

The assumption that the parental population is made up of identically distributed random variables is closely related to the assumption of independence and to the justification of asymptotic approximations. To tackle the problem by analyzing different seasons with consideration only of floods above a given level may allow the hydrologist to deal with an independent sample, but will result in a small number of events, which will affect the reliability of the results. The assumption of an identical distribution in the parent population may be a good approximation when made only on a monthly basis. However, from the perspective of physical justification, the Gumbel distribution is no longer applicable, because the 30 daily flows in each month are not independent. One cannot find among them a group of values which can be considered independent and numerous enough so that the asymptotic condition does not become critical. There has been interesting work on how to deal with the effect of seasonal variations and serial correlation on the distribution of extreme values (e.g. Buishand, 1983; Leadbetter et al., 1983). Nevertheless, most of the theory is of an asymptotic character.

The ideal tool for the modeling of floods would be a nonasymptotic theory which would allow for a correlation structure in the process. This is an impressive problem even under drastic simplifications and has yet to be solved. As example, we mention the classical work of Cramér (1965, 1966), which deals with the maximum of Gaussian stationary processes in an asymptotic framework. More recently Ditlevsen (1966, 1971) dealt with the same problem without asymptotic restrictions. Research along these lines is of great interest in hydrology, and it may well be that this is the most appropriate direction to take in providing a coherent and theoretically sound framework for the analysis of extreme streamflow.

The lognormal distribution arises as the result of a multiplicative mechanism acting on a number of factors. The interpretation of the multiplicative action is not obvious in the case of floods. The lognormal density function is

$$f_Q(q) = \frac{1}{q\sqrt{2\pi}\,\sigma_{\ln Q}} \exp\left\{-\frac{1}{2}\left[\frac{1}{\sigma_{\ln Q}} \ln\left(\frac{q}{m_Q}\right)\right]^2\right\}, \qquad q > 0, \tag{2.4}$$

where m_Q is the median of the floods and $\sigma_{\ln Q}$ is the standard deviation of $\ln Q$. The mean value is

$$\mu_Q = m_Q \exp(\tfrac{1}{2}\sigma_{\ln Q}^2). \tag{2.5}$$

and the skewness coefficient

$$\gamma_1 = 3V_Q + V_Q^3, \tag{2.6}$$

where V_Q is the coefficient of variation of the floods.

We mentioned before that besides the Gumbel and lognormal distributions, the Pearson type III and log–Pearson type III are also frequently used in flood frequency analysis. The probability density function for the Pearson III is

$$f_Q(q) = \frac{|\alpha|}{\Gamma(\lambda)}[\alpha(q - m)]^{\lambda-1} e^{-\alpha(q-m)}, \tag{2.7}$$

where α, λ, and m are parameters (λ is always positive). Two cases need to be considered:

$$\alpha > 0, \qquad m \le q < \infty,$$

$$\alpha < 0, \qquad -\infty < q \le m.$$

The mean, variance and skewness coefficient are

$$\mu_Q = m + \frac{\lambda}{\alpha} \tag{2.8}$$

$$\sigma_Q^2 = \frac{\lambda}{\alpha^2} \tag{2.9}$$

$$\gamma_1 = \frac{\alpha}{|\alpha|}\frac{2}{\lambda^{1/2}}. \tag{2.10}$$

Thus if $\alpha > 0$ the distribution is positively skewed with a lower bound of m, whereas if $\alpha < 0$ the distribution is negatively skewed with an upper bound equal to m.

The random variable Q has a log–Pearson III distribution if $y = \log_a Q$ has a Pearson III distribution. The density function of the log–Pearson III distribution can be written as

$$f_Q(q) = \frac{|\alpha| k}{\Gamma(\lambda)} \frac{e^{\alpha m}}{q^{1+\alpha k}}[\alpha(k \ln q - m)]^{\lambda-1}, \tag{2.11}$$

where $k = \log_a e = (\ln a)^{-1}$ and α, λ, and m are parameters of the Pearson III distribution. Two cases must be considered:

$$\alpha > 0, \qquad q \ge e^{m/k} = a^m, \tag{2.12}$$

$$\alpha < 0, \qquad 0 \le q \le e^{m/k} = a^m. \tag{2.13}$$

Usually only two values are considered for k: first, $a = e$ and then $k = 1$; second, $a = 10$ and then $k = 0.434$. The log–Pearson III distribution is extremely flexible in the shape it may take according to the values of the parameters α, m, and λ.

The moment of order r about the origin (μ_r^1) of the log–Pearson III distribution is given by Bobée (1975) as

$$\mu_r^1 = \int_D q^r f(q)\, dq, \tag{2.14}$$

where D is the interval defined through equations (2.12) and (2.13). It can be shown that if $\beta = \alpha k < 0$, the moment μ_r^1 is always defined, and if $\beta = \alpha k > 0$, then μ_r^1 is only defined up to order $r < \beta$. When μ_r^1 is defined it is expressible as

$$\mu_r^1 = \frac{e^{mr/k}}{(1 - r/\beta)^\lambda}. \tag{2.15}$$

The estimation of the parameters of the four distributions mentioned above is usually carried out in hydrological analysis either by the method of moments or by maximum likelihood. We will not describe details of the estimation methods here, as they are found in many standard references. Nevertheless, the following comments for the case of the log–Pearson type III distribution, based on Bobée (1975), are of interest.

The application of the method of maximum likelihood is problematic when the range of the variate depends upon an unknown parameter.[1] One case is the parameter m for the Pearson III and log–Pearson III distributions, for which the maximum likelihood estimates are not necessarily functions of the sufficient statistics. Besides, the maximum likelihood method depends on very laborious computations and sometimes does not lead to any solution. Note that when $\lambda < 1$ the likelihood function has a pole at m. Furthermore, even if λ is greater than one, but still small, the estimation via maximum likelihood is very unstable (Johnson and Kotz, 1970, p. 185). For this problem it is worth considering Bayesian estimation which is based on the area below the likelihood times the prior density rather than on the maximum of the likelihood function. Even when one has only a very rough idea of what the prior density is, a Bayesian procedure will give a sensible answer. A convenient reference for a different but similar situation is Hill (1963). Nevertheless, Matalas and Wallis (1973) have shown for the Pearson III distribution that, when the maximum likelihood method leads to a solution, it is preferable to the method of moments.

The method of moments is the most commonly applied technique in the case of the log–Pearson III distribution. As a matter of fact, it is the one requested by Federal agencies in the United States. Again, that may solve a legal and bureaucratic problem, but hides in the closet the scientific aspects of the problem. The truth is that quite frequently maximum likelihood and the method of moments result in very different estimates of flood frequencies. Moreover, in the case of the log–Pearson III distribution, the matter is even more complex because the method of moments can be applied in two different manners. The U.S. Water Resources Council (1981) specified the application of the method of moments to the logarithms of the observed data. In this method the values of the mean, the standard deviation, and the coefficient of skewness of the logarithms of the data are calculated, and then one proceeds to estimate the logarithm of an event with a given return period.

Bobée (1975) proposes the application of the method of moments preserving the mean, variance, and coefficient of skewness of the observed data rather than those of the logarithms of the data. His technique, which we will call the

[1] A discussion of these problems and a list of references are given by Cheng and Amin (1983).

direct method of moments, resembles the one used for the lognormal distribution (Aitchison and Brown, 1957). Unfortunately for those seeking a consistent theory, this technique yields in many cases flood frequencies quite different from those obtained with the method of moments as proposed by the U.S. Water Resources Council.

Bobée has argued that the direct method of moments is more attractive than the method proposed by the U.S. Water Resources Council (which can be called the indirect method of moments) because it retains the moments of the observed sample and thus gives the same weight to each observation. The indirect method gives the same weight to the logarithms of the observed values and consequently reduces in importance the larger elements of the sample (Bobée, 1975).

Notice that when $\alpha < 0$ the log–Pearson III has an upper bound equal to $e^{m/k}$, which is without physical meaning for extreme flows. To make things more exciting, there are cases in which α changes sign when estimated by different methods. Reich (1972) showed that in certain cases several observed values obtained from annual flood series were greater than the theoretical upper bound. This may be because of the shortness of the records, but needless to say it is an unacceptable feature whose occurrence in a particular case would argue against the application of the log–Pearson III.

3. A Glimpse of the Real World

This section presents flood frequency analyses for two rivers in the United States that are of importance for the regions in which they are located and whose floods have been the object of previous calculations for engineering works. They are the Feather River at Oroville, California, and the Blackstone River at Woonsocket, Rhode Island.

3.1. Feather River

The 59 years of data (1902–1960) taken from Benjamin and Cornell (1970) are given in Table 1. Figures 1 and 2 display the annual flood data on Gumbel and lognormal probability paper. Goodness of fit statistics such as the χ^2 and the Kolmogorov–Smirnov were calculated by Benjamin and Cornell (1970) for both distributions; these authors show that neither of these distributions can be rejected as representative models for the record. The goodness of fit statistics can therefore not be used successfully in this example to help in choosing between the models. Notice that the values of the coefficients of skewness for the original record and the logarithmic series are 1.04 and -0.35, reasonably close to the theoretical values of 1.14 and zero required by the Gumbel distribution and lognormal distribution respectively.

Table 1. Annual Floods of the Feather River (1902–1960)[a]

Year	Flood (ft^3/s)	Year	Flood (ft^3/s)
1907	230,000	1935	58,600
1956	203,000	1926	55,700
1928	185,000	1954	54,800
1938	185,000	1946	54,400
1940	152,000	1950	46,400
1909	140,000	1947	45,600
1960	135,000	1916	42,400
1906	128,000	1924	42,400
1914	122,000	1902	42,000
1904	118,000	1948	36,700
1953	113,000	1922	36,400
1942	110,000	1959	34,500
1943	108,000	1910	31,000
1958	102,000	1918	28,200
1903	102,000	1944	24,900
1927	94,000	1920	23,400
1951	92,100	1932	22,600
1936	85,400	1923	22,400
1941	84,200	1934	20,300
1957	83,100	1937	19,200
1915	81,400	1913	16,800
1905	81,000	1949	16,800
1917	80,400	1912	16,400
1930	80,100	1908	16,300
1911	75,400	1929	14,000
1919	65,900	1952	13,000
1925	64,300	1931	11,600
1921	62,300	1933	8,860
1945	60,100	1939	8,080
1952	59,200		

[a] Statistics of the record: mean = 70,265 ft^3/s; standard deviation = 51,581 ft^3/s; coefficient of skewness = 1.04; coefficient of skewness of the logarithms = −0.35.

Tables 2 and 3 give the discharges estimated by the different models for return periods of 50 years and 1000 years, respectively.

The first interesting thing to be noticed in this example is the very wide range of values covered in a smooth manner by the series of annual floods. In effect, the discharges in the record vary between 8000 ft^3/s and 230,000 ft^3/s without any abrupt change in type of behavior. This immediately shows the danger of outlier type considerations in flood analyses; with a shorter series (say 20 years), it is likely we would have found one particular value much above (or below) the second highest (or lowest) discharge of the record. One common temptation is to treat such a value as coming from a different type of population. If deletion of "outliers" is allowed, the statistical method becomes

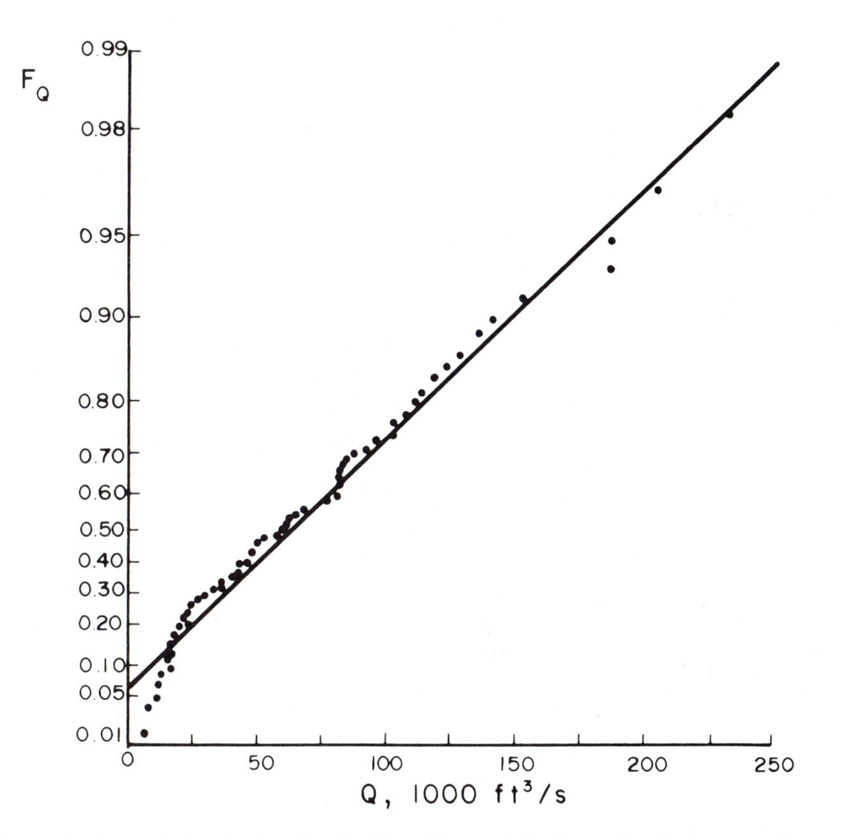

Figure 1. Annual floods for the Feather River plotted on Gumbel paper (from Benjamin and Cornell, 1970).

more adjustable and a better describer of the data. This should not be done except if there is considerable physically based evidence that the rare event was indeed due to a different kind of mechanism in nature which makes it part of another population.

The estimation of the 50 year flood—with a 59 year record—shows in this case a span of values which goes between 193,000 ft^3/s and 243,000 ft^3/s. The estimated 1000 year floods vary between 275,000 ft^3/s and 425,000 ft^3/s, which indeed poses a serious problem for the hydrologist. There is a very large difference in this example between the 1000 year floods estimated with the log–Pearson III distribution depending on how the parameters were estimated; the reason lies in the skewness coefficient. The coefficient of skewness of the original record is approximately 1, but the skewness of the logarithmic series is approximately -0.4. A negative value of α implies an upper bound in the log–Pearson III distribution; the direct method of moments estimates an upper bound for the Feather River of 355,058 ft^3/s. It is remarkable that the 500 year flood estimated by the indirect method in the log–Pearson III, 381,223 ft^3/s, already exceeds this upper bound. It is obvious that a change in

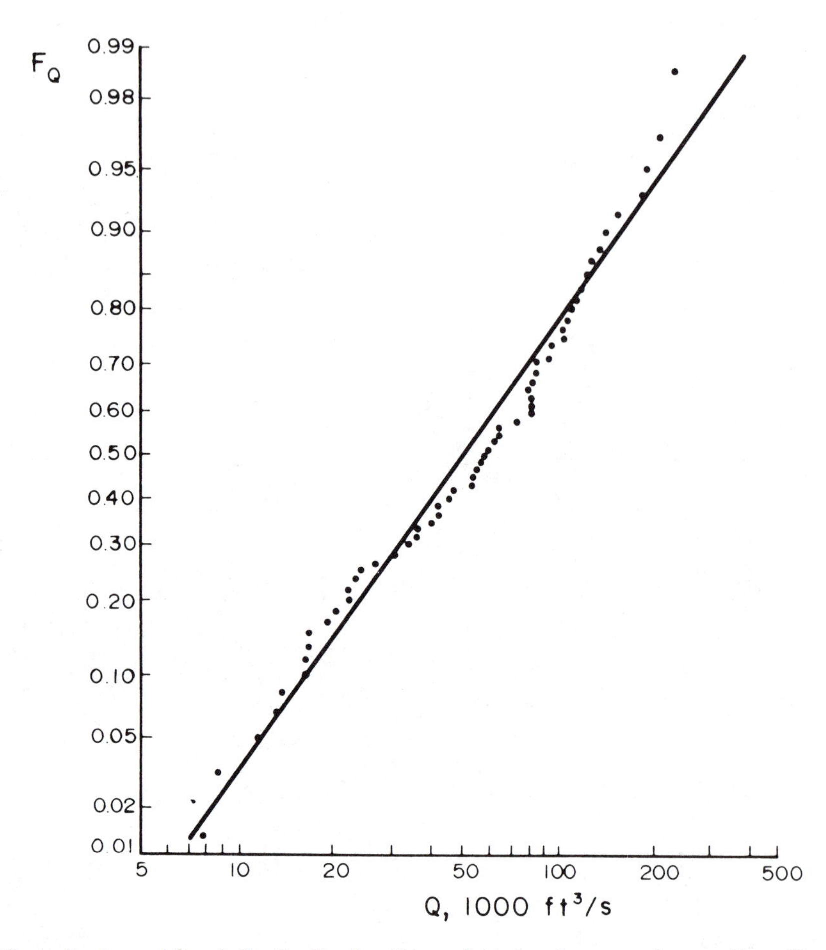

Figure 2. Annual floods for the Feather River plotted on lognormal paper (from Benjamin and Cornell, 1970).

the sign of the skewness, accompanied by the appearance of a different upper bound, will totally change any inference one wishes to make regarding extreme values in the upper tail of the distribution.

3.2. Blackstone River

The 37 years of data (1929–1965) taken from Wood, Rodríguez-Iturbe, and Schaake (1974) are given in Table 4. Tables 5 and 6 give the discharges estimated by the different models for return periods of 50 years and 1000 years, respectively.

Unlike the previous example, the range of observed floods is not covered smoothly in the historical record. In effect, the observed floods vary between

Table 2. Estimation of the 50 Year Flood for the Feather River Using Different Models

Distribution	Discharge (ft^3/s)
Gumbel[a]	192,887
Gumbel[b]	218,792
Lognormal	219,541
Pearson III[a]	243,363
Pearson III[b]	207,979
Log–Pearson III[a]	No solution
Log–Pearson III[c]	204,259
Log–Pearson III[d]	239,517

[a] Parameters estimated by the method of maximum likelihood.

[b] Parameters estimated by the method of moments.

[c] "Direct" method of moments.

[d] "Indirect" method of moments.

Table 3. Estimation of the 1000 Year Flood for the Feather River Using Different Models

Distribution	Discharge (ft^3/s)
Gumbel[a]	305,013
Gumbel[b]	352,031
Lognormal	435,687
Pearson III[a]	429,303
Pearson III[b]	328,348
Log–Pearson III[a]	No solution
Log–Pearson III[c]	274,670
Log–Pearson III[d]	425,087

[a] Parameters estimated by the method of maximum likelihood.

[b] Parameters estimated by the method of moments.

[c] "Direct" method of moments.

[d] "Indirect" method of moments.

2000 ft^3/s and 33,000 ft^3/s with only three values above 10,000 ft^3/s. Moreover, the highest value is more than twice the second largest one. Such extreme empirical distributions are not as infrequent as one might think, and, from a certain perspective, it is all a matter of time scale. An event that would be unthinkable in 20 or 50 years may be almost inevitable in ten thousand. The earth sciences have recorded many examples of major catastrophes in the time scale that such disciplines cover. The kind and magnitude of these extreme events make it difficult to evaluate them in the time scale to which human events are commonly related, and thus there is the temptation to consider natural catastrophes as somewhat beyond reality. The 1955 flood of the

Table 4. Annual Floods of the Blackstone River (1929–1965)[a]

Year	Flood discharge (ft^3/s)
1929	4,570
1930	1,970
1931	8,220
1932	4,530
1933	5,780
1934	6,560
1935	7,500
1936	15,000
1937	6,340
1938	15,100
1939	3,840
1940	5,860
1941	4,480
1942	5,330
1943	5,310
1944	3,830
1945	3,410
1946	3,830
1947	3,150
1948	5,810
1949	2,030
1950	3,620
1951	4,920
1952	4,090
1953	5,570
1954	9,400
1955	32,900
1956	8,710
1957	3,850
1958	4,970
1959	5,398
1960	4,780
1961	4,020
1962	5,790
1963	4,510
1964	5,520
1965	5,300

[a] Statistics of the record: mean = 6373 ft^3/s; standard deviation = 5276.6 ft^3/s; coefficient of skewness = 3.61; coefficient of skewness of the logarithms = 1.15.

Table 5. Estimation of the 50 Year Flood for the Blackstone River Using Different Models

Distribution	Discharge (ft^3/s)
Gumbel[a]	13,595
Gumbel[b]	22,009
Lognormal	21,650
Pearson III[a]	17,265
Pearson III[b]	21,817
Log–Pearson III[a]	17,158
Log–Pearson III[c]	21,298
Log–Pearson III[d]	22,546

[a] Parameters estimated by the method of maximum likelihood.

[b] Parameters estimated by the method of moments.

[c] "Direct" method of moments.

[d] "Indirect" method of moments.

Table 6. Estimation of the 1000 Year Flood for the Blackstone River Using Different Models

Distribution	Discharge (ft^3/s)
Gumbel[a]	20,404
Gumbel[b]	35,994
Lognormal	45,790
Pearson III[a]	28,308
Pearson III[b]	54,221
Log–Pearson III[a]	37,037
Log–Pearson III[c]	48,168
Log–Pearson III[d]	88,982

[a] Parameters estimated by the method of maximum likelihood.

[b] Parameters estimated by the method of moments.

[c] "Direct" method of moments.

[d] "Indirect" method of moments.

Blackstone River is a remarkable event—not uncommon in hydrology—in the sense that its occurrence makes one aware of the existence of major catastrophes. The existence of an artificial upper bound for extreme floods is thus made very questionable.

The estimated 1000 year floods for the Blackstone river vary between 20,000 ft^3/s and 90,000 ft^3/s. Notice that in this case the coefficient of skewness for the data (3.6) is larger than the theoretical skewness of the Gumbel distribution (1.13). This indicates that the Gumbel model may be inadequate to describe floods with high return periods. Similarly, the coefficient of skewness for the logarithms (1.15) makes one doubt that the lognormal distribution is an appropriate model in this case.

When the log–Pearson III distribution is used with parameters estimated using the indirect method of moments (as required by the U.S. Water Resources Council), the estimated 1000 year flood is considerably larger than in the other cases.

It is important to note that under the log–Pearson type III model, for the Blackstone river, all the methods of estimation yielded positives estimates of α. Thus the estimated densities do not have an upper bound.

4. The "Flood Rate" Criterion

We now propose a procedure that may give useful guides to the model selection problem in flood frequency analyses and that can be tested. Let us denote by $Q_1, \ldots, Q_n$ a sample of maximum yearly floods in a given river basin, which we suppose to be independent and identically distributed according to the distribution F with $F(0) = 0$. Denote by $Q_{(i)}$, $i = 1, \ldots, n$, the ordered values:

$$Q_{(1)} \leq Q_{(2)} \leq \cdots \leq Q_{(n)}.$$

We define the *flood rate* by

$$r(q) = \frac{f(q)}{1 - F(q)}, \tag{4.1}$$

where $f(q)$ is the probability density function, which we assume to exist. The interpretation of $r(q)\,dq$ is the probability that a flood of size between q and $q + dq$ is going to take place, given that a flood smaller than q has not occurred.

The function r has been found extremely useful in reliability analysis (among other fields), where it is called the hazard rate and the role of q is taken by t, the times of occurrences of failures of components.

The advantages of focusing a study on $r(q)$ are threefold:

(i) We have found it more natural to give subjective assessments about the shape of r than about that of f.

(ii) The flood rate separates drastically some of the usual candidates for models of floods, whereas inspection of the density shape does not give clear-cut discrimination.

(iii) There exists an extensive literature on estimation and test procedures for r, from the life testing theory, which applies to the present case.

To get an insight into what sort of behavior may be represented, it is important to focus on the study of the shapes of the flood rates of the four models mentioned throughout this paper.

We propose first that the selected model should represent the flood rate shape that the subjective belief of the research workers asserts for the river basin under study. As will be seen below, under certain models, the behavior of r is very sensitive to the parameter values. Since it was seen in the previous section that different methods of estimation yield substantially different estimates, we propose to choose only among the methods of estimation (if any) that produce estimates that, when substituted for the model parameters, resemble the proposed shape of r. It would also be useful to check the subjective beliefs about r, against the historical data, for example, by testing procedures such as will be outlined below. It is important to note that if $T(q)$ is the return period associated with q, i.e. $1/(1 - F(q))$, then from (4.1) we get

$$T(q) = \exp\left[\int_0^q r(y)\,dy\right], \tag{4.2}$$

and thus the return period, f, and F are all uniquely determined by r.

Secondly, we postulate tentatively that for a river basin, the flood rate ought to be nondecreasing. Under the assumption that all the yearly floods arise from the same population model, if $q_0 < q_1$, then it is natural to believe that in a given year

$$\text{Prob[flood is in } (q_0, q_0 + dq) \text{ given that } q > q_0]$$
$$\leq \text{Prob[flood is in } (q_1, q_1 + dq) \text{ given that } q > q_1]. \tag{4.3}$$

More formally, note that also from (4.1) we have that for $s > 0$, $q_0 > 0$

$$\text{Prob}\{q > q_0 + s \text{ given that } q > q_0\} = \exp\left(-\int_{q_0}^{s+q_0} r(x)\,dx\right). \tag{4.4}$$

Therefore, for $q_0 < q_1$ and $r(x)$ nondecreasing,

$$\text{Prob}[q \text{ is in } (q_0, q_0 + s] \text{ given that } q > q_0]$$
$$\leq \text{Prob}[q \text{ is in } (q_1, q_1 + s] \text{ given that } q > q_1], \tag{4.5}$$

and the inequality is reversed for $r(x)$ nonincreasing. Hence the postulate for a river basin implies that the conditional probability of getting a flood in an interval of length s, given that the flood is not smaller than the lowest point at that interval, is a nondecreasing function at that lowest point.

Of course we should be careful about the blanket application of this postulate. It is conceivable, for example, that the existence of a small subset of observations extremely far from the remaining observations will produce a nonmonotonic estimate of r. Whether to consider this set of "atypical" observations as outliers coming from a different population, due for example to hurricanes, prompting a separate analysis, or whether to remain with a unified analysis, is a difficult choice that should be decided on a case by case basis. But unless there is strong evidence that these "atypical" values are due to gross errors of recording, we believe they should not be rejected. And as we argued in Section 3, the situation could well be a matter of time scale. An event that in 30 years of recording seems to be far from the bulk of the data could be almost inevitable in 300 years. As the record gets larger, it is our experience that the

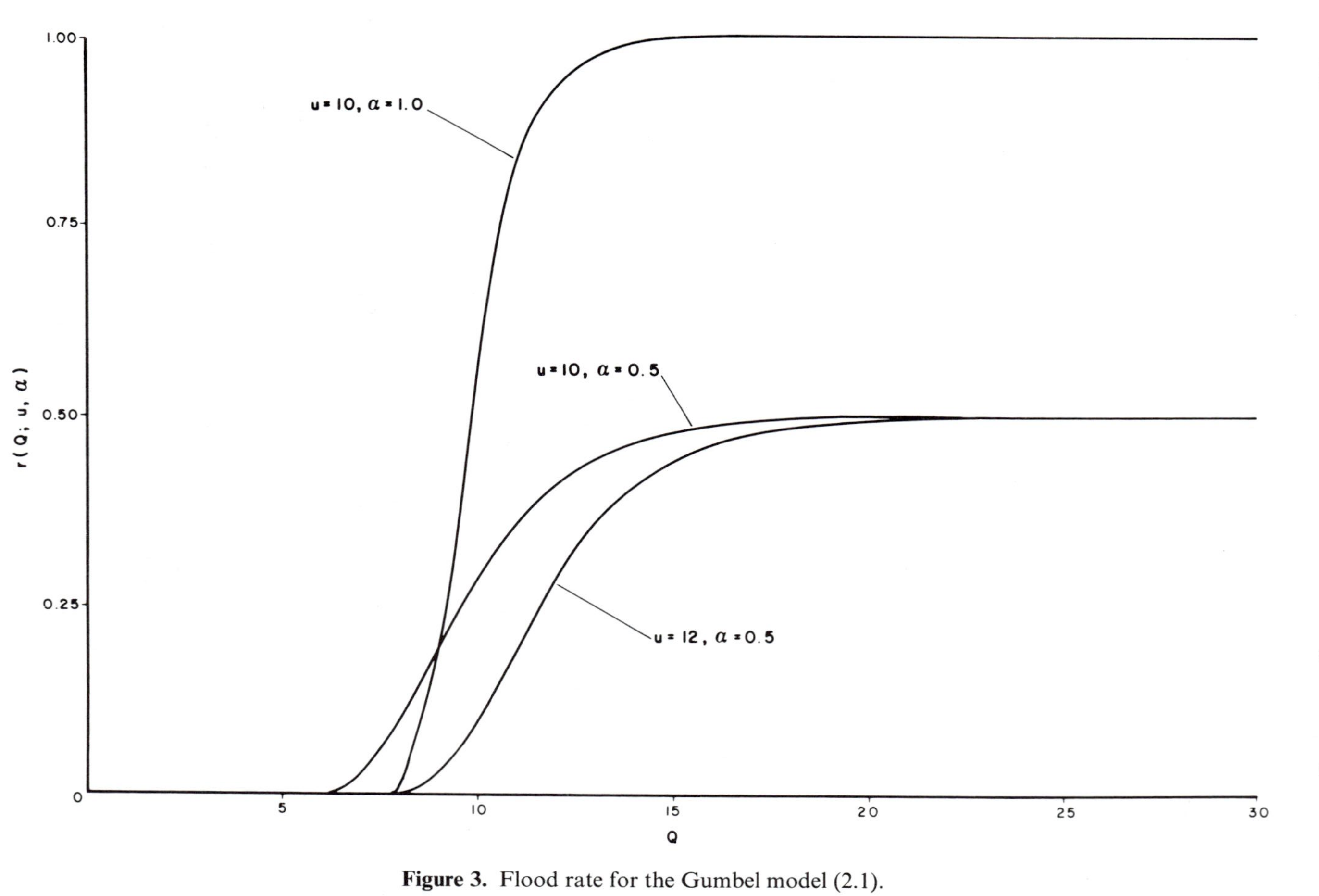

Figure 3. Flood rate for the Gumbel model (2.1).

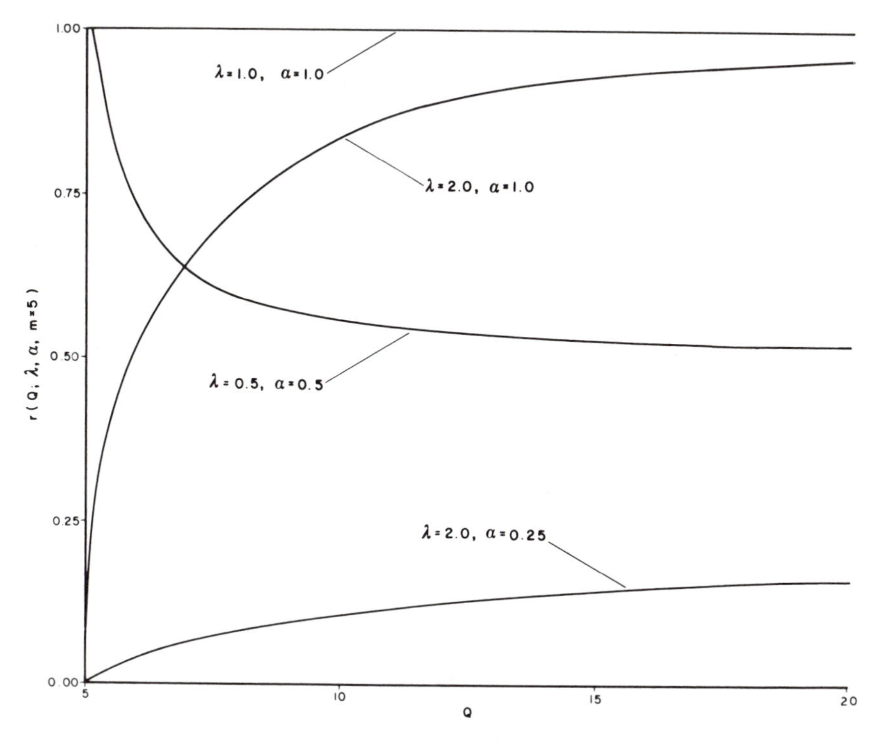

Figure 4. Flood rate for the Pearson type III model (2.7).

data cover the range more smoothly. In any event, the postulated flood rate ought to be checked against the data.

The flood rates for the four models and some parameter values are shown in Figures 3 to 7.

For the Gumbel distribution r is always increasing with a bound at α. Also the Pearson type III is increasing but bounded for $\lambda > 1$, constant for $\lambda = 1$, and decreasing but bounded for $\lambda < 1$. We remark that for the Gumbel model and the Pearson type III model (with $\lambda > 1$), the postulate of increasing r applies. But it is interesting to note also that equality almost occurs in equation (4.5) for very large q_0. This seems to us reasonable. Given that a flood is going to occur above q_0 or above q_1, $q_1 > q_0$, for q_0 extremely far from the span of the data, these models given almost the same conditional probability in $(q_0, q_0 + s]$ and in $(q_1, q_1 + s]$, respectively. The Gumbel model has in its favor its characteristic extreme value distribution to which there is an asymptotic convergence if the assumptions are met by the parent population.

On the other hand, the Pearson type III is more flexible, especially in the coefficient of skewness. Fuller (1914), one of the first engineers to make a systematic study of floods in the U.S.A., claimed on an empirical basis that floods should increase as the logarithm of the return period. Fuller's claim could be written for large q as:

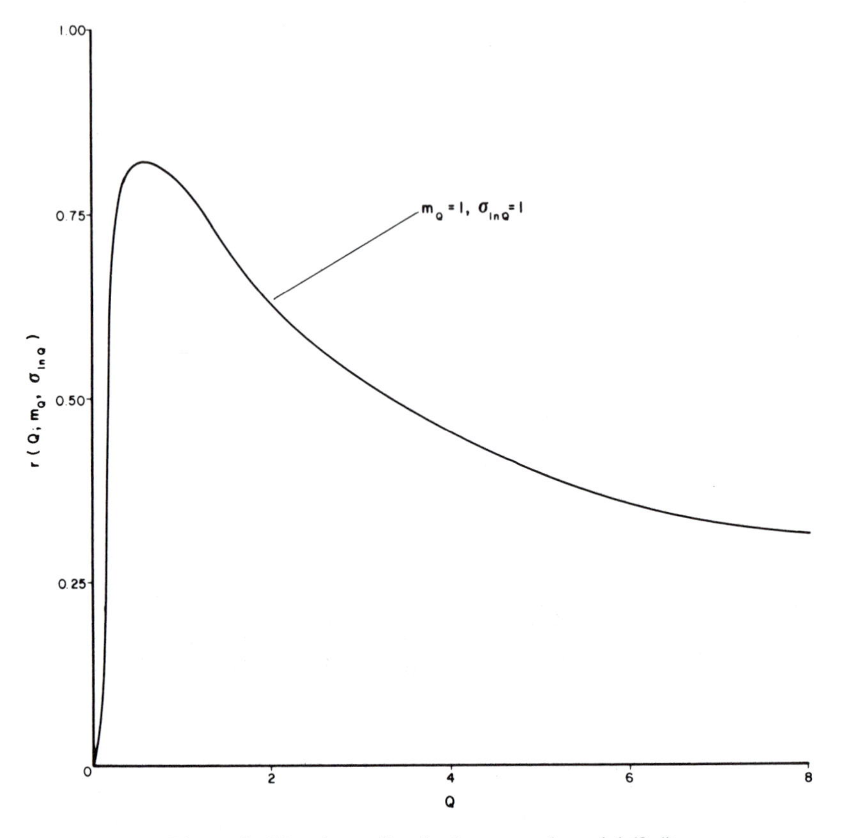

Figure 5. Flood rate for the lognormal model (2.4).

$$\ln T(q) \simeq aq + b.$$

On the other hand, note that from (4.2)

$$\frac{d \ln T(q)}{dq} = r(q). \tag{4.6}$$

Therefore, the behavior of both the Gumbel and Pearson type III distributions is in agreement with Fuller's claim, namely that for large q, the flood rate is approximately constant. Incidentally, equation (4.6) provides another way of expressing our postulate about the flood rate, i.e., the rate of change of the logarithm of the return periods ought to be nondecreasing in q.

Also, from (4.6) it is clear what the implications are for the return periods of r increasing towards infinity or decreasing towards zero.

The lognormal flood rate initially increases and then decreases, approaching zero in the limit, as shown in Goldthwaite (1961). If $r(q)$ decreases towards zero, then for very large q a large increment in the size of the flood will produce a very small increase in the return period, which seems rather odd. A plausible

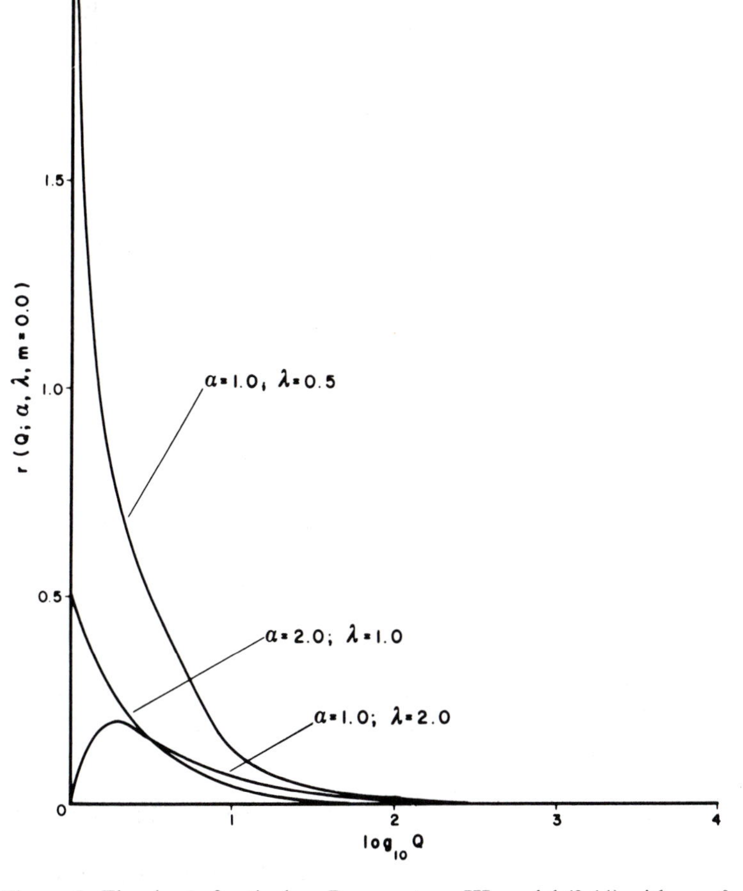

Figure 6. Flood rate for the log–Pearson type III model (2.11) with $\alpha > 0$.

alternative to the lognormal distribution for the representation of floods is the inverse Gaussian distribution,

$$f_Q(q) = \left(\frac{\lambda}{2\pi q^3}\right)^{1/2} \exp\left\{-\frac{\lambda}{2\mu^2 q}(q-\mu)^2\right\}, \qquad q, \mu, \lambda > 0.$$

This distribution has the same flood rate behavior as the lognormal, except that $r(q)$ approaches $\lambda/2\mu^2$ as $q \to \infty$ (Chhikara and Folks, 1977). The log–Pearson III flood rate depends on the value of α. For $\alpha > 0$ there is no upper bound to the distribution and the flood rate either is monotonically decreasing or initially increases and then decreases, approaching zero. For $\alpha < 0$ the distribution has an upper bound and the flood rate either is monotonically increasing or has a "bathtub" shape.

We consider such facts a strong shortcoming of the log–Pearson type III density for the prediction of floods. Despite its flexibility, in order to have an

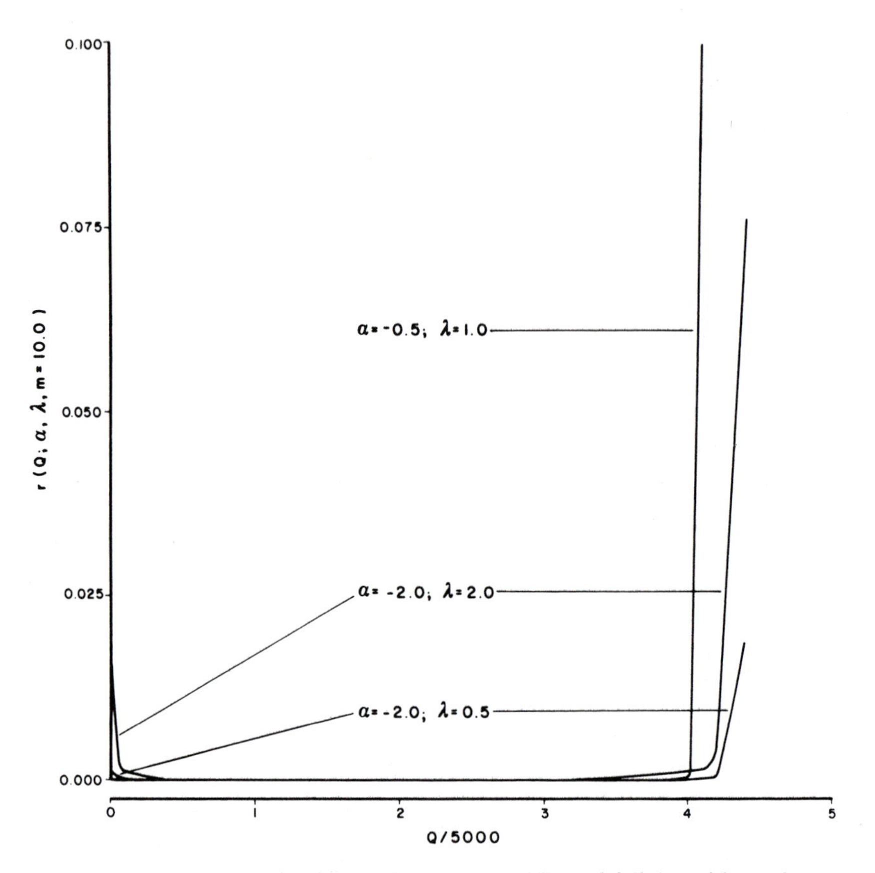

Figure 7. Flood rate for the log–Pearson type III model (2.11) with $\alpha < 0$.

increasing flood rate it is necessary that $\alpha < 0$, which in turn implies the existence of an absolute upper bound for q at e^m. The physical meaning of this upper bound is not at all clear to us. Incidentally, in reliability and survival analysis among other fields (see for example Martz and Waller, 1982; Miller, 1981), the "bathtub" shape of the hazard rate is often postulated. The log–Pearson III with $\alpha < 0$ may be a plausible model in those subjects, where the existence of an upper bound has some physical justification.

We next make use of the "total time on test transformation" in order to check if a given set of ordered data shows an increasing, decreasing, or constant r. We first examine graphically the behavior of r and then use two tests for constant r against increasing or decreasing alternatives. The theory and various optimal properties of the total time on test statistic are given in Barlow, Bartholomew, Bremner, and Brunk (1972). Let

$$D_i = (n - i + 1)(Q_{(i)} - Q_{(i-1)}), \qquad i = 1, \ldots, n, \quad Q_{(0)} = 0,$$

be the normalized sample spacings. If F has constant flood rate, then $\text{Prob}(D_i > D_j) = \frac{1}{2}$, $i \neq j$, but if F has an increasing (decreasing) r, then D_i tends to be

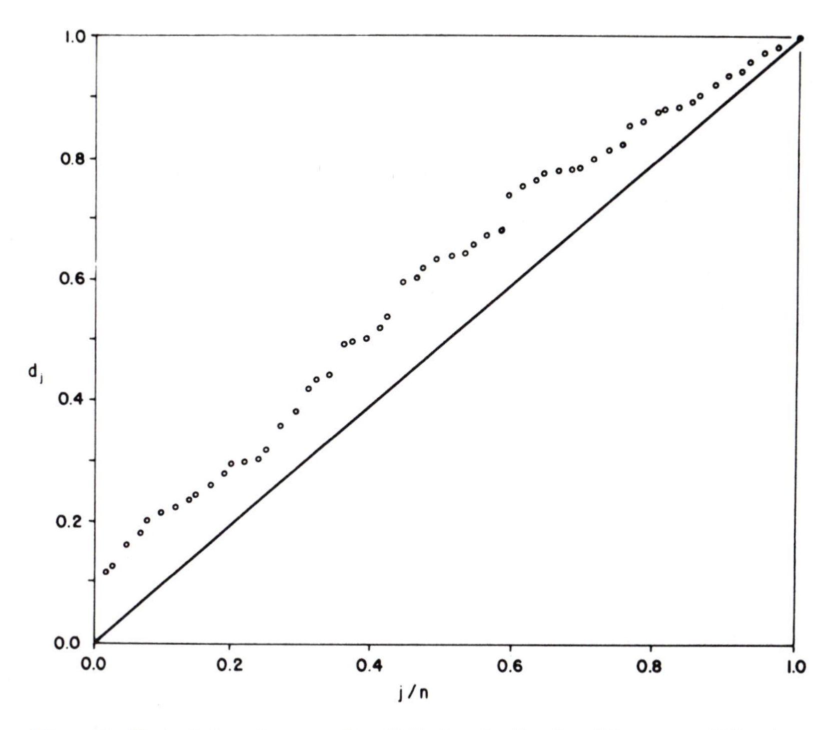

Figure 8. Plot of d_j vs j/n, equation (4.7), for the Feather River annual floods.

larger (smaller) than D_j for $i < j$ (Barlow et al., 1972). Hence, if r is increasing, this should be reflected in a positive slope for the regression of D_{n-i+1} on $i - 1$. In order to make the plot scale-invariant, we divide by the sum of D_i's:

$$d_j = \frac{\sum_{i=1}^{j} D_i}{\sum_{i=1}^{n} D_i}, \tag{4.7}$$

and plot d_j vs j/n for $j = 1, \ldots, n$. If r is increasing, constant, or decreasing, then the line tends to be concave, straight (45° slope), or convex respectively.

Figures 8 and 9 display the plot of d_j vs j/n for the Feather and Blackstone rivers respectively. We also give the plot of d_j vs j/n for the rivers Thames, Manawatu (New Zealand), and Neuquén (Argentina) in Figures 10, 11, and 12 respectively. The clear concave shapes of Figures 8, 10, 11, and 12 suggest an increasing flood rate over the whole span of the data. The situation for the Blackstone river is not so clear cut. After an initial strong concavity, it has a mild convexity suggesting a possible modality on the flood rate.

We now consider two tests put forward in the literature for constant vs increasing hazard rate. Barlow and Proschan (1964) proposed the following procedure: Let

$$V_{ij} = \begin{cases} 1 & \text{if } D_i > D_j, \\ 0 & \text{otherwise,} \end{cases}$$

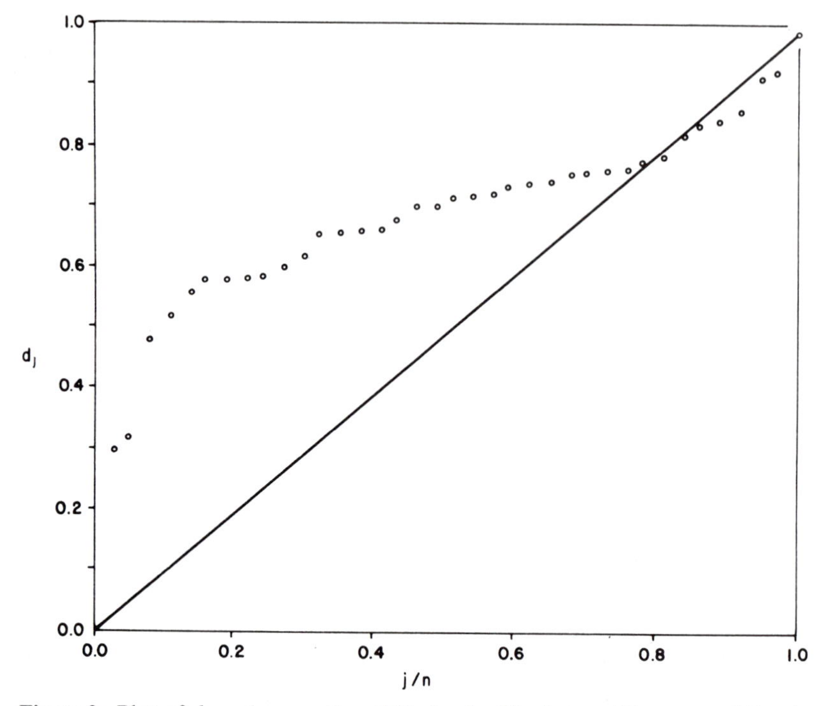

Figure 9. Plot of d_j vs j/n, equation (4.7), for the Blackstone River annual floods.

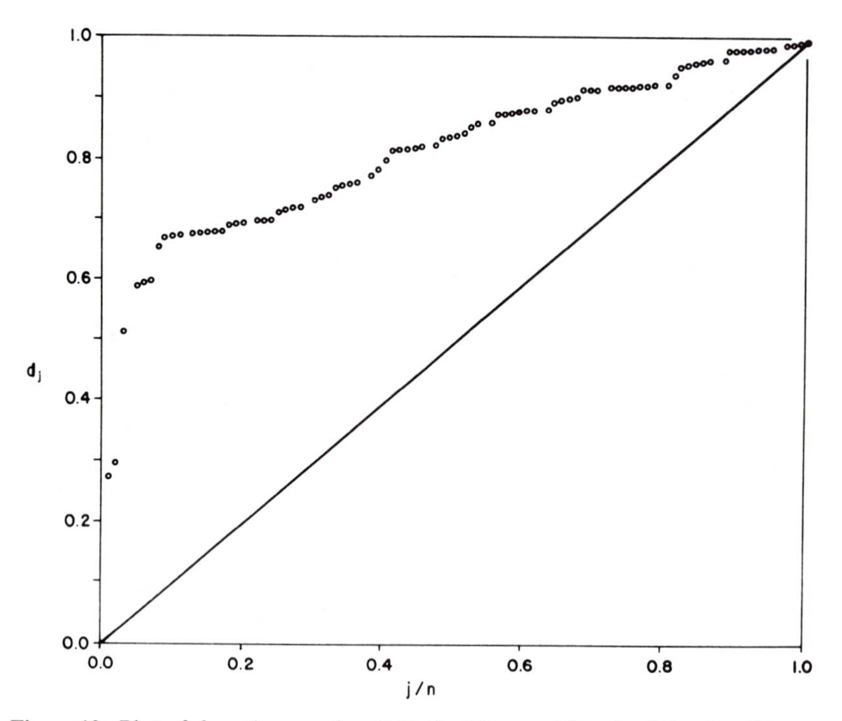

Figure 10. Plot of d_j vs j/n, equation (4.7), for 88 annual floods of the river Thames.

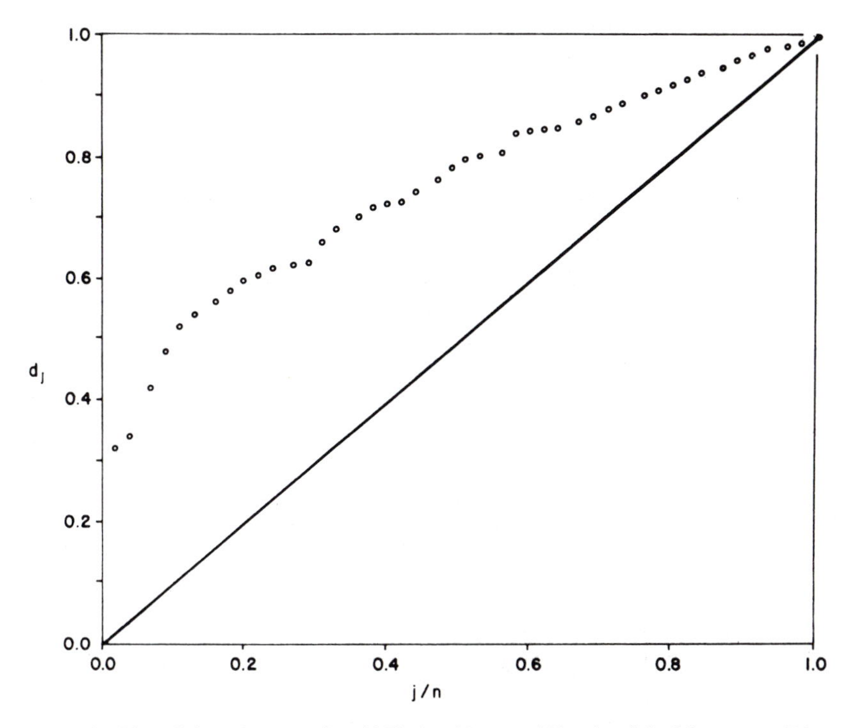

Figure 11. Plot of d_j vs j/n, equation (4.7), for 45 annual floods of the Manawatu River.

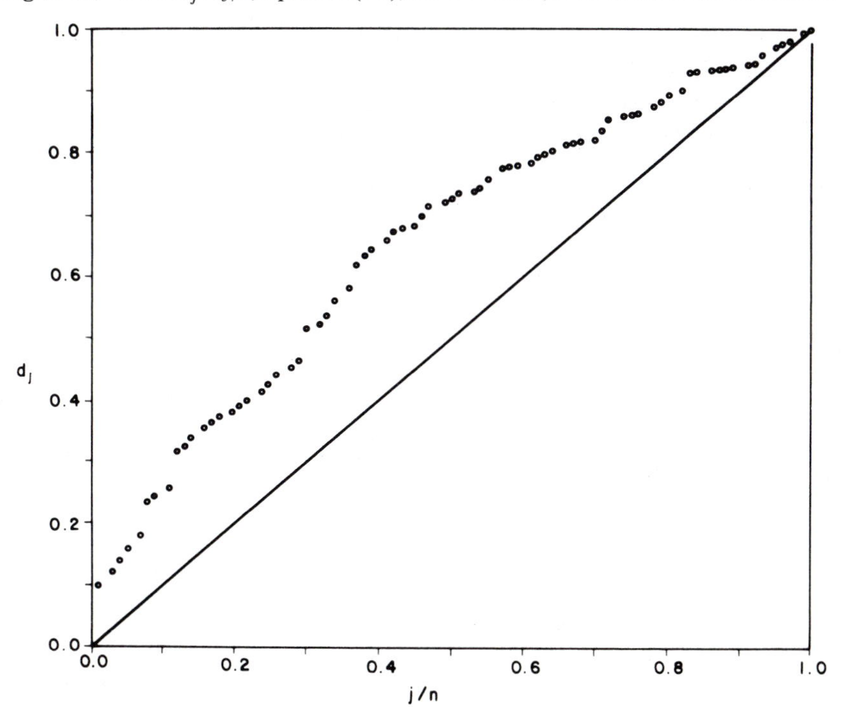

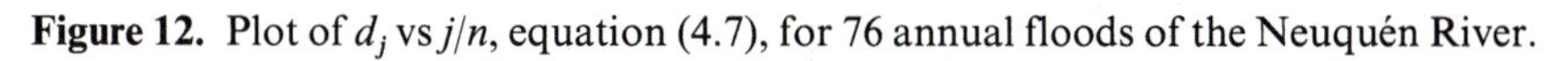

Figure 12. Plot of d_j vs j/n, equation (4.7), for 76 annual floods of the Neuquén River.

and let

$$V_n = \sum_{i<j=1}^{n} V_{ij}.$$

Large values of V_n are regarded as significant, since under the alternative, D_i tends to be larger than D_j, $i < j$. Under H_0, V_n is asymptotically normal with mean and variance $n(n-1)/4$ and $(n-1)n(n+6)/72$ respectively. Thus

$$Z_1 = \frac{V_n - n(n-1)/4}{[(n-1)n(n+6)/72]^{1/2}} \tag{4.8}$$

will be asymptotically standard normal under H_0.

Alternatively, Barlow et al. (1972) suggest a test based on

$$W = \frac{\sum_{i=1}^{n-1}\left[\sum_{j=1}^{i} D_j\right]}{\sum_{i=1}^{n} D_i}.$$

Under the null hypothesis the statistic W is stochastically equivalent to the sum of $n-1$ independent uniform variates in [0, 1]. Hence,

$$Z_2 = [12(n-1)]^{1/2}[(n-1)^{-1}W - \tfrac{1}{2}] \tag{4.9}$$

is asymptotically standard normal, and large values of Z_2 are regarded as significant.

The observed values for the Feather River are $Z_1 = 1.79$ and $Z_2 = 2.51$, which correspond to p-values, (i.e. the minimum level of the test for which the hypothesis would have been rejected) of 0.0367 and 0.0062 respectively. We feel this is a reasonable amount of evidence against constant r, the departure being in the direction of increasing flood rate. For the river Thames, $Z_1 = 4.44$ and $Z_2 = 10.2$; for the Manawatu, $Z_1 = 3.58$ and $Z_2 = 5.7$; and for the Neuquén, $Z_1 = 5.73$ and $Z_2 = 4.65$, yielding strong evidence against H_0 in favor of increasing r. But again, the situation for the Blackstone River is not so clear cut. The observed values of Z_1 and Z_2 are respectively 0.851 and 3.81, showing a strong discrepancy. The p-values are 0.2 and 0.0, respectively. We should recall that the number of years analyzed for the river is just 37, a modest number in view of the asymptotic character of the two tests described above.

We finally examine a method proposed by Cox (1979) to estimate $r(q)$ nonparametrically. Let m be a small integer, and calculate $S_1^{(m)}, S_2^{(m)}, \ldots$ as

$$S_1^{(m)} = Q_{(1)} + Q_{(2)} + \cdots + (n-m+1)Q_{(m)},$$

$$S_2^{(m)} = (Q_{(m+1)} - Q_{(m)}) + \cdots + (n-2m+1)(Q_{(2m)} - Q_{(m)}),$$

and so on. If r_j denotes the average flood rate on the range defining $S_j^{(m)}$, then $2r_jS_j^{(m)}$ is distributed as χ^2 with $2m$ degrees of freedom, assuming that the variations in $r(q)$ are very small on the range defining $S_j^{(m)}$. Thus, by the properties of the log χ^2 distribution,

$$Z_j^{(m)} = -\log\left(\frac{S_j^{(m)}}{m}\right) - \frac{1}{2m - \frac{1}{3}},$$

has mean $\log r_j$ and variance $1/(m - \frac{1}{2})$ approximately. Furthermore, these properties hold for each j independently of the preceeding $Z_j^{(m)}$. A natural plot is then $Z_j^{(m)}$ versus $\log \bar{q}_j^{(m)}$, where $\bar{q}_j^{(m)}$ is the sample mean of the relevant interval. Figures 13 and 14 show Cox's plots for the Feather and Blackstone rivers respectively. We also plotted the fitted values of the Gumbel and lognormal flood rates, where the parameters were estimated by the maximum likelihood method for each river. For the Feather River, $n = 59$ and $m = 6$, except for the last point, where $m = 11$. For the Blackstone River $n = 37$ and $m = 4$, except for the last point, where $m = 5$. It seems that the local variations in the plot are too big. Perhaps the irregular grouping and rounding is affecting the theoretical error.

There is considerable evidence in both plots against constant flood rate. The estimated flood rate of the Feather river appears to be consistent with the tests and with the postulate of ever increasing flood rate with bounded asymptotic value, and thus the Gumbel distribution (if any) performs better, particularly for the last points. The situation for the Blackstone river is quite different. The nonmonotonic behavior of the flood rate is quite clear, namely, an initial

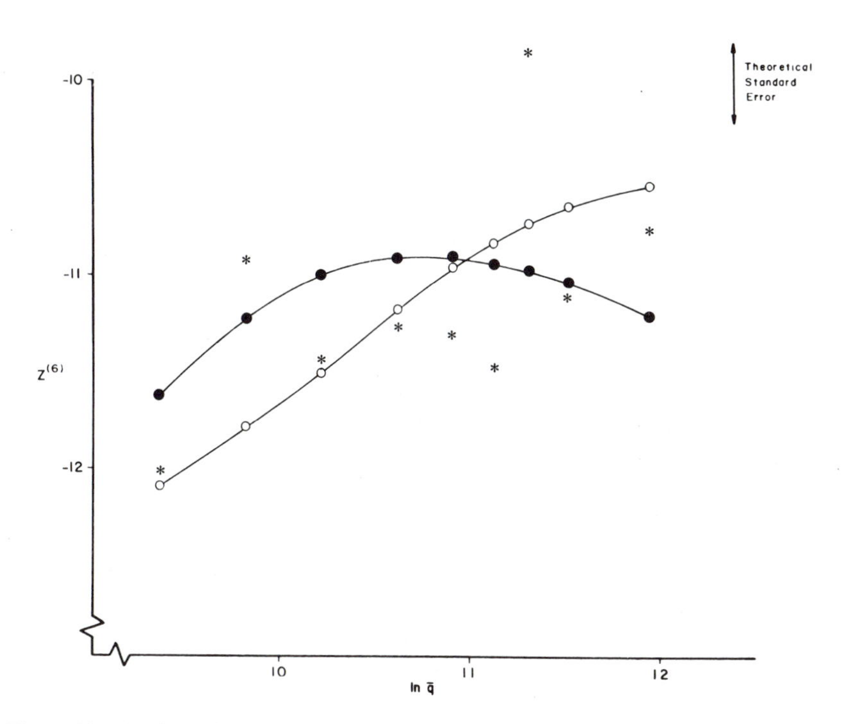

Figure 13. Feather River. $n = 59$, $m = 6$, except last point, where $m = 11$. Theoretical variance: 0.1818; estimated variance: 0.3652. $*$: estimates of $\log r(q)$ by Cox's procedure; o: fitted values under Gumbel model; $\bullet$: fitted values under lognormal model.

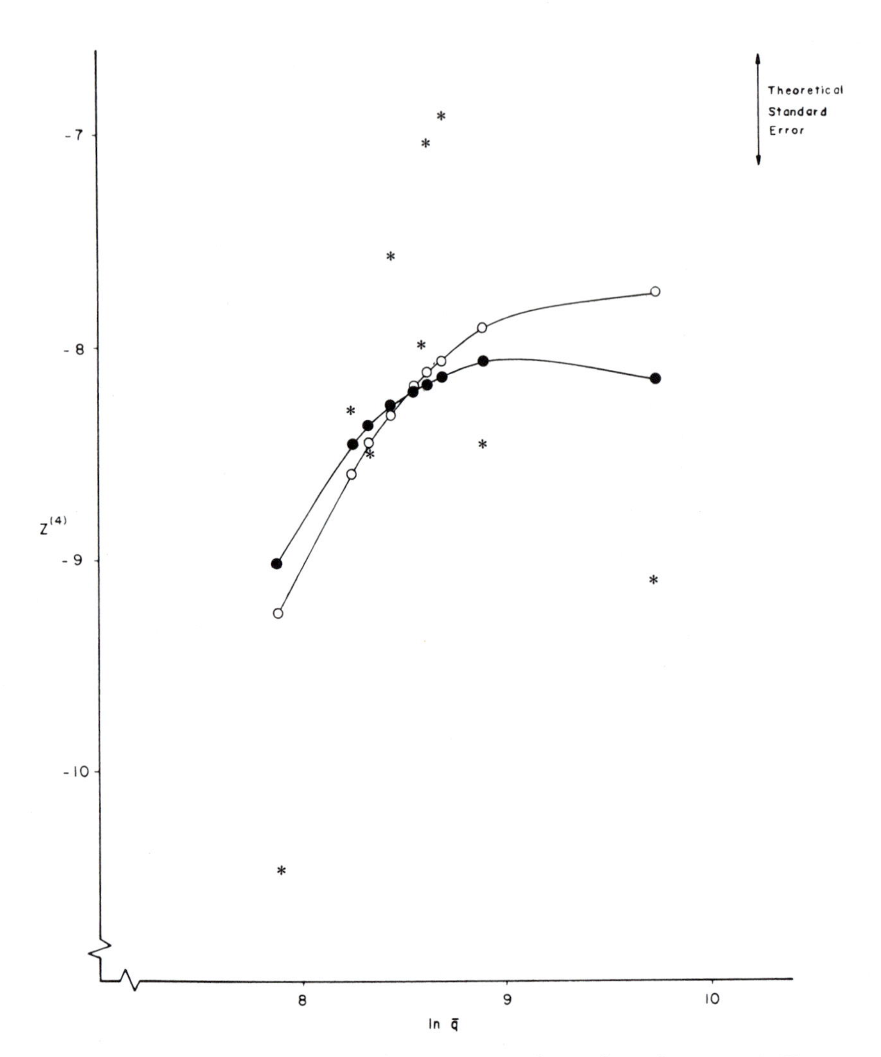

Figure 14. Blackstone River. $n = 37$, $m = 4$, except last point, where $m = 5$. Theoretical variance: 0.2857; estimated variance: 1.189. $*$: estimates of $\log r(q)$ by Cox's procedure; o: fitted values under Gumbel model; •: fitted values under lognormal model.

increase of $r(q)$ followed by a decrease towards an asymptotic value greater than zero, as in the inverse Gaussian, although this is not completely apparent from the picture because of the last, very small value. But a flood rate moving towards zero will pose strong difficulties of interpretation and will be against important empirical evidence for longer records. Although none of the models seem to fit well, the lognormal is slightly better over the span of the data, due to the nonmonotonic behavior of $r(q)$.

Different models could be fitted to the plots, which have the advantage that the $Z_j^{(m)}$'s, are independent. This makes the interpretation of the plots easier.

5. Conclusions

We have commented the necessity of model choice procedures for the representation of floods and the inconvenience of the blanket selection of the log–Pearson III model, for which some undesirable features have been noted. We have put forward simple procedures based on the flood rate, in order to make an initial screening of models, which seem to be successful in a case where classical goodness of fit procedures were indecisive.

From the cases studied, there appear to be two classes of rivers:

(1) Rivers with ever increasing flood rates and bounded asymptotic value.
(2) Rivers that show an initial increase in the flood rate and after that a decrease towards an asymptotic value greater than zero.

We then propose that plausible candidate models should be able to display the observed behavior of the flood rate.

For more specific, refined, and complex methods of separate model discrimination, the reader is referred to Cox (1961, 1962) for a classical framework or to the outline in Pericchi and Rodríguez-Iturbe (1983) for a Bayesian approach.

The discrepancy between the different methods of estimation remains open, but this may be solved numerically, by a Bayesian procedure of estimation that incorporates data-based priors obtained from regional analysis, or non-data-based priors which represent the subjective knowledge of experts. In any case, the fitted flood rate should be plotted against the estimated flood rate, in order to check if it is well represented.

Acknowledgements

We wish to express our gratitude for the help and efforts of Mr. Rafael Seoane and Mrs. Lelys Bravo de Guenni in the calculations of the examples presented in this paper, as well as for many stimulating discussions. We also gratefully acknowledge the enjoyable and stimulating interaction we have had with Professor D. R. Cox, from Imperial College, both on the scientific and on the human side, during his recent visit to our University.

Bibliography

Aitchison, J. and Brown, J. A. C. (1957). *The Log Normal Distribution*. London: Cambridge U.P.

Barlow, R. E. and Proschan, F. (1964). *Mathematical Theory of Reliability*. New York: Wiley.

Barlow, R. E., Bartholomew, D. J., Bremner, J. M., and Brunk, H. D. (1972). *Statistical Inference under Order Restrictions*. New York: Wiley.

Benjamin, J. R. and Cornell, C. A. (1970). *Probability, Statistics and Decision for Civil Engineers*. New York: McGraw-Hill.

Benson, M. A. (1968). "Uniform flood frequency estimating methods for the Federal agencies." *Water Resources Res.*, **4**, 891–908.

Bobée, B. (1975). "The log Pearson type 3 distribution and its application in hydrology." *Water Resources Res.*, **11**, 681–689.

Buishand, T. A. (1983). "The effect of seasonal variation and serial correlation on the extreme value distribution of rainfall data." In *Proceedings II International Meeting on Statistical Climatology*, 10.4.1–10.4.6.

Cheng, R. C. H. and Amin, N. A. K. (1983). "Estimating parameters in continuous univariate distributions with a shifted origin." *J. Roy. Statist. Soc. Ser. B*, **45**, 394–403.

Chhikara, R. S. and Folks, J. L. (1977). "The inverse Gaussian distribution as a lifetime model." *Technometrics*, **19**, 461–468.

Cox, D. R. (1961). "Test of separate families of hypotheses." In J. Neyman (ed.), *Proceedings of the Fourth Berkeley Symposium on Mathematical Statistics and Probability*, **1**. Berkeley: Univ. of California Press, 105–123.

Cox, D. R. (1962). "Further results on test of separate families of hypotheses." *J. Roy. Statist. Soc. Ser. B*, **24**, 406–424.

Cox, D. R. (1979). "A note on the graphical analysis of survival data." *Biometrika*, **66**, No. 1, 188–190.

Cramér, H. (1965). "A limit theorem for the maximum values of certain stochastic processes." *Teoriya Veroyatnostie i ee Primeniya*, **10**, 137–139.

Cramér, H. (1966). "On the intersections between the trajectories of a normal stationary process and a high level." *Ark. Mat.*, **6**, 337–349.

David, H. A. (1970). *Order Statistics*. New York: Wiley.

Ditlevsen, O. (1966). "Extremes of realizations of continuous time stationary stochastic processes on closed intervals." *J. Math. Anal. Appl.* **14**, 463–474.

Ditlevsen, O. (1971). *Extremes and First Passage Times with Applications in Civil Engineering*. Doctoral Thesis. Copenhagen: Technical Univ. of Denmark, 414.

Fuller, W. E. (1914). "Flood flows." *Trans. Amer. Soc. Civ. Eng.*, 77.

Goldthwaite, L. (1961). "Failure rate study for the lognormal lifetime model." In *Proceedings of the Seventh National Symposium on Reliability and Quality Control*, 208–213.

Gumbel, E. J., (1958). *Statistics of Extremes*. New York: Columbia U.P.

Hill, B. M. (1963). "The three parameter lognormal distribution and Bayesian analysis of a point source epidemic." *J. Amer. Statist. Assoc.*, **58**, 72–84.

Jenkinson, A. F., (1955). "The frequency distribution of annual maximum (or minimum) values of meteorologial elements." *Quart. J. Roy. Meteorol. Soc.*, **81**, 158–171.

Johnson, N. L. and Kotz, S. (1970). *Continuous Univariate Distributions—1*. New York: Wiley.

Leadbetter, M. R., Lindgren, G., and Rootzén, H. (1983). *Extremes and Related Properties of Random Sequences and Processes*. New York: Springer.

Martz, H. F. and Waller, R. A. (1982). *Bayesian Reliability Analysis*. New York: Wiley.

Matalas, N. C. and Wallis J. R. (1973). "Eureka! It fits a Pearson type 3 distribution." *Water Resources Res.*, **9**, 281–289.

Miller, R. G. (1981). *Survival Analysis.* New York: Wiley.

Pericchi, L. R. and Rodriguez-Iturbe, I. (1983). "On some problems in Bayesian model choice in hydrology." The *Statistician*, **32**.

Reich, B. M. (1972). "Log–Pearson type 3 and Gumbel analysis of floods." In *Second International Symposium in Hydrology.* Fort Collins, CO: Water Resources Publications.

Tiago de Oliveira, J. (1982). "Decision and modelling for extremes." In J. Tiago de Oliveria and B. Epstein (eds.), *Some Recent Advances in Statistics.* New York: Academic.

United Nations–WMO (1967). *Assessment of the Magnitude and Frequency of Flood Flows.* Water Resources Ser., No. 30, 206.

U.S. Water Research Council (1981). *Guidelines for Determining Flood Flow Frequency.* Bulletin No. 17B. Washington, DC: U.S. Government Printing Office.

Wood, E. F., Rodríguez-Iturbe, I., and Schaake, J. C. (1974). *The Methodology of Bayesian Inference and Decision Making Applied to Extreme Hydrologic Events.* Report No. 178. Ralph M. Parsons Lab., M. I. T.

CHAPTER 24

Weighted Distributions Arising Out of Methods of Ascertainment: What Population Does a Sample Represent?

C. Radhakrishna Rao

Abstract

The concept of weighted distributions can be traced to the study of the effects of methods of ascertainment upon the estimation of frequencies by Fisher in 1934, and it was formulated in general terms by the author in a paper presented at the First International Symposium on Classical and Contagious Distributions held in Montreal in 1963. Since then, a number of papers have appeared on the subject. This paper reviews some previous work, points out, through appropriate examples, some situations where weighted distributions arise, and discusses the associated methods of statistical analysis.

Weighted distributions occur in a natural way in specifying probabilities of events as observed and recorded by making adjustments to probabilities of actual occurrence of events taking into account methods of ascertainment. Failure to make such adjustments can lead to wrong conclusions.

1. Sample and Population

Statisticians are often required to work with data provided by customers and answer questions raised by them. The questions relate to a real or hypothetical population which gave rise to the data, usually referred to as a "sample from a population." The role of statistical methodology is to extract the relevant information from a given sample to answer specific questions about the parent population (which the sample is presumed to represent). For this purpose, it is necessary to identify all possible samples that can be observed from a population (sample space) and to provide a stochastic model for attaching probabilities to different sets of samples (specification). The link between sample and population is specification, and the question "What population does a sample represent?" is technically equivalent to "What is the appropriate specification?" Wrong specification can lead to invalid inference, which is sometimes referred to as the third kind of error, the first two being the

Key words and phrases: damage models, nonresponse, probability sampling, quadrat sampling, size biased sampling, truncation, weighted distributions.

familiar two kinds of errors associated with the Neyman–Pearson theory of testing of hypotheses.

How does a statistician decide on an appropriate specification for a given sample? Unfortunately, there is not much discussion of this basic question in statistical literature, although any statistical inference presupposes some kind of specification. The present paper is primarily addressed to the problem of specification based on how a sample is drawn from a population. It illustrates through live examples in some areas of applied research the use of what are called weighted distributions in choosing the appropriate specification and the associated statistical methodology.

The problem of specification is not a simple one. A detailed knowledge of the procedure actually employed in acquiring data is an essential ingredient in arriving at a proper specification. The situation is more complicated with field observations and nonexperimental data, where nature produces events according to a certain stochastic model, which are observed and recorded by investigators. There does not always exist a suitable sampling frame for observing events and applying the classical sampling theory. In practice, it is not always possible to observe and record all events which occur. For instance, certain events may not be observable by the method we follow and therefore be missed in the record (truncated, censored, and incomplete samples). Or an event may be observable only with a certain probability depending on the characteristics of the event, such as its conspicuousness and the procedure employed to observe it (unequal probability sampling). Or an event may change in a random way by the time or during the process of observation, so that what comes on record is a modified event (damage models). Sometimes, events produced under two or more different mechanisms with unspecified relative frequencies get mixed up and brought into the same record (outliers, contaminated samples). In all these cases, the specification for the original events (as they occur) may not be appropriate for the events as they are recorded (observed data) unless it is suitably modified.

In a classical paper, Fisher (1934) demonstrated the need for such adjustment in specification depending on the way the data are ascertained. In extending the basic ideas of Fisher, the author (Rao, 1965) introduced the concept of a weighted distribution as a method of adjustment applicable to many situations. In the present paper we discuss, through live examples, some procedures for making adjustments in specification based on methods of ascertaining data.

Although I have mentioned only field observations which are collected without the help of a suitable sampling frame, I must emphasize that similar problems of specification arise with data collected in large scale sample surveys and also with data acquired through field and laboratory experiments. Survey practitioners are faced with problems of incomplete frame, which raise questions of the representativeness of a sample for a given population (see Kruskal and Mosteller, 1980, and references therein); nonresponse, which raises questions of repeated visits to sampled units; replacement of nonresponding units by others with possibly similar characteristics, and imputation of values (Fienberg and Tanur, 1983; Fienberg and Stasny, 1983; Rubin, 1976, 1980); and nonsampling errors, which raise questions about their recognition, detection, and measurement, and lead to making adjustments in expressing the precision of estimates (Mahalanobis, 1944; Mosteller, 1978). Similarly, in the design of experiments, difficulties in random allocation of treatments and choice of controls in field trials, pooling of evidence from different experiments conducted over space and time, and missing values (dropouts) introduce additional uncertainties in statistical inference and the interpretation of results for practical use or policy purposes (for typical problems and references see Fienberg, Singer, and Tanur, 1985, Chapter 12 in this volume; Neyman, 1977).

2. Truncation and Censoring

Some events, although they occur, may be unascertainable, so that the observed distribution is truncated to a certain region of the sample space. An example is the frequency of families with both parents heterozygous for albinism but having no albino children. There is no evidence that the parents are heterozygous unless they have an albino child, and the families with such parents and having no albino children get confounded with normal families. The actual frequency of the event "zero albino children" is thus not ascertainable. Adjustment to the probability distribution applicable to observable events in such a case is simple.

In general, if $p(x, \theta)$ is the pdf (probability density function), where θ denotes unknown parameters, and the rv X is truncated to a specified region $T \subset \mathscr{X}$, the sample space, then the pdf of the truncated random variable X^w is

$$p^w(x, \theta) = \frac{w(x, T)p(x, \theta)}{u(T, \theta)}, \tag{2.1}$$

where $w(x, T) = 1$ if $x \in T$ and $= 0$ if $x \notin T$, and $u(T, \theta) = E[w(X, T)]$. If $x_1, \ldots, x_n$ are independent observations subject to truncation, then the likelihood is

$$\frac{p(x_1, \theta) \cdots p(x_n, \theta)}{[u(T, \theta)]^n}. \tag{2.2}$$

In some cases we may have independent observations $x_1, \ldots, x_n$ arising from a truncated distribution in addition to a number m (and not the actual values) of observations falling outside T. Then the likelihood is

$$\frac{(n + m)!}{m!} p(x_1, \theta) \cdots p(x_n, \theta)[1 - u(T, \theta)]^m. \tag{2.3}$$

A more complicated case is the following. Suppose that we have a measuring device which records the time at which a bulb fails. If we are experimenting with n bulbs in a life testing problem using a measuring device which may itself fail at a random time, then the observations would be of the type

$$x_1, \ldots, x_{n_1}, n_2, n_3, \tag{2.4}$$

where $x_1, \ldots, x_{n_1}$ are the lifetimes of n_1 bulbs recorded before an unknown time point T at which the measuring device failed, n_2 is the number of bulbs that failed between T and T_0, the known time at which the experiment was terminated, and n_3 is the number of bulbs still burning after T_0. Let

$$w_1(T,\theta) = P(x \le T), \qquad w_2(T,\theta) = P(T < x \le T_0),$$
$$w_3(T,\theta) = 1 - w_1(T,\theta) - w_2(T,\theta).$$

Then the likelihood based on the data (2.4) is

$$\frac{n!}{n_2!n_3!} p(x_1,\theta)\cdots p(x_{n_1},\theta)[w_2(T,\theta)]^{n_2}[w_3(T,\theta)]^{n_3}, \tag{2.5}$$

where T is unknown as well as the basic parameters θ. Inference on T and θ based on (2.5) does not seem to have been fully worked out, but could be developed on standard lines.

The expressions (2.2), (2.3), and (2.5) are simple examples of weighted distributions, whose general definition is given in Section 3.

3. Weighted Distributions

In Section 2, we have considered situations where certain events are unobservable. But a more general case is where an event that occurs has a certain probability of being recorded (or included in the sample). Let X be a rv with $p(x,\theta)$ as the pdf, and suppose that when $X = x$ occurs, the probability of recording it is $w(x,\alpha)$, depending on the observed value x and possibly also on an unknown parameter α. Then the pdf of the resulting rv X^w is

$$p^w(x,\theta,\alpha) = \frac{w(x,\alpha)p(x,\theta)}{E[w(X,\alpha)]}. \tag{3.1}$$

Although in deriving (3.1) we chose $w(x,\alpha)$ such that $0 \le w(x,\alpha) \le 1$, we can define (3.1) for any arbitrary nonnegative weight function $w(x,\alpha)$ for which $E[w(X,\alpha)]$ exists. The distribution (3.1) obtained by using any nonnegative weight function $w(x,\alpha)$ is called (see Rao, 1965) a weighted version of $p(x,\theta)$. In particular, the weighted distribution

$$p^w(x,\theta) = \frac{|x|p(x,\theta)}{E[|x|]}, \tag{3.2}$$

where $|x|$ is the norm or some measure of size of x, is called the size biased distribution. When x is univariate and nonnegative, the weighted distribution

$$p^w(x, \theta) = \frac{xp(x, \theta)}{E(X)} \tag{3.3}$$

is called length (size) biased distribution. For example, if X has the logarithmic series distribution

$$\frac{\alpha^r}{-r\log(1-\alpha)}, \qquad r = 1, 2, \ldots, \tag{3.4}$$

then the distribution of the size biased variable is

$$\alpha^{r-1}(1-\alpha), \qquad r = 1, 2, \ldots, \tag{3.5}$$

which shows that $X^w - 1$ has a geometric distribution. A truncated geometric distribution is sometimes found to provide a good fit to an observed distribution of family size (Feller, 1968). But, if the information on family size has been ascertained from school children, then the observations will have a size biased distribution. In such a case a good fit of the geometric distribution to the observed family sizes would indicate that the underlying distribution of family size is, in fact, a logarithmic series.

Table 1 gives a list of some basic distributions and their size biased forms. It is seen that the size biased form belongs to the same family as the original distribution in all cases except the logarithmic series [see Rao (1965), Patil and Ord (1975), Janardhan and Rao (1983) for characterizations and examples of size biased distributions].

An example of weighted distributions arises in sample surveys when unequal probability sampling or pps (probability proportional to size) sampling is employed. A general version of the sampling scheme involves two rv's X and Y with pdf $p(x, y, \theta)$ and a weight function $w(y)$ which is a function of y only, giving the weighted pdf

$$p^w(x, y, \theta) = \frac{w(y)p(x, y, \theta)}{E[w(Y)]}. \tag{3.6}$$

In sample surveys we obtain observations on (X^w, Y^w) from the pdf (3.6) and draw inference on the unknown parameter θ.

It is of interest to note that the marginal pdf of X^w is

$$p^w(x, \theta) = \frac{w(x, \theta)p(x, \theta)}{E[w(X, \theta)]}, \tag{3.7}$$

which is a weighted version of $p(x, \theta)$ with the weight function

$$w(x, \theta) = \int p(y \mid x)w(y)\, dy, \tag{3.8}$$

which may involve the unknown parameter θ.

Table 1. Certain Basic Distributions and Their Size-Biased Forms

Random variable (rv)	pf (pdf)	Size-biased rv
Binomial, $B(n,p)$	$\binom{n}{x}p^x(1-p)^{n-x}$	$1 + B(n-1,p)$
Negative binomial, $\mathrm{NB}(k,p)$	$\binom{k+x-1}{x}q^x p^k$	$1 + \mathrm{NB}(k+1,p)$
Poisson, $\mathrm{Po}(\lambda)$	$e^{-\lambda}\lambda^x/x!$	$1 + \mathrm{Po}(\lambda)$
Logarithmic series, $L(\alpha)$	$\{-\log(1-\alpha)\}^{-1}\alpha^x/x$	$1 + \mathrm{NB}(1,\alpha)$
Hypergeometric, $H(n,M,N)$	$\binom{n}{x}\frac{M^x(N-M)^{n-x}}{N^n}$	$1 + H(n-1,M-1,N-1)$
Binomial beta, $\mathrm{BB}(n,\alpha,\gamma)$	$\binom{n}{x}\frac{\beta(\alpha+x,\gamma+n-x)}{\beta(\alpha,\gamma)}$	$1 + \mathrm{BB}(n-1,\alpha,\gamma)$
Negative binomial beta, $\mathrm{NBB}(k,\alpha,\gamma)$	$\binom{k+x-1}{x}\frac{\beta(\alpha+x,\gamma+k)}{\beta(\alpha,\gamma)}$	$1 + \mathrm{NBB}(k+1,\alpha,\gamma)$
Gamma, $G(\alpha,k)$	$\alpha^k x^{k-1}e^{-\alpha x}/\Gamma(k)$	$G(\alpha,k+1)$
Beta first kind, $B_1(\delta,\gamma)$	$x^{\delta-1}(1-x)^{\gamma-1}/\beta(\delta,\gamma)$	$B_1(\delta+1,\gamma)$
Beta second kind, $B_2(\delta,\gamma)$	$x^{\delta-1}(1+x)^{-\gamma}/\beta(\delta,\gamma-\delta)$	$B_2(\delta+1,\gamma-\delta-1)$
Pearson type V, $\mathrm{Pe}(k)$	$x^{-k-1}\exp(-x^{-1})/\Gamma(k)$	$\mathrm{Pe}(k-1)$
Pareto, $\mathrm{Pa}(\alpha,\gamma)$	$\gamma\alpha^\gamma x^{-(\gamma+1)}, x \geq \alpha$	$\mathrm{Pa}(\alpha,-1)$
Lognormal, $\mathrm{LN}(\mu,\sigma^2)$	$(2\pi\sigma^2)^{-1/2}x^{-1}\exp - \left(\frac{\log x - \mu}{\sigma\sqrt{2}}\right)^2$	$\mathrm{LN}(\mu+\sigma^2,\sigma^2)$

An extensive literature on weighted distributions has appeared since the concept was formalized in Rao (1965); it is reviewed with a large number of references in a paper by Patil (1984) with special reference to ecological work. Reference may also be made to two earlier contributions by Patil and Rao (1977, 1978), and Patil and Ord (1976) which contain reviews of previous work and details of some new results.

In the next sections, we consider several examples where weighted distributions are used in the analysis of data.

4. Are Only Small Skulls Well Preserved?

The following problem arose in the analysis of cranial measurements. A sample of skulls dug out from ancient graves in Jebel Moya, Africa, consisted of some well-preserved skulls and the rest in a broken condition (see Mukherji, Trevor, and Rao, 1955). On each well-preserved skull it was possible to take four measurements, C (capacity), L (length), B (breadth), and H (height), while on a broken skull only a subset of L, B, and H and *not* C could be measured. The observed data, thus, consisted of samples from a four variate population with several observations missing. There were some sets with all the four measurements C, L, B, H, and some with one or two or three of the measurements L, B, and H only. The problem was to estimate the mean values of C, L, B, and H in the *original* population of skulls from the recovered fragmentary samples. In a number of papers which appeared in the early issues of *Biometrika*, it was the practice to estimate the unknown population mean value of any characteristic, say C, by taking the mean of all the available measurements on C. An alternative to this, which is often recommended, is to compute maximum likelihood estimates of the unknown mean values, variances, and covariances by writing down the likelihood function based on all the available data assuming a four variate normal distribution for C, L, B, and H and using the derived marginal distribution for an incomplete set of measurements. This is based on the *assumption* that each skull admitting all the four measurements or any subset of the four can be considered as a random sample from the *original* population of skulls. Is this assumption valid?

It is common knowledge that a certain proportion of the original skulls get broken, depending on the length of time and depth at which they lay buried. Let $w(c)$ be the probability that a skull of capacity c is not broken, and $p(c, \theta)$ be the pdf of C in the original population. Then the pdf of C measured on well-preserved skulls is

$$\frac{w(c)p(c, \theta)}{E[w(C)]}. \tag{4.1}$$

If $w(c)$ depends on c, then the *observed* measurements on C cannot be considered as a random sample of C from the *original* population. Further, if $w(c)$ is a decreasing function of c, then there will be a larger representation of small skulls among the unbroken skulls, and therefore the mean of the available measurements on C will be an underestimate of the mean capacity of the original population.

Is there any evidence that $w(c)$ depends on c? To answer this question, the regression of C on L, B, and H (in terms of logarithms) was estimated from the data sets where all the four measurements were available and used to predict the mean capacity of broken skulls by substituting the observed averges $\bar{L}$, $\bar{B}$, and $\bar{H}$ of broken skulls in the regression equation. At least in two series of cranial measurements (see Rao and Shaw, 1948; Rao, 1973, p. 280), it was found that the average measured capacity of unbroken skulls was smaller than the estimated average capacity of broken skulls. This provided some evidence about the differential preservation of skulls, with smaller skulls having a higher chance of remaining unbroken.

This finding invalidates the assumption that skulls providing all four measurements constitute a random sample from the original population of skulls. The pdf associated with these measurements is more appropriately (4.1), which is a weighted version of the original pdf with an unknown weight function. Presumably, the pdf associated with observations on any subset of L, B, and H will again be a weighted pdf with a weight function depending on the degree of damage to a skull. The expression for the correct likelihood will then depend on the original pdf and the probabilities of different degrees of damage as assessed by subsets of measurements that can be taken on a skull, which are likely to be unknown. Is there a reasonable solution to the problem of estimation of mean values in a situation like the above?

There are several possibilities, of which the following procedure for estimating the mean of C appears to be a natural one. We use the complete sets of measurements, C, L, B, and H, on unbroken skulls to compute the regressions of C on different subsets of L, B, and H. Using the appropriate regression function, we estimate (predict) the missing value of C for each broken skull. Then an average is taken of all the measured and estimated values of C. Such an average is likely to be a valid estimate of the mean of C. The estimation is based on the assumption that the complete sets of measurements (C, L, B, H) can provide valid estimates of relationships like the regression functions of C on L, B, H and its subsets, although they are biased samples from the original population. Similar methods can be used to estimate the mean values of L, B, and H.

Paleontologists compare the characteristics of fossils of long bones and cranial material discovered in different parts of the world to trace the evolutionary history of hominids. Such studies based on physical measurements may be misleading, as the discovered fossils may not be representative samples from the original populations due to differential preservation of skeletal material. It is gratifying to note that attempts are being made to compare the fossils in terms of some basic chemical measurements which are not likely to be subject to the phenomenon of differential preservation.

5. Enquiry Through an Offspring

In genetic and sociopsychological studies it is the common practice to locate an abnormal individual and through him or her collect information on the status of brothers and sisters, parents, uncles, and aunts. From such data estimates are made of the incidence of abnormality in families by sex and parity of birth. A family is the basic unit whose characteristics may have a specified distribution. But our method of ascertainment gives unequal probabilities to families depending on the mechanism inherent in the selection of an abnormal family member. Thus, the distribution applicable to observed data on families is a weighted version of the distribution specified for the families. We consider some examples, discuss the nature of the problems involved in each case, and suggest possible solutions.

5.1. Too Many Males?

During the last few years, while lecturing to students and teachers in different parts of the world, I collected data on the numbers of brothers and sisters in the family of each individual in the audience. The results are summarized in Tables 2, 3, and 4. The data from the male respondents given in Tables 2 and 4 show that the ratio of B, the total number of brothers including the respondents, to $B + S$, the total number of brothers and sisters, is much larger than one-half in each case, indicating a preponderence of male children in the families of male members of my audiences.

The number of male children in a family of a given size has a binomial distribution, and this would have been the specification if the sib compositions had been ascertained from families selected at random from the population of families.

Table 2. Data on Male Respondents (Students)[a]

Place and year	k	B	S	$\frac{B}{B+S}$	$\frac{B-k}{B+S-k}$ [b]	χ^2
Bangalore (India, 75)	55	180	127	.586	.496	.02
Delhi (India, 75)	29	92	66	.582	.490	.07
Calcutta (India, 63)	104	414	312	.570	.498	.04
Waltair (India, 69)	39	123	88	.583	.491	.09
Ahmedabad (India, 75)	29	84	49	.632	.523	.35
Tirupati (India, 75)	592	1902	1274	.599	.484	.50
Poona (India, 75)	47	125	65	.658	.545	1.18
Hyderabad (India, 74)	25	72	53	.576	.470	.36
Tehran (Iran, 75)	21	65	40	.619	.500	.19
Isphahan (Iran, 75)	11	45	32	.584	.515	.06
Tokyo (Japan, 75)	50	90	34	.725	.540	.49
Lima (Peru, 82)	38	132	87	.603	.519	.27
Shanghai (China, 82)	74	193	132	.594	.474	.67
Columbus (USA, 75)	29	65	52	.556	.409	2.91
College St. (USA, 76)	63	152	90	.628	.497	.01
Total	1206	3734	2501	.600	.503	0.14

[a] k = number of students, B = total number of brothers including the respondent, S = total number of sisters.

[b] Estimate of π under size biased binomial distribution.

Table 3. Data on Female Respondents (Students)

Place and year	k	B	S	$\frac{S}{B+S}$	$\frac{S-k}{B+S-k}$	χ^2
Lima (Peru, 82)	16	37	48	.565	.464	.36
Los Banos (Philippines, 83)	44	101	139	.579	.485	.18
Manila (Philippines, 83)	84	197	281	.588	.500	.00
Bilbao (Spain, 83)	14	19	35	.576	.525	.10
Total	158	354	503	.587	.493	.11

Table 4. Data on Male Respondents (Professors)

Place and year	k	B	S	$\frac{B}{B+S}$	$\frac{B-k}{B+S-k}$	χ^2
State College (USA, 75)	28	80	37	.690	.584	2.53
Warsaw (Poland, 75)	18	41	21	.660	.525	2.52
Poznan (Poland, 75)	24	50	17	.746	.567	1.88
Pittsburgh (USA, 81)	69	169	77	.687	.565	2.99
Tirupati (India, 76)	50	172	132	.566	.480	0.39
Maracaibo (Venezuela, 82)	24	95	56	.629	.559	1.77
Richmond (USA, 81)	26	57	29	.663	.517	0.03
Total	239	664	369	.642	.535	3.95

In the case of the data reported in Tables 2 and 4, a male student is located first and the sib composition in his family is ascertained; in such a case, each family included in the sample has at least one male child, which indicates a departure from the full binomial distribution. What population then does our sample represent? It is clear that the effective population is the *subset* of families having a male child of a particular description, such as a specified age group and qualifications which gave him a chance of being included in the enquiry. Rao (1977) argued that the distribution of brothers and sisters in such a subset of families of a given size is likely to be size biased binomial, so that the probability of r brothers and $n-r$ sisters in a family of size n is the weighted binomial

$$\frac{r}{E(r)}\binom{n}{r}\pi^r(1-\pi)^{n-r} = \binom{n-1}{r-1}\pi^{r-1}(1-\pi)^{n-r}, \tag{5.1.1}$$

where π is the probability of a male child. Under this hypothesis we find that

$$E\left(\frac{B-k}{B+S-k}\right) = \pi, \tag{5.1.2}$$

where k is the number of male respondents, so that $(B-k)/(B+S-k)$ is an estimate of π, and

$$\frac{[B - k - (B + S - k)\pi]^2}{(B + S - k)\pi(1 - \pi)} \tag{5.1.3}$$

has an asymptotic chi-square distribution on 1 degree of freedom. Similar results hold for the data from female respondents in Table 3. It is seen from the chi-square values in Tables 2 and 3 that the data collected from the students are consistent with the hypothesis of a size biased binomial with $\pi = \frac{1}{2}$. (Actually the chi-squares are too small. This needs an investigation).

The situation is somewhat different in Table 4, relating to data from the professors. The estimated π is more than one-half in each case, and the chi-square values are high. This implies that the weight function appropriate for these data is of a higher order than r, the number of brothers. A possible sociological explanation for this is that a person coming from a family with a larger number of brothers tends to acquire better academic qualifications to compete for jobs.

The following example on observed sex ratio is illuminating. In a survey of fertility and mortality, Dandekar and Dandekar (1953) gave the distribution of brothers (excluding the informant), sisters, sons, and daughters as reported by 1115 "male heads," contacted through households chosen with equal probability for each household. It may be observed that in a survey of this type, a family with r brothers gets a chance nearly proportional to r, and the conditions for a weighted binomial with $w(r) = r$ hold for the number of brothers in a family. Yet we find from Table 5 that the total number of brothers, 1325 (excluding the informants) is far in excess of the number of sisters, 1014, giving

$$\chi^2 = \frac{(1325 - 1014)^2}{1325 + 1014} = 41.35,$$

which is very high on 1 degree of freedom. Is the theory of the size biased binomial wrong?

Table 5. Distribution by Age of Brothers, Sisters, Sons, and Daughters[a]

Age group	Brothers	Sisters	Sons	Daughters
0–4	5	10	357	348
5–9	27	31	330	354
10–14	63	62	305	226
15–19	87	85	208	190
20–24	155	100	167	130
25–29	181	130	85	63
30–34	156	130	29	33
35–39	179	123	18	16
40–44	146	105	13	5
Rest	336	228	21	10
Total	1325	1014	1533	1375

[a] Dandekar and Dandekar (1953).

It is clear from Table 5 that the disproportionate sex ratio is confined to the age groups above 15–19 years, and the same phenomenon seems to occur in the case of sons and daughters. There is perhaps an underreporting of sisters and daughters who are married off, due to a superstitious custom of not including them as members of the household. Underreporting of female members is a persistent feature in data on fertility and mortality collected in developing countries.

5.2. Albinism

In studies of the inheritance of rare diseases, it is more convenient to collect family data by first locating an affected individual and then enquiring about the status of each of his or her brothers and sisters. While the different categories of children classified by disease, sex, etc., may have a multinomial distribution among families of given size, the numbers as ascertained do not have the same distribution, due to unequal probabilities of selection of families. In the previous section we have encountered a situation where the probability of selection of a family was proportional to the number of male children. In the case of rare diseases, the probability of selection of a family through an affected child may be not a linear but a more complicated function of the number of affected children. In this section we propose a model for selection probabilities in such cases and develop the appropriate methodology.

Let π_1 be the probability that a male child is an albino, and π_2 that a female child is an albino. Then the probability that a family of n children has r_1 males of whom t_1 are albinos and r_2 females of whom t_2 are albinos is

$$p(r_1, t_1; r_2, t_2) = \binom{n}{r_1}\left(\frac{1}{2}\right)^n \binom{r_1}{t_1} \pi_1^{t_1} \phi_1^{r_1 - t_1} \binom{r_2}{t_2} \pi_2^{t_2} \phi_2^{r_2 - t_2}, \tag{5.2.1}$$

where $\phi_1 = 1 - \pi_1$ and $\phi_2 = 1 - \pi_2$, and the probability of a child being a male or a female is taken as one-half.

There are a number of ways in which we can introduce probabilities of selection of affected families. We consider some models which are extensions of those suggested by Fisher (1934) and Haldane (1938).

Introducing α and $\beta = 1 - \alpha$ as relative probabilities of observing a male or a female albino, we may consider a mixture of two size biased distributions,

$$p^w(r_1, t_1; r_2, t_2) = \left(\frac{2\alpha t_1}{n\pi_1} + \frac{2\beta t_2}{n\pi_2}\right) p(r_1, t_1; r_2, t_2), \tag{5.2.2}$$

as the appropriate distribution of the observed vector (r_1, t_1, r_2, t_2). If we have data on (r_1, t_1, r_2, t_2) from N ascertained families, we can write down the likelihood using the expression (5.2.2) and estimate the parameters α, π_1, and π_2. Alternatively, we can use the method of moments, using the statistics $\sum t_1$, $\sum t_2$, and $\sum r_1$ to estimate the unknown parameters.

If $\pi_1 = \pi_2 = \pi$, the expression (5.2.2) reduces to

$$\frac{2}{n\pi}(\alpha t_1 + \beta t_2) p(r_1, t_1; r_2, t_2), \tag{5.2.3}$$

and the estimates of α and π can be obtained from the equations

$$\begin{aligned} \bar{t}_1 &= \alpha + \frac{\pi}{2k}\sum(n_i - 1), \\ \bar{t}_2 &= \beta + \frac{\pi}{2k}\sum(n_i - 1), \end{aligned} \tag{5.2.4}$$

where k is the number of families, n_i is the number of children in the ith family, and $\bar{t}_1$ and $\bar{t}_2$ are the average numbers of male and female albino children in a family.

Another model is as follows. Let ρ_1 and ρ_2 be the probabilities of observing a male and a female albino respectively. Then the probability that a family with n children having t_1 male albinos and $r_1 - t_1$ normal males, and t_2 female albinos and $r_2 - t_2$ normal females, is investigated s_1 times by observing a male albino and s_2 times by observing a female albino is

$$\binom{t_1}{s_1} \rho_1^{s_1}(1 - \rho_1)^{t_1 - s_1} \binom{t_2}{s_2} \rho_2^{t_2}(1 - \rho_2)^{t_2 - s_2} p(r_1, t_1; r_2 t_2). \tag{5.2.5}$$

Since a family is not investigated unless at least one of t_1 and t_2 is different from zero, the effective distribution for the observed data is (5.2.5) normalized by the quotient $1 - (1 - \rho)^n$, where $\rho = (\rho_1 \pi_1 + \rho_2 \pi_2)/2$. The method of estimation of ρ_1, ρ_2, π_1, and π_2 when we have the additional information on the number of times each family is investigated is discussed in detail in Rao (1965).

In case a family is investigated only once although more than one abnormal child in the family is observed, the appropriate distribution is

$$\frac{[1 - (1 - \rho_1)^{t_1}(1 - \rho_2)^{t_2}] p(r_1, t_1; r_2, t_2)}{1 - (1 - \rho)^n}, \tag{5.2.6}$$

where $\rho = (\pi_1 \rho_1 + \pi_2 \rho_2)/2$. If $\rho_1 = \rho_2 = \rho$ and $\pi_1 = \pi_2 = \pi$, then the expression (5.2.6) reduces to

$$\frac{1 - (1 - \rho)^{t_1 + t_2}}{1 - (1 - \pi\rho)^n} \frac{n!}{t_1!(r_1 - t_1)! t_2!(r_2 - t_2)!} \left(\frac{\pi}{2}\right)^{t_1 + t_2} \left(\frac{\phi}{2}\right)^{n - t_1 - t_2}. \tag{5.2.7}$$

If sex is ignored, then(5.2.7) becomes

$$\frac{1 - (1 - \rho)^t}{1 - (1 - \pi\rho)^n} \frac{n!}{t!(n - t)!} \pi^t \phi^{n-t}, \tag{5.2.8}$$

where $t = t_1 + t_2$, which is the expression used by Haldane (1938).

We have considered three different models (5.2.2), (5.2.5), and (5.2.6) for the probability of selection of a family. In the case where we have information only on the number r of abnormal children in a family of size n without any sex distinction, we may consider the weighted binomial distribution

$$\frac{w(r)}{E[w(r)]}\binom{n}{r}\pi^r\phi^{n-r}, \tag{5.2.9}$$

where $\phi = 1 - \pi$, with three possible alternatives for $w(r)$:

$$w(r) = r \tag{5.2.10}$$

$$= r^\alpha \qquad (\alpha \text{ unknown}) \tag{5.2.11}$$

$$= 1 - (1 - \rho)^n \qquad (\rho \text{ unknown}). \tag{5.2.12}$$

The maximum likelihood method of estimating α and π under the model (5.2.9), (5.2.11) is discussed in Rao (1965), and of ρ and π under the model (5.2.9), (5.2.12) in Haldane (1938). To demonstrate the relevance of the weight function (5.2.11), we compare in Table 6 the observed data on frequencies of albino children in families of different sizes with the expected values under the two different weight functions $w(r) = r$ and $w(r) = r^{1/2}$, choosing $\pi = \frac{1}{4}$. It is seen that the weight function $w(r) = r^{1/2}$ provides a better fit.

For a general discussion of the type of problems discussed in this section, and a few other models for selection probabilities, the reader is referred to Stene (1981) and other references mentioned in that paper. For estimation of α and π in the model (5.2.9), (5.2.11), reference may be made to Rao (1965).

5.3. Alcoholism, Family Size, and Birth Order

Smart (1963, 1964) and Sprott (1964) examined a number of hypotheses on the incidence of alcoholism in Canadian families using the data on family size and birth order of 242 alcoholics admitted to three alcoholism clinics in Ontario. The method of sampling is thus of the type discussed in Sections 5.1 and 5.2.

One of the hypotheses tested was that "larger families contain larger numbers of alcoholics than expected." The null hypothesis was interpreted to imply that the observations on family size as ascertained arise from the weighted distribution

$$np(n)/E(n), \qquad n = 1, 2, \ldots, \tag{5.3.1}$$

where $p(n)$, $n = 1, 2, \ldots$, is the distribution of family size in the general population. Smart and Sprott used the distribution of family size as reported in the 1931 census of Ontario for $p(n)$ in their analysis. It is then a simple matter to test whether the observed distribution of family size in their study is in accordance with the expected distribution (5.3.1).

It may be noted that the distribution (5.3.1) would be appropriate if we had chosen individuals (alcoholic or not) at random from the general population (of individuals) and ascertained the sizes of the families to which they belonged. But it is not clear whether the same distribution (5.3.1) holds if the

Table 6. Observed and Expected Frequencies of Albino Children for Each Family Size n

No. of albinos	$n = 2$ Observed	$n = 2$ Expected[a] (1)	$n = 2$ Expected[a] (2)	$n = 3$ Observed	$n = 3$ Expected[a] (1)	$n = 3$ Expected[a] (2)	$n = 4$ Observed	$n = 4$ Expected[a] (1)	$n = 4$ Expected[a] (2)	$n = 5$ Observed	$n = 5$ Expected[a] (1)	$n = 5$ Expected[a] (2)
1	31	30.00	32.37	37	30.93	35.81	22	21.10	26.07	25	19.00	24.93
2	9	10.00	7.63	15	20.63	16.88	21	21.09	18.43	23	25.31	23.50
3				3	3.44	2.30	7	7.03	5.02	10	12.65	9.59
4							0	0.78	0.48	1	2.81	1.85
5										1	0.23	0.13

No. of albinos	$n = 6$ Observed	$n = 6$ Expected[a] (1)	$n = 6$ Expected[a] (2)	$n = 7$ Observed	$n = 7$ Expected[a] (1)	$n = 7$ Expected[a] (2)	Total Observed	Total Expected[a] (1)	Total Expected[a] (2)
1	18	12.58	17.46	16	8.21	11.98	149	121.82	148.62
2	13	20.96	20.58	10	16.37	16.94	96	114.36	103.98
3	18	13.98	11.20	14	13.64	11.53	47	50.74	39.64
4	3	4.66	3.23	5	6.06	4.43	9	14.31	10.00
5	0	0.77	0.48	1	1.51	0.99	1	2.51	1.61
6	1	0.05	0.03	0	0.20	0.12	1	0.25	0.15
				0	0.01	0.01	0	0.01	0.01

[a] (1) for $w_r = r$; (2) for $w_r = r^{1/2}$.

enquiry is restricted to alcoholic individuals admitted to a clinic, as assumed by Smart and Sprott. This could happen, as demonstrated below, under an interpretation of their null hypothesis that the number of alcoholics in a family has a binomial distribution (like failures in a sequence of independent trials), and a further assumption that every alcoholic has the same independent chance of being admitted to a clinic.

Let π be the probability of an individual becoming an alcoholic, and suppose that the probability that a member of a family becomes an alcoholic is independent of whether another member is alcoholic or not. Further let $p(n)$, $n = 1, 2, \ldots$, be the probability distribution of family size (whether a family has an alcoholic or not) in the general population. Then the probability that a family is of size n and has r alcoholics is

$$p(n)\binom{n}{r}\pi^r\phi^{n-r}, \qquad r = 0, \ldots, n; \qquad n = 1, 2, \ldots, \tag{5.3.2}$$

where $\phi = (1 - \pi)$. From (5.3.2), it follows that the distribution of family size in the general population, given that a family has at least one alcoholic, is

$$\frac{(1 - \phi^n)p(n)}{1 - E(\phi^n)}, \qquad n = 1, 2, \ldots. \tag{5.3.3}$$

If we had chosen households at random and recorded the family sizes in households containing at least one alcoholic, then the null hypothesis on the excess of alcoholics in larger families could be tested by comparing the observed frequencies with the expected frequencies under the model (5.3.3). However, under the sampling scheme adopted, the weighted distribution of (n, r),

$$p^w(n, r) = rp(n)\binom{n}{r}\frac{\pi^r\phi^{n-r}}{\pi E(n)}, \tag{5.3.4}$$

is more appropriate. If we had information on the family size n as well as on the number of alcoholics (r) in the family, we could have compared the observed joint frequencies of (n, r) with those expected under the model (5.3.4).

From (5.3.4), the marginal distribution of n alone is

$$np(n)/E(n), \qquad n = 1, 2, \ldots, \tag{5.3.5}$$

which is used by Smart and Sprott as a model for the observed frequencies of family sizes. It is shown in (5.3.3) that in the general population, the distribution of family size in families with at least one alcoholic is

$$\frac{(1 - \phi^n)p(n)}{1 - E(\phi^n)},$$

which reduces to (5.3.5) if ϕ is close to unity. In other words, if the probability of an individual becoming an alcoholic is small, then the distribution of family size as ascertained is close to the distribution of family size in families with at least one alcoholic in the general population. This is not true if ϕ is not close to unity.

Smart and Sprott found that the distribution (5.3.5) did not fit the observed frequencies, which had heavier tails. What conclusions can we draw from this test? It is seen that the weighted distribution (5.3.5) is derived under two hypotheses. One is that the distribution of family size in the subset of families having at least one alcoholic in the general population is of the form (5.3.3) which is implied by the original null hypothesis posed by Smart. The other is that the method of ascertainment is equivalent to pps sampling of families, with probability proportional to the number of alcoholics in a family. The rejection of (5.3.5) would imply the rejection of the first of these two hypotheses if the second is assumed to be correct. There are no *a priori* grounds for such an assumption, and in the absence of an objective test for this, some caution is needed in accepting Smart's conclusions.

An alternative to (5.3.4) is obtained by assuming that each alcoholic has a chance θ of being admitted to a clinic independently of other alcoholic family members. In such a case, the probability that a family of size n has r alcoholics and a member has been admitted to a clinic is

$$p(n)\binom{n}{r}\pi^r\phi^{n-r}(1-(1-\theta)^r). \tag{5.3.6}$$

The marginal distribution of n with the normalizing factor is then

$$p(n)\frac{1-(1-\pi\theta)^n}{E(1-(1-\pi\theta)^n)}, \qquad n = 1, 2, \ldots. \tag{5.3.7}$$

The distribution (5.3.7) involves one unknown parameter $\pi\theta$ which needs to be estimated in fitting to the observed frequencies of family sizes. Some examples of distributions of the type (5.3.7) have been considered by Barrai, Mi, Morton, and Yasuda (1965). The distribution (5.3.7) is close to (5.3.5) if $\pi\theta$ is small.

We may also consider a more complicated model by assuming different probabilities π_1 and π_2 respectively for a male and a female becoming alcoholic, and also different probabilities θ_1 and θ_2 for male and female alcoholics being referred to a clinic. In such a case, the probability of inclusion of a family of size n with r_1 males, s_1 male alcoholics, r_2 females, and s_2 female alcoholics is

$$p(n)\binom{n}{r}\left(\frac{1}{2}\right)^n\binom{r_1}{s_1}\pi_1^{s_1}\phi_1^{r_1-s_1}\binom{r_2}{s_2}\pi_2^{r_2}\phi_2^{r_2-s_2}(1-(1-\theta_1)^{s_1}(1-\theta_2)^{s_2}), \tag{5.3.8}$$

where $\phi_1 = 1 - \pi_1$ and $\phi_2 = 1 - \pi_2$. This gives the marginal distribution of n as

$$p(n)\frac{1-2^{-n}(2-\pi_1\theta_1-\pi_2\theta_2)^n}{E(1-2^{-n}(2-\pi_1\theta_1-\pi_2\theta_2)^n)}, \tag{5.3.9}$$

which again involves one unknown parameter, $(\pi_1\theta_1 + \pi_2\theta_2)/2$. The marginal distribution of r_1 and r_2 obtained from (5.3.8) is

$$p(n)\binom{n}{r_1}\left(\frac{1}{2}\right)^n \frac{1-(1-\pi_1\theta_1)^{r_1}(1-\pi_2\theta_2)^{r_2}}{E(1-2^{-n}(2-\pi_1\theta_1-\pi_2\theta_2)^n)}, \tag{5.3.10}$$

where $n = r_1 + r_2$. If $\pi_1\theta_1$ and $\pi_2\theta_2$ are small, then (5.3.10) becomes

$$p(n)\binom{n}{r_1}\left(\frac{1}{2}\right)^n \frac{r_1\pi_1\theta_1 + r_2\pi_2\theta_2}{2^{-1}(\pi_1\theta_1 + \pi_2\theta_2)E(n)}. \tag{5.3.11}$$

If we had the joint frequencies of males and females in the observed families of alcoholics, we could have fitted distributions of the type (5.3.10) and (5.3.11) to test the null hypothesis of larger numbers of alcoholics in larger families.

It is seen from (5.3.10) and (5.3.11) that the distribution of (r_1, r_2) will not be symmetric unless $\pi_1\theta_1 = \pi_2\theta_2$. This may result in an excess of males or females in observed families. Such an effect (with an excess of males) can be seen in similar data studied by Freire-Mala and Chakraborty (1975) and Rao, Mazumdar, Waller, and Li (1973); these authors have not, however, commented on this phenomenon.

Another hypothesis considered by Smart was that the later-born children have a greater tendency to become alcoholic than the earlier-born. The method used by Smart may be somewhat confusing to statisticians. Some comments were made by Sprott criticizing Smart's approach. We shall review Smart's analysis in the light of the model (5.3.4). If we assume that birth order has no relationship to becoming an alcoholic, and the probability of an alcoholic being referred to a clinic is independent of the birth order, then the probability that an observed alcoholic belongs to a family with n children and r alcoholics and has given birth order $s \le n$ is, using the model (5.3.4),

$$\frac{rp(n)}{nE(n)}\binom{n}{r}\pi^{r-1}\phi^{n-r}, \qquad s = 1, \ldots, n, \quad r = 1, \ldots, n, \quad n = 1, 2, \ldots. \tag{5.3.12}$$

Summing over r, we find that the marginal distribution of (n, s), the family size and birth order, applicable to the observed distribution, is

$$p(n)/E(n), \qquad s = 1, \ldots, n, \quad n = 1, 2, \ldots, \tag{5.3.13}$$

where it may be recalled that $p(n)$, $n = 1, 2, \ldots$, is the distribution of family size in the general population. Smart gave the observed bivariate frequencies of (n, s), and since $p(n)$ was known, the expected values could have been computed and compared with the observed. But, he did something else.

From (5.3.13), the marginal distribution of birth rank is

$$\frac{1}{E(n)}\sum_{i=r}^{\infty} p(i), \qquad r = 1, 2, \ldots. \tag{5.3.14}$$

Smart's (1963) analysis in his Table 2 is an attempt to compare the observed distribution of birth ranks with the expected under the model (5.3.14) with $p(i)$ itself estimated from data using the model (5.3.1).

A better method is as follows: from (5.3.13) it is seen that for given family size, the expected birth order frequencies are equal as computed by Smart (1963) in Table 1, in which case individual chi-squares comparing the expected and observed frequencies for each family size would provide all the information about the hypothesis under test. Such a procedure would be independent of any knowledge of $p(n)$. But it is not clear whether a hypothesis of the type posed by Smart can be tested on the basis of the available data without further information on the other alcoholics in the family, such as their ages, sexes, etc.

Table 7 reproduces a portion of Table 1 in Smart (1963) relating to families up to size 4 and birth ranks up to 4. It is seen that for family sizes 2 and 3, the observed frequencies seem to contradict the hypothesis, and for family sizes above 3 (see Smart's Table 1), birth rank does not have any effect. It is interesting to compare the above data with a similar type of data (Table 8) collected by the author on birth rank and family size of the staff members in two departments at the University of Pittsburgh. It appears that there are too many earlier-borns among the staff members, indicating that becoming a professor is an affliction of the earlier born! It is expected that in data of the kind we are considering there will be an excess of the earlier born without implying an implicit relationship between birth order and a particular attribute, especially when it is age dependent.

Table 7. Distribution of Birth Rank s and Family Size n[a]

	$n = 1$		2		3		4	
s	O	E	O	E	O	E	O	E
1	21	21	22	16	17	13.3	11	11.75
2			10	16	14	13.3	10	11.75
3					9	13.3	13	11.75
4							13	11.75

[a] Smart (1963, Table 1). O = observed, E = expected.

Table 8. Distribution of Birth Rank s and Family Size $n \leq 4$ Among Staff Members (University of Pittsburgh)

s	$n = 1$	2	3	4
1	7	14	9	6
2		6	4	2
3			2	0
4				0

6. Quadrat Sampling with Visibility Bias

For estimating wildlife population density,quadrat sampling has been found generally convenient. Quadrat sampling is carried out by first selecting at random a number of quadrats of fixed size from the region under study and ascertaining the number of animals in each. The following assumptions are made:

A_1. Animals are found in groups within each quadrat, and the number of animals X in a group follows a specified distribution.

A_2. The number of groups N within a quadrat has a specified distribution.

A_3. The number of groups within a quadrat and the numbers of animals within groups are independent.

Let the method of sampling be such that the probability of sighting (or recording) a group of x animals is $w(x)$. If X^w and N^w represent the rv's of the number of animals in a group and number of groups within a quadrat as ascertained, then we have the following results:

(i)

$$P(N^w = m \mid N = n) = \binom{n}{m} w^m (1 - w)^{n-m}, \tag{6.1}$$

where

$$w = \sum_{1}^{\infty} w(x)\, P(X = x) \tag{6.2}$$

is the visibility factor (or the probability of recording a group).

(ii)

$$P(N^w = m) = \sum_{n=m}^{\infty} \binom{n}{m} w^m (1 - w)^{n-m} P(N = n). \tag{6.3}$$

(iii)

$$P(N^w = m, X_1^w = x_1, \ldots, X_m^w = x_m) = w^{-m} P(N^w = m) \prod_{i=1}^{m} w(x_i)\, P(X = x_i). \tag{6.4}$$

(iv) Let $S^w = X_1^w + \cdots + X_m^w$. Then

$$P(S^w = y) = \sum_{m=1}^{\infty} P(N^w = m) P(S^w = y \mid m) \tag{6.5}$$

and

$$P(S^w = y \mid m) = \sum_{\Sigma x_i = y}^{\infty} \frac{w(x_1)}{w} \cdots \frac{w(x_m)}{w} P(X_1 = x_1) \cdots P(X = x_m). \quad (6.6)$$

The formulae listed above are useful in many practical situations. Usually the sighting probability is of the form

$$w(x) = 1 - (1 - \beta)^x. \quad (6.7)$$

For some applications, the reader is referred to papers by Cook and Martin (1974) and Patil and Rao (1977, 1978).

7. Waiting Time Paradox

Patil (1984) reported a study conducted in 1966 by the Institute National de la Statistique et de l'Economie Appliquee in Morocco to estimate the mean sojourn time of tourists. Two types of surveys were conducted, one by contacting tourists residing in hotels and another by contacting tourists at frontier stations while leaving the country. The mean sojourn time as reported by 3000 tourists in hotels was 17.8 days, and by 12321 tourists at frontier stations was 9.0. Suspected by the officials in the department of planning, the estimate from the hotels was discarded.

It is clear that the observations collected from tourists while leaving the country correspond to the true distribution of sojourn time, so that the observed average 9.0 is a valid estimate of the mean sojourn time. It can be shown that in a steady state of flow of tourists, the sojourn time as reported by those contacted at hotels has a size biased distribution, so that the observed average will be an overestimate of the mean sojourn time. If X^w is a size biased random variable, then

$$E(X^w)^{-1} = \mu^{-1} \quad (7.1)$$

where μ is the expected value of X, the original variable. The formula (7.1) shows that the harmonic mean of the size biased observations is a valid estimate of μ. Thus the harmonic mean of the observations from the tourists in hotels would have provided an estimate comparable with the arithmetic mean of the observations from the tourists at the frontier stations.

It is interesting to note that the estimate from hotel residents is nearly twice the other, a factor which occurs in the waiting time paradox (see Feller, 1966; Patil and Rao, 1977) associated with the exponential distribution. This suggests, but does not confirm, that the sojourn time distribution may be exponential.

Suppose that the tourists at hotels were asked how long they had been staying in the country up to the time of enquiry. In such a case, we may assume that the pdf of the rv Y, the time a tourist has been in a country up to the time of enquiry, is the same as that of the product $X^w R$, where X^w is the size biased version of X, the sojourn time, and R is an independent rv with a uniform distribution on [0, 1]. If $F(x)$ is the distribution function of X, then the pdf of Y is

$$\mu^{-1}[1 - F(y)]. \tag{7.2}$$

The parameter μ can be estimated on the basis of observations on Y, provided the functional form of $F(y)$, the distribution function of the sojourn time, is known.

It is interesting to note that the pdf (7.2) is the same as that obtained by Cox (1962) in studying the distribution of failure times of a component used in different machines from observations on the ages of the components in use at the time of investigation.

8. Damage Models

Let N be a rv with probability distribution, $p_n, n = 1, 2, \ldots$, and R be a rv such that

$$P(R = r \mid N = n) = s(r, n). \tag{8.1}$$

Then the marginal distribution of R truncated at zero is

$$p'_r = (1 - p)^{-1} \sum_{n=r}^{\infty} p_n s(r, n), \qquad r = 1, 2, \ldots, \tag{8.2}$$

where

$$p = \sum_{1}^{\infty} p_i s(0, i). \tag{8.3}$$

The observation r represents the number surviving when the original observation n is subject to a destructive process which reduces n to r with probability $s(r, n)$. Such a situation arises when we consider observations on family size counting only the surviving children (R). The problem is to determine the distribution of N, the original family size, knowing the distribution of R and assuming a suitable survival distribution.

Suppose that $N \sim P(\lambda)$, i.e., distributed as Poisson with parameter λ, and let $R \sim B(\cdot, \pi)$, i.e., binomial with parameter π. Then

$$p'_r = e^{-\lambda\pi} \frac{(\lambda\pi)^r}{r!} \frac{1}{1 - e^{-\lambda\pi}}, \qquad r = 1, 2, \ldots. \tag{8.4}$$

It is seen that the parameters λ and π get confounded, so that knowing the distribution of R, we cannot find the distribution of N. Similar confounding occurs when N follows a binomial, negative binomial, or logarithm series distribution. When the survival distribution is binomial, Sprott (1965) gives a general class of distributions which has this property. What additional information is needed to recover the original distribution? For instance, if we know which of the observations in the sample did not suffer damage, then it is possible to estimate the original distribution as well as the binomial parameter π.

It is interesting to note that observations which do not suffer any damage have the distribution

$$p_r^u = c p_r \pi^r, \tag{8.5}$$

which is a weighted distribution. If the original distribution is Poisson, then

$$p_r^u = e^{-\lambda\pi} \frac{(\lambda\pi)^r}{r!} \frac{1}{1 - e^{-\lambda\pi}}, \tag{8.6}$$

which is same as (8.4). It is shown in Rao and Rubin (1964) that the equality $p_r^u = p_r'$ characterizes the Poisson distribution.

The damage models of the type described above were introduced in Rao (1965). For theoretical developments on damage models and characterization of probability distributions arising out of their study, the reader is referred to Alzaid, Rao, and Shanbhag (1984).

9. Nonresponse: The Story of an Extinct River

Sample survey practitioners define nonresponse as a missing observation or nonavailability of measurements on a unit included in a sample. It is clear that if the missing values can be considered as a random sample from the population under survey, then the observed values constitute a representative sample of the whole population (Rubin, 1976). Usually this is not the case, and special techniques are developed in sample surveys to cope with such situations.

In general, nonresponse poses serious issues, such as the problem of broken skulls not providing direct measurements on capacity (see Section 4 of this paper). More complex cases are as follows.

For instance, we may try to estimate the underground resources in a given region by making borings at a randomly chosen set of points and taking some measurements. But it may so happen that borings cannot be made at some chosen points, for example because of the presence of rocks. The measurements at such points may be of a different type from the rest, in which case the observed sample will not be a representative sample from the whole region.

Such a problem arose in an investigation by geologists at the Indian Statistical Institute to estimate the mean direction of flow of an extinct river of geological times in a certain region (see Sengupta, 1966; J. S. Rao and Sengupta, 1966). The geologists collected a series of observations on direction cosines of flow (two dimensional vector data), which seemed ideal for an application of Fisher's (1953) distribution and the associated theory for estimation of the mean direction of flow. Then the question arose as to what the hypothetical population was from which the observations could be considered as a random sample. The measurements on direction cosines could not be made at any chosen point, but only at certain points where there were outcrops. The geologists walked along the region under exploration and made measurements wherever they came across outcrops. If the outcrops had been uniformly distributed over space, then it might have been possible to define a population of which the observations made by the geologists could be a representative sample. The locations at which observations were made, when plotted on a topographical map of the region, showed an unequal distribution of outcrops in different areas of the region, indicating the nonrandom nature of the occurrence of outcrops. In such a case the estimate of mean direction assuming that each observation is an independent sample with a common expectation will be biased. In order to minimize the bias in estimation, the following method of estimation was adopted. A square lattice was imposed on the topographical map, and the measurements in each grid were replaced by their average. Then a simple average of these averages was taken as an estimate of the mean direction of flow. This estimate differed somewhat from the average of all the measurements and was considered to have less bias.

This study points out the need for a reexamination of the data on directions of rock magnetism collected by geologists and analysed by Fisher (1953), who developed a special theory for that purpose. If the outcrops at which measurements of direction are possible are not uniformly distributed over space, then there will be some difficulty in interpreting the observed mean direction as an estimate of some specific parameter.

10. Conclusions

Some of the broad conclusions that emerge from the discussion of the live examples in the paper are as follows:

Specification, or the choce of a model, is of great importance in data analysis. An appropriate specification for given data can be arrived at on the basis of past experience, information on the stochastic nature of events, a detailed knowledge of how observations are ascertained and recorded, and an exploratory analysis of current data itself using graphical displays, preliminary tests, and cross validation studies.

Inaccuracies in specification can lead to wrong inference. It is therefore worthwhile to review the data under different possible specifications (models) to determine how variant the conclusions could be.

What population does an observed sample represent? What is the widest possible universe to which the conclusions drawn from a sample apply? The answers depend on how the observations are ascertained and what the deficiencies in data are in terms of nonresponse, measurement errors, and contamination.

Every data set has its own unique features which may be revealed in an initial scrutiny of data and/or during statistical analysis, which may have to be taken into account in interpreting data. Routine data analysis based on textbook methods or software packages can be misleading.

Generally in scientific investigations, a question cannot be answered without knowing the answers to several other questions. It often pays to analyse the data to throw light on a broader set of relevant and related questions.

What data should be collected to answer a given question? Lack of information on certain aspects may create undue complications in applying statistical methods and/or restrict the nature of conclusions drawn from available data. Attempts should be made to collect information on concomitant variables to the extent possible, whose use can enhance the precision of estimators of unknown parameters, and provide broader validity to statistical inference.

Acknowledgements

The work is supported by the Air Force Office of Scientific Research under Contract F49620-82-K-0001. Reproduction in whole or in part is permitted for any purpose of the United States Government.

Bibliography

Alzaid, A. H., Rao, C. R., and Shanbhag, D. N. (1984). *Solutions of Certain Functional Equations and Related Results on Probability Distributions*. Technical Report. Univ. of Sheffield.

Barrai, I., Mi, M. P., Morton, N. E., and Yasuda, N. (1965). "Estimation of prevalence under incomplete selection." *Amer. J. Hum. Genet.*, **17**, 221–236.

Cook, R. D. and Martin, F. B. (1974). "A model for quadrat sampling with visibility bias." *J. Amer. Statist. Assoc.*, **69**, 345–349.

Cox, D. R. (1962). *Renewal Theory*. London: Chapman and Hall.

Dandekar, V. M. and Dandekar, K. (1953). *Survey of Fertility and Mortality in Poona District*. Publication No. 27. Poona, India: Gokhale Institute of Politics and Economics.

Feller, W. (1966). *Introduction to Probability Theory and its Applications, Vol. 2*, New York: Wiley.

Feller, W. (1968). *An Introduction to Probability Theory and its Applications, Vol. 1* (3rd edn.) New York: Wiley.

Fienberg, S. E., Singer, B., and Tanur, J. M. (1985). "Large scale social experimentation." In this volume, Chapter 12.

Fienberg, S. E. and Stasny, E. A. (1983). "Estimating monthly gross flows in labor force participation." *Survey Methodology*, **9**, 77–98.

Fienberg, S. E. and Tanur, J. M. (1983). "Large scale social surveys: perspectives, problems and prospects." *Behavioral Sci.*, **28**, 135–153.

Fisher, R. A. (1934). "The effect of methods of ascertainment upon the estimation of frequencies." *Ann. Eugen.*, **6**, 13–25.

Fisher, R. A. (1953). "Dispersion on a sphere." *Proc. Roy. Soc. London Ser. A*, **217**, 295–305.

Freire-Mala, A. and Chakraborty, R. (1975). "Genetics of archeiropodia." *Ann. Hum. Genet. London*, **39**, 151–161.

Haldane, J. B. S. (1938). "The estimation of the frequency of recessive conditions in man." *Ann. Eugen*, (*London*), **7**, 255–262.

Janardhan, K. G. and Rao, B. R. (1983). "Lagrange distributions of the second kind and weighted distributions." *SIAM J. Appl. Math.*, **43**, 302–313.

Kruskal, W. and Mosteller, F. (1980). "Representative sampling IV: The history and the concept in statistics, 1815–1939." *Internat. Statist. Inst. Rev.*, **48**, 169–195.

Mahalanobis, P. C. (1944). "On large scale sample surveys." *Philos. Trans. Roy, Soc. Ser. B*, **231**, 329–451.

Mosteller, F. (1978). "Errors: Nonsampling errors." In W. H. Kruskal and J. M. Tanur (eds.), *The International Encyclopedia of Statistics.* New York: Free Press, 208–229.

Mukherji, R. K., Trevor, J. C., and Rao, C. R. (1955). *The Ancient Inhabitants of Jebel Moya.* London: Cambridge U. P.

Neyman, J. (1977). "Experimentation with weather control and statistical problems generated by it." In P. R. Krishnaiah (ed.), *Applications of Statistics.* Amsterdam: North-Holland, 1–26.

Patil, G. P. and Ord, J. K. (1976). "On size-biased sampling and related form-invariant weighted distributions." *Sankhyā B*, **38**, 48–61.

Patil, G. P. (1984). "Studies in statistical ecology involving weighted distributions." In *Statistics: Applications and New Directions*, Calcutta: Indian Statistical Institute, 478–503.

Patil, G. P. and Rao, C. R. (1977). "The weighted distributions: A survey of their applications." In P. R. Krishnaiah (ed.), *Applications of Statistics*, Amsterdam: North Holland, 383–405.

Patil, G. P. and Rao, C. R. (1978). "Weighted distributions and size biased sampling with applications to wildlife populations and human families." *Biometrics*, **34**, 179–189.

Rao, B. R., Mazumdar, S., Waller, J. H., and Li, C. C. (1973). "Correlation between the numbers of two types of children in a family." *Biometrics*, **29**, 271–279.

Rao, C. R. (1965). "On discrete distributions arising out of methods of ascertainment." In G. P. Patil (ed.), *Classical and Contagious Discrete Distributions*, Calcutta: Statist. Publ. Soc., 320–333. Reprinted in *Sankhyā A*, **27**, 311–324.

Rao, C. R. (1973). *Linear Statistical Inference and its Applications*, 2nd edn. New York: Wiley.

Rao, C. R. (1977). "A natural example of a weighted binomial distribution." *Amer. Statist.*, **31**, 24–26.

Rao, C. R. (1975). "Some problems of sample surveys." *Suppl. Adv. Appl. Probab.*, **7**, 50–61.

Rao, C. R. and Rubin, H. (1964). "On a characterization of the Poisson distribution." *Sankhyā A*, **25**, 295–298.

Rao, C. R. and Shaw, D. C. (1948). "On a formula for the prediction of cranial capacity." *Biometrics*, **4**, 247–253.

Rao, J. S. and Sengupta, S. (1966). "A statistical analysis of cross-bedding azimuths from the Kamthi formation around Bheemaram, Pranhita-Godavari Valley." *Sankhyā B*, **28**, 165–174.

Rubin, D. B. (1976). "Inference and missing data." *Biometrika*, **63**, 581–592.

Rubin, D. B. (1980). *Handling Nonresponse in Sample Surveys by Multiple Imputations*. A Census Bureau Monograph. Washington.

Sengupta, S. (1966). "Studies on orientation and imbrication of pebbles with respect to cross-stratification." *J. Sed. Petrology*, **36**, 362–369.

Smart, R. G. (1963). "Alcoholism, birth order, and family size." *J. Abnorm. Soc. Psychol.*, **66**, 17–23.

Smart, R. G. (1964). "A response to Sprott's 'Use of chi square'." *J. Abnorm. Soc. Psychol.*, **69**, 103–105.

Sprott, D. A. (1964). "Use of chi square." *J. Abnorm. Soc. Psychol.*, **69**, 101–103.

Sprott, D. A. (1965). "Some comments on the question of identifiability of parameters raised by Rao." In G. P. Patil (ed.), *Classical and Contagious Discrete Distributions*. Calcutta: Statist. Publ. Soc., 333–336.

Stene, Jon (1981). "Probability distributions arising from the ascertainment and analysis of data on human families and other groups." In C. Taille, G. P. Patil and B. Baldessari (eds.), *Statistical Distributions in Scientific Work, Vol. 6*. Dordrecht: Reidel, 51–62.

CHAPTER 25

Statistical Problems in Crop Forecasting

D. Singh and M. P. Jha

Abstract

A reliable estimate of a crop yield well before harvest is of considerable importance in formulating policy. Usually, such preharvest estimates of the yield rate of a crop are obtained on the basis of visual observations of crop reporters, which is subjective. An objective method of preharvest forecasting, based on observations on biometrical characters (viz. plant population, plant height, number of leaves, etc.) as well as on weather parameters such as rainfall, temperature and humidity, is discussed in this paper. Several statistical problems are briefly examined, for example, the choice of appropriate variables, sampling design, and prediction model. Empirical studies have shown that the explanatory variables included in these models explain a large part of the variation in yield; however, for further improvement, it is necessary to use biometrical variables in conjunction with weather factors in prediction models. Further, research is needed on weather modeling, including weather forecasting and the use of satellite imagery, as well as on sample design for the collection of data for modeling purposes, and compartmental analysis of plant process models.

1. Introduction

An estimate of crop production is customarily obtained by multiplying the area under the crop and its yield rate per hectare. The area estimate in several countries is obtained by complete enumeration through administrative agencies or through special surveys, normally conducted before the harvest of the crop. The estimate of the yield rate, on the other hand, which is obtained using the objective method of random sampling in crop-cutting surveys, becomes available only after harvest. Thus the objective estimate of crop production also becomes available after harvest only. However, a reasonably good forecast of crop production in advance of harvest is of immense value in the planning of food procurement, storage, distribution, price fixation, movement of food and other major agricultural products, import–export plans, and marketing, and in some cases for taking timely action to increase production. Since crop acreages can be accurately estimated through objective sample surveys, the problem of forecasting crop production, therefore, may be

Key words and phrases: estimated yields; prediction models; regression models; sample surveys; visual estimates.

confined to the development of an objective procedure for forecasting the yield rate.

The Food and Agriculture Organisation of the United Nations has recently started exploring the possibility of developing a procedure which would enable it to monitor, as much in advance as possible, the anticipated production of major crops in the world with a particular reference to possible shortages in specific countries or regions within countries. The FAO has named this programme the Global Information and Early Warning System on Food and Agriculture.

The visual method of estimation for forecasting production of standing crops is still followed in many countries. An advance estimate of production is usually obtained by crop reporters through a subjective, visual estimation technique which quite often produces dubious results. It has been observed in some countries that this method leads generally to overestimation of yield in a bad crop year, but to underestimation in a good crop year. In view of these deficiencies, the necessity of developing an objective method of preharvest crop forecasting assumes importance. Research efforts are being made by the FAO and several countries in this direction.

2. Crop Forecasting and Estimation

2.1. Crop Area Statistics

The crop acreages are estimated mostly through sample surveys. However, in some countries, an annual crop enumeration scheme on a complete census basis is implemented for obtaining land utilization and crop statistics. Unfortunately, these statistics usually become available long after the harvest of the crop, thus reducing their utility. To reduce the time lag in the availability of acreage figures, attempts are being made in some countries to organize crop area estimation surveys immediately after sowing or planting. Any scheme of crop area estimation conducted prior to maturity or harvesting of the crop will provide only the sown or standing crop area. Thus to obtain the harvested area an appropriate correction factor is needed. The conduct of the crop area surveys while the crop is standing in the field (not necessarily mature for harvest) has practical advantages. Firstly, a longer time is available for field work, and this may facilitate the enumeration of more samples by the same set of field enumerators. Secondly, sowing and growth periods of several crops may coincide (but not necessarily the harvesting period), and this may make possible the coverage of several crops in the same survey program, if the domain of the survey is the same. In countries with multiple cropping systems, such surveys will often turn out to be multipurpose, in that acreages under different crops are estimated from the same survey. Sometimes the scope of such surveys is expanded for economic or other reasons to cover many other agricultural items, such as farm characteristics, livestock, etc.

The design of such multipurpose sample surveys is beset with a number of technical problems. Grouping the sampling units to form homogeneous strata is not easy. Determination of probabilities for the selection of units to increase the efficiency of estimates becomes difficult. Even the preparation of a sampling frame for the selection of units is sometimes difficult. An extensive literature on the planning of multi-purpose surveys is available in the statistical journals and books on sampling. It is beyond the scope of this paper to describe them here. Fred A. Vogel (1983) of the Statistical Reporting Service, USDA, has discussed the experiences of the USA in a bulletin *Sample Designs for Multi-purpose Agricultural Surveys—Problems in Application*. In the FAO's report (1965) the procedures followed by different countries in the estimation of crop area are discussed.

Advances in remote sensing, especially the successful use of satellite imagery and of aerial photography for the estimation of crop areas in some countries, may have an important bearing in the future on programs of agricultural statistics. However, the high costs involved in these techniques have discouraged their use in the developing countries, and it is hoped that further research in this field will in due course reduce costs and also overcome other technical difficulties, e.g. those due to cloud cover, small sizes of fields, the presence of mixed crops, and varying crop densities.

2.2. Estimation of Yield Rate

The estimate of the yield rate of a crop is obtained according to two methods, referred to as the traditional method and the method of crop-cutting surveys. The traditional method also provides the preharvest estimates, and this technique is being followed in several countries with modifications to suit the local conditions.

The other method of obtaining the final estimate of the yield rate of a crop is that of crop-cutting surveys. Over the last four decades this scientific method has been increasingly used in several countries. It involves the selection of a representive sample of a crop over the area for which estimate is required, and harvesting and weighing the produce from each of the several units constituting the sample. The sample units actually harvested are plots of prescribed dimensions located and marked according to clearly defined procedures. These plots may be located by a single stage of random selection over the entire area of the crop or, more commonly, by the following series of steps: first, selecting segments of area clearly marked with well-defined boundaries, or farms at random out of all segments, or farms in the area which grow the crop; then, selecting one or more random fields growing the crop within these primary units; and finally, marking out one or more random plots in the selected fields for harvesting. In some cadastrally surveyed countries with up-to-date village maps, a list of villages yields the primary sample units.

Considerable research has been done in India and elsewhere on the determination of the optimum size and shape of plots for yield estimation. The size and shape of the plot will no doubt depend upon the stand of the crop and method of its sowing. It has been observed, however, that yield rates based on very small plots are generally overestimated due to border effects. Therefore, a large size plot of say 25 to 50 square meters is recommended, particularly in developing countries where well-trained field staff in adequate numbers is not available for conducting random crop-cutting surveys. That this procedure is practicable and capable of giving yield estimates free from bias and providing a higher degree of accuracy is well recognized. The procedure adopted for harvesting and processing the product of the crop cutting is generally similar to that prevalent among the farmers of the locality.

3. Techniques of Crop Forecasting

Primarily three methods have been tried for forecasting crop yields, with varying success in different countries: (i) based on visual observation of the crop during its growth period, (ii) based on biometrical variables recorded during various stages of crop growth, and (iii) based on weather variables. Of these methods, the most widely used is the one based on visual observations, which, although subjective, is generally practicable. The various approaches to crop forecasting are briefly discussed in the following paragraphs.

3.1. Forecasts Based on Visual Observation of the Crop During Its Growth Period

The visual method of forecasting may be resorted to for obtaining usable forecasts where more precise quantitative observations are either impossible or impracticable. The reasons why visual estimation, as practiced, fails to give reliable estimates are:

(i) These estimates are vague general impressions of the crop prospects by various officials. A more effective procedure might be to obtain individual estimates for a representative set of fields and then to combine them statistically. The fields selected for yield estimation surveys of a crop provide a representative set of the fields under the crop, and hence visual estimates based on these fields would give far more reliable forecasts than those obtained by the current method.

(ii) With respect to the scale for measuring the condition of the crop, eye estimates suffer from severe limitations, since the officials concerned usually tend to give estimates which are within a fairly narrow range on either side of the normal yield observed in the past. As a result, crop yield is overestimated when the crop is poor and underestimated when the crop is especially good.

Considerable research is being carried out to make improvements in this subjective method through the use of sampling methods and regression techniques.

3.1.1. *Graphical Methods*. Wherever possible, historical data are used as a guide to establish a relationship between condition reports and final yields. If such historical series are available, then a graph of the condition reports versus final outcome may indicate a functional relationship that will allow probable yield corresponding to the condition figure for the current year to be read immediately off of the graph. Obviously, such relationships may change with technology, and may therefore need to be adjusted. Though in principle it may be possible to derive a mathematical relationship between the final yield, on the one hand, and the condition and other variables such as pest damage or weather conditions, on the other, it has been found usually convenient to use the graphical approach for a rough and ready indication of the likely harvest level.

3.1.2. *Regression of Estimated Yield on Condition*. Under the Statistical Reporting System in the United States, a simple regression of estimated yield on condition or a regression chart with time as an additional variable is followed in forecasting the yield of a number of crops. The feasibility of using weather data to forecast or to estimate crop yields has also been investigated on a number of occasions in the United States; e.g., rainfall data have proved useful in estimating the yield of winter wheat and soyabean crops. A multiple regression equation of the form has been used:

$$Y_c = a + b_1X_1 + b_2X_2 + b_3X_3 + b_4X_4,$$

where
Y_c = computed yield per acre,
X_1 = reported condition or probable yield per acre,
X_2 = precipitation for specified months prior to the date of forecasting,
X_3 = forecasted precipitation for specified months after date of forecasting
X_4 = time,
and b_1, b_2, b_3, b_4 are multiple regression coefficients. It is presumed that weather conditions and the damage from insects or other causes will be around normal during the period after the forecast is made.

3.2. Forecasts Based on Biometrical Variables

Preharvest forecasting of crop yield based on observations on biometrical variables recorded during various stages of crop growth involves consideration of such aspects as the choice of biometrical variables, sampling design for collection of data, and appropriateness of the chosen model. These aspects are briefly described hereunder.

3.2.1. *Choice of Biometrical Characteristics.* The forecasting of the yield of a crop at periodic intervals during the growing season is more difficult than estimating the yield at harvest time. Since the crop yield is a function of biometrical characteristics, an appropriate set of such characteristics having a profound influence on crop yield should be carefully selected for inclusion in the forecasting model. Apart from being useful indicators of the final outcome of the crop yield, these should be such as can be easily measured without much error well in advance of the harvest. To minimize multicollinearity, the biometrical characteristics selected should not have substantial correlation among themselves. Further, a decision in regard to the usefulness of a biometrical characteristic to serve as an indicator of a crop yield should be made on the basis of a study of the relationship between the yield and different biometrical characteristics. Finally, the cost of collection of observations on the biometrical characteristics selected on this basis should also be taken into consideration to ensure that their use is economically feasible on a large scale.

The problem of choosing a suitable sampling design for the collection of data on crop yield and different biometrical characteristics has to be viewed in the light of some practical considerations. If there is variation in the dates of sowing or planting within any given area, the sample selected from the area will not be uniform in this respect. A similar difficulty also arises from variation in the crop maturity dates. One way to deal with this problem would be to poststratify the sample, say, according to early and late sowing in case of wide differences in such dates.

Another problem concerns the multivariate nature of the sample survey, since data on both the crop yield and various plant biometrical characteristics have to be secured from the same survey. The sampling design usually adopted for collection of data on biometrical characteristics in such surveys is one of stratified multistage sampling, with stratification generally made on the basis of agroclimatic and geographical factors.

A decision in regard to the size of the sample and the frequency of recording data is generally made on the basis of availability of resources and operational feasibility. For this purpose, it is also necessary at the same time to take into account the variability associated with both crop yield and different biometrical characteristics. In case of biometrical characteristic the time of recording should coincide with the critical stage of crop growth.

To establish an effective statistical relationship between biometrical variables as explanatory variables and crop yields as dependent variables, it may be desirable to make observations on biometrical variables from the same plots which are selected for crop-cutting experiments for yield data. Since frequent observations on biometrical characteristics during the growth period of the crop will be costly and time-consuming, it may be economical and practical to use only a subsample of plots selected for collecting data on the crop yield. However, if a biometrical characteristic is such that its measurement during the early season also involves destruction, a plot (perhaps in the neighborhood) different from what is actually harvested will have to be selected. This procedure will obviously introduce errors in variables used for prediction.

Since the sample design plays an important role in the collection of relevant data, it thus seems necessary to study in depth the effect of sampling design on regression models.

3.2.2. *Prediction Models.* Various types of functional relationships, such as simple linear and log–linear, can be considered as models capable of describing adequately the relationship between crop yield and various biometrical characteristics. Before adoption for general use, however, the suitability and appropriateness of such models in a given situation should be first established. This could be done through studies on the basis of data collected through pilot sample surveys specially designed for the purpose. The criterion currently most widely used to choose a specific forescasting model among several alternatives is the mean squared error (MSE) between the actual and predicted values for past data. However, the ability of the model to forecast future values that have not been included in developing the model is more important and needs attention. The stability of regression coefficients involved should be examined on the basis of data pertaining to a fairly long period of time, so that the model is capable of providing realistic forecasts.

Another aspect of the use of prediction models that deserves attention is that of updating the model to take account of the rapid changes in agricultural technology and farm practices evolving through agricultural research and experimentation. Such changes may alter the coefficients of variables in the model, and it will be useful to study the extent of such changes.

Complications can arise due to errors of measurement in both crop yield and various explanatory biometrical variables. Such errors in general tend to depress both multiple correlation and regression coefficients, the extent of depression being dependent on the reliability coefficients of the variables and the level of correlation among them. The main sources of such errors are instrument bias, enumerator bias, and their interaction. In this regard, methods of measurement need to be standardized and field staff need to be thoroughly trained to minimize the extent of such errors.

It may be useful here to mention a few pilot studies recently conducted in India for developing statistical methodology on several aspects of crop forecasting using biometrical characteristics. Pilot studies on crop forecasting methodology using biometrical variables have been carried out recently by Indian Agricultural Statistics Research Institute, New Delhi. These studies were conducted with a view to evolving a suitable methodology for preharvest forecasts of crop yield on the basis of observations on biometrical characteristics such as plant density, plant height, and plant girth recorded periodically during crop growth. The crops covered under these studies included rice, wheat, sorghum, cotton, jute, tobacco, and sugarcane. The data on crop yields and biometrical characteristic were secured through specially designed sample surveys on farmers' fields in the country. The survey for each crop was carried out for a period of 3 to 4 years and covered crops on farmers' fields in one to three agroclimatically homogeneous zones. An appropriate sampling design was used in each case, usually a stratified multistage random-sample design.

The general approach of these studies was to fit a multiple-regression equation for the yield to biometrical observations at different stages of crop growth. Typical equations used were of the form

Model I: $$Y = a_0 + \sum_{i-1}^{k} a_i X_i + \varepsilon,$$

Model II: $$\log Y = b_0 + \sum_{i-1}^{k} b_i \log X_i + \varepsilon,$$

Model III: $$\sqrt{Y} = c_0 + \sum_{i=1}^{k} c_i \sqrt{X_i} + \varepsilon,$$

Model IV: $$\frac{1}{Y} = d_0 + \sum_{i-1}^{k} \frac{d_i}{X_i} + \varepsilon,$$

where Y denotes the crop yield and $\{X_i : i = 1, \ldots, k\}$ denote the respective biometrical characteristics included as explanatory variables in the regression models. Further, according to another set of models tried, the crop yield Y was not subjected to any transformation, and the transformations on the X's, if any, were made with the idea of transforming their respective distributions to an approximately normal form. Still another model for cotton crops involved the use of "first picking" yield as an additional explanatory variable.

The results showed that no single model is consistently superior to the others. Thus, a linear model using both Y and the X_i's in the original scale may be preferred for its simplicity and ease in interpretation.

The amount of variation in crop yield explained by the fitted regression equations in terms of R^2 values differed from crop to crop. For cereal crops, rice, wheat, and sorghum the biometrical characteristics at the stage of 2 to 3 months after planting or sowing of the crop explained about 50 to 60 per cent of the variation in crop yield. For cotton, it ranged from 30 to 40 per cent, but by including the "first-picking" yield as an additional explanatory variable in the regression equation, the extent of explained variation increased to as much as 80 per cent. Since the "first-picking" yield becomes available at least two months before final harvesting, the yield forecast can be made about 2 months before harvesting. For sugarcane, biometrical characteristics such as the number of millable canes and their height and girth explained over 70 per cent of the variation in yield at a crop age of 6 to 7 months. Finally, the R^2 values range from 50 to 60 per cent for jute and 40 to 50 per cent for tobacco.

In view of the small extent of variation in crop yield explained by the fitted regression equations, which for a number of crops (except cotton and sugarcane) is of the order of 50 per cent, there is a need to include factors such as weather and crop inputs as additional explanatory variables in the forecasting models.

Another approach has been developed by using growth indices of biometrical characteristic based on two or more periods simultaneously (Jain et al., 1981). the growth indices are obtained as weighted accumulations of observations of biometrical characteristic in different periods, the weights being respective correlation coefficients between yield and biometrical characteristics A model using principal components of biometrical characteristics in two or more periods was also developed, which tackles the problem of multicollinearity (Jain et al., 1984).

In the U.S.A., yield-forecasting methodology based on the use of the biometrical charactersitics has been applied to corn, cotton, soyabeans, wheat, rice, sunflowers, cherries, peaches, grapes, almonds, walnuts, lemons, and oranges. The approach involves forecasting the number of fruits and their weights separately, using appropriate biometrical characteristics through simple linear regression. If more than one character is considered important, separate forecasts are obtained considering regression with one variable at a time, and then these are combined using coefficients of determination as weights. These forecasts of number of fruits and average fruit weight are multiplied to give the final yield (USDA, 1975).

3.3. Forecasts Based on Weather Parameters

Weather exerts a profound influence on agricultural production, especially in countries which depend heavily on rain for timely agricultural operations and for proper growth of crops. Weather may influence production either directly through its effect on growth and structural characteristics of the crop such as plant population, number of tillers, and leaf area, or indirectly through its effect on the incidence of pests and diseases. Even the extent of cultivation is influenced by the time of onset of rain. Several studies of this aspect of crop forecasting have been carried out both in India and abroad. These fall broadly into three groups: (i) fitting of distributions to weather variables, (ii) studies on correlation between crop yield and weather variables, and (iii) regression analysis of yield on weather variables. The results of studies in the first category provide a basis for phasing agricultural operations. For yield forecasting purposes, however, such knowledge could be utilized in two ways: firstly to transform the weather variables to an approximately normal form if necessary, while using them as explanatory variables in the regression model; and secondly in using the untransformed distributions of these weather variables for developing stochastic models for predicting the crop yield. The other two groups of studies provide a measure of the degree of association of different weather variables with crop yield and help to identify the variables that affect the yield. Such studies also provide estimates of the yield response to unit change in different weather variables, such as rainfall, but the regression equations developed under these studies are of limited utility for predicting the crop yield. To be useful as explanatory variables for prediction in regression models, weather variables should be observed at different stages of crop growth prior to harvest. This approach will obviously increase the number of explanatory variables in the prediction model. To estimate the corresponding regression coefficients will require a long series of data, which may not be available in practice.

To deal with this problem of large numbers of variables, the sample size is often increased artificially through the application of the cross-sectional method, which involves taking several sets of weather variables from the same (usually geographically homogeneous) region. This approach has shortcomings. If several sets of weather data are matched with the same yield value, the resulting increase in the degrees of freedom is fictitious, and there is no real improvement in prediction and estimation precision. The elements of the sample formed by combining several subsamples are highly interdependent due to the strong correlation between weather variables measured at neighboring locations. This independence contradicts the assumption of statistically independent elements which is made in standard regression analysis. In addition, the statistical information contained in such a sample does not increase in proportion to the number of observations in the sample. The estimated model coefficients based on this method will always have inflated values, and this approach may not be considered the most efficient way of increasing the size of a sample. Another approach is to decrease the number of variables in the model by taking weather variables during some periods only when these variables show significant correlation with yield. But in this method, the information over the complete growing season is not utilized in the model. Fisher (1924) tackled this problem by fitting distribution constants. He assumed that the effects of change in weather variables in successive weeks would not be abrupt or erratic, but orderly, in accord with some mathematical law. He expressed these effects, as well as weather factors, in terms of orthogonal polynomials in the time, and he developed the model using those polynomials.

Hendrick and Scholl (1943) modified Fisher's technique. They assumed that a second-degree polynomial in the week number would be sufficiently flexible to express the effects. Their model was:

$$Y = A_0 + a_0 \sum_W X_W + a_1 \sum_W W X_W + a_2 \sum_W W^2 X_W.$$

In this model the number of constants to be determined reduced to four, irrespective of the number of weeks. The model was extend to two weather variables in order to study joint effects:

$$\begin{aligned} Y = A_0 &+ a_0 \sum_W X_{1W} + a_1 \sum_W W X_{1W} + a_2 \sum_W W^2 X_{1W} \\ &+ b_0 \sum_W X_{2W} + b_1 \sum_W W X_{2W} + b_2 \sum_W W^2 X_{2W} \\ &+ c_0 \sum_W X_{1W} X_{2W} + c_1 \sum_W W X_{1W} X_{2W} + c_2 \sum_W W^2 X_{1W} X_{2W}. \end{aligned}$$

Since the data required for such studies extended over a long period of years, an additional variable T representing the year was included to make allowance for a time trend.

Agrawal et al. (1980) and Jain et al. (1980) have further modified Hendrick and Scholl's model by expressing the effects of weather variables on the yield in the wth week as second-degree polynomial in the correlation coefficients between the yield and weather variables of the respective weeks. This approach explains the relationship in a better way, as it gives appropriate weight to different periods. A forecast model considering all weather variables through this model has been developed. Further, weather indices have been constructed and principal components obtained to be used in the model in place of the original weather variables, thereby removing multicollinearity among the variables in the model. The study revealed that the rice yield in Raipur district, India, can be forecast by weekly climatic variables $2\frac{1}{2}$ months after sowing for a crop of 5-month duration, using this model. Simulated forecasts of the subsequent years not included for obtaining regression equations showed that, in most of the cases, the difference between the forecasts and the actual values was within 5% of the latter.

Another line of work has been reported by Sakamato (1977) on the basis of the Z index. The Z index is defined as the product of the difference between the observed and the climatically appropriate precipitation, d, and an empirical weight, k, which is the average demand-and-supply coefficient and is a measure of the local significance of the moisture departure. The monthly Z index and the algebraic temperature departures were used as variables in multiple-regression models. The Z values were combined in some areas so that the effect of the moisture anomaly was aggregated over a period of more than one month. This procedure permitted the use of this effect as a single variable. The Z index also combines temperature and precipitation, thus minimizing the collinearity problem which often exists with these two variables in a single month. A quadratic term was included wherever appropriate because the response of crop to weather may not always be linear.

In the FAO (1976) report several techniques for forecasting crop production using historical meteorological data are discussed. The development of a methodology for forecasting area and yield, the product of the two providing production, based on weather parameters is illustrated with the help of data from Turkey. A multiple regression equation was used to relate the yield rate per unit area with meteorological factors. A time-trend variable was introduced to represent a secular yearly change in the yield due to the development of agricultural technology, and time was defined by the calendar year in which the crop was harvested. The following formula was used to calculate the precision of the forecast:

$$\sum W_i^2(F_i - A_i)^2,$$

where F_i and A_i are the forecast and the actual yield/area/production figures respectively for the ith region, and W_i is the crop area when yield forecast was obtained and is equal to one when either the area or the production forecast is made. The method generally gave satisfactory results, with close agreement between the actual and the forecast figures of crop areas, yield rates, and production.

The crop forecasting technique based on agrometeorological information utilizes a cumulative water balance established over the whole growing season for the given crop and established for successive periods of a week or 10 days. The water balance is the difference between the precipitation received by the crop and the water lost by the crop and the soil. The water retained by the soil is also taken into account in the calculation. The data needed are the actual rainfall and climatological records of temperature, relative humidity, sunshine hours, and wind velocity. Based on the data, an agrometeorological index is worked out which gives a good indication of the satisfaction of the crop water requirements in several areas where water represents the main constraint for crops. Direct relations with yield have been demonstrated (FAO, 1979).

Related to weather is the incidence of crop pests and diseases, which affect the crop yield adversely. The extent of loss in yield, however, depends on the severity of the attack, e.g., the loss in yield may be nominal in case of light to moderate incidence of pests and diseases, but it may be heavy to the extent of complete destruction in case of very heavy and severe incidence. A realistic yield prediction model should take cognizance of the incidence of pests and diseases. Since the observations on pests and diseases recorded at different stages of crop growth are expected to be correlated, one could use an overall index based on these observations or alternatively use principal components constructed therefrom as explanatory variables in the regression model. The effects of light incidence of pests and diseases would be reflected through the estimates provided by the forecast model, and no adjustment need be made therein. The loss in yield due to severe attacks, if any, should be assessed separately, and then a suitable adjustment made in the predicted yield.

4. Some Suggestions for Further Work

As the precision of the forecast depends on the models used, there is also need to study the various aspects of statistical models, including nonlinear and stochasti models. Further research is needed on compartmental modeling and analysis techniques to predict the yield, as in the plant process simulation (pps) model developed for corn data by Matis et al. (1984). The compartmental model in the form of a discrete-time Markov chain can be applied to a pps model. It is expected that the best results could be obtained by combining observations on biometrical characteristics with weather factors. As the several variables, both weather and biometrical, will show considerable intercorrelations, they have to be suitably reduced through appropriate multivariate techniques. The system of crop forecasting should be developed so that there is a regular flow of data from the field in a form which can be fed easily to a computer in order to get a forecast figure in time for use.

The following are some of the areas on which research is underway and further work needed:

(a) Plant-process modeling.
(b) Weather modeling, including weather forecasting.
(c) Use of satellite imagery.
(d) Time-series and regression diagnostic procedures.
(e) Sample design to collect data for modeling purposes.

Acknowledgements

The authors are thankful to the referees for their valuable suggestions for improving the presentation of the paper, and to Dr. Ranjana Agrawal for her assistance in revising the manuscript.

Bibliography

Agrawal, R., Jain, R. C., Jha, M. P., and Singh, D. (1980). "Forecasting of rice yield using climatic variables." *Ind. J. Agric. Sci.*, **50**, 680–684.

Agrawal, R., Jain, R. C., and Jha, M. P. (1983). "Joint effects of weather variables on rice yield." *Mausam*, **34**, 189–194.

FAO (1965). *Estimation of Areas in Agricultural Statistics.* S. S. Zarkovich (ed.). Statistics Division.

FAO (1976). *Report on Crop Production Forecasting.*

FAO (1979). *Agrometeorological Crop Monitoring and Forecasting.* FAO Plant Production and Protection Paper 17.

Fisher, R. A. (1924). "The influence of rainfall on the yield of wheat at Rothamsted." *Philos. Trans. Roy. Soc. London Ser. B*, **213**, 89–142.

Hendrick, W. A. and Scholl, J. C. (1943). "Technique in measuring joint relationship. The joint effect of temperature and precipitation on corn yields." *N. Carolina Agric. Exp. Sta. Tech. Bull.*, **74**, 1–34.

Houseman, E. E. and Huddleston, H. F. (1966). "Forecasting and estimating crop yields from plant measurements." *Monthly Bull. Agric. Econ. Statist. F.A.O.*, **15**, No. 10.

Jain, R. C., Agrawal, R., and Jha, M. P. (1980). "Effect of climatic variables on rice yield and its forecast." *Mausam*, **31**, 591–596.

Jain, R. C. Agrawal, R., and Jha, M. P. (1981). "Models for forecasting crop yields." Paper presented at the Symposium on Quantitative Methods in Environmental and Biological Sciences, 68th Session, Indian Science Congress Association.

Jain, R. C., Sridharan, H., and Agrawal, R. (1984). "Use of principal component technique in yield forecast." *Ind. J. Agric. Sci.*, **54**, 467–470.

Jha. M. P., Jain, R. C., and Singh, D. (1981). "Pre-harvest forecasting of sugarcane yield." *Ind. J Agric. Sci.*, **51**, 757–761.

Matis, J. H., Saito, T., and Grant, W. E. (1984). *Compartmental Analysis of a Plant Process Model.* College Station, TX: Dept. of Statistics and Dept. of Wildlife & Fisheries Sciences, Texas A&M Univ.

Sakamato, C. M. (1977). "The *Z* index as a variable for crop estimation." *Agric. Meteorol.*, **18**, 305–313.

Singh, D., Singh, H. P., and Singh, P. (1976). "Pre-harvest forecasting of wheat yield." *Ind. J. Agric. Sci.*, **8**, 445–450.

Singh, D., Singh, H. P., Singh, P., and Jha, M. P. (1979). "A study of pre-harvest forecasting of yield of jute." *Ind. J. Agric. Res.*, **13**, 167–179.

United States Department of Agriculture (1975). *Scope & Methods of the Statistical Reporting Service.* Miscellaneous publication No. 1308.

Vogel, F. A. (1983). *Sample Designs for Multi-purpose Agricultural Surveys–Problems in Application.* Arcata, CA: Statistical Reporting Service, U.S. Dept. of Agriculture.

Author Index

Subject Index